AF545035

Numerical Methods in *HEAT TRANSFER*

VOLUME II

WILEY SERIES IN
NUMERICAL METHODS IN ENGINEERING

Consulting Editors
R. H. Gallagher, *College of Engineering,*
University of Arizona
and
O. C. Zienkiewicz, *Department of Civil Engineering,*
University College Swansea

Rock Mechanics in Engineering Practice
Edited by K. G. Stagg and O. C. Zienkiewicz

Optimum Structural Design: Theory and Applications
Edited by R. H. Gallagher and O. C. Zienkiewicz

Finite Elements in Fluids
Vol. 1 Viscous Flow and Hydrodynamics
Vol. 2 Mathematical Foundations, Aerodynamics and Lubrication
Edited by R. H. Gallagher, J. T. Oden, C. Taylor, and O. C. Zienkiewicz

Finite Elements in Geomechanics
Edited by G. Gudehus

Numerical Methods in Offshore Engineering
Edited by O. C. Zienkiewicz, R. W. Lewis, and K. G. Stagg

Finite Elements in Fluids, Vol. 3
Edited by R. H. Gallagher, O. C. Zienkiewicz, J. T. Oden, M. Morandi Cecchi and C. Taylor

Energy Methods in Finite Element Analysis
Edited by R. Glowinski, E. Rodin, and O. C. Zienkiewicz

Finite Elements in Electrical and Magnetic Field Problems
Edited by M. V. K. Chari and P. Silvester

Numerical Methods in Heat Transfer
Edited by R. W. Lewis, K. Morgan, and O. C. Zienkiewicz

Finite Elements in Biomechanics
Edited by R. H. Gallagher, B. R. Simon, P. C. Johnson, and J. F. Gross

Soil Mechanics—Transient and Cyclic Loads
Edited by G. N. Pande and O. C. Zienkiewicz

Finite Elements in Fluids, Vol. 4
Edited by R. H. Gallagher, D. Norrie, J. T. Oden, and O. C. Zienkiewicz

Foundations of Structural Optimization: A Unified Approach
Edited by A. J. Morris

Creep and Shrinkage in Concrete Structures
Edited by Z. Bazant and F. Wittmann

Hybrid and Mixed Finite Element Methods
Edited by S. N. Atluri, R. H. Gallagher, and O. C. Zienkiewicz

Numerical Methods in Heat Transfer, Vol. II
Edited by R. W. Lewis, K. Morgan, and B. A. Schrefler

Numerical Methods in HEAT TRANSFER

VOLUME II

Edited by
R. W. Lewis
K. Morgan
Department of Civil Engineering,
University of Wales,
Swansea, U.K.
B. A. Schrefler
Istituto di Costruzioni,
Universita di Padova,
Italy

A Wiley–Interscience Publication

JOHN WILEY & SONS
Chichester . New York . Brisbane . Toronto . Singapore

***Library of Congress Cataloging in Publication Data*:**

(Revised for vol. 2)
Main entry under title:
Numerical methods in heat transfer.
(Wiley series in numerical methods in engineering)
'A Wiley–Interscience publication.'
Papers from an international conference held in Swansea, July 1979.
Vol. 2: edited by R. W. Lewis, K. Morgan, B. A. Schrefler.
Includes indexes.
1. Heat—Transmission. 2. Numerical analysis.
I. Lewis, Roland, Wynne. II. Morgan, Kenneth,
1945– . III. Zienkiewicz, O. C.
QC320.N86 1981 536'.2'0015194 80-49973
ISBN 0 471 27803 3 (v. 1)

***British Library Cataloguing in Publication Data*:**

Numerical methods in heat transfer.—(Wiley series in numerical methods in engineering)
Vol. 2
1. Heat Transmission—Mathematics—Congresses
2. Numerical calculations—Congresses
I. Lewis, R. W. II. Morgan, K. 1945–
III Schrefler, B. A.
536'.2.001511 QC320.2

ISBN 0 471 90064 8

Printed in Great Britain by J. W. Arrowsmith Ltd., Bristol
Bound by Pitman Press Ltd., Bath, Avon

Contributing Authors

R. L. AKAU	*Purdue University, West Lafayette, IN 47907 U.S.A.*
J. T. BERRY	*School of Mechanical Engineering, Georgia Institute of Technology, Atlanta, Georgia, U.S.A.*
P. L. BETTS	*Department of Mechanical Engineering, UMIST, Sackville Street, Manchester, U.K.*
P. BONTOUX	*Institut de Mécanique des Fluides, 1 rue Honnorat, 13003 Marseille, France.*
M. BORSETTO	*ISMES (Esperimental Institute for Models and Structures) Viale Giulio Cesare 29, 24100 Bergamo, Italy.*
G. CARRADORI	*ISMES, Viale Giulio Cesare 29, 24100 Bergamo, Italy.*
P. DECHAUMPHAI	*Mechanical Engineering & Mechanics Department, Old Dominion University, Norfolk, VA 23508, U.S.A.*
M. FANELLI	*Center for Hydraulic & Structural Research, R & D Dept. ENEL (Italian National Power Agency) Milano, Italy.*
D. F. FISCHER	*Voest-Alpine AG, Linz, Austria.*
F. K. FONG	*Department of Chemical Engineering, The Ohio State University, Columbus, Ohio, 43210 U.S.A.*
C. R. GANE	*Central Electricity Research Laboratories, Kelvin Avenue, Leatherhead, Surrey, U.K.*
P. GERREKENS	*Aerospace Laboratory of the University of Liege R. E. Solvay, 21, B-4000 Liege, Belgium.*

D. J. GILBY — *Polydynamics Limited, Zurich, Switzerland.*

B. GILLY — *Institut de Mécanique des Fluides, 1 rue Honnorat, 13003 Marseille, France.*

G. GIUSEPPETTI — *Center for Hydraulic & Structural Research, R & D Dept, ENEL Milano, Italy.*

O. GRUNDLER — *Vereinigte Edelstahlwerke AG (VEW) Kapfenberg, Austria.*

P. N. HANSEN — *Department of Civil Engineering, Technical University of Denmark, Lyngby, Denmark.*

M. HOGGE — *Aerospace Laboratory of the University of Liege R. E. Solvay, 21, B-4000 Liege, Belgium.*

R. J. HOPKIRK — *Polydynamics Ltd., Zurich, Switzerland.*

F. HOISLBAUER — *Voest-Alpine AG, Linz, Austria.*

L. IMRE — *Technical University, Budapest, Hungary.*

C. JAQUEMAR — *Voest Alpine AG, Linz, Austria.*

J. A. JOHNSON — *Weyerhaeuser Company, Tacoma, Washington, U.S.A.*

J. S. KANG — *Department of Mechanical Engineering, UMIST, Sackville Street, Manchester, U.K.*

F. KAYIHAN — *Weyerhaeuser Company, Tacoma, Washington, U.S.A.*

G. A. KERAMIDAS — *Fluid Dynamics Branch, Naval Research Laboratory, Washington D.C. 20375, U.S.A.*

L. I. KISS — *Technical University, Budapest, Hungary.*

G. W. KRUTZ — *Purdue University, West Lafayette, IN 47907 U.S.A.*

W. MITTER — *Vereinigte Edelstahlwerke AG (VEW) Kapfenberg, Austria.*

K. MORGAN — *Department of Civil Engineering, University College, Swansea, Singleton Park, Swansea SA2 8PP, U.K.*

A. J. OLIVER — *Central Electricity Research Laboratories, Kelvin Avenue, Leatherhead, Surrey, U.K.*

A. PEANO — *ISMES, Viale Giulio Cesare 29, 24100 Bergamo, Italy.*

P. J. PRALONG — *Polydynamics Limited, Zurich, Switzerland.*

J. RAE — *AERE, Harwell, Didcot, Oxfordshire, U.K.*

F. G. RAMMERSTORFER — *Voest-Alpine AG, Linz, Austria.*

J. N. REDDY — *Department of Engineering Science and Mechanics, Virginia Polytechnic Institute and State University, Blacksburg, Virginia, 24061 U.S.A.*

P. C. ROBINSON — *AERE, Harwell, Didcot, Oxfordshire, U.K.*

B. ROUX — *Institut de Mécanique des Fluides, 1 rue Honnorat, 13003 Marseille, France.*

D. SHARMA — *Polydynamics Limited, Zurich, Switzerland.*

D. R. SKIDMORE — *Department of Chemical Engineering, The Ohio State University, Columbus, Ohio 43210 U.S.A.*

D. R. SOULSBY — *Central Electricity Research Laboratories, Kelvin Avenue, Leatherhead, Surrey, U.K.*

P. L. STEPHENSON — *Central Electricity Research Laboratories, Kelvin Avenue, Leatherhead, Surrey, U.K.*

K. K. TAMMA — *Mechanical Engineering and Mechanics Department, Old Dominion University, Norfolk VA. 23508, U.S.A.*

C. TAYLOR — *Department of Civil Engineering, University College of Swansea, Singleton Park, Swansea, SA2 8PP, U.K.*

C. E. THOMAS — *Department of Civil Engineering, University College of Swansea, Singleton Park, Swansea SA2 8PP, U.K.*

E. A. THORNTON — *Mechanical Engineering and Mechanics Department, Old Dominion University, Norfolk, VA. 23508, U.S.A.*

E. T. Till — *Voest-Alpine AG, Linz, Austria.*

C. Wei — *School of Mechanical Engineering, Georgia Institute of Technology, Atlanta, Georgia, U.S.A.*

L. M. Wickens — *AERE, Harwell, Didcot, Oxfordshire, U.K.*

K. H. Winters — *Theoretical Physics Division, AERE, Harwell, Oxfordshire, U.K.*

Contents

Preface

The Second International Conference on Numerical Methods in Thermal Problems was held in Venice, Italy during July 1981. Scientists from many countries, employed in research institutions, universities and research organizations, attended the conference. They all shared the common interest of being actively engaged in the development or application of computational methods for the solution of heat transfer problems. The editors have attempted to capture the flavour of the conference in this volume by including, in expanded form, twenty of some of the most interesting papers presented at the meeting.

The papers will be found to consist of either a significant new development in the modelling area or an important industrial application of traditional methods. Among the practical problem areas considered are modelling the effects of fire, ablation, heat flow in porous rock, thermal stress and dissolving coal. In addition, papers concerned with the problems of simulating new energy sources such as geothermal reservoirs and solar radiation are included. The important area of natural convection receives an appropriate amount of attention while the numerical techniques expounded range from the use of influence functions to the use of finite elements, boundary elements and finite differences.

The editors wish to thank the authors for giving their time to make this publication possible and hope that this book will serve as a valuable source of reference to all those engaged in the numerical simulation of thermal problems.

R. W. Lewis
K. Morgan
B. A. Schrefler

Numerical Methods in Heat Transfer, Volume II
Edited by R. W. Lewis, K. Morgan, and B. A. Schrefler

Chapter 1

The Computation of Time-dependent Thermal Structural Effects by Means of Influence Functions Pertaining to 'Unit Temperature Distributions': A Unified Approach Free from Restrictive Hypotheses

Michele Fanelli and Gabriella Giuseppetti

1.1 ORIGIN AND MOTIVATION OF THE PRESENT STUDY

Quite frequently, in the field of structural analysis, the necessity arises of assessing the structural effects of a time-dependent thermal distribution inside a solid elastic body, starting from very sketchy information about the thermal field itself.

In some cases, the temperature-vs.-time behaviour of the medium (or the media) in which the structure is immersed may be known; in other cases the temperatures of a few points lying on the surfaces of the body may be sampled in time at regular intervals (and this is the case we shall be concerned with); only seldom, the temperatures of a few inner points of the body are also monitored in time with the help of suitably embedded sensors.

Of course, the problem of assessing the structural effects of a thermal field so scantily defined would be an insoluble one, were we not able to reconstruct in a reliable way the 'complete' thermal field (throughout the volume of the body and over all the time span of interest) from the available information.

This reconstruction is in principle feasible thanks to an additional piece of knowledge: the fact that inside the body the temperature distribution, $\vartheta(x, y, z, t)$, has to satisfy—at least to a first approximation—the Fourier heat conduction equation ($\dot{\vartheta} = \partial\vartheta/\partial t$):

$$a\nabla^2\vartheta = \dot{\vartheta}, \tag{1.1}$$

where a = diffusivity constant of the material. This can be a function of space and temperature, but for certain problems it may be supposed—again, at least to a first approximation—to be a *constant*. We will stick to this hypothesis; in our developments, however, the more general assumption of a known variation of a with the space coordinates:

$$a = a(x, y, z), \tag{1.2}$$

would entail no great difficulties; but usually no information is available about such space variations of a, if any. Equation (1.1), together with complete boundary conditions on the surfaces:

$$\vartheta(P_s, t) = T(P_s, t) \tag{1.3}$$

known for all points P_s lying on the surface of the body, S, and with complete *initial conditions*:

$$\vartheta(x, y, z, t_0) = \vartheta_0(x, y, z) \tag{1.4}$$

known for *all* $P(x, y, z)$ inside the body, would allow one, in principle, to determine

$$\vartheta(x, y, z, t) \tag{1.5}$$

for all points $P(x, y, z)$ inside the body and for every $t \geq t_0$.

But in practice, suitable and economical numerical procedures to achieve this aim must be found. And these procedures should not only circumvent the fact that the available information is always *far less complete* than (1.3) and (1.4); they should also be so contrived as to make as easy and fast as possible the following step: the computation of structural effects from the temperature distribution (1.5).

This poses particular constraints in those cases (viz. the continuous surveillance of big structures, such as concrete dams) where this appraisal of thermal structural effects must be effected in a repetitive way for a very great number of time instants:

$$t = t_1, t_2, \ldots, t_N \qquad (N \text{ in the order of some thousands}). \tag{1.6}$$

The hypothesis of elastic behaviour of the solid body, whenever acceptable (and we may mention that, for the interpretation of concrete dam behaviour, the acceptability of this hypothesis has been proved beyond doubt), suggests that we operate with the technique of *influence functions*.

In other words, the 'thermal load' represented by a temperature distribution inside the solid body at time t is suitably approximated by a linear operator $\mathscr{L}$ of the Q known temperatures, $T_i(\tau)$, $1 \leq i \leq Q$, up to time t, plus another linear operator $\mathscr{L}_0$ of the known initial temperatures at time $t_0 < t$:

$$\vartheta(x, y, z, t) \cong \mathscr{L}[T_i(\tau)] + \mathscr{L}_0[\vartheta(t_0)] \qquad \text{with } \tau \leq t, \tag{1.7}$$

where T_i = measured temperatures† ($1 \leq i \leq Q$), and $\vartheta(t_0)$ = thermal distribution at initial time $t_0 < t$ (usually defined in a very imperfect way through a few embedded thermometers). Hence also any linear structural effect of interest becomes a linear operator of these same temperatures, and its coefficients can be computed once and for all.

In the past,[1–4] advantage was taken of the fact that for dams the thermal field admits of decomposition into a sum of partial fields, each obeying the Fourier equation (1.1) and each satisfying a restricting hypothesis; e.g.

$$\vartheta = \vartheta_0 + \vartheta_1 + \vartheta_2 + \vartheta_3 + \vartheta_4, \tag{1.8}$$

with

ϑ_0 = time-independent (average) component;
ϑ_1 = transient due to initial conditions;
ϑ_2 = periodic component of yearly period (fundamental plus a few harmonics: usually only the semestral harmonic is considered);
ϑ_3 = periodic component of daily period;
ϑ_4 = quasi-random residual.

This approach yielded very acceptable results in a number of cases[5,6] and has the advantage that (due to the hypothesis of periodicity about ϑ_2 and ϑ_3) the linear operator $\mathscr{L}$ can be made formally to depend only on *present* values of measured temperatures (instead of on the whole previous thermal history). But there were also drawbacks, among them:

(i) the necessity of operating a quite cumbersome decomposition of the thermal field‡ and of treating each component with specialized analytical and numerical tools;
(ii) slow (long-term) variations in the average temperatures or sudden changes in the shape of the successive cycles for the 'periodic' components would not be accommodated by this scheme and would thus introduce apparent 'errors' into the computed structural effects;§
(iii) the effects of the quasi-random residuals ϑ_4 are not explicity taken into account and therefore add to the amplitude of the apparent error; etc.

† Usually only on the exposed surfaces of the body: this particular disposition of the Q thermometers is retained throughout the present work.

‡ Because of this influence functions had to be multiplied, not into the measured temperatures themselves, but into the respective components (and time-derivatives thereof for the periodic components).

§ In effect, if the decomposition (1.8) allowed perfectly periodic components ϑ_2 and ϑ_3, the corresponding effects would also be exactly periodic and it would be possible to write them down as explicit functions of time. The operative solution adopted (to compute structural influence functions as if the periodicity hypothesis were true and use them, with suitable arrangements, with components ϑ_2 and ϑ_3 not exactly periodic) is not a satisfactory one from a formal standpoint, even if in practice is has usually proved successful.

It was deemed, thus, interesting to inquire about the possibility of developing a 'general' formulation of the problem endowed with the following requisites:

—no restricting hypothesis about the time behaviour of thermal variations (arbitrarily irregular boundary thermal histories should be accommodated by the formal structure of the analytical treatment);
—the technique of the 'influence functions' should be retained for the computation of structural thermal effects; i.e., the treatment should isolate certain special 'unit' (or 'base') thermal distributions, onto which any actual thermal situation $\vartheta(x, y, z, t)$ should be developed by simple linear operators using directly the measured temperatures, T_i, as weights. Then any structural effect of interest could be first computed for each of the unit distributions ('influence functions') and then the running value of that structural effect would be synthesized by applying the same linear operators to the products of the influence functions into the same factors that effect the development of $\vartheta(x, y, z, t)$ into the base distributions, again using directly the measured temperatures, T_i, as weights;
—the procedure adopted to synthesize 'complete' thermal distributions from 'incomplete' information should always conform to Fourier's equation (1.1);
—it should be easy to translate the mathematical formulation into efficient, fast, numerical computation procedures.

We intend to demonstrate in the remainder of this paper that such an approach can be found and developed, and to discuss the features, advantages and drawbacks of the ensuing numerical procedures for automatic computation.

1.2 STEP 1: THE SYNTHESIS OF 'COMPLETE' BOUNDARY CONDITION INFORMATION FROM ACTUALLY MEASURED BOUNDARY TEMPERATURES. STEP 2: THE DECOMPOSITION OF INPUT AND RESPONSE INTO A SUM OF TIME-STAGGERED STEP FUNCTIONS

As already stated, boundary temperatures are in practice known only in a limited number of points lying on the boundary surfaces:

$$T_i(t) \qquad \text{at } P_i(S) \tag{1.9},$$

see Figure 1.1 (for simplicity a two-dimensional case is illustrated);

$$1 \leqslant i \leqslant Q. \tag{1.10}$$

For the moment each $T_i(t)$ is considered as a function of t defined for all values of time t itself, even though it is usually sampled at regular or irregular time intervals.

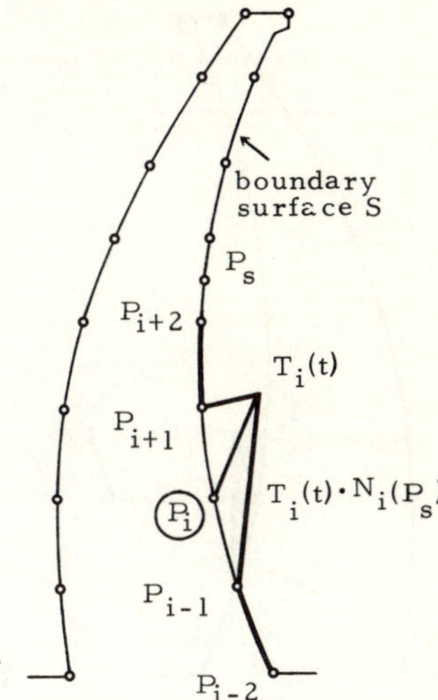

Figure 1.1 Typical boundary conditions for a two-dimensional case

The problem addressed in this step is to define in an acceptable way a surface distribution

$$\vartheta(P_s) \tag{1.11}$$

valid for *all* points P_s lying on the boundary S, even if not coincident with thermometric points P_i.

This can be effected without difficulty thanks to suitable 'shape functions'

$$N_i(P_s), \tag{1.12}$$

suitably defined for every value of i over all of S (obviously $N_i = 1$ for $P_s = P_i$, $N_i = 0$ for $P_s = P_{i-1}$, $P_s = P_{i+1}$ and outside the boundary interval P_{i-1} to P_{i+1}).

The simplest choice for $N_i(P_s)$ is the piecewise-linear function in the case of two-dimensional domains (see Figure 1.2). This obviously gives a linear interpolation (space-wise) for $\vartheta(P_s)$, between T_i and T_{i+1}, when P_s lies between P_i and P_{i+1}.

Thus, the general expression for $\vartheta(P_s)$ is

$$\vartheta(P_s, t) = \sum_{i=1}^{Q} T_i(t) N_i(P_s). \tag{1.13}$$

Let us now consider the time behaviour of T_i, $T_i(t)$: it is possible to express it as follows:

$$T_i(t) \equiv T_i(t_0) g(t - t_0) + \int_{t_0}^{t} \frac{dT_i(\tau)}{d\tau} g(t-\tau)\, d\tau, \qquad \text{for } t \geqslant t_0, \tag{1.14}$$

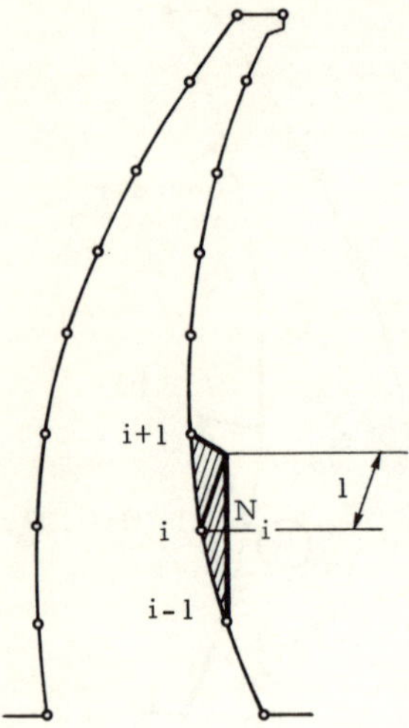

Figure 1.2 Features of the shape functions $N_i(P_s)$

where $g(t-\tau)$ is the well-known 'unit step function':

$$g(t-\tau)\equiv 0 \quad \text{for } t-\tau<0, \quad \text{i.e. for } t<\tau; \quad \text{(see Figure 1.3a)}$$
$$g(t-\tau)\equiv 1 \quad \text{for } t-\tau>0, \quad \text{i.e. for } t>\tau. \tag{1.15}$$

(Obviously, $T_i(t)$ is known only through a discrete sampling, so that (1.14) will be replaced by a sum:

$$T_i(t)\cong T_i(t_0)g(t-t_0)+\sum_{t>t_0}\Delta T_i(\tau)g(t-\tau) \qquad \text{(see Figure 1.3b).} \tag{1.16}$$

Now (1.13) and (1.14) allow us to express the 'complete' surface temperature distribution, $\vartheta(P_s, t)$ at time t, i.e. the instantaneous boundary conditions:

$$\begin{aligned}\vartheta(P_s, t) &= \sum_i T_i(t_0)g(t-t_0)N_i(P_s)+\sum_i N_i(P_s)\int_{t_0}^{t}\frac{\mathrm{d}T_i(\tau)}{\mathrm{d}\tau}g(t-\tau)\,\mathrm{d}\tau \\ &= \sum_{i=1}^{Q} N_i(P_s)\left[T_i(t_0)g(t-t_0)+\int_{t_0}^{t}\frac{\mathrm{d}T_i}{\mathrm{d}t}g(t-\tau)\,\mathrm{d}t\right]\end{aligned} \tag{1.17}$$

or, see (1.16),

$$\vartheta(P_s, t)\cong\sum_{i=1}^{Q} N_i(P_s)\left\{T_i(t_0)g(t-t_0)+\sum_{t>t_0}\Delta T_i(\tau)g(t-\tau)\right\}. \tag{1.18}$$

It is now evident that, due to the linear structure of (1.18) in the T_is, it will be sufficient to find the responses of the temperature of an arbitrary inner point to special unit step functions of time:

$$w_i(x, y, z, t-\tau) = \text{response of } \vartheta(x, y, z, t), \tag{1.19}$$

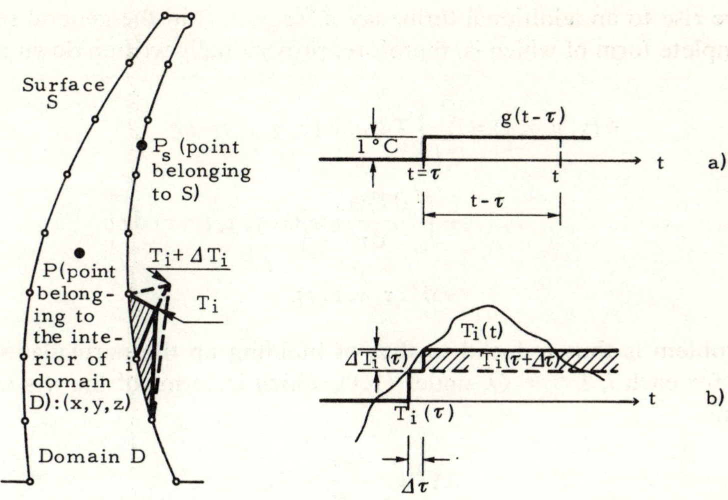

Figure 1.3 Decomposition of $T_i(t)$ into a sum of staggered step functions

according to Fourier's equation (1.1), to a boundary condition defined as follows:

$$\vartheta_i(P_s) = N_i(P_s)g(t-\tau). \tag{1.20}$$

In fact, it is immediate to see that, once the w_is are known, and with *homogeneous initial conditions,*†

$$\vartheta(x, y, z, t_0) = \vartheta_0(x, y, z) \equiv 0 \tag{1.21}$$

for any $P(x, y, z)$ within D, the sought-after solution,

$$\vartheta(x, y, z, t) \qquad \text{for } t > t_0 \tag{1.22}$$

will be given by

$$\vartheta(x, y, z, t) = \sum_{i=1}^{Q} T_i(t_0)w_i(x, y, z, t-t_0) + \sum_{i=1}^{Q} \int_{t_0}^{t} \frac{\mathrm{d}T_i(\tau)}{\mathrm{d}\tau} w_i(x, y, z, t-\tau)\,\mathrm{d}\tau. \tag{1.23}$$

The influence of non-homogeneous initial conditions:

$$\vartheta(x, y, z, t_0) = \vartheta_0(x, y, z) \neq 0 \tag{1.24}$$

† The influence of actual (non-homogeneous) initial conditions will be dealt with later.

will give rise to an additional term, say $\vartheta^*(x, y, z, t)$ in the general solution, the complete form of which is, therefore, provisionally written down as

$$\vartheta(x, y, z, t) = \sum_{i=1}^{Q} \left[T_i(t_0) w_i(x, y, z, t-t_0) + \int_{t_0}^{t} \frac{\mathrm{d}T_i(\tau)}{\mathrm{d}\tau} w_i(x, y, z, t-\tau)\, \mathrm{d}\tau \right] + \vartheta^*(x, y, z, t). \tag{1.25}$$

The problem is thus reduced to that of building up the particular solution (1.19), for each i, $1 \leqslant i \leqslant Q$, under (1.1), which in terms of the w_is takes on the form

$$a\nabla^2 w_i = \frac{\partial w_i}{\partial t} = \dot{w}_i, \tag{1.26}$$

and with boundary conditions (1.20), besides finding the term ϑ^* (transient due to non-homogeneous initial conditions).

In the following section it will be shown how to build up ϑ^* as well as the desired solutions (1.19), starting from a knowledge of the *natural cooling modes* (N.C.M.s) of the solid body.

1.3 STEP 3: TRANSFORMATION OF THE GENERAL SOLUTION (1.25) IN TERMS OF THE NATURAL COOLING MODES OF THE SOLID BODY

1.3.1 Essential notions about the 'natural cooling modes' (N.C.M.s)

It will be seen in the following that all the essential information about the thermal behaviour of the body can be related to the knowledge of certain particular solutions of the Fourier equation (1.1). Each of these particular solutions is characterized by a shape that at different times is not altered but for a factor of proportionality that is simply an exponentially decaying function of time t (the decay constant being in principle different for each particular solution). Indeed, if we look for solutions of (1.1), within domain D and with homogeneous boundary conditions, of the form

$$\vartheta(x, y, z, t) = \psi_j(x, y, z)\mathrm{e}^{-K_j t}, \tag{1.27}$$

a classical eigenvalue problem obtains, in which the time variable t is formally eliminated:

$$a\nabla^2\psi_j + K_j\psi_j = 0, \qquad \text{with } \psi_j(P_s) \equiv 0. \tag{1.28}$$

The constants K_j form an infinite, discrete series of *eigenvalues*; and to each of them there corresponds a distribution ψ_j,

$$K_j \rightarrow \psi_j(x, y, z), \tag{1.29}$$

termed an *eigenfunction*, which is determined up to an arbitrary multiplicative factor.†

The distributions $\psi_j(x, y, z)$, each together with its constant K_j, are the so-called 'natural cooling modes' (N.C.M.s) of the solid body.

The K_js and the corresponding distributions $\psi_j(x, y, z)$ can be determined numerically, once the geometry of domain D is defined and the thermal diffusivity constant a is known, thanks to suitable F. E. numerical computation programs.

Also, any initial distribution (provided it vanishes at the boundary S, $\vartheta_0(P_s) = 0$):

$$\vartheta_0(x, y, z) \qquad \text{with } \vartheta_0(P_s) = 0, \tag{1.30}$$

can be developed into a series of the N.C.M.s:

$$\vartheta_0(x, y, z) = \sum_{j=1}^{\infty} A_j \psi_j(x, y, z), \tag{1.31}$$

so that the solution of (1.1) with

$$\begin{aligned} &\vartheta(x, y, z, t_0) = \vartheta_0(x, y, z) \qquad \text{in } D, \\ &\vartheta(P_s, t) \equiv 0 \qquad \text{[homogeneous boundary conditions‡]} \end{aligned} \tag{1.32}$$

is given, for any $t \geq t_0$, by

$$\vartheta^*(x, y, z, t) = \sum_{j=1}^{\infty} A_j \psi_j(x, y, z)\, \mathrm{e}^{-K_j(t-t_0)}. \tag{1.33}$$

The numerical coefficients A_j can be easily computed thanks to a well-known property of the orthonormal functions ψ_j:

$$\iiint_D \psi_j(x, y, z) \psi_k(x, y, z)\, \mathrm{d}x\, \mathrm{d}y\, \mathrm{d}z = \begin{cases} \equiv 0 & \text{for } k \neq j, \\ = 1 & \text{for } k = j. \end{cases} \tag{1.34}$$

so that from (1.33) and (1.34):

$$A_j = \iiint_D \vartheta_0(x, y, z) \psi_j(x, y, z)\, \mathrm{d}x\, \mathrm{d}y\, \mathrm{d}z. \tag{1.35}$$

† This arbitrary numerical factor is usually fixed due to a 'normalization condition'; see the second equation of (1.34).

‡ Note that in our formulation the initial non-homogeneous boundary conditions $\vartheta(P_s, t_0) = \vartheta_0(P_s) \neq 0$ is separately treated through the first term in the square bracket in (1.25), second member, which in its turn presupposes homogeneous inner initial conditions. So now it is correct to assume homogeneous boundary conditions for $\vartheta_0(x, y, z)$.

Now let us see how the response (1.19) to a boundary input (1.20) which is step-like in time and has the spatial distribution $N_i(P_s)$ on the boundary, can be computed for all inner points once the N.C.M.s are known.

1.3.2 How to build up the particular solutions $w_i(x, y, z, t)$, $1 \leq i \leq Q$

Let us begin by finding Q *time-independent* solutions $s_i(x, y, z)$ (see Figure 1.4) of the Fourier equation (1.1), which in this case becomes a Laplace-type equation:

$$\nabla^2 s_i = 0, \tag{1.36}$$

each with boundary conditions

$$s_i(P_s) = N_i(P_s). \tag{1.37}$$

The solutions s_i can be found numerically thanks to suitable F.E. programs.
Let us also define the Q auxiliary functions $s_i^*(x, y, z)$ as follows:

$$s_i^*(x, y, z) = \\ = \begin{cases} = s_i(x, y, z) & \text{for } P(x, y, z) \textit{ inside } D \ (\textit{not} \text{ on the boundary}); \\ = 0 & \text{for } P(x, y, z) \text{ on the boundary } S \text{ (i.e. for } P = P_s). \end{cases} \tag{1.38}$$

(The functions s_i^* present a discontinuity at the boundary.) It is quite obvious that the auxiliary functions $s_i^*(x, y, z)$ can be developped into a series of the

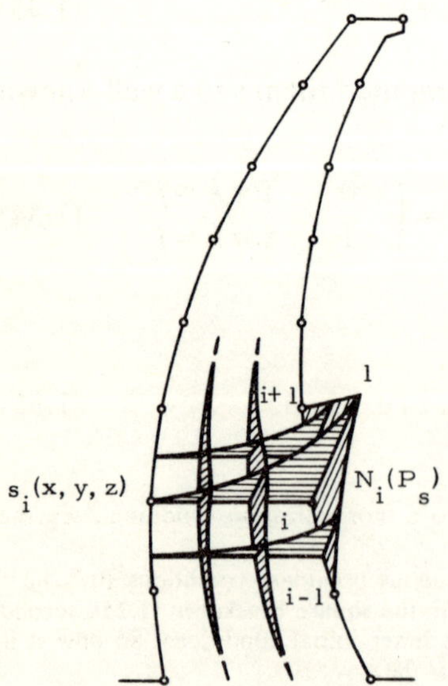

Figure 1.4 Time-independent solutions for $s_i(x, y, z)$

N.C.M.s:

$$s_i^*(x, y, z) = \sum_{j=1}^{\infty} S_{ij}\psi_j(x, y, z). \tag{1.39}$$

The coefficients S_{ij} will in fact be computed quite simply with the following formula, using (1.34):

$$S_{ij} = \iiint_D s_i^*(x, y, z)\psi_j(x, y, z)\,\mathrm{d}x\,\mathrm{d}y\,\mathrm{d}z. \tag{1.40}$$

It is now evident that the response (1.19) could be computed theoretically as

$$w_i(x, y, z, t-\tau) = s_i(x, y, z) - \sum_{j=1}^{\infty} S_{ij}\psi_j(x, y, z)\,\mathrm{e}^{-K_j(t-\tau)} \tag{1.41}$$

for every $t-\tau>0$.

Practical difficulties arise from the fact that the summation with respect to j should be extended to infinity; a possible device to circumvent this difficulty will be discussed in Section 1.6.

1.3.3 Formulation of the general solution

Let us now take up again the general solution in its initial form (1.25), but taking into account (1.33) and (1.41). From (1.25), integrating by parts,

$$\begin{aligned}\vartheta(x, y, z, t) = \sum_{i=1}^{Q} \Big[& T_i(t_0)w_i(x, y, z, t-t_0) + T_i(t)w_i(x, y, z, 0) \\ & - T_i(t_0)w_i(x, y, z, t-t_0) \\ & - \int_{t_0}^{t} T_i(\tau)\frac{\partial w_i(x, y, z, t-\tau)}{\partial \tau}\,\mathrm{d}\tau \Big] \\ & + \vartheta^*(x, y, z, t) = [\text{taking into account (1.41)}] \\ = \sum_{i=1}^{Q} \Big[& T_i(t)(s_i - s_i^*) + \sum_{j=1}^{\infty} S_{ij}\psi_j(x, y, z)K_j \\ & \times \int_{t_0}^{t} T_i(\tau)\,\mathrm{e}^{-K_j(t-\tau)}\,\mathrm{d}\tau \Big] \\ & + \sum_{j=1}^{\infty} A_j\psi_j(x, y, z)\,\mathrm{e}^{-K_j(t-t_0)}. \end{aligned} \tag{1.42}$$

It is worth noticing that

$$\left.\begin{aligned} & s_i - s_i^* = 0 && \text{for all points } P(x, y, z) \textit{ inside } D; \\ & s_i - s_i^* = N_i(P_s) && \text{for all points } P_s \text{ belonging to } S. \end{aligned}\right\} \tag{1.43}$$

We see that—apart from initial conditions, reflected in the coefficients A_j, and points on the boundary where $s_i - s_i^* = N_i(P_s)$—the general solution hinges on knowledge of the following sets of quantities:

—the Q stationary solutions $s_i(x, y, z)$;
—the N.C.M.s $\psi_j(x, y, z)$, and the associated constants K_j;
—the coefficients S_{ij} (see (1.40));
—the *convolution integrals*

$$C_{ij}(t) = \int_{t_0}^{t} T_i(\tau)\, e^{-K_j(t-\tau)}\, d\tau. \tag{1.44}$$

It is worth stressing the point that (1.42) is, up to now, a merely formal development, insofar as it involves the evaluation of infinite series. In practice, it will be necessary to arrest the development at a certain value, say J, of the subscript j, and to assess the error thus introduced. This question will be dealt with in the following and it will be seen that the convergence properties of these series are rather poor, so that in principle one should take into account a quite large value of J. However, a particular device (see further on) was found useful in improving the approximation reached in truncating the series at an acceptable value of J (say not more than $J \cong 20$ for two-dimensional domains).

It is also to be hoped that mathematicians can indicate other more rational ways of improving the convergence of the truncated series to their exact values.

With this proviso in mind, it appears that the solution (1.42) possesses in principle all the desired properties as enumerated toward the end of Section 1.1.

Besides, it can be verified by substituting (1.42) into (1.1) that (1.42) satisfies the Fourier heat conduction equation. Simple inspection of (1.42) shows, furthermore, that the proposed solution satisfies boundary and initial conditions.

Finally, if we calculate the inner temperature distribution $\vartheta(x, y, z, t_1)$ at a time $t_1 > t_0$, and take this as a new initial distribution, in the place of ϑ_0, it can be shown that identical results obtain for the inner distribution $\vartheta(x, y, z, t_2)$ at all times $t_2 > t_1$, irrespective of whether it is calculated by (1.42) assuming either t_0 or t_1 as initial instant.

Other 'necessary' tests of internal coherence for the proposed solution also yield positive results. For instance:

(a) if all initial and boundary conditions are operated on by a linear operator, the new inner distribution that obtains by introducing the new initial and boundary conditions into (1.42) is the product of the old solution and of the chosen linear operator;

(b) if we start with an initial distribution

$$\vartheta_0(x, y, z) = \sum_{i=1}^{Q} T_i(t_0)s_i(x, y, z), \tag{1.45}$$

which is obviously complying with a stationary regime defined by boundary conditions

$$T_i(t) = T_i(t_0), \tag{1.46}$$

and we keep all T_i actually constant in time, application of (1.42) easily yields:

$$\vartheta(x, y, z, t) = \vartheta_0(x, y, z), \tag{1.47}$$

as it should be in this case; and so on.

1.4 THE CONCEPT OF 'MODAL MEMORY MATRIX' FOR THE THERMAL BEHAVIOUR OF A SOLID BODY

The main time-dependent quantities that appear in (1.42) and that need a continuous updating based on experimental data $T_i(t)$ are the convolution integrals $C_{ij}(t)$ (see (1.44)).

The double suffix ij and the structure of the convolution integrals (1.44) suggest a very interesting schematization.

All the essential information about the varying thermal régime of the solid body can be organized into an ideal 'matrix' (with Q rows and infinitely many columns, see Figure 1.5), at least as pertains to the effects of boundary inputs.

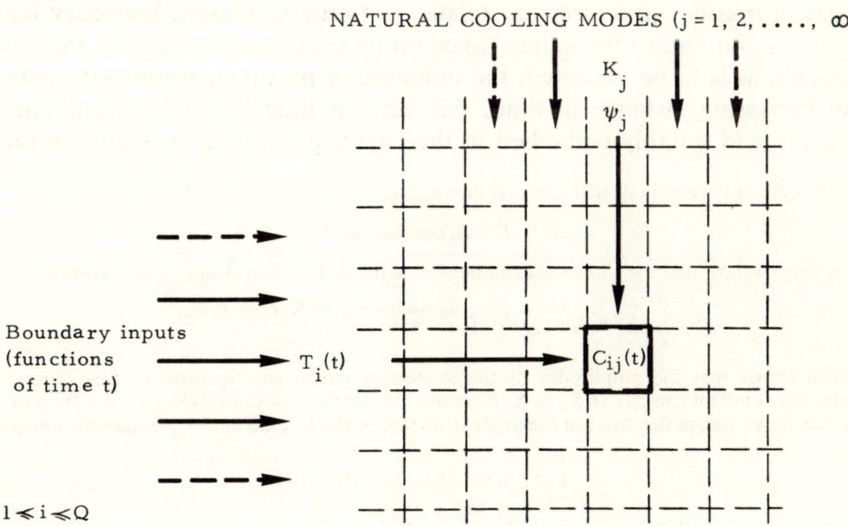

Figure 1.5 'Modal memory matrix' for boundary inputs

The elements of this matrix are precisely the varying values (continuously updated) of the convolution integrals $C_{ij}(t)$.

If we now rewrite the solution (1.42) as follows:†

$$\vartheta(x, y, z, t) = \sum_{i=1}^{Q} [T_i(t)(s_i - s_i^*) + \sum_{j=1}^{\infty} S_{ij}\psi_j(x, y, z)K_jC_{ij}(t)]$$
$$+ \sum_{j=1}^{\infty} A_j\psi_j(x, y, z)\,e^{-K_j(t-t_0)}, \tag{1.48}$$

it is seen that (apart from the time-independent first term and the last contribution, given by the initial conditions) the distribution $\vartheta(x, y, z, t)$ can be viewed as the sum of the 'modal shapes' $\psi_j(x, y, z)$, multiplied each into the coefficients $S_{ij}K_jC_{ij}(t)$, and then summed over j and i.

If we now analyse how these multiplicative coefficients

$$S_{ij}K_jC_{ij}(t) \tag{1.49}$$

vary with time t, we see easily on the basis of (1.44) that each of them would tend to decay exponentially (with time constant $1/K_j$), were it not for new contributions coming from the non-zero values, if any, taken on successively by $T_i(t)$. So, if a particular 'input excitation' $T_i(t)$ should abruptly cease, the cells of the corresponding row of the 'modal memory matrix' would thenceforth fade away exponentially. This fact justifies the name 'modal memory matrix' now introduced and throws a suggestive light on the physical meaning of the mathematical decomposition adopted in our formulation.

From the physical point of view, it was well known that structural thermal effects in massive structures are related not only to present boundary temperatures, but also to the values taken on by those temperatures in the past, although, as is to be expected, the influence of past temperatures decreases with increasing distance in time; this kind of time lag and attenuation is explicitly and suitably embodied in the convolution integrals $C_{ij}(t)$. In fact,

† In the case of purely sinusoidal external inputs, e.g.:

$$T_i = B_i \cos(\omega t + \varphi_i),$$

it is easy to show that the sinusoidal component of the distribution $\vartheta(x, y, z, t)$ is given by:

$$\sum_{i=1}^{Q} \sum_{j=1}^{\infty} S_{ij}\psi_j(x, y, z)K_jB_i \frac{\omega \sin(\omega t + \varphi_i) + K_j \cos(\omega t + \varphi_i)}{\omega^2 + K_j^2}$$

which shows that the amplitudes of the successive modal contributions to this sinusoidal component fall off roughly as S_{ij} as K_j increases substantially beyond $\omega (K_{j\max} \gg \omega)$. (It is to be kept in mind that, as the order of the mode, j, increases, the S_{ij} tend to zero because the integrals

$$\int_D s_i^*(x, y, z)\psi_j(x, y, z)\,dx\,dy\,dz$$

assign to an essentially symmetrical function, s_i^*, weight functions ψ_j that comprise an ever-increasing number of positive and negative 'waves'.)

although up to now we spoke only of the temperature distribution, it is quite obvious that in the case of a linearly elastic body very similar developments and expressions can be found for any structural effect of interest caused by these temperature distributions (see Section 1.5).

Moreover, since our whole formulation hinges on two kinds of time-independent distributions, which in a sense form the 'base space' of our decomposition, namely:

$$\begin{cases} s_i(x, y, z) & \text{(and } s^*\text{)}, \\ \psi_i(x, y, z), & \end{cases} \tag{1.50}$$

it is evident that it will be possible to operate through the technique of the 'influence functions', by pre-computing once and for all, for any effect of interest, the value it assumes under thermal distributions (1.50), and then combining linearly these 'influence coefficients' with weights identical to those multiplying each of the (1.50) in the expression of $\vartheta(x, y, z, t)$ (see (1.48)).

The concept will be developed in detail in the following paragraph.

1.5 EVALUATION OF THE THERMAL STRUCTURAL EFFECTS

The closing remarks of the preceding paragraph can be developed formally without difficulty.

Let us consider a given type of structural effect, say $\eta(x, y, z, t)$. We shall consider those F. E. structural-analysis programs, where the effect distribution

$$\eta(x, y, z, t) \tag{1.51}$$

due to a thermal distribution

$$\vartheta(x, y, z, t) \tag{1.52}$$

can be *directly* determined, this last distribution being known through (1.48).

But we can also proceed in an indirect way. Suppose we precompute a set of influence coefficient for the effect η. In particular, let us use our F.E. structural analysis with the following inputs (the outputs are correspondingly indicated in the scheme):

$$\begin{array}{ccc} \text{INPUT} & & \text{OUTPUT} \\ \text{(thermal distr.)} & & \text{(effect } \eta\text{)} \\ s_i(x, y, z) & \rightarrow & \lambda_i(x, y, z) \\ s_i^*(x, y, z) & \rightarrow & \lambda_i^*(x, y, z) \\ \psi_j(x, y, z) & \rightarrow & E_j(x, y, z) \end{array} \tag{1.53}$$

Looking at the expression (1.48) of $\vartheta(x, y, z, t)$ as a linear function of s_i, s_i^*, ψ_j, we may immediately write down $\eta(x, y, z, t)$ as follows:

$$\eta(x, y, z, t) = \sum_{i=1}^{Q}\left[T_i(t)(\lambda_i - \lambda_i^*) + \sum_{j=1}^{\infty} S_{ij}K_jE_j(x, y, z)C_{ij}(t)\right]$$

$$+ \sum_{j=1}^{\infty} A_jE_j(x, y, z)\,\mathrm{e}^{-K_j(t-t_0)}. \tag{1.54}$$

Since λ_i, λ_i^*, E_j are now known quantities, evaluation of formula (1.54) for any time t is in principle a straightforward matter. Of course, also in this case the difficulty of evaluating to an acceptable degree of approximation the sum of the infinite series appearing in (1.54) (without having to treat numerically a forbiddingly high number, J, of N.C.M.s) must be reckoned with; this particular aspect will be dealt with in the next paragraph.

If, however, a satisfactory solution to this problem can be found (so that the number J of N.C.M.s to be explicitly treated is a reasonable one—say no more than 20), then it is evident that the indirect approach indicated by (1.54) is preferable to the direct one as soon as the number of time instants t to be considered is large, and it becomes necessary to use this approach if the appraisal of $\eta(x, y, z, t)$ is to be made in real time.

In fact, the direct approach would entail, for large structures (such as dams) the use of the very heavy, time-consuming numerical computation program (e.g. a full three-dimensional stress analysis program) for each new time instant t to be considered.

On the other hand, (1.54) entails simple multiplications and sums, with known numerical coefficients (λ_i, λ_i^*, E_j, S_{ij}, K_j, A_j) of quantities either directly measured (T_i) or readily amenable to continuous updating (C_{ij}).

Provided the infinite summations over subscript j can be replaced by finite sums, this kind of numerical treatment requires neither heavy programs nor appreciable computer time. It could, therefore, be considered for on-line monitoring, implementing all operations on an on-site, inexpensive micro-computer.

1.6 A POSSIBLE DEVICE TO REPLACE THE INFINITE SUMMATIONS OVER SUBSCRIPT *j* WITH FINITE (TRUNCATED) SERIES

Suppose the summations over subscript j indicated in (1.50) and in (1.54) are truncated at $j = J$. Formally we could then write:

$$s_i^*(x, y, z) = \sum_{j=1}^{J} S_{ij}\psi_j(x, y, z) + F_i(x, y, z) \tag{1.55}$$

in place of (1.39). As s_i^* is known [see (1.38)], and likewise the coefficients

S_{ij} [see (1.40)] and the distributions $\psi_j(x, y, z)$, (1.55) allows us to define Q distributions $F_i(x, y, z)$, $1 \leq i \leq Q$, as

$$F_i(x, y, z) = s_i^*(x, y, z) - \sum_{j=1}^{J} S_{ij}\psi_j(x, y, z). \tag{1.56}$$

A distribution like $F_i(x, y, z)$ existing at time τ (notice that, on the surface S, $F_i(P_s) \equiv 0$), subject to the heat-conduction law with homogeneous boundary conditions, would decay in time certainly without keeping the initial shape,† but certainly faster than $e^{-K_J(t-\tau)}$. This is because by comparison with (1.39) we can write

$$F_i(x, y, z) = \sum_{j=J+1}^{\infty} S_{ij}\psi_j(x, y, z), \tag{1.57}$$

and we suppose that all the positive constants K_j are ordered with j increasing, so that

$$K_j > K_J \qquad \text{for any } j > J. \tag{1.58}$$

On the other hand, in any development like (1.55), the omission of terms like F_i would entail serious errors if J were kept to reasonable values, due to the very slow convergence of the series

$$\sum_j S_{ij}\psi_j(x, y, z). \tag{1.59}$$

From the above considerations, the idea arose of treating formally each of the Q distributions F_i as if it were a $(J+1)$th term of the summation over subscript j, of the type $S_{i,J+1}\psi_{J+1}$, associated with a suitable time constant $1/K^*$, where $K^* > K_J$. With this provisional device in mind, let us retrace formally all developments subsequent to (1.39). In particular

$$w_i(x, y, z, t-\tau) \cong s_i(x, y, z) - \sum_{j=1}^{J} S_{ij}\psi_j(x, y, z)\, e^{-K_j(t-\tau)}$$
$$-F_i(x, y, z) e^{-K^*(t-\tau)} \tag{1.60}$$

in place of (1.41); and it follows that in place of (1.42) we have

$$\vartheta(x, y, z, t) \cong \sum_{i=1}^{Q} \Big[T_i(t)(s_i - s_i^*) + \sum_{j=1}^{J} S_{ij}K_j\psi_j(x, y, z)C_{ij}(t)$$
$$+K^*F_i(x, y, z)C_{i,J+1}(t) \Big] + \sum_{j=1}^{J^*} A_j\psi_j(x, y, z)\, e^{-K_j(t-t_0)} \tag{1.61}$$

† On the contrary, the fundamental property of the N.C.M.s ψ_j, as already stated, is that under similar conditions each of them decays keeping its characteristic modal shape. In the analytical developments, each ψ_j is affected by a factor $e^{-K_j(t-\tau)}$.

for $t \geq t_0$; in the last term obviously the symbol $C_{i,J+1}$ stands for

$$C_{i,J+1}(t) = \int_{t_0}^{t} T_i(\tau)\, \mathrm{e}^{-K^*(t-\tau)}\, \mathrm{d}\tau, \tag{1.62}$$

and, as for the last term of (1.61), the summation has been truncated at a suitable value J^* of subscript j (not necessarily coincident with J, although in practice it is usually convenient to assume $J^* = J$), knowing that the error thus introduced on the evaluation of this last term will decay exponentially in time faster than $\mathrm{e}^{-K_{J^*}(t-t_0)}$, until it eventually vanishes.

Likewise, when evaluating structural thermal effects, (1.54) will be replaced by

$$\begin{aligned}\eta(x, y, z, t) = \sum_{i=1}^{Q} \Big[T_i(t)(\lambda_i - \lambda_i^*) + \sum_{j=1}^{J} S_{ij} K_j E_j(x, y, z) C_{ij}(t) \\ + K^* E_{i,J+1}(x, y, z) C_{i,J+1}(t) \Big] \\ + \sum_{j=1}^{J^*} A_j E_j(x, y, z)\, \mathrm{e}^{-K_j(t-t_0)},\end{aligned} \tag{1.63}$$

where the $E_{i,J+1}$ are defined as the structural effects (*outputs*) pertaining to distributions $F_i(x, y, z)$ taken as thermal *inputs* in the structural analysis program.

It is to be remarked that the structure of (1.61) and (1.63) ensures that the effects of a given $\Delta T_i(\tau)$ will be correctly assessed for $t = \tau$ and $t - \tau \to \infty$.

On the other hand, it is also to be kept in mind that the *computed* N.C.M.s tend to differ from the actual N.C.M.s more and more for $j \to \infty$, due e.g. to the inevitable inhomogeneities present in the material, so that the point of pursuing an 'exact' appraisal of the infinite summation with respect to subscript j is more an apparent than a real one. Also, the inevitable errors affecting the measured T_is, as well as the approximations introduced by the boundary interpolation procedure (via the N_is) and by the imperfect knowledge of initial conditions, all tend to make rather pointless the quest for an 'exact' appraisal of the above-cited summation.

It now remains to be seen if the approximation introduced by formulae such as (1.61), (1.63) is acceptable and, if so, what is the most suitable value to be attributed to the constant K_{J+1}.

A series of numerical tests was performed on the reconstruction of the thermal distribution by (1.61). As a reference, a thermal distribution due to exactly time-sinusoidal boundary conditions was chosen, arranging things so as to exclude the effects of initial transients, if any; in other words, only the steady-state, periodic-régime distribution was considered. This can be computed independently by very accurate specialized F.E. programs (BIREPE for two-dimensional domains, TRIREPE for three-dimensional domains) and the results of (1.58) can be compared with this 'reference solution' for different values of t.

The above-cited numerical results will be illustrated in Section 1.8. However, even without resorting to numerical checking it is quite evident that, if the shortest time-constants coming into play for boundary conditions are of the order Δt, it will be necessary to consider an upper value J of the subscript j such that

$$\frac{1}{K_J} < \Delta t. \tag{1.64}$$

1.7 F.E. FORMULATION OF THE MAIN COMPUTATIONAL STEPS. NUMERICAL EVALUATION AND UPDATING OF THE CONVOLUTION INTEGRALS, $C_{ij}(t)$

1.7.1 F.E. Formulation for basic functions $s_i(x, y, z)$, $\psi_j(x, y, z)$ with their associated K_js.

For numerical determination of the Q steady-state distributions $s_i(x, y, z)$, see (1.36), (1.37), the F.E. formulation is the classical one adopted for Laplace-type problems, namely:

$$[C]\{s_i\} = \{0\} \tag{1.65}$$

sub boundary conditions

$$s_i \text{ (boundary nodes)} = N_i \text{ (boundary nodes)}. \tag{1.66}$$

Obviously, in (1.65) $[C]$ is the so-called 'conductivity matrix'; $\{s_i\}$ is the vector of *nodal* values of s_i.

Many good F.E. programs are available for numerical computation of $\{s_i\}$. The writers use their own versions, which are called BICAMP and TRICAMP for two- and three-dimensional problems respectively.

As for numerical determination of the J N.C.M.s (natural cooling modes) ψ_j, see (1.27), (1.28), with their associated time constants $1/K_j$, the F.E. formulation is likewise a well-known one:

$$(a[C] - K_j[m])\{\psi_j\} = \{0\}, \tag{1.67}$$

sub boundary conditions:

$$\psi_j \text{ (boundary nodes)} \equiv 0. \tag{1.68}$$

$[C]$ is, as above, the 'conductivity matrix' and $[m]$ the 'capacity', or 'mass', matrix.†

† Since we have used the value a of diffusivity as in (1.67), the conductivity matrix $[C]$ has to be computed with a unit value of the conductivity constant (k), and the mass matrix $[m]$ likewise has to be computed with a unit value of the thermal capacity per unit volume (c). (As is well known, $a = k/c$.)

It is evident that a non-trivial solution

$$\psi_j(x, y, z) \qquad \text{(not everywhere zero)} \tag{1.69}$$

can be obtained only for those particular values (*eigenvalues*) K_j such that

$$\det(a[C]-K_j[m])=0. \tag{1.70}$$

(1.70) yields numerical estimation of the eigenvalues K_j (up to a certain order), and for each of these (1.67) then yields the corresponding modal distribution ψ_j, apart from an arbitrary multiplicative factor. This arbitrariness is removed by the normalization condition (1.34).

The F.E. program used by the authors to solve (1.67) has been called DETER; it uses the well-known Rutishauser technique to determine numerically the series of the eigenvalues K_j and the associated spatial distributions $\psi_j(x, y, z)$.

1.7.2 Numerical evaluation and continuous updating of the convolution integrals, $C_{ij}(t)$

Reference is made to (1.44). Temperatures $T_i(\tau)$, $1 \leq i \leq Q$, are sampled at time instants t_l. Figure 1.6 shows how the value $C_{ij}(t_{n+1})$ can be gleaned from the previously computed value, $C_{ij}(t_n)$.

Analytically, one can obviously write, according to Figure 1.6:

$$C_{ij}(t_{n+1}) = C_{ij}(t_n)\,\mathrm{e}^{-K_j(t_{n+1}-t_n)} + \int_{t_n}^{t_{n+1}} T_i(\tau)\,\mathrm{e}^{-K_j(t_{n+1}-\tau)}\mathrm{d}\tau. \tag{1.71}$$

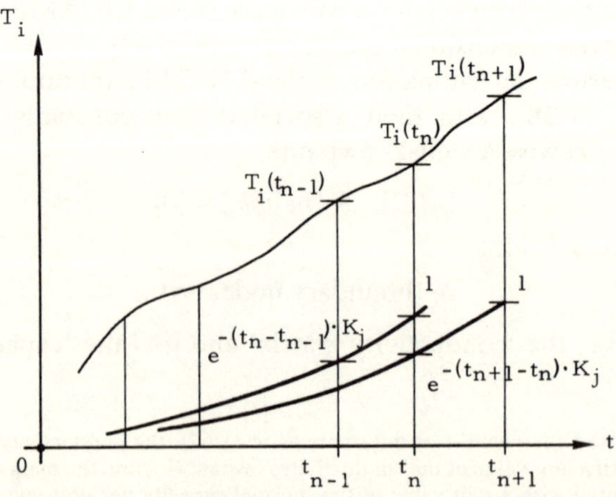

Figure 1.6 Evaluation of convolution integrals $C_{ij}(t)$

Now, in the absence of any specific information about the behaviour of $T_i(t)$ for $t_n \leqslant t \leqslant t_{n+1}$, the only reasonable assumption for numerical computation is that of a linear variation of T_i with time:

$$T_i = T_i(t_n) + \frac{T_i(t_{n+1}) - T_i(t_n)}{t_{n+1} - t_n}(\tau - t_n) \qquad \text{for } t_n \leqslant \tau \leqslant t_{n+1}; \tag{1.72}$$

under this hypothesis one gets

$$\begin{aligned}
\int_{t_n}^{t_{n+1}} T_i(\tau)\,\mathrm{e}^{-K_j(t_{n+1}-\tau)}\,\mathrm{d}\tau &= \int_{t_n}^{t_{n+1}} T_i(t_n)\,\mathrm{e}^{-K_j(t_{n+1}-\tau)}\,\mathrm{d}\tau \\
&\quad + \frac{T_i(t_{n+1}) - T_i(t_n)}{t_{n+1} - t_n}\left(\int_{t_n}^{t_{n+1}} (\tau - t_n)\,\mathrm{e}^{-K_j(t_{n+1}-\tau)}\,\mathrm{d}\tau\right) \\
&= \left[\frac{T_i(t_n)}{K_j}\,\mathrm{e}^{-K_j(t_{n+1}-\tau)}\right]_{\tau=t_n}^{\tau=t_{n+1}} + \frac{T_i(t_{n+1}) - T_i(t_n)}{t_{n+1} - t_n} \\
&\quad \times \left[\frac{1}{K_j}(\tau - t_n)\,\mathrm{e}^{-K_j(t_{n+1}-\tau)} - \frac{1}{K_j^2}\,\mathrm{e}^{-K_j(t_{n+1}-\tau)}\right]_{\tau=t_n}^{\tau=t_{n+1}} \\
&= \frac{T_i(t_n)}{K_j}\left[1 - \mathrm{e}^{-K_j(t_{n+1}-t_n)}\right] \\
&\quad + \frac{T_i(t_{n+1}) - T_i(t_n)}{t_{n+1} - t_n}\left[\frac{1}{K_j}(t_{n+1} - t_n) - \frac{1}{K_j^2}\left(1 - \mathrm{e}^{-K_j(t_{n+1}-t_n)}\right)\right] \\
&= \frac{T_i(t_{n+1})}{K_j} - \frac{T_i(t_n)}{K_j}\,\mathrm{e}^{-K_j(t_{n+1}-t_n)} \\
&\quad + \frac{T_i(t_{n+1}) - T_i(t_n)}{K_j^2(t_{n+1} - t_n)}\left[\mathrm{e}^{-K_j(t_{n+1}-t_n)} - 1\right]
\end{aligned} \tag{1.73}$$

Recursive use of (1.71), together with (1.73) and with the obvious equation for initialization:

$$C_{ij}(t_0) = 0, \tag{1.74}$$

enables one to evaluate and update $C_{ij}(t)$ in a continuous way as new values of T_i come in at successive times, without introducing any unnecessary hypothesis or approximation.

1.8 NUMERICAL RESULTS OF THE PROPOSED PROCEDURE

1.8.1 Comparison of temperature distributions

In order to check the practical applicability of the proposed numerical procedure, a two-dimensional domain of simple shape was chosen.

The domain is a rectangle, 500×2000 cm, with thermal diffusivity $40\ \text{cm}^2\ \text{h}^{-1}$. The boundary conditions are homogeneous on the two short sides and on one of the long sides; the other long side is subjected to a temperature distribution sinusoidal in time, with period $T = 8760$ h (1 yr), and triangular along the side, with zero value at the ends and maximum amplitude 10 °C at mid-point (see Figure 1.7). The total number, Q, of 'thermometric points' along the boundary was 5. The F.E. mesh used to compute the Q steady-state distributions s_i† and the N.C.M.s ψ_j included 128 elements and 297 nodes.

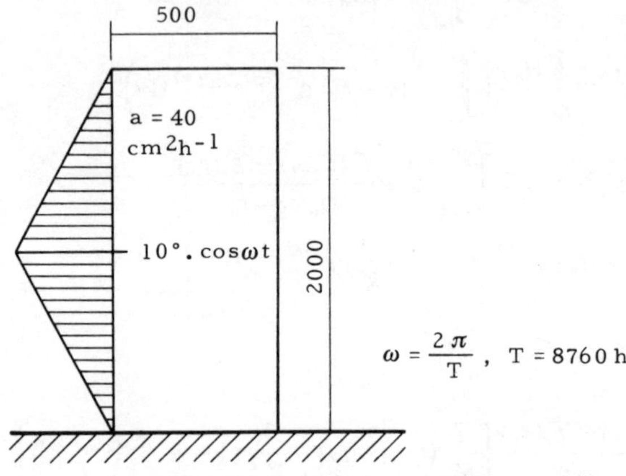

Figure 1.7 Two-dimensional example

The inner distribution of temperatures under régime conditions, i.e. after dying down of the initial transient, was computed on one hand using the BIREPE F.E. program, which by independent checking is known to give excellent results (for purely sinusoidal time variations), and on the other hand with a numerical program embodying the developed formulae above. These last computations were effected for different values of J, i.e. $J = 5$, 10, 17 and 40, and for 12 equidistant time instants covering one period.‡

Figures 1.8–15 summarize the essential results of the comparison, for $J = 17$ and $J = 40$ and for 7 of the 12 time instants referred to above.

It is seen that

(a) the approximation reached increases with increasing J, as was to be expected;

† By the BICAMP program.
‡ The boundary cycle on the long side was sampled at regular intervals, equally spaced along the cycle. The comparison was effected for a subdivision of the whole cycle into 12 intervals ($t_{n+1} - t_n = 730$ h).

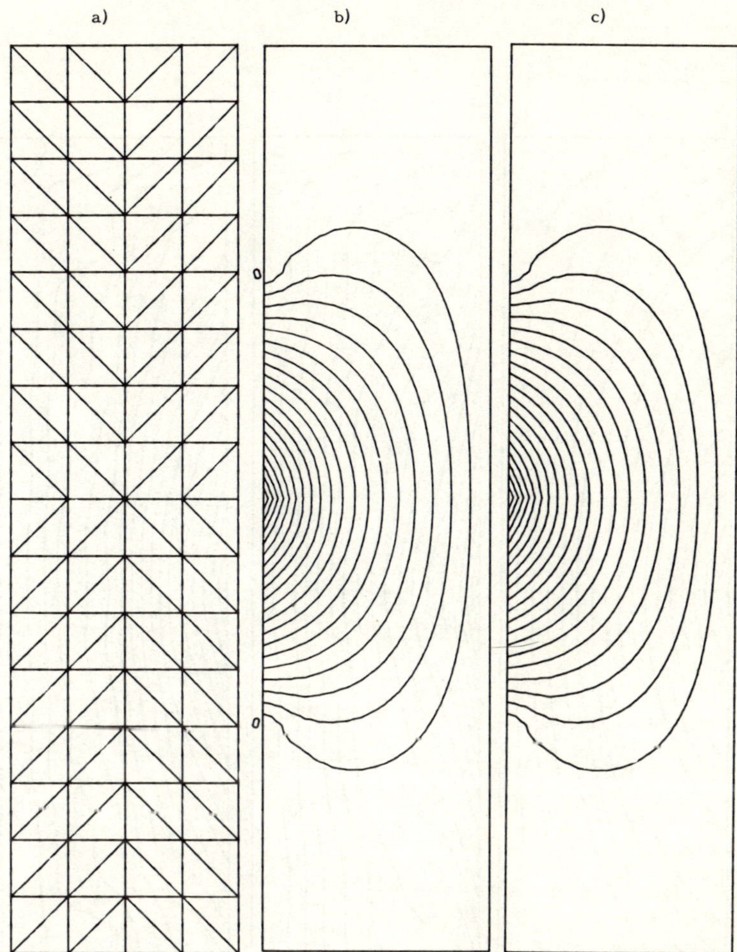

Figure 1.8 Steady-state distribution for central thermometer: (a) F.E. mesh used; (b) steady-state distribution according to BICAMP; (c) steady-state distribution according to superposition of 17 N.C.M. plus correcting function

(b) for J as small as 17 the results appear to be acceptable, keeping into account all the uncertainties that are bound to occur in practice;

(c) if the practical device alluded to under Section 1.6 is not used, i.e. if the summations over subscript j is simply truncated at $j = J$ without any further ado, the quality of the results is markedly worse (these bad results are not shown).

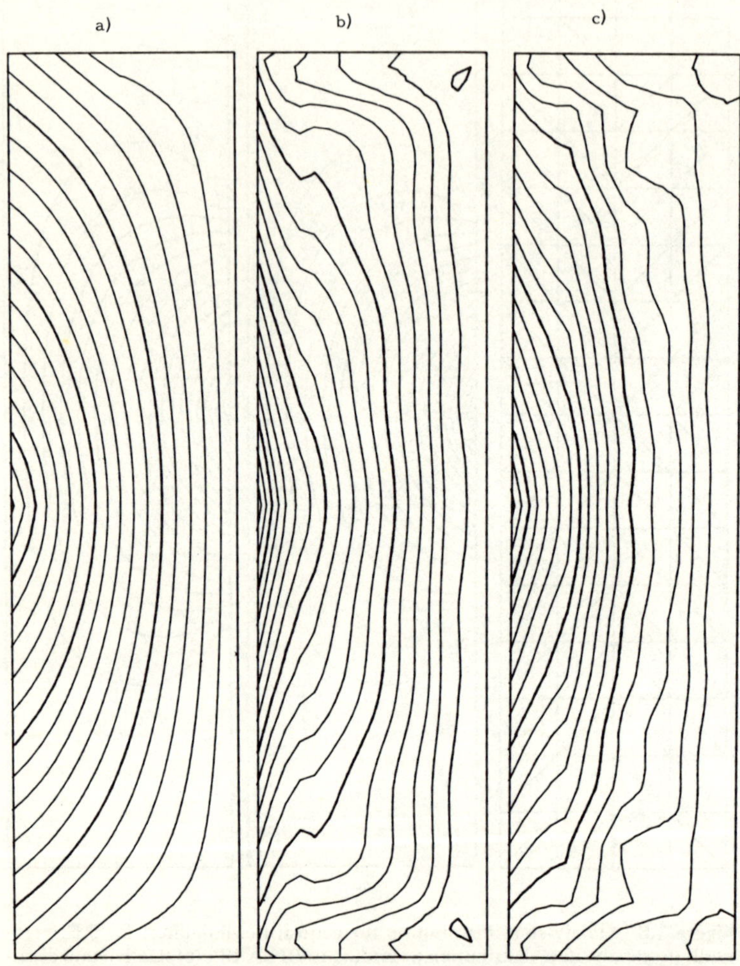

Figure 1.9 Sinusoidal régime—comparison between: (a) reference temperature distribution at $t = 730$ h ($\frac{1}{12}$ of period); (b) Temperature distribution according to proposed formula (Section 1.6) at $t = 730$ h with 17 N.C.M.; (c) temperature distribution according to proposed formula (Section 1.6) at $t = 730$ h with 40 N.C.M.

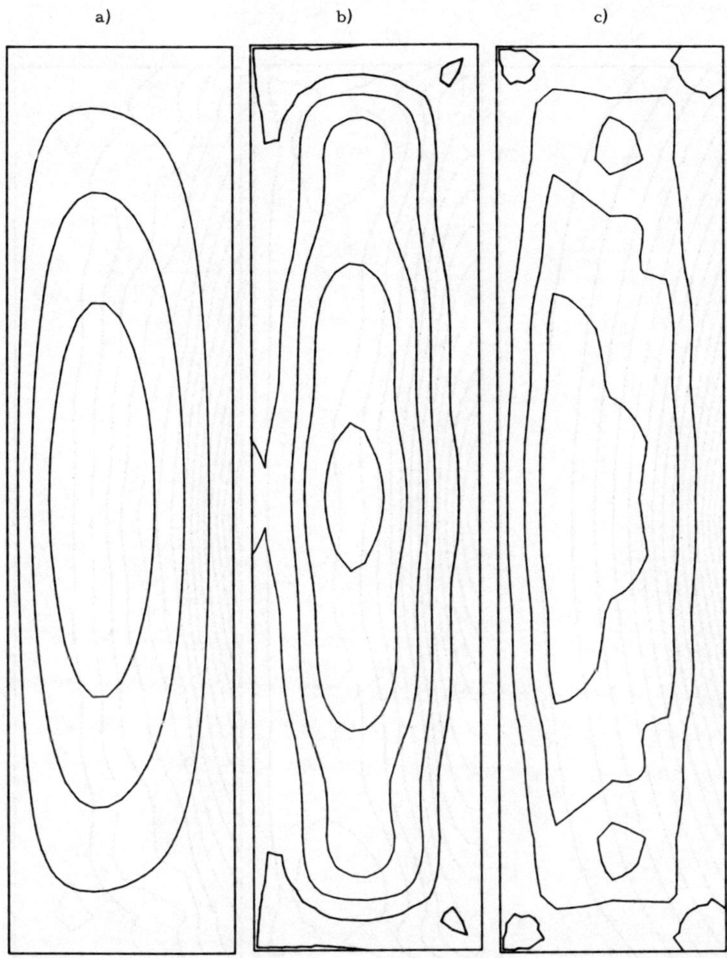

Figure 1.10 Sinusoidal régime—comparison between: (a) reference temperature distribution at $t = 2190$ h ($\frac{1}{4}$ of period), (b) temperature distribution according to proposed formula (Section 1.6) at $t = 2190$ h with 17 N.C.M.; (c) temperature distribution according to proposed formula (Section 1.6) at $t =$ 2190 h with 17 N.C.M.; (c) temperature distribution according to proposed formula (Section 1.6) at $t = 2190$ h with 40 N.C.M.

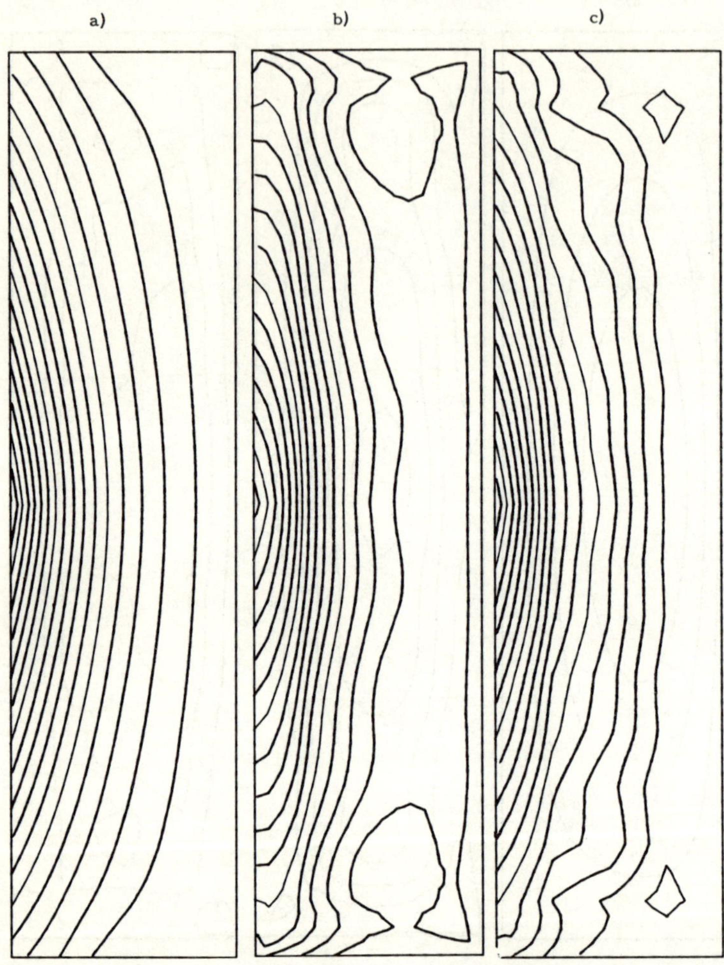

Figure 1.11 Sinusoidal régime—comparison between: (a) reference temperature distribution at $t = 3650$ h ($\frac{5}{12}$ of period); (b) temperature distribution according to proposed formula (Section 1.6) at $t = 3650$ h with 17 N.C.M.; (c) temperature distribution according to proposed formula (Section 1.6) at $t = 3650$ h with 40 N.C.M.

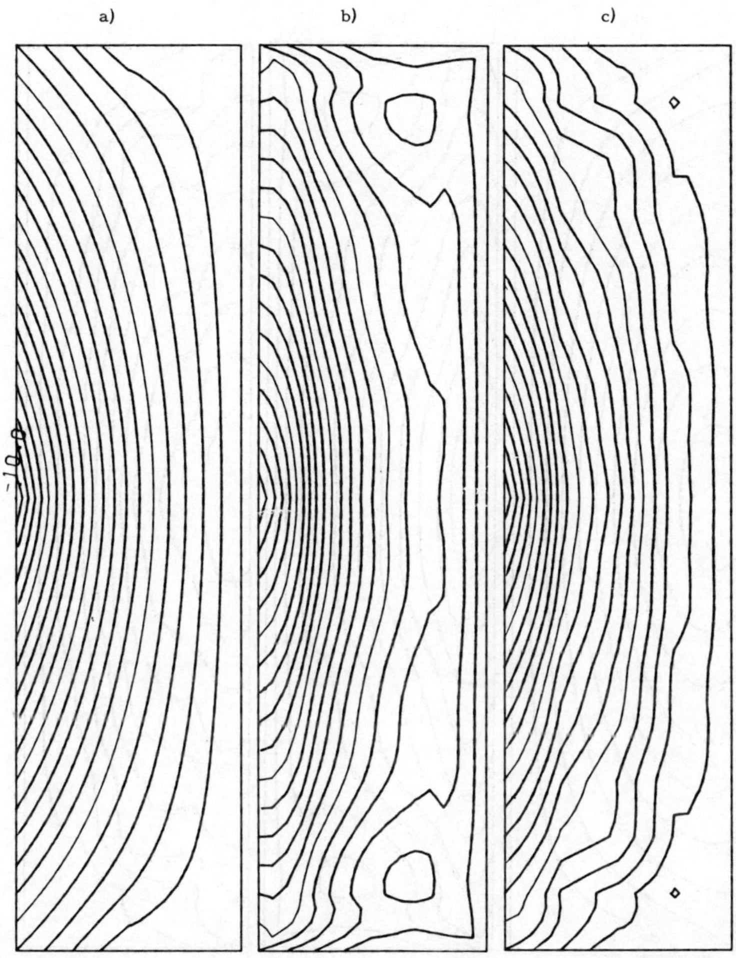

Figure 1.12 Sinusoidal régime—comparison between: (a) reference temperature distribution at $t = 4380$ h ($\frac{1}{2}$ of period); (b) temperature distribution according to proposed formula (Section 1.6) at $t = 4380$ h with 17 N.C.M.; (c) temperature distribution according to proposed formula (Section 1.6) at $t = 4380$ h with 40 N.C.M.

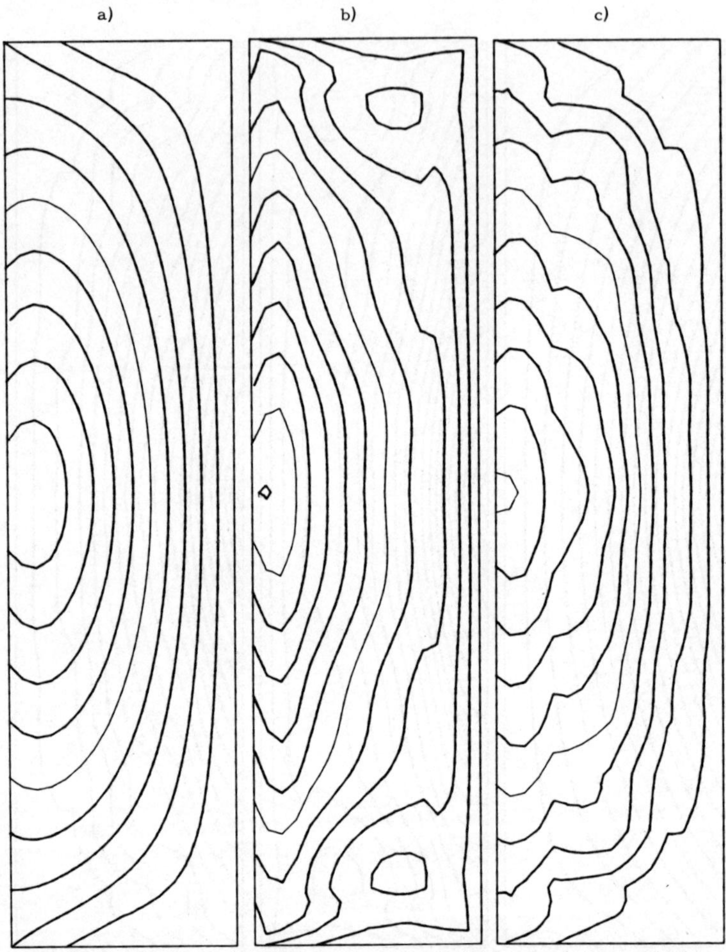

Figure 1.13 Sinusoidal régime—comparison between: (a) reference temperature distribution at $t = 5840$ h ($\frac{2}{3}$ of period); (b) temperature distribution according to proposed formula (Section 1.6) at $t = 5840$ h with 17 N.C.M.; (c) temperature distribution according to proposed formula (Section 1.6) at $t = 5840$ h with 40 N.C.M.

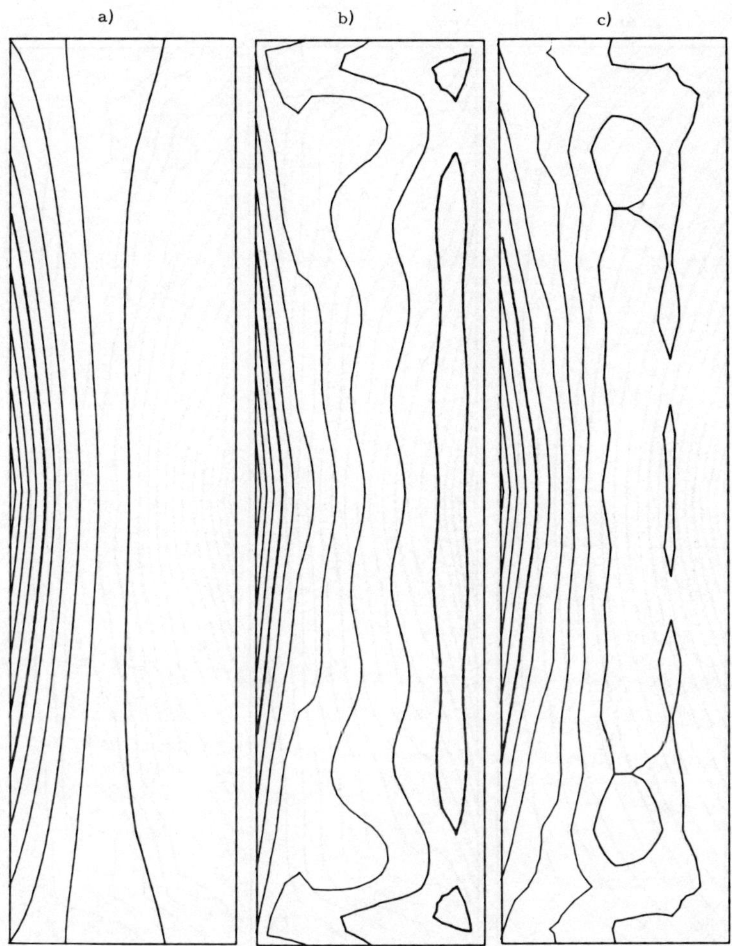

Figure 1.14 Sinusoidal régime—comparison between: (a) reference temperature distribution at $t = 7300$ h ($\frac{5}{6}$ of period); (b) temperature distribution according to proposed formula (Section 1.6) at $t = 7300$ h with 17 N.C.M.; (c) temperature distribution according to proposed formula (Section 1.6) at $t = 7300$ h with 40 N.C.M.

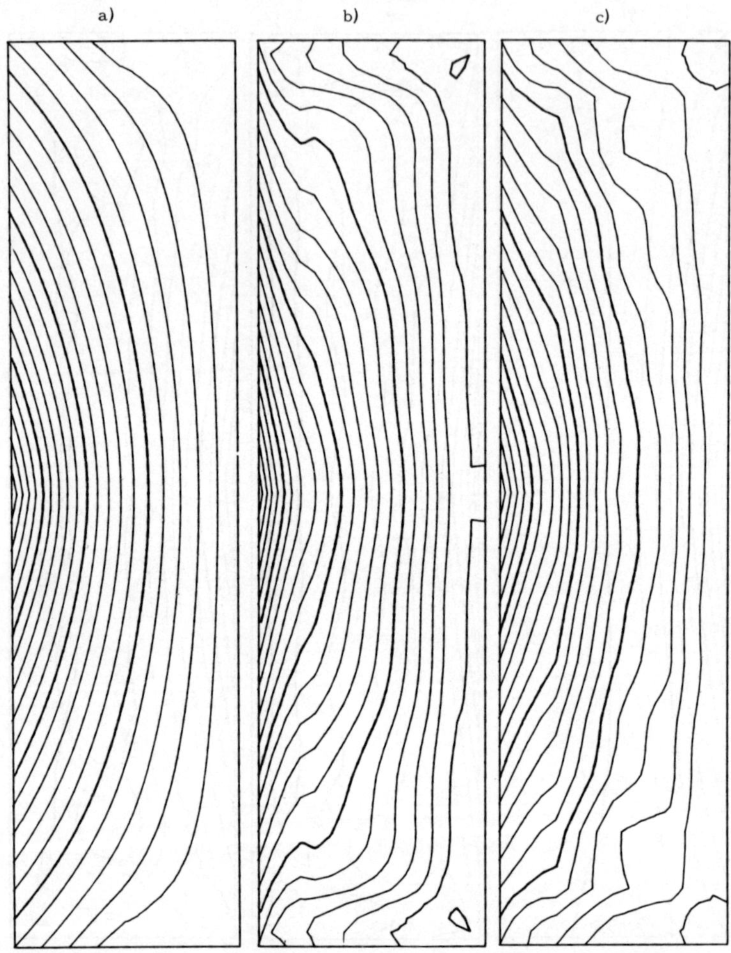

Figure 1.15 Sinusoidal régime—comparison between: (a) reference temperature distribution at $t = 8760$ h (full period); (b) temperature distribution according to proposed formula (Section 1.6) at $t = 8760$ h with 17 N.C.M.; (c) temperature distribution according to proposed formula (Section 1.6) at $t =$ 8760 h with 40 N.C.M.

1.8.2 Comparison of thermal displacements

For the same thermal régime above described, the corresponding displacement fields:

$$\delta_x(x, y, z) \qquad \text{and} \qquad \delta_y(x, y, z) \tag{1.75}$$

were computed for a value of the thermal dilatation coefficient

$$\alpha = 10^{-5}(°\text{C})^{-1}$$

and for built-in constraint conditions at the lower end of the cantilever (see Figure 1.7).

The computations were effected using on the one hand the 'reference' temperature distributions—at each of the 12 time instants along one cycle—as inputs to the PLATEN F.E. stress-analysis program; on the other hand, computing the influence functions E_j as described in Section 1.5 and using them together with the convolution integrals $C_{ij}(t)$ and the corrective functions F_i (see (1.63)).

The comparisons were carried out as for the thermal distributions (see Section 1.8.1), for $J = 17$, $J = 40$ respectively, and they are illustrated by Figures 1.16–19 for 7 time instants along the cycle. It is seen that:

(a) also in this case the approximation reached increases with increasing J; however, again the results are quite acceptable for J as small as 17;
(b) for a given value of J, the approximation reached for displacements appears to be better than that for temperatures. This was in a sense to be expected, because displacements are integral effects (as opposed to temperatures which are local variables) and thus the approximations introduced in evaluating the thermal field should be rendered insignificant in computing the displacements.

1.9 CONCLUSIONS

The above-described procedure is potentially adequate to satisfy the requirements put forward at the end of Section 1.1. Of course, there is a price to pay for the greater generality of this procedure as compared with previously proposed and tested ones.

The main drawback appears to be the necessity to go to a quite substantial order of the N.C.M.s (at least of the order of 20 for two-dimensional bodies, probably considerably more for three-dimensional geometries).

A practical rule of thumb for fixing this order (J) is to be sure that N.C.M.s are included possessing at least an inner zero line across the thickness (i.e. with at least one positive and one negative 'half-wave' across the thickness).

It would be interesting to develop efficient methods of truncating the summations over subscript j at a reasonably low value J while keeping an

Figure 1.16 Sinusoidal régime—comparison between: (a) reference distribution of displacements along x-axis at $t = 2190$ h ($\frac{1}{4}$ of period); (b) distribution of displacements along x-axis according to proposed formula (Section 1.6) at $t = 2190$ h with 17 N.C.M.; (c) distribution of displacements along x-axis according to proposed formula (Section 1.6) at $t = 2190$ h with 40 N.C.M.

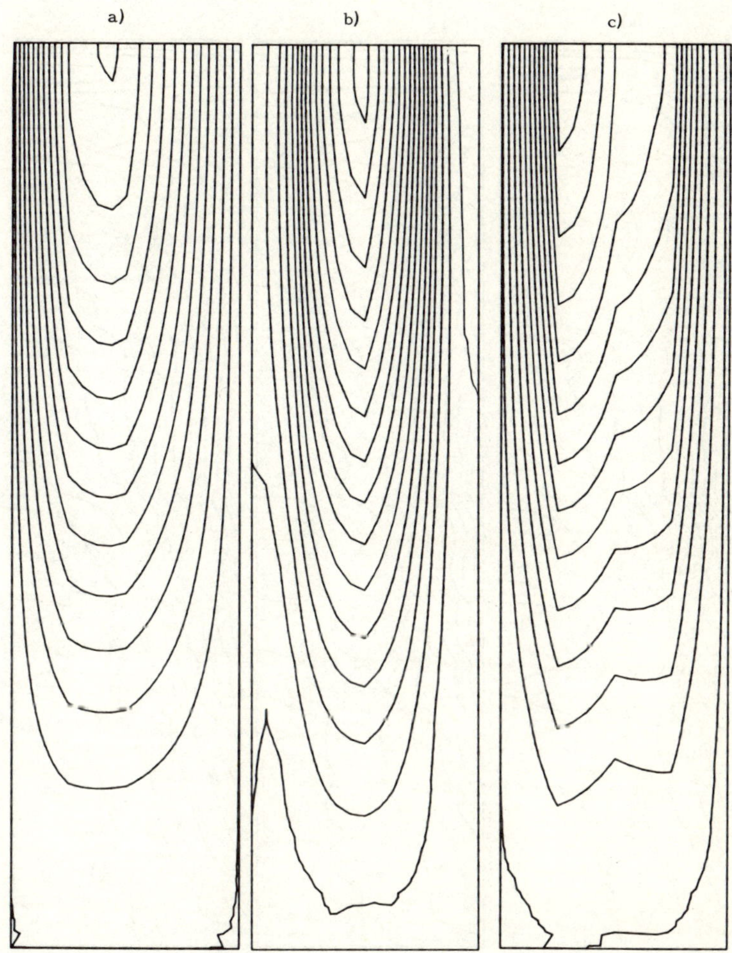

Figure 1.17 Sinusoidal régime—comparison between: (a) reference distribution of displacements along y-axis at $t = 2190$ h ($\frac{1}{4}$ of period); (b) distribution of displacements along y-axis according to proposed formula (Section 1.6) at $t = 2190$ h with 17 N.C.M.; (c) distribution of displacements along y-axis according to proposed formula (Section 1.6) at $t = 2190$ h with 40 N.C.M.

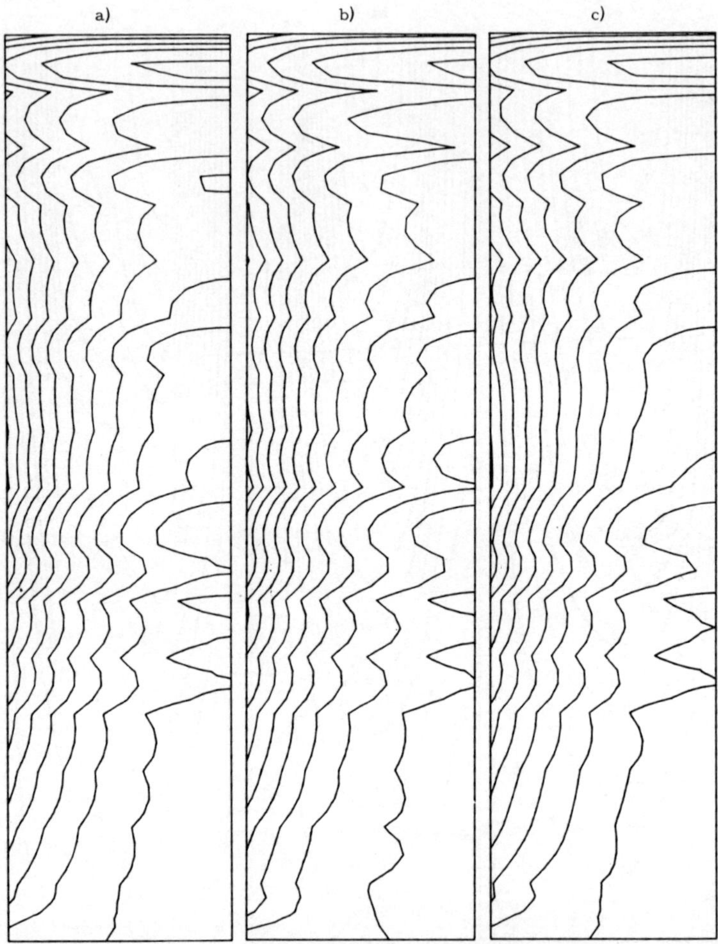

Figure 1.18 Sinusoidal régime—comparison between: (a) reference distribution of displacements along x-axis at $t = 4380$ h ($\frac{1}{2}$ of period; (b) distribution of displacements along x-axis according to proposed formula (Section 1.6) at $t = 4380$ h with 17 N.C.M.; (c) distribution of displacements along x-axis according to proposed formula (Section 1.6) at $t = 4380$ h with 40 N.C.M.

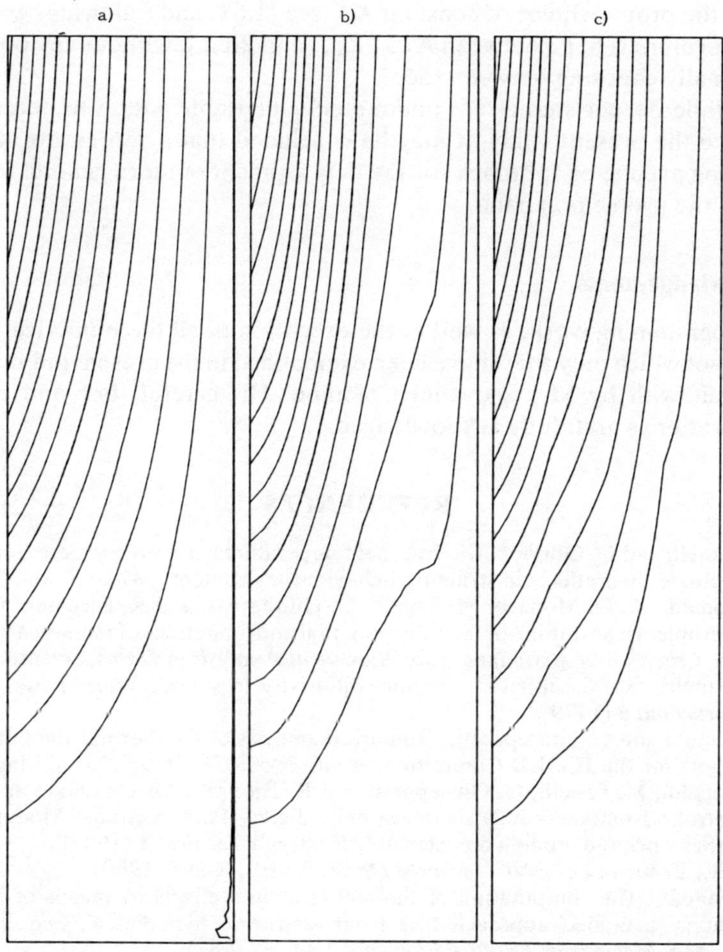

Figure 1.19 Sinusoidal régime—comparison between: (a) reference distribution of displacements along y-axis at $t = 4380$ h ($\frac{1}{2}$ of period), (b) distribution of displacements along y-axis according to proposed formula (Section 1.6) at $t = 4380$ h with 17 N.C.M.; (c) distribution of displacements along y-axis according to proposed formula (Section 1.6) at $t = 4380$ h with 40 N.C.M.

acceptably high degree of approximation in the evaluation of the relevant series. In this connection, the practical device proposed in Section 1.6 is no more, for the moment being, than an empirical short-cut.

Also the proper choice of constant K^* (see (1.61) and following) was found by practical trials to lie between K_{J+1}, K_{J+2} and $2K_{J+1}$, without any possibility of rationally choosing between them.

So, while deeper studies are undoubtedly advisable and even necessary to complete the present study, it may be concluded that it represents at least a promising avenue of approach for the development of more general formulations of the problem in hand.

Acknowledgements

The programming work, as well as the execution of all the numerical explorations—of which only a few have been exemplified in the present publication—was dealt with by Mr. Giovanni Colombo. His careful, fast and effective collaboration is gratefully acknowledged.

REFERENCES

1. M. Fanelli and G. Giuseppetti, 'Internal temperatures in mass concrete: techniques to evaluate their effects in structure behaviour assessment', *Water Power*, **7** (1975).
2. P. Bonaldi, A. Di Monaco, M. Fanelli, G. Giuseppetti and R. Riccioni, 'Concrete dam problems: an outline of the role, potential and limitations of numerical analysis', *Symp. Criteria and Assumptions for Numerical Analysis of Dams*, Swansea (1975).
3. M. Fanelli, G. Giuseppetti, 'Thermal diffusivity in solids', *Water Power & Dam Construction*, **5** (1979).
4. M. Fanelli and G. Giuseppetti, 'Numerical analysis of the thermal state of a dam', A report for the ICOLD Committee, *Analysis and Design of Dams*, May (1980).
5. P. Bonaldi, M. Fanelli, G. Giuseppetti and R. Riccioni, 'Osservazione automatica e controllo istantaneo della sicurezza delle dighe—Parte seconda: Modelli deterministici a priori e modelli a posteriori', *Rassegna Tecnica*, **1** (1981).
6. ENEL, *Behaviour of ENEL's Large Dams*, ENEL, Rome (1980).
7. M. Fanelli, 'The computation of thermal structural effects by means of influence functions: a unified approach free from restrictive hypotheses', 2*nd Int. Conf. Numerical Methods in Thermal Problems*, Venice (1981).

Numerical Methods in Heat Transfer, Volume II
Edited by R. W. Lewis, K. Morgan, and B. A. Schrefler

Chapter 2

Exact Finite Element Solutions for Linear Steady-state Thermal Problems

Earl A. Thornton, Pramote Dechaumphai, and Kumar K. Tamma

SUMMARY

An approach for developing exact conduction–convection finite elements is presented. For one-dimensional problems and a quasi two-dimensional problem interpolation functions are derived based on solutions to the governing differential equations. Exact interpolation functions are presented for combined heat transfer in several solids of different shapes and for combined heat transfer in a flow passage. Numerical results demonstrate that exact elements offer significant advantages over elements based on approximate interpolation functions.

2.1 INTRODUCTION

A significant development in finite element methodology for forced convection analysis is the concept of upwind weighting functions.[1] For the one-dimensional convective-diffusion equation upwind weighting functions are expressed in terms of an upwind parameter which can be varied to adjust the accuracy of the finite element solution. Optimal values of the upwind parameter lead to exact values of the temperature at nodal points. Other elements in one-dimensional thermal and structural analysis also predict exact nodal values. For example, one-dimensional solutions[2] for heat conduction with internal heat generation by linear interpolation functions yield exact temperatures at the nodes. Formulations and examples of exact one dimensional structural elements appear in two recent papers.[3,4] The ability of finite elements to predict exact nodal temperatures has previously been regarded as a property of a few particular equations and to be of limited generality. There is, however, a broad class of one-dimensional problems governed by linear ordinary differential equations for which finite elements can be developed to yield exact solutions. The one-dimensional convective-diffusion equation and the one-dimensional conduction equation are special cases of this class of problems. In a previous paper,[5] the authors have presented two examples of one-dimensional thermal–structural finite elements which yield exact solutions for linear problems.

The purpose of the paper is to present an approach for developing conduction–convection finite elements which yield exact nodal temperatures and an exact variation of temperature within an element for steady-state linear analysis. One-dimensional conduction–convection problems are considered first. A nodeless parameter approach for deriving exact interpolation functions is presented and illustrated for several cases. The general form of the element equations is presented and discussed. An exact quasi two-dimensional conduction element for conduction in aerospace structures with thermal protection systems is then developed. Finally, the benefits of the exact finite elements are illustrated for four problems by comparing numerical solutions from finite elements with exact and linear interpolation functions.

2.2 EXACT ONE-DIMENSIONAL ELEMENTS

2.2.1 Governing equations and boundary conditions

The geometry and terminology for eight one-dimensional conduction–convection cases are shown in Figure 2.1. Cases 1–7 are solids of various shapes where heat transfer consists of: (a) conduction, which may be combined with, (b) surface convection, (c) internal heat generation, and (d) surface heating. Case 8 is a one-dimensional flow where the heat transfer consists of fluid conduction and mass-transport convection, which may be combined with: (b) surface convection, and (d) surface heating. In the figure, $\dot{Q}$ is the volumetric heat generation rate, $\dot{q}$ is the surface heating rate, h is the convective heat transfer coefficient and T_∞ is the environmental temperature for the convection heat exchange. Heat transfer in Cases 1–8 is described by differential equations of the form

$$a_0(x)\frac{\mathrm{d}^2T}{\mathrm{d}x^2}+a_1(x)\frac{\mathrm{d}T}{\mathrm{d}x}+a_2(x)T=a_3(x) \tag{2.1}$$

where $a_i = 0, 1, 2, 3$, are functions which depend on the geometry and thermal parameters of each case. Finite element equations for Equation (2.1) may be formulated by the method of weighted residuals, but due to the presence of the odd-order derivative an unsymmetrical coefficient matrix results. However, if Equation (2.1) is first cast into self-adjoint form, the differential operator is symmetric. Then, finite element matrices can be derived by the method of weighted residuals or a variational method, and coefficient matrices are symmetrical.[2] Equation (2.1) is written in self-adjoint form by multiplying

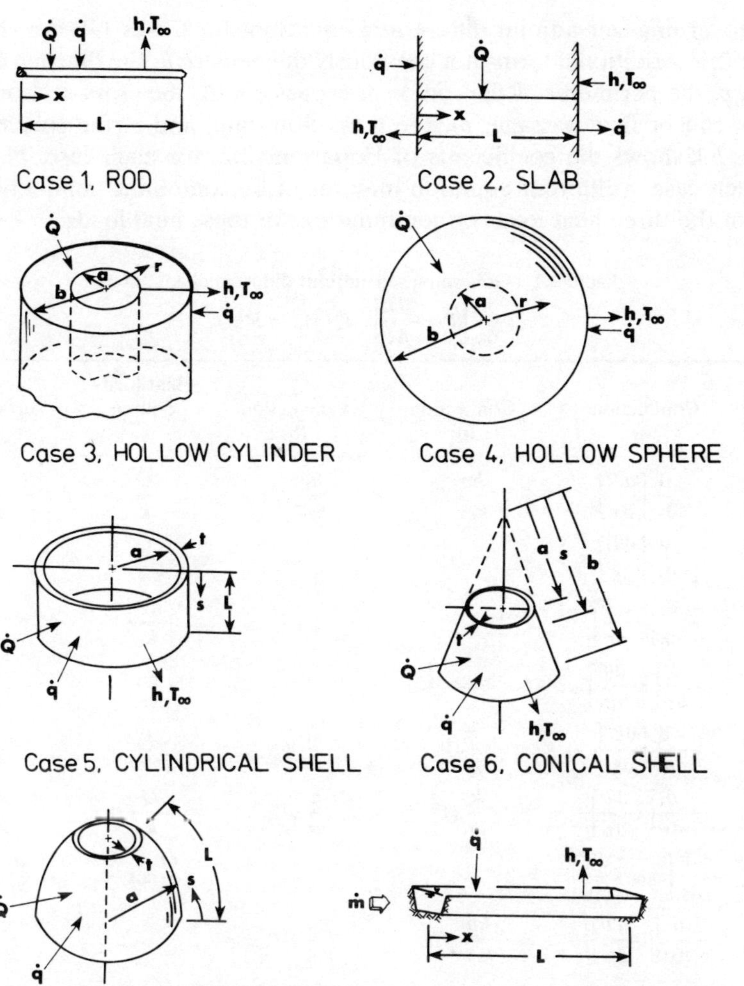

Figure 2.1 One-dimensional conduction and convection cases

it by the factor

$$P = \exp\left[\int (a_1/a_0)\,\mathrm{d}x\right]$$

so

$$\frac{\mathrm{d}}{\mathrm{d}x}\left[P(x)\frac{\mathrm{d}T}{\mathrm{d}x}\right] + Q(x)T = R(x) \qquad (2.2)$$

where

$$Q(x) = a_2P/a_0 \qquad \text{and} \qquad R(x) = a_3P/a_0.$$

The governing self-adjoint differential equations for Cases 1–8 are shown in Table 2.1. Additional terms not previously defined are k, the thermal conductivity, p, the perimeter of the rod or flow passage, A, the cross-sectional area of the rod or flow passage, $\dot{m}$, the mass flow rate, and c, the specific heat. Table 2.1 shows the coefficients of Equation (2.2) for each case. Note that for each case a different equation may result by combining conduction with each of the three heat loads or combinations of these heat loads.

Table 2.1 Governing self-adjoint differential equations

$$\frac{\mathrm{d}}{\mathrm{d}x}\left[P(x)\frac{\mathrm{d}T}{\mathrm{d}x}\right]+Q(x)T=R(x)$$

Case	Conduction (a)	Convection (b)	Convection (b)	Heat loads Source (c)	Surface flux (d)
1	$-\frac{\mathrm{d}}{\mathrm{d}x}\left[\frac{\mathrm{d}T}{\mathrm{d}x}\right]$	$\frac{hp}{kA}T$	$\frac{hp}{kA}T_\infty$	$\frac{\dot{Q}}{k}$	$\frac{\dot{q}p}{kA}$
2	$-\frac{\mathrm{d}}{\mathrm{d}x}\left[\frac{\mathrm{d}T}{\mathrm{d}x}\right]$	—	—	$\frac{\dot{Q}}{k}$	—
3	$-\frac{\mathrm{d}}{\mathrm{d}r}\left[r\frac{\mathrm{d}T}{\mathrm{d}r}\right]$	—	—	$\frac{\dot{Q}}{k}r$	—
4	$-\frac{\mathrm{d}}{\mathrm{d}r}\left[r^2\frac{\mathrm{d}T}{\mathrm{d}r}\right]$	—	—	$\frac{\dot{Q}}{k}r^2$	—
5	$-\frac{\mathrm{d}}{\mathrm{d}s}\left[\frac{\mathrm{d}T}{\mathrm{d}s}\right]$	$\frac{h}{kt}T$	$\frac{h}{kt}T_\infty$	$\frac{\dot{Q}}{k}$	$\frac{\dot{q}}{kt}$
6	$-\frac{\mathrm{d}}{\mathrm{d}s}\left[s\frac{\mathrm{d}T}{\mathrm{d}s}\right]$	$\frac{h}{kt}sT$	$\frac{h}{kt}sT_\infty$	$\frac{\dot{Q}}{k}s$	$\frac{\dot{q}}{kt}s$
7	$-\frac{\mathrm{d}}{\mathrm{d}s}\left[\cos\frac{s}{a}\frac{\mathrm{d}T}{\mathrm{d}s}\right]$	—	—	$\frac{\dot{Q}}{k}\cos\frac{s}{a}$	$\frac{\dot{q}}{kt}\cos\frac{s}{a}$
8	$-\frac{\mathrm{d}}{\mathrm{d}x}\left[P\frac{\mathrm{d}T}{\mathrm{d}x}\right]^{\dagger}$	$\frac{hp}{kA}PT$	$\frac{hp}{kA}PT_\infty$	—	$\frac{\dot{q}p}{kA}P$

† Combined conduction and mass transport convection where $P=\exp(-\dot{m}cx/kA)$

The boundary conditions for Equation (2.2) consist of specifying the temperature or the temperature gradient at the endpoints of the solution domain, $a \leqslant x \leqslant b$. For cases 1–8, the boundary conditions considered herein are:

$$T=\text{constant},\quad \text{or}$$

$$-k\frac{\mathrm{d}T}{\mathrm{d}x}=q,\quad \text{or} \tag{2.3}$$

$$-k\frac{\mathrm{d}T}{\mathrm{d}x}=h(T-T_\infty)$$

2.2.2 Interpolation functions

Exact interpolation functions for finite element formulations of Equation (2.2) can be derived from the general solution to the differential equation. The general solution has the form

$$T(x)=c_1f_1(x)+c_2f_2(x)+G(x) \tag{2.4}$$

where $f_1(x)$ and $f_2(x)$ are linearly independent solutions of the homogeneous equation, c_1 and c_2 are constants of integration, and $G(x)$ is a particular solution. A finite element with two nodes is formulated by imposing the conditions

$$T(x_1)=T_1, \qquad T(x_2)=T_2 \tag{2.5}$$

where x_i, $i=1,2$ are nodal coordinates and T_i, $i=1,2$ are the nodal temperature values. To accommodate the particular solution $G(x)$, the element temperature interpolation is written as

$$T(x)=N_0(x)T_0+N_1(x)T_1+N_2(x)T_2 \tag{2.6}$$

where T_0 is a nodeless parameter. The nodeless parameter T_0 is similar to a nodeless variable[2] except that the nodeless parameter is uniquely determined for each differential equation. Note that $N_0(x)$ is zero at the nodes to satisfy Equation (2.5). The interpolation functions N_i, $i=0, 1, 2$, are determined by imposing the boundary conditions, Equation (2.5), on Equation (2.4) and then writing the result in the form of Equation (2.6). The nodeless parameter is identified by writing the particular solution $G(x)=T_0g(x)$. Thus, in general,

$$N_0(x)=g(x)+\left[\frac{f_2(x_1)g(x_2)-f_2(x_2)g(x_1)}{D}\right]f_1(x) + \left[\frac{f_1(x_2)g(x_1)-f_1(x_1)g(x_2)}{D}\right]f_2(x) \tag{2.7a}$$

$$N_1(x)=\frac{f_2(x_2)f_1(x)-f_1(x_2)f_2(x)}{D} \tag{2.7b}$$

$$N_2(x)=\frac{f_1(x_1)f_2(x)-f_2(x_1)f_1(x)}{D} \tag{2.7c}$$

where

$$D=f_1(x_1)f_2(x_2)-f_1(x_2)f_2(x_1).$$

Element interpolation functions derived in the form of Equation (2.6) are not generally the same as used in the conventional finite elements since they result from the differential equation solution. Interpolation functions, in fact, may range from simple polynomials to higher transcendental functions depending on the equation. The exact interpolation functions, Equations (2.7), are derived for specified temperature boundary conditions, but they also yield exact solutions for the gradient boundary conditions, Equation (2.3), when these boundary conditions are consistently incorporated in the finite element equations. Since the interpolation functions are exact, temperature gradients and fluxes computed from these functions are also exact.

Nodeless parameters and element interpolations functions have been derived for Cases 1–8 for most of the relevant heat loads (b)–(d). Nodeless parameters appear in Table 2.2 and interpolation functions are presented in the Appendix.

Table 2.2 Nodeless parameter, T_0

Case	Heat loads Convection (b)	Source (c)	Surface flux (d)
1	T_∞	$\frac{\dot{Q}L^2}{2k}$	$\frac{\dot{q}pL^2}{2kA}$
2	—	$\frac{\dot{Q}L^2}{2k}$	—
3	—	$\frac{\dot{Q}b^2}{4kw}$	—
4	—	$\frac{\dot{Q}}{6k}$	—
5	T_∞	$\frac{\dot{Q}L^2}{2k}$	$\frac{\dot{q}L^2}{2kt}$
6	T_∞	$\frac{\dot{Q}b^2}{4kw}$	$\frac{\dot{q}b^2}{4ktw}$
7	—	$\frac{\dot{Q}a^2}{k}$	$\frac{\dot{q}a^2}{kt}$
8	T_∞	—	$\frac{\dot{q}pL}{\dot{m}c}$

where $w = \ln(b/a)$

2.2.3 Element matrices

For general interpolation functions, element matrices for equation (2.2) are

$$K_{ij} = \int_{x_1}^{x_2} -P\frac{dN_i}{dx}\frac{dN_j}{dx}\,dx + \int_{x_1}^{x_2} QN_iN_j\,dx \tag{2.8a}$$

$$F_i = \int_{x_1}^{x_2} RN_i\,dx \tag{2.8b}$$

where K_{ij} is a typical term in the symmetric conductance matrix, and F_i is a typical term in the heat load vector. For a node on the boundary additional terms are required to represent surface heating or a convective heat exchange. For the exact interpolation functions, element matrix equations have the general form

$$\begin{bmatrix} K_{00} & K_{01} & K_{02} \\ K_{10} & K_{11} & K_{12} \\ K_{20} & K_{21} & K_{22} \end{bmatrix} \begin{Bmatrix} T_0 \\ T_1 \\ T_2 \end{Bmatrix} = \begin{Bmatrix} F_0 \\ F_1 \\ F_2 \end{Bmatrix}$$

since $i = 0, 1, 2$. Because T_0 is known the first equation may be uncoupled from the nodal unknowns in the second and third equations. Thus, the exact element matrices have the same size as a conventional linear element and can be written as

$$\begin{bmatrix} K_{11} & K_{12} \\ K_{21} & K_{22} \end{bmatrix} \begin{Bmatrix} T_1 \\ T_2 \end{Bmatrix} = \begin{Bmatrix} F_1 \\ F_2 \end{Bmatrix} - T_0 \begin{Bmatrix} K_{10} \\ K_{20} \end{Bmatrix} \tag{2.9}$$

Equation (2.9) can be simplified further by noting that $K_{10} = K_{20} = 0$ for self-adjoint equations. The last result can be shown by observing that N_i, $i = 1, 2$ are solutions of the ordinary differential equation, (2.2). Multiplying this equation by N_0 and integrating by parts shows that $K_{10} = K_{20} = 0$. Thus, for self-adjoint equations, element matrices need to be evaluated only for $i = 1, 2$. As emphasized previously, the development of the element matrices, Equation (2.8), and element equations, Equation (2.9), are based upon the differential equation being written in the self-adjoint form, Equation (2.2). As an alternative approach, exact finite element matrices may also be derived from the standard form of the differential equation. Equation (2.1). However, element matrices are not symmetrical, due to the odd-order derivative, and the exact element equations have $K_{0i} \neq 0$, $i = 1, 2$.

The principal advantage of the exact one-dimensional finite elements is the superior accuracy in comparison to elements based on approximate interpolation functions. The principal disadvantage of the exact elements is the additional effort required to form the more complex interpolation functions and to evaluate the element matrices. This disadvantage had been overcome, in part, by using a computer-based symbolic manipulation language MACSYMA to perform the algebra and calculus required for these derivations. There is also the potential for computer underflow in computations with some of the transcendental functions for large arguments, but such difficulties do not arise on most modern computers.

2.3 AN EXACT QUASI TWO-DIMENSIONAL ELEMENT

In this section, an exact quasi two-dimensional element is developed. As with the family of one-dimensional elements previously discussed, the element interpolation functions are derived from solutions to the governing differential equations. The element is presented to illustrate one example of other possibilities of exact heat transfer finite elements.

2.3.1 Governing equations

The exact element presented in this section is an example of new finite element capability under development with support of NASA for more effective modelling of heat transfer in aerospace structures with thermal protection systems (TPS). The geometry and terminology for the quasi two-dimensional conduction element are shown in Figure 2.2. Aerodynamic heating on the

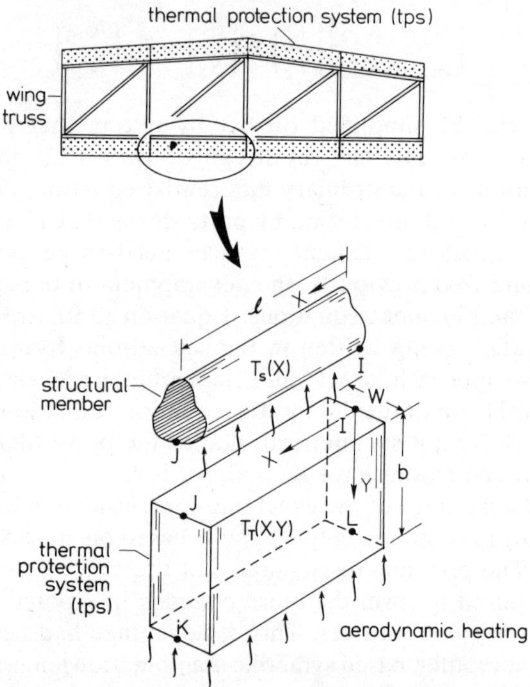

Figure 2.2 Quasi two-dimensional conduction element for a structure with thermal protection system (TPS)

exposed surface of the TPS is conducted through the thickness of the thermal protection system and is transferred nonuniformly to the supporting structural member, where it is conducted along the length of the structural member to connecting members. For steady-state heat transfer, energy balances on small segments of the TPS and attached structure give the governing differential equations for the temperature distributions $T_T(x, y)$ and $T_S(x)$ for the TPS and structure, respectively.

TPS:
$$\frac{\partial}{\partial x}\left(K_{Tx}W\frac{\partial T_T}{\partial x}\right)+\frac{\partial}{\partial y}\left(K_{Ty}W\frac{\partial T_T}{\partial y}\right)=0 \tag{2.10a}$$

Structure:

$$-\frac{\partial}{\partial x}\left(K_s A_s\frac{\partial T_s}{\partial x}\right)=K_{Ty}W\frac{\partial T_T}{\partial y}(x, 0) \tag{2.10b}$$

In the above coupled TPS–structure equations, K_{Tx} is the TPS thermal conductivity in the x-direction, K_{Ty} is the TPS thermal conductivity in the y-direction, K_S is the thermal conductivity of the structural member, A_S is the cross-sectional area of the structural member. The boundary condition for the coupled equations (2.10) consists of specifying the heating or temperature on the aerodynamic surface of the TPS.

2.3.2 Interpolation functions

Detailed studies of heat transfer in the TPS show that the heat flux in the x-direction of the TPS is small in comparison to the y-direction and therefore can be neglected. This assumption makes it possible to obtain a closed-form solution to the coupled Equations (2.10a) and (2.10b). The boundary conditions used in the solution of the governing equations specify a linear variation of temperature on the TPS surface and continuity of temperature at the TPS and structure interface:

$$T_T(x, b)=T_K\frac{x}{l}+T_I\left(1-\frac{x}{l}\right) \tag{2.11a}$$

$$T_T(x, 0)=T_S(x) \tag{2.11b}$$

$$T_S(0)=T_I \tag{2.11c}$$

$$T_S(l)=T_J \tag{2.11d}$$

Element interpolation functions can then be written in the form,

TPS:
$$T_T(x, y)=[N_T(x, y)]\{T\} \tag{2.12a}$$

Structure:

$$T_S(x)=[N_S(x)]\{T\} \tag{2.12b}$$

where $[N_T(x, y)]$ and $[N_S(x)]$ denote the interpolation functions for the TPS and structure, respectively; {T} denotes the vector of element nodal temperatures. The interpolation functions are combinations of polynomials in x and y and hyperbolic functions of x which arise in the solutions to the differential equations (2.10).

2.3.3 Element matrices

Following the method of weighted residuals, the finite element matrices for the coupled equations (2.10a) and (2.10b) are derived in the form

$$[K_T+K_S]\{T\}=\{Q\} \tag{2.13}$$

with

$$[K_T]=\int_A K_{Ty}W\left\{\frac{\partial N_T}{\partial y}\right\}\left[\frac{\partial N_T}{\partial y}\right]dA \tag{2.14a}$$

$$[K_S]=\int_0^l K_S A_S\left\{\frac{\partial N_S}{\partial x}\right\}\left[\frac{\partial N_S}{\partial x}\right]dx \tag{2.14b}$$

where the integration for the elements of $[K_T]$ is over the area A of the TPS in the x–y plane. These element matrices may be evaluated numerically for quadrilateral TPS shapes or evaluated explicitly for a rectangular TPS shape. When evaluated explicitly, the finite element is exact within, of course, the assumptions leading to the mathematical model.

2.4 NUMERICAL EXAMPLES

Four numerical examples are presented to illustrate the benefits of the exact elements; other examples for one-dimensional elements appear in reference 5. The solution of problems with exact elements typically requires only a few elements because mesh refinement is not required in regions of large temperature gradients. The number of elements used is determined by changes in geometry, material properties or heating variations. After nodal temperatures are computed, temperatures are computed at several points within each element by using the element interpolation functions.

2.4.1 Coffee spoon with conduction and convection

Rod elements with conduction and convection (Table 2.1, Case 1, (a)–(b)) are used to model one-dimensional heat transfer in a coffee spoon,[6] Figure 2.3. The lower one-half of the spoon is convectively heated by the coffee at a specified temperature of 150 °F, and the upper one-half of the spoon is convectively cooled by the atmosphere at a specified temperature of 50 °F. The ends of the one-dimensional spoon model are assumed to have negligible heat transfer. Element matrices for the exact finite elements appear in reference 5. Temperatures for the spoon model are computed from: (1) two exact finite elements; (2) two elements with linear temperature interpolation; (3) ten elements with linear temperature interpolation.

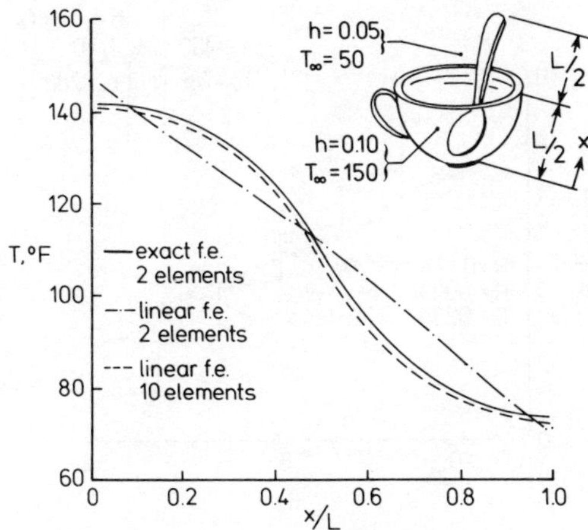

Figure 2.3 Rod element analysis of coffee spoon with conduction and convection

The temperature distributions (Figure 2.3) show the exact temperature distribution computed by the two exact elements with three nodal unknown temperatures compared with predictions from the elements with linear temperature interpolation. Two linear elements predict the temperatures at the three nodes with fair accuracy, but the linear interpolation functions are not capable of representing the zero temperature gradient boundary conditions. Temperatures computed from ten linear elements, however, show excellent agreement with the exact finite element solution. Solutions for similar problems have also shown this trend; typically about five to ten elements with linear interpolation per exact element are required for acceptable accuracy.

2.4.2 Layered pressure vessel

Axisymmetric heat conduction in a layered pressure vessel (Figure 2.4) is modelled with each layer represented by a hollow cylinder element (Table 2.1, Case 3, (a)). Contact resistance between contiguous layers is represented by a conductance matrix,

$$[K]=\frac{r}{R}\begin{bmatrix} 1 & -1 \\ -1 & 1 \end{bmatrix}$$

where r denotes the radius of the interface between layers and R denotes the thermal contact resistance. The pressure vessel has convective boundary conditions on its inner and outer surfaces. Steady-state temperature distributions calculated from a model with six elements for three values of contact resistance are shown in Figure 2.4. The temperatures are identical to results computed from an analytical solution.[7]

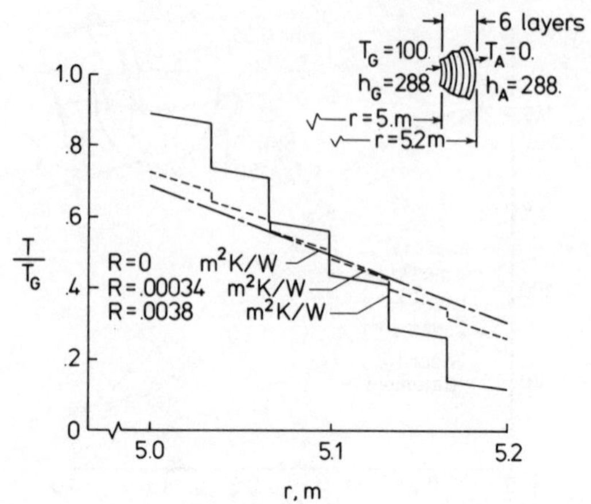

Figure 2.4 Exact finite element analysis for axisymmetric heat transfer in a layered pressure vessel

2.4.3 Heat transfer in merging flows

In merging flows, significant temperature gradients may occur at the flow confluence; therefore, merging flows provide a good test for one-dimensional convective finite element evaluations. In reference 8, the authors compared conventional (Bubnov–Galerkin) finite element solutions with upwind (Petrov–Galerkin) solutions for steady-state and transient merging flows. Figure 2.5 presents the geometry and terminology of a merging flow where conduction is combined with mass-transport and surface convection.

Temperatures for the merging flow were computed from: (1) three exact finite elements with four nodes, and (2) 75 conventional elements with 76

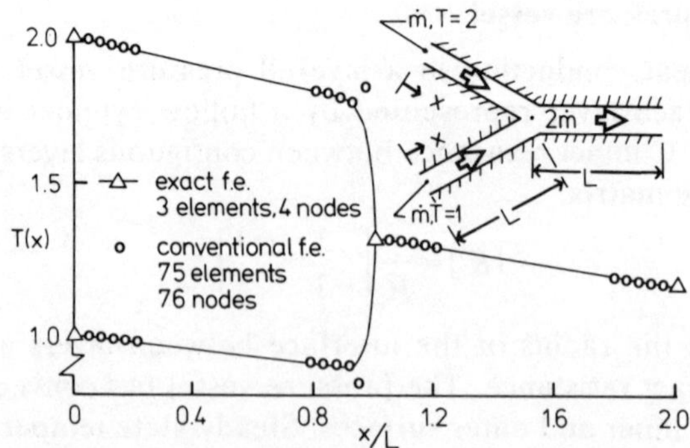

Figure 2.5 Exact and conventional finite element analysis of merging flow with conduction, mass transport convection, and surface convection

nodes. There is good agreement between the two solutions except in the vicinity of the flow confluence, where the conventional element shows oscillations indicating a need for mesh refinement in the region of large temperature gradients. The clear superiority of the exact elements is demonstrated. Exact nodal temperatures are also predicted in problems of this type (without surface convection) by optimum upwind elements, but a larger number of nodes are required since upwind elements do not predict exact temperatures within an element.

2.4.4 Thermal analysis of space shuttle wing section

Figure 2.6(a) shows two thermal models of a section of the space shuttle wing. The bottom exposed portion of the thermal protection system has a specified temperature of 2000 °R and the top ends of typical structural members have specified temperatures of 530 °R. To predict the structural temperature $T_S(x)$ the conventional approach requires 50 bilinear quadrilateral elements and ten rod elements while only 1 TPS element is sufficient. A comparison of the conventional and exact element solutions for temperatures is shown in Figure

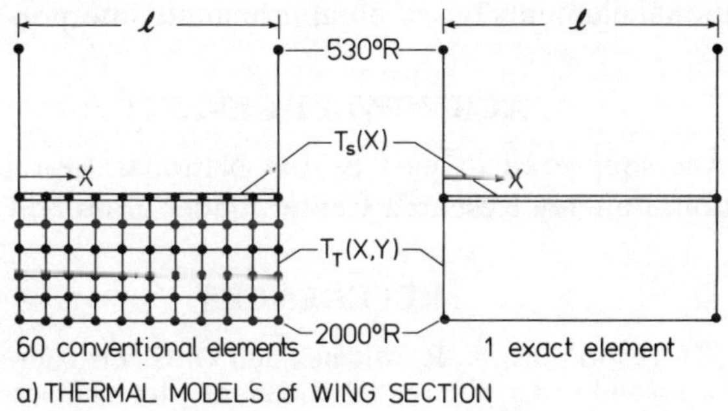

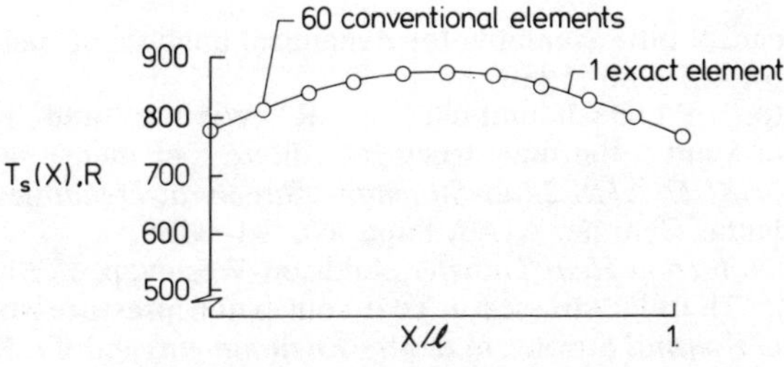

Figure 2.6 Conventional and exact element conduction analyses of structure with thermal protection system (TPS)

2.6(b). Since the TPS element is formulated using interpolations functions based on the governing heat conduction equations, it gives the exact solution.

2.5 CONCLUDING REMARKS

An approach for developing exact finite elements for linear, steady conduction–convection analysis is presented. For one-dimensional equations, the element interpolation functions employ the solution from the governing second-order differential equation by utilizing a nodeless parameter approach. The elements require two nodes and yield exact nodal temperatures and an exact variation of the temperature within an element. Exact nodal interpolation functions are presented for several cases of conduction and convection. An exact conduction element is formulated for a quasi two-dimensional problem to illustrate the possibility of developing other exact elements. Numerical results are presented for conduction and surface convection in a rod element model, conduction in a layeted pressure vessel, conduction, mass-transport convection and surface convection in a merging one-dimensional flow model and conduction in a space shuttle wing section. The examples demonstrate that the exact elements offer significant advantages over conventional elements based on approximate interpolation functions.

ACKNOWLEDGEMENT

This study was supported in part by the National Aeronautics and Space Administration, Langley Research Center, under grant NSG 1321.

REFERENCES

1. I. Christie, D. F. Griffiths, A. R. Mitchell and O. C. Zienkiewicz, 'Finite element methods for second order differential equations with significant first derivatives', *Int. J. Num. Meth. Eng.*, **10**, 1389–1396 (1976).
2. O. C. Zienkiewicz, *The Finite Element Method*, 3rd edn, McGraw-Hill (1977).
3. E. Lightfoot, 'Exact straight-line elements', *J. Strain Anal. Eng. Des.* **15**, (2), 89–96 (1980).
4. J. E. Mottershead, 'Finite elements for dynamical analysis of helical rods', *Int. J. Mech. Sci.*, **22**(5), 267–283 (1980).
5. E. A. Thornton, P. Dechaumphai, A. R. Wieting, and K. K. Tamma, 'Integrated transient thermal-structural finite element analysis', *Proc. AIAA/ASME/ASCE/AHS 22nd Structures, Structural Dynamics, and Materials Conference*, Atlanta, Georgia, AIAA Paper No. 81–0480.
6. V. S. Arpaci, *Conduction Heat Transfer*, Addison-Wesley, p. 173 (1966).
7. T. R. Tauchert, 'Thermal stresses in coal conversion pressure vessels of layered construction', in *Thermal Stresses in Severe Environments* (ed. D. P. H. Hasselman and R. A. Heller), Plenum Press, pp. 183–205 (1980).
8. E. A. Thornton and P. Dechaumphai, 'Convective heat transport in merging flows, in *Finite Elements in Water Resources* (ed. S. Y. Wang, C. A. Brebbia, C. V. Alonso, W. G. Gray, and G. F. Pinder), The University of Mississippi (1980).

APPENDIX

Element interpolation functions in the form of equation (2.6) for Cases 1–8 (Figure 2.1 and Table 2.1) are presented. Nodeless parameters are shown in Table 2.2. The lower case letters in parentheses denote heat load cases defined in Table 2.1.

Case 1 (Rod)

$$N_1 = 1 - \frac{x}{L} \qquad N_2 = \frac{x}{L}$$

$$N_0 = \frac{x}{L}\left(1 - \frac{x}{L}\right) \qquad \text{(a, c, d)}$$

$$N_1 = \frac{\sinh m(L-x)}{\sinh mL} \qquad N_2 = \frac{\sinh mx}{\sinh mL}$$

$$N_0 = 1 - N_1 - N_2 \qquad \text{(b)}$$

where $m = \sqrt{(hp/kA)}$

Case 2 (Slab)

$$N_1 = 1 - \frac{x}{L} \qquad N_2 = \frac{x}{L}$$

$$N_0 = \frac{x}{L}\left(1 - \frac{x}{L}\right) \qquad \text{(a, c)}$$

Case 3 (Hollow cylinder)

$$N_1 = \frac{1}{w}\ln\left(\frac{b}{r}\right) \qquad N_2 = \frac{1}{w}\ln\left(\frac{r}{a}\right)$$

$$N_0 = \ln\left(\frac{r}{a}\right) + \frac{a^2}{b^2}\ln\left(\frac{b}{r}\right) - \frac{r^2}{b^2}w \qquad \text{(a, c)}$$

where $w = \ln(b/a)$

Case 4 (Hollow sphere)

$$N_1 = \frac{a(b-r)}{r(b-a)} \qquad N_2 = \frac{b(r-a)}{r(b-a)}$$

$$N_0 = \frac{1}{r}(r-a)(b-r)(r+a+b) \qquad \text{(a, c)}$$

Case 5 (Cylindrical shell)

$$N_1 = 1 - \frac{s}{L} \qquad N_2 = \frac{s}{L}$$

$$N_0 = \frac{s}{L}(1 - \frac{s}{L}) \qquad \text{(a, c, d)}$$

$$N_1 = \frac{\sinh m(L-s)}{\sinh mL} \qquad N_2 = \frac{\sinh ms}{\sinh mL}$$

$$N_0 = 1 - N_1 - N_2 \qquad \text{(b)}$$

where $m = \sqrt{(h/kt)}$

Case 6 (Conical shell)

$$N_1 = \frac{1}{w}\ln\left(\frac{b}{s}\right) \qquad N_2 = \frac{1}{w}\ln\left(\frac{s}{a}\right)$$

$$N_0 = \ln\left(\frac{s}{a}\right) + \frac{a^2}{b^2}\ln\left(\frac{b}{s}\right) - \frac{s^2}{b^2}w \qquad \text{(a, c, d)}$$

$$N_1 = \frac{I_0(ms)K_0(mb) - I_0(mb)K_0(ms)}{I_0(ma)K_0(mb) - I_0(mb)K_0(ma)}$$

$$N_2 = \frac{I_0(ma)K_0(ms) - I_0(ms)K_0(ma)}{I_0(ma)K_0(mb) - I_0(mb)K_0(ma)}$$

$$N_0 = 1 - N_1 - N_2 \qquad \text{(b)}$$

where $w = \ln(b/a)$, $m = \sqrt{(h/kt)}$; I_0 and K_0 are modified Bessel functions of the first and second kind of order zero, respectively.

Case 7 (Spherical shell)

$$\mathrm{N}_1 = 1 - \mathrm{N}_2 \qquad \mathrm{N}_2 = \frac{\log\left[\frac{1+\sin(s/a)}{1-\sin(s/a)}\right]}{\log\left[\frac{1+\sin(L/a)}{1-\sin(L/a)}\right]}$$

$$N_0 = \log[\cos(s/a)] - N_2 \log[\cos(L/a)] \qquad \text{(a, c, d)}$$

Case 8 (Flow passage)

$$N_1 = 1 - \frac{1 - \mathrm{e}^{2\alpha x}}{1 - \mathrm{e}^{2\alpha L}} \qquad N_2 = \frac{1 - \mathrm{e}^{2\alpha x}}{1 - \mathrm{e}^{2\alpha L}}$$

$$N_0 = \frac{x}{L} - N_2 \qquad \text{(a, d)}$$

$$\mathrm{N}_1 = \mathrm{e}^{\alpha x}\frac{\sinh\beta(L-x)}{\sinh\beta L} \qquad N_2 = \mathrm{e}^{\alpha(x-L)}\frac{\sinh\beta x}{\sinh\beta L}$$

$$N_0 = 1 - N_1 - N_2 \qquad \text{(b)}$$

where $\alpha = \dot{m}c/2kA$, $\beta = \sqrt{(\alpha^2 + m^2)}$, and $m = \sqrt{(hp/kA)}$

Numerical Methods in Heat Transfer, Volume II
Edited by R. W. Lewis, K. Morgan, and B. A. Schrefler

Chapter 3

Finite Element with Measurements: A Predictive Simulation

Gary W. Krutz and R. L. Akau

3.1 INTRODUCTION

Advancements in finite element analysis have been increasing at an exponential rate if one judges the increase by the number of publications. Its value to mankind for solving complex problems has been improving as computer systems expand in their computational power.

The main concern among users of the finite element technique is the lack of accurate physical property data and load information. Accessibility of this data for input might be very costly or unobtainable with current experimental techniques. The finite element method can be used to estimate physical property data, the only requirement being a limited number of easily measured variables. This process is called the 'inverse' technique.

Details on the use of measurements to determine force vector quantities will be discussed for the welding thermal model. Once heat flux values are known, then reliable predictions of metallurgical structure and residual stress may be simulated by a finite element model. The welding process has temperatures in the 3000 °C range, steep thermal gradients, turbulent flow in the liquid region creating great difficulty in simulating exact transient temperature distributions. Other problems described in this chapter include the simulation of overland flooding in the real time domain. Again the flow input into the watershed needs to be determined in order for the model to predict areas of flooding. Benefits are longer lead times for flooding area evacuation, saving human life and property destruction.

Finally, dynamic physical properties during impact are nearly impossible to determine. Using the time-dependent measurements that are available during impact (i.e. strain gauge data), and applying a finite element simulation, properties such as the dynamic Young's modulus can be determined, and this numerical process could be less costly than elaborate experimental testing.

3.2 APPLICATION OF THE INVERSE FINITE ELEMENT METHOD TO HEAT TRANSFER PROBLEMS

3.2.1 One-dimensional—variable heat flux

The finite element method has proved to be an effective means for solving heat transfer problems. Some of its advantages include: utilizing unequally spaced nodes, handling propery variations and nonlinear boundary conditions. The classical 'direct' type of transient heat transfer problem specifies the boundary conditions and calculates the temperature distribution in the medium. A special type of approach is the 'inverse' method whereby the boundary condition at the surface is determined knowing the time–temperature history at an interior point. Many references[1] are available on the inverse method. The inverse method is a helpful tool in determining boundary conditions in high-temperature environments such as nuclear reactors[2] and welding processes[3] when direct measurements are impractical.

Krutz *et al.*[4] used the inverse method to determine the boundary heat flux for a one-dimensional finite element model as shown in Figure 3.1. The

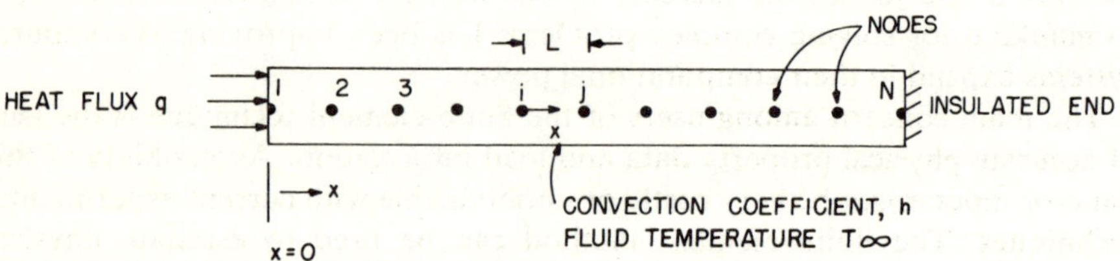

Figure 3.1 One-dimensional finite element model.[4] (Reproduced by permission of Hemisphere Publishing Corporation)

differential equation for the one-dimensional model is

$$k\frac{\partial^2 T}{\partial x^2} - \frac{hP}{A}(T - T_\infty) = \rho c_\mathrm{p}\frac{\partial T}{\partial t}, \tag{3.1}$$

where k = thermal conductivity (W/m K)
T = temperature (K)
h = convective heat transfer coefficient (W/m^2 K)
P = element perimeter (m)
A = element cross sectional area (m^2)
T_∞ = ambient temperature (K)
ρ = density (kg/m^3)
c_p = specific heat (KJ/kg K)
t = time (s).

The initial and boundary conditions are

$$T = T_\infty, \qquad \text{when } t = 0$$

$$-k\frac{\partial T}{\partial x} = q \qquad \text{at } x = 0, \quad \text{when } t > 0$$

$$\frac{\partial T}{\partial x} = 0 \qquad \text{at } x = NL, \quad \text{when } t > 0$$

The physical properties of the model are assumed to be constant.

Using variational calculus the differential equation and the boundary and initial conditions are transformed into a functional I defined as

$$I = \int_V \left[\tfrac{1}{2}k\left(\frac{\partial T}{\partial x}\right)^2 - \left(\rho c_p \frac{\partial T}{\partial t}\right)T\right] \mathrm{d}V + \int_{S_1} qT\,\mathrm{d}S + \int_{S_2} \tfrac{1}{2}h_c(T - T_\infty)^2\,\mathrm{d}S \tag{3.2}$$

The solution is achieved by minimizing Equation (3.2) with respect to temperature, which yields

$$[K]\{T\} + [C]\frac{\partial\{T\}}{\partial t} + \{F\} = 0, \tag{3.3}$$

where $[K]$ = global stiffness matrix (W/K)
$[C]$ = global capacitance matrix (W/K)
$\{T\}$ = global temperature vector (K)
$\{F\}$ – global force vector (W).

The time derivative Equation (3.3) is approximated using a central difference scheme and the result is

$$[A]\{T\}^{t+1} = [P]\{T\}^t - [\{F\}^{t+1} + \{F\}^t] = \{H\}, \tag{3.4}$$

where $[A] = \left[[K] + \frac{2}{\Delta t}[C]\right]$

$[P] = \left[\frac{2}{\Delta t}[C] - [K]\right]$

$\{H\} = [P]\{T\}^t - [\{F\}^{t+1} + \{F\}^t]$

Δt = time step (s).

For a simple one-dimensional linear element under the defined boundary and initial conditions, the matrices are defined as

$$[K^e] = \frac{Ak}{L}\begin{bmatrix} 1 & -1 \\ -1 & 1 \end{bmatrix} + \frac{PhL}{6}\begin{bmatrix} 2 & 1 \\ 1 & 2 \end{bmatrix}$$

$$[C^e] = \frac{\rho c_p AL}{6}\begin{bmatrix} 2 & 1 \\ 1 & 2 \end{bmatrix}$$

The value of the force vector is dependent on the boundary conditions imposed on that particular element. For example, the force vector for the element containing nodes 2 and 3 is

$$\{F^e\} = \frac{h_c T_\infty PL}{2} \begin{Bmatrix} 1 \\ 1 \end{Bmatrix}$$

The inverse method used by Krutz *et al.*[4] involved first calculating the time–temperature history at node 2 by the direct method. This information is then used as input into the inverse program. Next an initial boundary heat flux is guessed at node 1 and the inverse algorithm

$$(\{H_1\}^{m+1})_t = (\{H_1\}^m)_t \left[1 + \frac{(T_{2,\text{known}})_t^2 - (T_{2,\text{calc}})_t^2}{(T_{2,\text{known}})_t^2} \right]$$

is performed at each time step by the difference of squares between the input temperature and the calculated temperature at node 2 until the actual and computed temperatures are within a specified error factor. Results for a step change in heat flux is graphically shown in Figure 3.2. The results exhibit excellent determination of the boundary heat flux and good comparison with a convolution integral presented by Beck[5] and a finite difference method by D'Souza.[6]

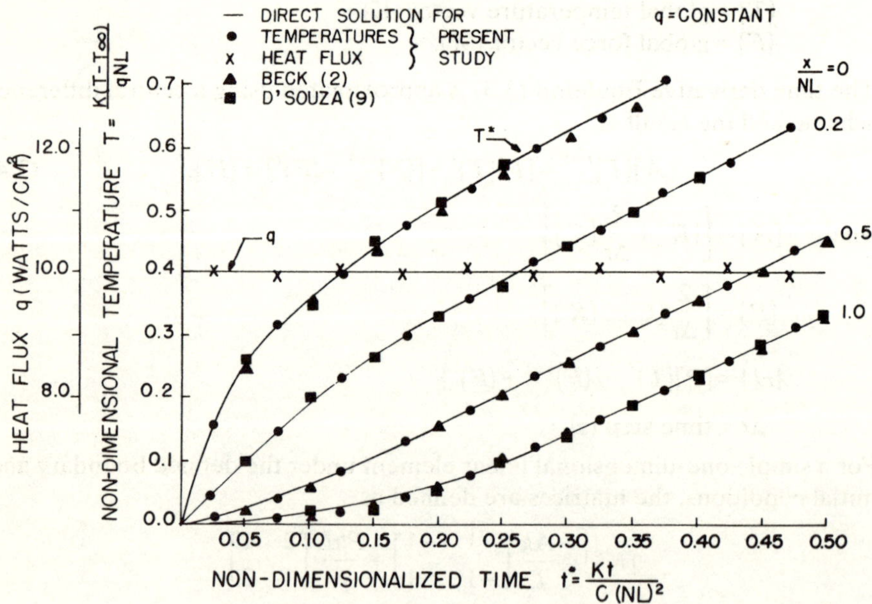

Figure 3.2 Comparison of results for step change in heat flux at node 1.[4] (Reproduced by permission of Hemisphere Publishing Corporation)

An added degree of difficulty is imposed by specifying a periodic square wave at node 1 with the values for the third cycle given in Figure 3.3. The results illustrate an oscillatory behaviour for the predicted heat flux but good comparison between the computed temperatures and an exact solution. Further analysis shows that the value of the convergence criteria has an effect on the oscillatory behaviour of the predicted heat flux. But the calculated temperatures are still within a small percentage of the expected value. The results and conclusions formulated by Krutz *et al.*[4] for solving the inverse problem by finite elements proved to be very successful in predicting surface heat flux. Their motivation provided incentive for further application to more practical heat transfer problems.

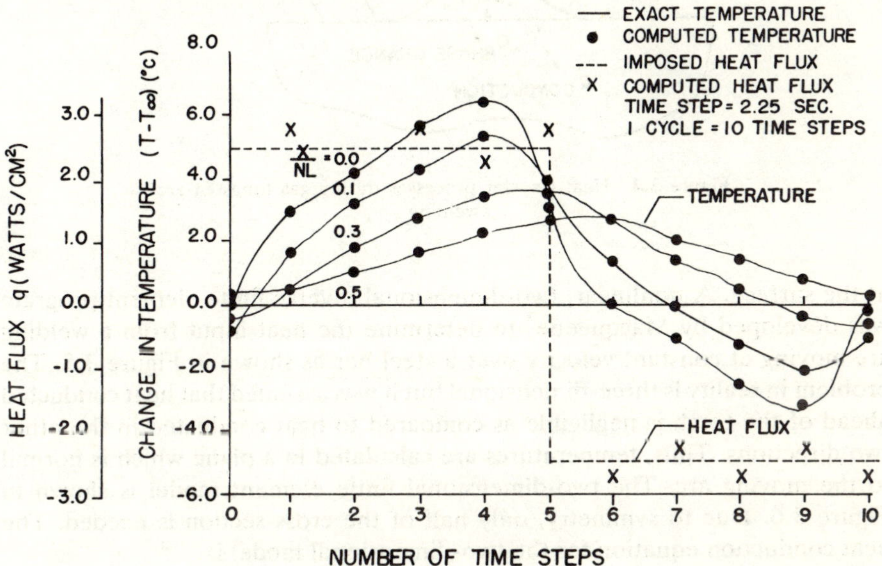

Figure 3.3 Comparison of predicted surface heat flux (node 1) and temperature variations for an imposed square wave heat flux (third cycle).[4] (Reproduced by permission of Hemisphere Publishing Corporation)

3.2.2 Two-dimensional–welding

Macqueene *et al.*[7,8] developed a nondestructive procedure for calculating energy efficiencies for a gas tungsten arc welding process using the inverse finite element method. Arc welding is used extensively in the manufacturing industry because of its versatility for both manual and automatic welding and the intense and concentrated heat which promotes fast welding speeds. The metallurgical characteristics and strength of the weld is determined primarily by the heat transfer occurring during the welding process. The welding arc is generated by passing an electric current from a nonconsumable electrode to the base material across a gap of high resistance.

The relevant heat transfer modes occurring during the gas tungsten arc welding process are shown in Figure 3.4 and include phase change, heat conduction in the medium, convection, radiation, and an imposed heat flux

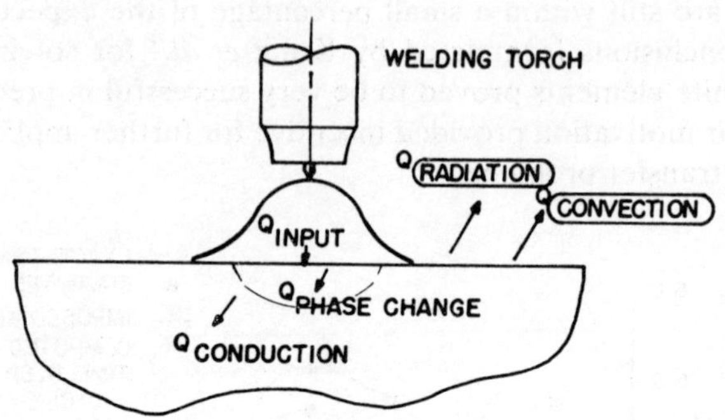

Figure 3.4 Heat transfer processes during gas tungsten arc welding

at the surface. A nonlinear, two-dimensional inverse finite element program was developed by Macqueene[3] to determine the heat input from a welding arc moving at constant velocity over a steel bar as shown in Figure 3.5. The problem in reality is three-dimensional but it was assumed that heat conducted ahead of the torch is negligible as compared to heat conducted in the other two directions. Thus, temperatures are calculated in a plane which is normal to the moving arc. The two-dimensional finite element model is shown in Figure 3.6. Due to symmetry, only half of the cross-section is needed. The heat conduction equation for the two-dimensional model is

$$k\frac{\partial^2 T}{\partial x^2}+k\frac{\partial^2 T}{\partial y^2}=\rho c_p\frac{\partial T}{\partial t} \tag{3.5}$$

where the medium is assumed to be isotropic. The initial and boundary conditions are

$$T=T_0, \qquad \text{when } t=0$$

$$k\frac{\partial T}{\partial x}l_x+k\frac{\partial T}{\partial y}l_y+q_c+q_r-q=0, \qquad \text{when } t>0$$

$$T=T_s, \qquad \text{when } t>0,$$

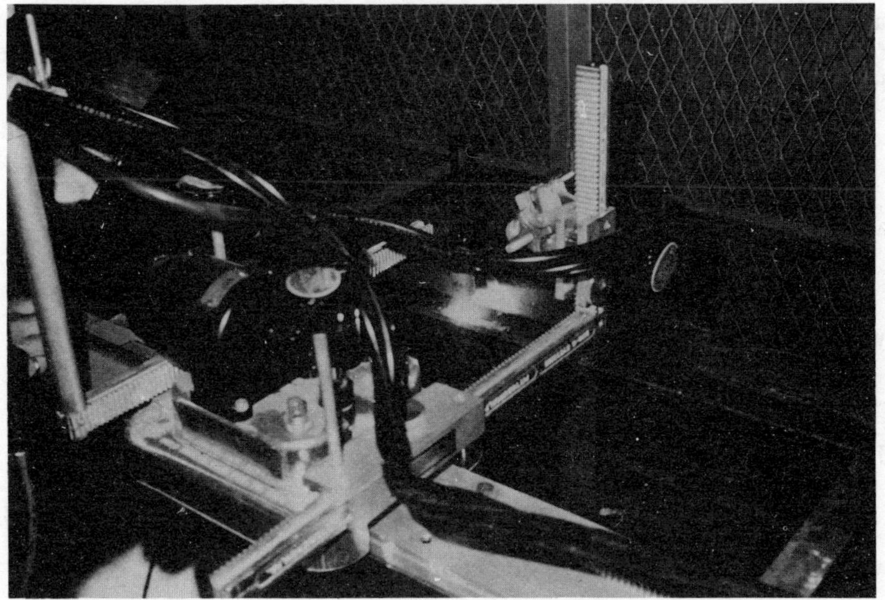

Figure 3.5 Welding arc experiment

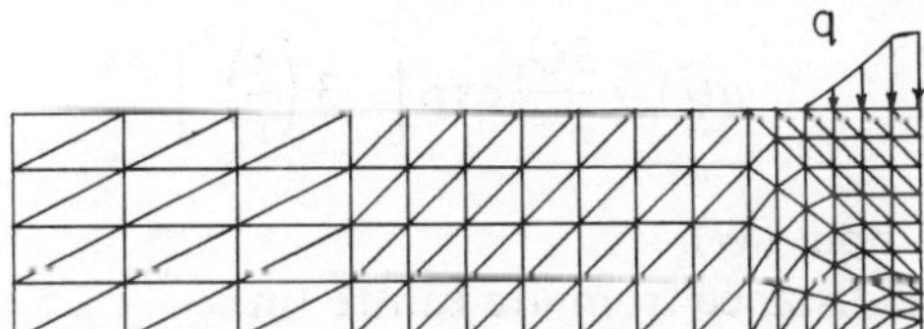

Figure 3.6 Two-dimensional finite element model for welding process

where T_0 = initial temperature (K)
l_x, l_y = direction cosines
q_c = convection heat transfer flux at surface (W/m^2)
q_r = radiation heat transfer flux at surface (W/m^2)
q = imposed heat flux (W/m^2)
T_s = constant surface temperature (K).

The functional I for the two-dimensional model is

$$I = \int_V \left[\tfrac{1}{2}k\left(\frac{\partial T}{\partial x}\right)^2 + \tfrac{1}{2}k\left(\frac{\partial T}{\partial y}\right)^2 - \left(\rho c_p \frac{\partial T}{\partial t}\right) T \right] dV - \int_{S_1} qT \, dS$$

$$+ \int_{S_2} \tfrac{1}{2}(T - T_\infty)^2 \, dS + \sigma\varepsilon \int_{S_3} \tfrac{1}{2}(T^4 - T_{\text{eff}}^4)^2 \, dS \tag{3.6}$$

and is similiar to the one-dimensional formulation (Equation (3.2)) except for the terms added for heat conduction in the y-direction and the radiation heat flux.

The convection and radiation heat fluxes are defined as

$$q_c = h(T - T_\infty) \tag{3.7}$$

and

$$q_r = \sigma\varepsilon(T^4 - T_{eff}^4),$$

where σ = Stefan–Boltzman constant ($W/m^2\ K^4$)
ε = total hemispherical emmissivity
T_{eff} = effective temperature for radiation (K).

In order to linearize the radiation heat flux, q_r, the equation is rewritten

$$q_r = \sigma\varepsilon(\bar{T}^3 T - T_{eff}^4) \tag{3.8}$$

where $\bar{T}$ = average element side temperature (K)

The incident heat flux (q) was modelled to be normally distributed and is defined by Pavelic *et al.*[9] as

$$q(r) = \frac{3Q_i}{\pi\bar{r}^2}\exp\left[-3\left(\frac{r}{\bar{r}}\right)^2\right], \tag{3.9}$$

where Q_i = heat input (W)
r = radial distance from arc centre (m)
$\bar{r}$ = radius within which 95% of heat input occurs (m).

Note that Equation (3.9) is for a stationary arc whereas the arc is actually moving. Thus the observed temperatures were considered equivalent to a weaker pseudo-arc centred directly above the thermocouple. The discretization of the heat flux distribution for the finite element model is shown in Figure 3.7. For the purpose of the inverse program, $\bar{r}$ was given an assigned value and Q_i was calculated by an iterative procedure.

Because of the high temperature environment induced by the welding arc, both the thermal conductivity and specific heat are a function of temperature and are assumed to vary linearly with temperature. The properties for low carbon steel are:

thermal conductivity (W/m. K)

$$k = 77.634 - 0.0458\, T_{ave} \qquad \text{for } T_{ave} < 1300\ \text{K}$$

$$k = 30.0 \qquad \text{for } T_{ave} \geq 1300\ \text{K}$$

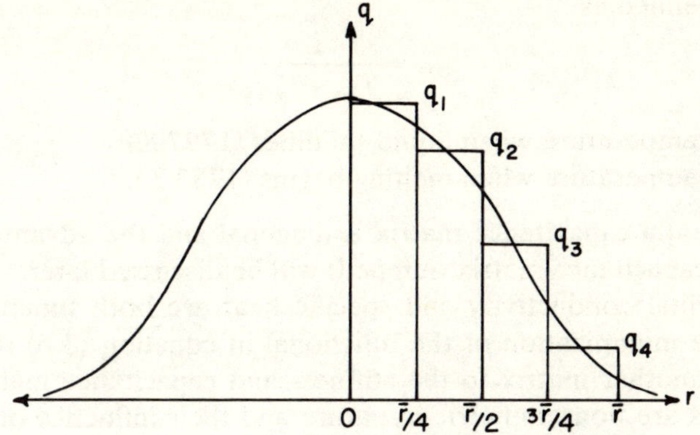

Figure 3.7 Discretization of heat flux distribution

Specific heat (W.s/kg. K)

$$c_{\mathrm{p}} = 253.91 + 0.588\, T_{\mathrm{ave}} \qquad \text{for } T_{\mathrm{ave}} < 775 \text{ K}$$

$$c_{\mathrm{p}} = 685.6 \qquad \text{for } T_{\mathrm{ave}} \geqslant 775 \text{ K}$$

Density (kg/m^3)

$$\rho = 7850.0$$

Latent heat of fusion (W.s/kg)

$$\Delta H = 271.440$$

The temperature T_{ave} is the average elemental temperature defined as

$$T_{\mathrm{ave}} = \frac{(T_i + T_j + T_k)}{3}$$

The subscripts i, j, and k are the nodal points for the particular two-dimensional simplex (triangular) element. Phase change for carbon steel occurs over a temperature range. The latent heat was defined in terms of an effective specific heat and treated as an increase in the capacitance matrix. During phase change the capacitance matrix for a two-dimensional simplex element is

$$[C^{\mathrm{e}}] = \frac{A\rho(c_{\mathrm{p}} + \alpha\, \Delta H)}{3} \begin{bmatrix} 1 & 0 & 0 \\ 0 & 1 & 0 \\ 0 & 0 & 1 \end{bmatrix}$$

where α is defined as

$$\alpha = \frac{1}{T_{\text{liq}} - T_{\text{sol}}}$$

and T_{liq} = temperature when liquid solidifies (1797 K)
T_{sol} = temperature when melting begins (1783 K).

The elemental capacitance matrix is diagonal and the advantage of using this type of capacitance matrix (lumped) will be discussed later.

Since thermal conductivity and specific heat are both functions of temperature, the minimization of the functional in equation (3.6) results in the addition of another matrix to the stiffness and capacitance matrices. These new matrices are nonsymmetric in nature and their influence on calculating temperatures has been investigated by Krutz[10] who illustrated that the incorporation of the nonsymmetric matrices only results in an increase in computer costs with no significant influence on the calculated temperatures, and thus can be neglected. After the minimization process is performed, the algebraic expression is identical to equation (3.4) but now the global matrices are defined in terms of two-dimensional simplex elements.

A plot of the thermocouple temperatures and the calculated heat inputs for an experimental test is shown in Figure 3.8. The temperature increases as the arc approaches the thermocouple, reaches a maximum value, and then slowly decreases when the arc passes by. The heat input at the surface also follows the same type of trend but decreases rapidly after the maximum heat input is reached. When the maximum value of Q_{i} is obtained, the welding efficiency is calculated by

$$\eta = \frac{Q_{\text{i}}}{VI} \times 100\%, \tag{3.10}$$

where V = measured welder supply voltage (V)
I = measured welder supply current (A).

Several welding efficiencies were calculated by Macqueene[3] from experimental results and these values compared well with those given by Christensen *et al.*[11] Macqueene[3] also did a sensitivity analysis on the effect of $\bar{r}$ on the maximum heat input ($Q_{\text{i,max}}$) as shown in Table 3.1. The results illustrate that

Table 3.1 Affect of $\bar{r}$ on the maximum heat input

Arc velocity (cm/s)	Welder voltage (V)	Welder current (A)	Welder power (W)	$\bar{r}$ (cm)	Max. heat input (W)	Eff. (%)
0.254	14	150	2100	0.47625	802	38.2
				0.635	1002	47.7
				0.79375	1328	63.2

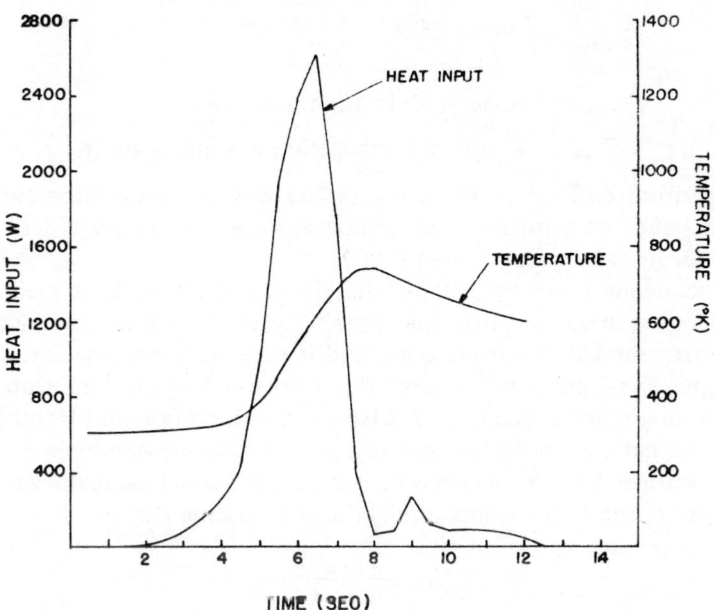

Figure 3.8 Typical curves of the transient thermocouple temperatures and calculated heat inputs (inverse method) from welding experiment

a 25% increase in $\bar{r}$ causes as much as a 35% change in the maximum heat input. Therefore, further investigation should be done on modelling the heat flux.

Macqueene's[3] inverse technique uses the idea introduced by Beck *et al.*[12] of incorporating one future time step in the convergence algorithm. This helps to reduce oscillatory behaviour of the heat input predicted at the surface and the time lag exhibited by the embedded thermocouple to a change in surface temperature. The time–temperature history from an embedded thermocouple located directly below the path of the moving arc (also below the molten pool) was determined experimentally for use as input into the inverse program. The iterative procedure that calculates the heat input at the surface is

$$Q^{t}_{i,n+1} = Q^{t}_{i,n}(1 + \gamma TE), \tag{3.11}$$

where $Q^t_{i,n+1}$ = heat input at time t and iteration $n+1$
$Q^t_{i,n}$ = heat input at time t and iteration n
γ = constant to speed convergence (Macqueene[3] used 5)
TE = total temperature error.

The total temperature error is the sum of the temperature error (T_{error}) over two consecutive time steps, where

$$T_{\text{error}} = \frac{T_{\text{known}} - T_{\text{cal}}}{T_{\text{known}}}$$

T_{known} = known nodal temperature (K)

T_{calc} = calculated temperature at same node (K).

The algorithm defined in Equation (3.11) was performed for each given iteration heat flux until the total temperature error (TE) was less than an error criterion (Macqueene[3] used 0.001).

In Macqueene's[3] investigations, stability problems such as mean square instability and maximum principle instability were encountered and proved to be detrimental to the inverse method if they were not handled properly. Mean square instability is primarily due to the size of the time step and the time-stepping scheme being used. Myers[13] and Lemmon and Heaton[14] have done investigations on the subject and have developed expressions to determine the critical time step which is the value above which oscillatory behaviour occurs. For a one-dimensional model the critical time step is

$$\Delta t_c = \frac{1}{6}\frac{\rho c_p (\Delta x)^2}{k} \tag{3.12}$$

where Δt_c = critical time step (s)
Δx = element length (m)

and for two-dimensional linear simplex (triangular) elements,

$$\Delta t_c = \frac{1}{18}\frac{\rho c_p L^2}{k}, \tag{3.13}$$

where L = minimum element side length (m).

On the other hand, maximum principle instability is a result of the construction of the capacitance matrix which causes a decrease in the internal temperature during the first few time steps when a heat flux is imposed at the surface. Maximum principle instability occurs when a consistent type of capacitance matrix is used. For a linear two-dimensional simplex element the consistent capacitance matrix is

$$[C^e] = \frac{\rho c_p A}{12}\begin{bmatrix} 2 & 1 & 1 \\ 1 & 2 & 1 \\ 1 & 1 & 2 \end{bmatrix} \tag{3.14}$$

In order to alleviate this problem, a lumped capacitance matrix is proposed, where

$$[C^e] = \frac{\rho c_p A}{3} \begin{bmatrix} 1 & 0 & 0 \\ 0 & 1 & 0 \\ 0 & 0 & 1 \end{bmatrix} \tag{3.15}$$

and now $[C^e]$ is a diagonal matrix. A comparison of results using a lumped and consistent capacitance matrix is illustrated in Figure 3.9. The results show that the consistent capacitance scheme undergoes a negative change in temperature at early times and some oscillation at later times.

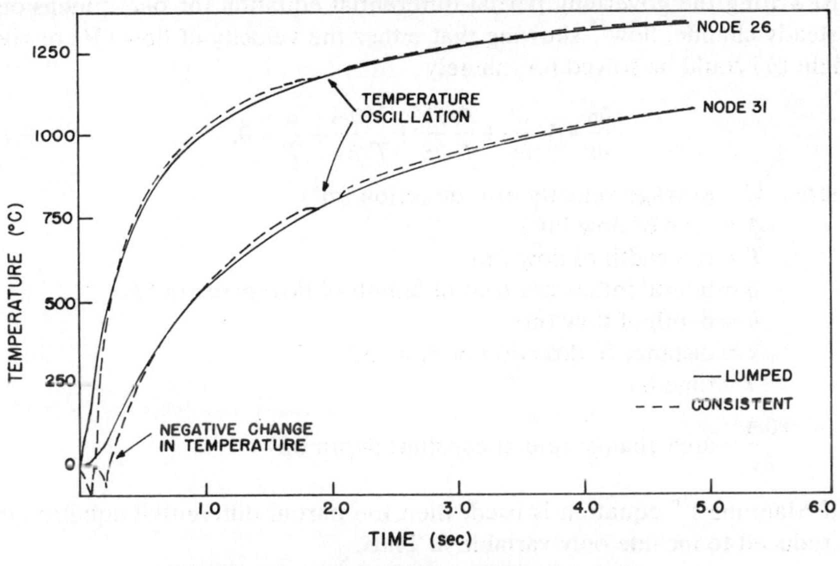

Figure 3.9 Comparison of temperatures using lumped capacitance and consistent capacitance

Macqueene[3] and Macqueene *et al.*[7,8] have developed a nondestructive type of procedure in determining arc welding efficiencies and have shown the important application of the inverse method to a practical heat transfer problem. But further experimentation and analysis needs to be done in improving the heat flux model.

3.3 FLOOD PREDICTION—FINITE ELEMENT MODEL

Recent work by agricultural engineers has provided time-dependent data of rainfall, silt and chemical runoff, and flooding information for specific watersheds that in the past have overflowed their river bottoms. With this data one can determine the parameter values that are in the stiffness matrix $[K]$, capacitance matrix $[C]$, and force vectors for a given soil structure. Then when a rainfall starts to occur real-time river levels can be recorded and used as input to an inverse finite element model that would determine the flux (rainfall) over the watershed. Once the rainfall (force vector) is known, the finite element method can be again used to predict river levels at future times. When flooding is projected evacuation procedures can then be implemented by local authorities which will potentially allow greater lead time for saving property and lives. This would be a real-time application of using an inverse finite element program to predict an occurrence.

Rewriting the governing partial differential equation for one-dimensional unsteady channel flow[15] showing that either the velocity of flow (V) or river height (h) could be solved for, namely

$$\frac{\partial h}{\partial t}+V\frac{\partial h}{\partial x}+\frac{A}{T}\frac{\partial V}{\partial x}+\frac{V}{T}\frac{\partial A}{\partial x}-\frac{q}{T}=0, \tag{3.16}$$

where V = average velocity in x-direction (m^2)
A = area of flow (m^2)
T = top width of flow (m)
q = lateral inflow per foot of length of flow plane (m^2/s)
h = depth of flow (m)
x = distance in direction of flow (m)
t = time (s)
$\frac{\partial A}{\partial x}$ = area change rate at constant depth (m).

If Manning's[16] equation is used, then the partial differential equation can be reduced to include only variable h. Thus,

$$V=\frac{1.49R^{2/3}S_f^{1/2}}{\beta}$$

where V = velocity in x-direction (m/s)
R = hydraulic radius (m)
S_f = slope of free surface
β = roughness coefficient (0.04).

Expanding for the one-dimensional case after applying Galerkin's method for linear elements, the following form results:

$$\begin{bmatrix}\frac{1}{3} & \frac{1}{6}\\ \frac{1}{6} & \frac{1}{3}\end{bmatrix}\begin{Bmatrix}\dot{h}_i\\ \dot{h}_j\end{Bmatrix}+\frac{h_j-h_i}{x_j-x_i}\begin{bmatrix}\frac{1}{3} & \frac{1}{6}\\ \frac{1}{6} & \frac{1}{3}\end{bmatrix}\begin{Bmatrix}V_i\\ V_j\end{Bmatrix}$$

$$+\frac{\partial A/\partial x}{T_m}\begin{bmatrix}\frac{1}{3} & \frac{1}{6}\\ \frac{1}{6} & \frac{1}{3}\end{bmatrix}\begin{Bmatrix}V_i\\ V_j\end{Bmatrix}=\frac{q_m}{2T_m}\begin{Bmatrix}1\\ 1\end{Bmatrix}-\frac{A_m(V_j-V_i)}{2T_m(x_j-x_i)}\begin{Bmatrix}1\\ 1\end{Bmatrix}, \tag{3.17}$$

where q_m = average lateral flow per foot (m^{-1})
A_m = average area $(A_i + A_j)/2$ (m^2)
T_m = average top width $(T_i + T_j)/2$ (m)

$$\frac{\partial A}{\partial x} = \text{average rate of cross-section change with distance (m)}$$

$$= \frac{1}{2}\left[\frac{A_j - A_i}{x_j - x_i}\bigg|_{h_i} + \frac{A_j - A_i}{x_j - x_i}\bigg|_{h_j}\right]$$

Figure 3.10 shows an example of a finite element model for a watershed. Since velocity is a function of height the following general matrix form will result:

$$[C]\{\dot{h}\} + [K]\{h\} + \{F\} = 0 \tag{3.18}$$

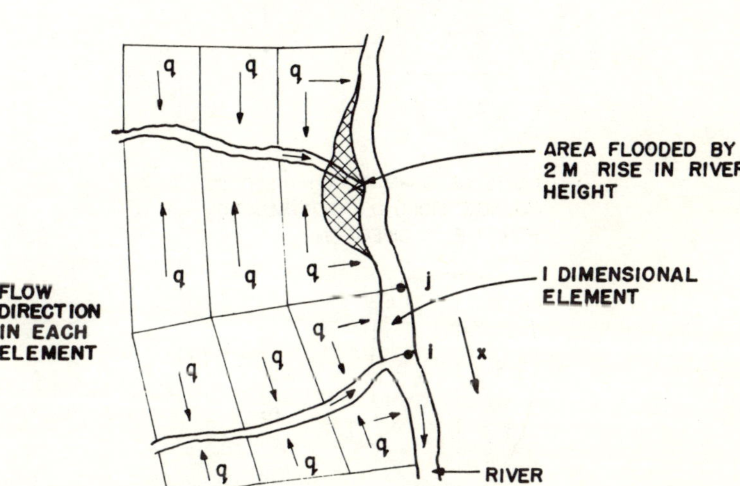

Figure 3.10 Example watershed with finite elements for overland water flow

A solution can be obtained by using a time-stepping scheme similiar to the weld thermal model.

3.4 DETERMINING PHYSICAL PROPERTY DATA

Finite elements can also be used as a nondestructive test procedure for static, dynamic, and vibrational loading conditions. The determination of impact load history or variant property data can be an expensive experimental process. Finite elements can reduce this cost by using the computer to determine variable properties. This simulation process has been used to solve the equilibrium equations similiar to the welding thermal problem discussed previously.

For any loading condition, the equilibrium equation is

$$[M]\{\ddot{x}\}+[C]\{\dot{x}\}+[K]\{x\}=\{F\} \tag{3.19}$$

In dynamic problems dampening ratios or dynamic modulus of elasticity, located in $[K]$ and $[C]$, are properties that are difficult to obtain experimentally.

One such problem discussed was the stress wave propagation in corn kernels subjected to threshing or dropped from great heights during the transportation process. Moreira *et al.*[17] discussed the current limitations of determining dynamic properties of various parts of the kernel (Figure 3.11). Table 3.2

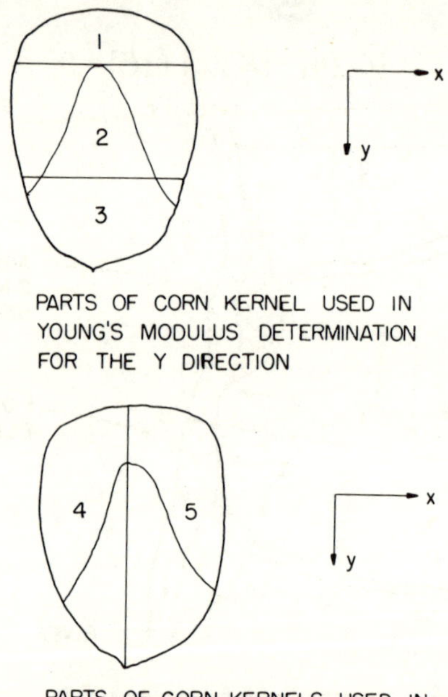

Figure 3.11 Kernel sections for modulus of elasticity

Table 3.2 Modulus of Elasticity (MPa) for parts of corn kernel at different levels of moisture content

Moisture content (%w.b.)	Position of specimen (Fig. 3.11) 1	2	3	4	5
13.4	92.3	290.1	119.7	150.0	120.1
20.0	61.4	135.7	91.3	26.7	30.8
29.3	20.0	33.1	21.9	11.6	13.9

depicts static test data for the modulus of elasticity which varies by kernel location and moisture content. Experimental difficulties in determining the dynamic values resulted from the specimen being physically too small to input controlled loads and measure small displacements without damaging transducers. Force–time plots and the effect of velocity were determined experimentally (Figure 3.12) for input into the equilibrium equation leaving property

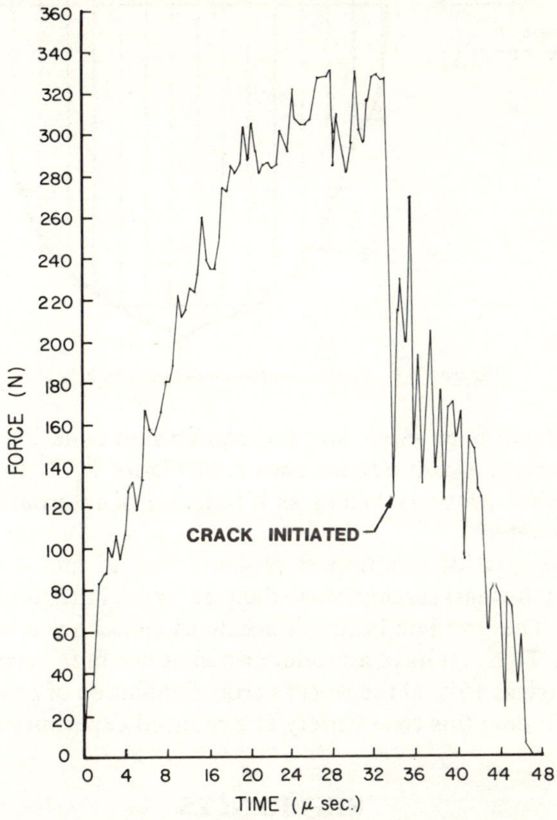

Figure 3.12 Kernel impact curve when cracking occurs

data as the only unknown variable. Figure 3.13 shows the one-dimensional model for the kernel. In Moreira's[17] work different ratios of the modulus of elasticity were used in a sensitivity analysis to find its effect on element stress. The sensitivity tests did not show significantly different levels of stress for various elasticity ratios. Therefore, it was deemed unnecessary to use the correct dynamic modulus value in each element. If significantly different stress levels had been present then the acceleration, velocity, and displacement

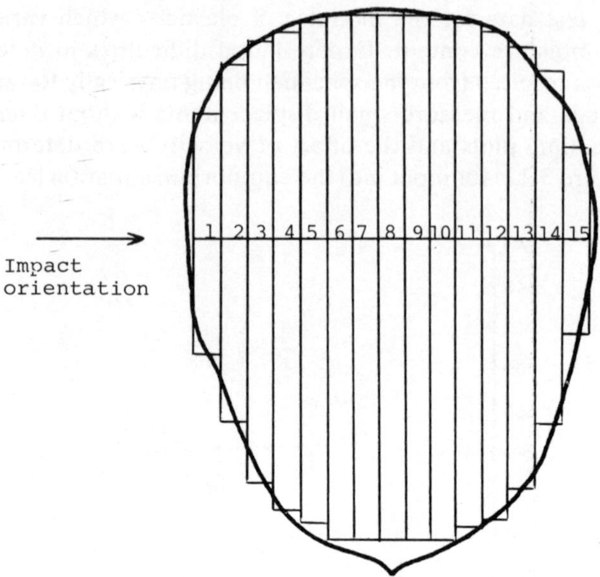

Figure 3.13 Finite element model of corn kernel

vectors could have been input into the equilibrium equation along with the known force-vector quantities (as shown in Figure 3.12) to determine the unknown physical property values as a function of internal kernel position and moisture content.

The ultimate goal of this impact problem was to find a variety of corn kernel that was the least susceptible to damage (crack initiation) during harvest and transport. The problem becomes accute as cracks occur which allows for mould growth. This results in a product which is not fit for consumption, now estimated as high as 15% of the world's crop. Simulation of physical properties could lead to finding this crop variety at a reduced experimental cost.

REFERENCES

1. J. V. Beck and J. Minhorn, *Inverse Heat Conduction Bibliography*, ASME Short Course Program–Solution Methods for Inverse Heat Conduction Problems, Washington D.C. (November 17, 1981).
2. B. R. Bass, 'Application of the finite element method to the nonlinear inverse heat conduction problem using beck's second method' *J. Heat Transfer, Trans ASME*, **102**, 168–176 (1980).
3. J. W. Macqueene, *Application of Finite Element Methods to the Inverse Heat Transfer Problem*, Masters Thesis, Purdue University (1981).
4. G. W. Krutz, P. Hore and R. J. Schoenhals 'Application of the finite element method to the inverse heat conduction problem', *Numerical Heat Transfer*, **1**, 489–498 (1978).

5. J. V. Beck, 'Non-linear estimation applied to the non-linear inverse heat conduction problem', *Int. J. Heat Mass Transfer*, **13**, 703–716 (1970).
6. N. D'Souza, 'Numerical solution of one-dimensional inverse transient heat conduction by finite-difference method', *ASME Paper* 75-WA/HT-81 (1975).
7. J. W. Macqueene, R. L. Akau, G. W. Krutz, and R. J. Schoenhals, 'Numerical methods and measurements related to welding processes', in *Numerical Methods in Thermal Problems* (ed. K. Morgan and B. A. Shrefler), Pineridge Press, Swansea, U.K., *Proc. 2nd Int. Conf.*, Venice, Italy (July 7–10 1981), pp. 152–167.
8. J. W. Macqueene, R. L. Akau, G. W. Krutz, and R. J. Schoenhals, 'Development of inverse finite element techniques for evaluation of measurements obtained from welding processes', presented at the 2nd Nat. Symp. Num. Meth. Heat Transfer, University of Maryland, College Park, Maryland (September 28–30, 1981).
9. V. Pavelic, R. Tanbakuchi, O. A. Uyehara, and P. S. Meyers, 'Experimental and computed temperature histories in gas tungsten arc welding of thin plates', *Welding J.* **48**(7), 295s–305s (1969).
10. G. W. Krutz, *A Thermo-metallurgical Model Predicting the Strength of Welded Joints Using the Finite Element Method* Ph.D. Thesis, Michigan State University, (1976).
11. N. Christensen, V. L. Davies, and K. Gjermundsen, 'Distribution of temperature in arc Welding', *Brit. Welding J.*, **12**, 54–75 (1965).
12. J. V. Beck, B. Litkouhi, and C. R. St. Clair, Jr., 'Efficient numerical solution on the non-linear inverse heat conduction problem', *ASME Paper* 80-*HT*-3, New York (1980).
13. G. E. Myers, *Numerically Induced Oscillation and Stability Characteristics of Finite Element Solutions to Two Dimensional Heat Conduction Transients*, Engineering Experiment Station Report No. 43, University of Wisconsin, Madison, Wisconsin (August 1977).
14. E. C. Lemmon, and H. S. Heaton, 'Accuracy, stability and oscillation characteristics of finite element method for solving heat conduction equation', *ASME Paper 69-WA/HT-35,* New York (1969).
15. O. M. Judah, V. O. Shanholtz, and D. N. Contractor, 'Finite element simulation of flood hydrographs', *Trans. ASAE*, **18**, 518–522 (1975).
16. *Agricultural Engineers Yearbook* 1981–1982, published by the American Society of Agricultural Engineers, St. Joseph, Michigan 49085.
17. S. M. C. Moreira, G. W. Krutz, and C. T. Sun, 'Simulation of stresses inside the corn kernel', *ASAE paper no.* 81-3043.

Numerical Methods in Heat Transfer, Volume II
Edited by R. W. Lewis, K. Morgan, and B. A. Schrefler

Chapter 4

Steep Gradient Modelling in Diffusion Problems

M. Hogge and P. Gerrekens

4.1 STEEP GRADIENT INACCURACY

Modelling of transient diffusion problems faces the classical difficulty in discretizing parabolic or hyperbolic partial differential equations: the choice of the spatial and the temporal path of discretization to achieve the best compromise with respect to accuracy and cost of the solution.

Contrary to finite difference methods, the finite element method applied to time-dependent problems usually proceeds by relatively uncoupled semi-discretization operations: i.e. first choose a finite element space grid which leaves a set of first-order differential equations, generally nonlinear, and second, integrate these equations by a time-marching scheme which yields finally a set of (nonlinear) algebraic equations for each time step considered. Very often in these operations space discretization is preponderant in the sense that the analyst has several tools at his disposal to qualify the spatial grid (error analysis, dual analysis, experience, . . .) and that the time steps elected for the transient response are merely chosen to insure global stability and the best achievable accuracy in the response for the given space discretization.[1]

In the present chapter we shall concentrate on the particular problem of heat transfer in the presence of steep gradients. This interest has been originally motivated by a broader study on appropriate modelling techniques for thermal ablation problems.[2,3] Inaccurate modelling of spatial gradients is, however, typical of all conduction problems (and diffusion problems in general) where the heat penetration is tracked inwards from the boundaries in isothermal media, and where the time steps are small for some reason while the corresponding spatial discretization is not fine enough in the vicinity of the appropriate boundary. This can be illustrated by considering the spatial temperature distribution within a one-dimensional medium initially at $T=0$ and exposed to a heat input $q(t)$ at its boundary $x=0$ for $t>0$. The boundary $x=l$ is insulated and the thermal characteristics (conductivity k and specific heat p.u. volume ρc) are constant. Integration of the heat equation over the domain when due account is taken of the boundary conditions yields the result

$$\int_0^l \rho c T \, dx = \int_0^t \int_0^L (kT_{,x})_{,x} \, dx \, d\tau = \int_0^t q(\tau) \, d\tau. \tag{4.1}$$

where a comma denotes spatial differentiation. In other words the total enthalpy of the medium (or the area between the spatial temperature curve and the x axis in the present case) at any time is completely determined by the total heat input at that time, irrespective of the conductivity properties of the medium. This is also true for the temperature distribution resulting from a finite element model with spatial path Δx. Suppose now that the conductivity value is such that the exact shape of the spatial temperature distribution after one time step be $E(\Delta t)$ as shown in Figure 4.1, i.e. a very

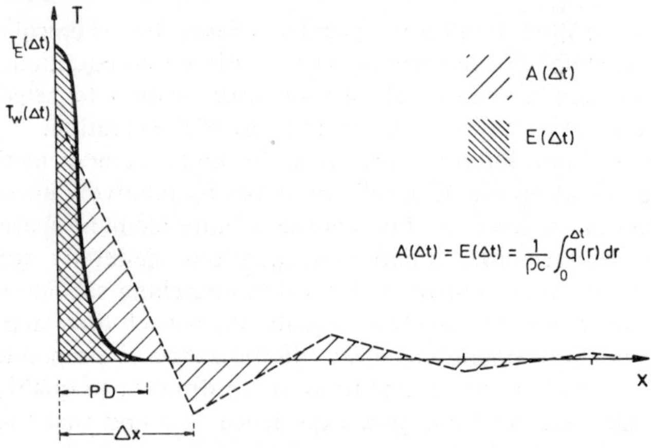

Figure 4.1 Steep gradient inaccurate modelling

thin penetration of heat in the medium measured by the penetration depth PD. If the spatial discretization path Δx happens to be larger than PD, the model exhibits inadequate spatial distribution of temperature: the wall temperature $T_w(\Delta t)$ is underestimated, interior nodes suffer temperatures below the initial value (zero here) and severe spatial oscillations arise to comply with both condition (4.1), i.e. $A(\Delta t) = E(\Delta t)$, and global heat balance in the mean for the heat flux in the elements.

Increasing the time step is an obvious remedy to these inaccuracies if the physics and the desired global accuracy allow it; otherwise mesh refinement must occur in the vicinity of the boundary, but this refinement usually becomes unnecessary as time proceeds. Deforming grid concepts seem thus a natural remedy for these situations if

(a) an exogeneously prescribed or automatic deformation process can be settled to track the steep gradients;
(b) accuracy is definitely improved at small cost, even in standard linear situations.

The aim of this chapter is precisely to answer these two questions by

(a) basing the explicit deformation of the mesh on penetration depth concepts;
(b) experimenting numerically on the procedures for model one-dimensional situations.

4.2 PENETRATION DEPTH CONCEPTS

4.2.1 The general concept

Parabolic partial differential equations behave mathematically as if infinite wave propagation speeds were implied: i.e. the influence of any perturbation at one point is felt instantaneously at all the other points of the domain. For instance, in heat transfer problems, heat input at a boundary immediately influences the temperature at every point within the medium. One can however decide on a threshold below which the temperature may be considered invariant during the period of observation, in keeping with the physical observation of finite penetration times of heat inwards into isothermal media.

The distance from the boundary beyond which the temperature within the domain is unaffected in that sense is the so-called penetration depth (PD).[4] There are other mathematical definitions of the PD.[5,6] Let us recall them by considering again the one-dimensional situation of a semi-infinite medium of thermal conductivity k and heat capacity ρc (both eventually space- and temperature-dependent), whose boundary $x = 0$ is exposed to a prescribed heat input $q(t)$ (Neumann boundary condition). The heat equation for $T(x, t)$ reads then, without volumetric heat sources:

$$(kT_{,x}) = \rho c\dot{T} \qquad x \in [0, \infty] \tag{4.2}$$

with the boundary conditions

$$-kT_{,x} = q(t) \qquad \text{at } x = 0 \tag{4.3}$$

$$T_{,x} = 0 \qquad \text{at } x = \infty \tag{4.4}$$

and the initial condition (with T_0 uniform):

$$T(x, 0) = T_0, \tag{4.5}$$

where a dot over a symbol stands for time rate of change.

In order to allow for nonlinear characteristics from the beginning, let us transform the problem in terms of volumetric enthalpy by introducing

$$h(x, t) = \int_0^T \rho c \, \mathrm{d}T, \tag{4.6}$$

so that we can write the initial-value problem (4.2)–(4.5) as

$$(\eta h_{,x})_{,x} = \dot{h} \qquad x \in [0, \infty] \tag{4.7a}$$

$$-\eta h_{,x} = q(t) \qquad \text{at } x = 0 \tag{4.7b}$$

$$h_{,x} = 0 \qquad \text{at } x = \infty \tag{4.7c}$$

$$h(x, 0) = h_0, \tag{4.7d}$$

where η denotes the instantaneous thermal diffusivity $\eta = k/\rho c$.

A first way to define the penetration depth magnitude is to introduce approximate enthalpy profiles[4,5]

$$h(x, t) = P(x, \delta(t)) + h_0 \qquad \text{for } x \in [0, \delta(t)], \tag{4.8a}$$

$$h(x, t) = h_0 \qquad \text{for } x \geqslant \delta(t), \tag{4.8b}$$

where P is a spatial polynomial of degree m, whose derivatives of order up to $m-1$ are zero at $x = \delta(t)$ and which is such that conditions (4.7(b)) and (4.7(d)) are satisfied if $\delta(0) = 0$.

The PD magnitude $\delta(t)$ is thus determined by Equation (4.7(a)) integrated over space, i.e.

$$\int_0^{\delta(t)} \dot{h} \, \mathrm{d}x = q(t) \tag{4.9}$$

and exhibits the required feature of being the instantaneous distance from the boundary beyond which the boundary condition (4.7(b)) has no influence over the initial enthalpy.

This absolute PD concept is however, in contradiction with the parabolic character of the partial differential equation; moreover, its magnitude is highly dependent on the degree m of the polynomial P[7] and can be shown to become arbitrarily large for any t if this degree is indefinitely increased.

It therefore seems preferable to introduce a relative PD concept by assuming the enthalpy distribution:[6]

$$h(x, t) = h_0 + 1/\eta_w q(t)\delta(t) \, \mathrm{e}^{-x/\delta(t)}, \tag{4.10}$$

where η_w is the boundary diffusivity. This profile satisfies conditions (4.7(b)–(4.7(d)) if $\delta(0) = 0$ and is such that

$$h[n\delta(t), t] - h_0 = \mathrm{e}^{-n}[h(0, t) - h_0]. \tag{4.11}$$

The associated PD magnitude reads thus via Equation (4.9):

$$\delta(t)=\left[\eta_{\mathrm{w}}/q(t)\int_0^t q(\tau)\,\mathrm{d}\tau\right]^{1/2}. \tag{4.12}$$

Note that it depends only on the boundary diffusivity and in no way on the diffusivity internal variation. This can be explained by the fact that heat balance is satisfied in the mean only via Equation (4.9). Inadequate evaluation of the effective PD is thus possible in critical cases.

4.2.2 Influence of the boundary condition history

The PD magnitude given by Equation (4.12) is obviously dependent on the particular heat input history $q(t)$. If the latter is, for instance, a step function $q(t)=qu(t)$, with $u(t)$ denoting the Heaviside function, we get

$$\delta_{\mathrm{s}}(t)=(\eta_{\mathrm{w}}t)^{1/2}. \tag{4.13}$$

If it is a ramp function $q(t)=qt$, we are led to

$$\delta_{\mathrm{r}}(t)=(\eta_{\mathrm{w}}t/2)^{1/2}. \tag{4.14}$$

This shows that the PD magnitude mainly depends on the shape of the heat input and not on its amplitude: step functions penetrate 140% deeper than ramp functions for a given propagation time.

In general all heat inputs which remain constant or increase monotonically with time are easily taken into account by the integral formula (4.12). However, heat input histories which vanish or change sign without their time integral doing the same produce infinite or imaginary PD magnitudes via that equation.

The assumed enthalpy profile (4.10) must thus be modified to allow for combination of heat inputs. Let us begin by changing the initial heat input into $q^*(t)=q(t)+p(t)$. It is thus natural to seek a modified enthalpy solution of type

$$h(x,t)=h_0+1/\eta_{\mathrm{w}}(q(t)\delta(t)\,\mathrm{e}^{-x/\delta(t)}+p(t)\xi(t)\,\mathrm{e}^{-x/\xi(t)}). \tag{4.15}$$

satisfying Equation (4.7(c))–(4.7(d)) and Equation (4.7(b)) modified if $\xi(0)=\delta(0)=0$.

If $\delta(t)$ remains defined by Equation (4.12), substitution of Equation (4.15) into the modified integral balance, Equation (4.9), yields

$$\xi(t)=\left[\eta_{\mathrm{w}}/p(t)\int_0^t p(\tau)\,\mathrm{d}\tau\right]^{1/2} \tag{4.16}$$

Note that if the thermal characteristics of the medium are temperature-dependent, the assumed decomposition (4.15) does not reduce to a superposition principle since the modified enthalpy profile leads to a different surface diffusivity η_w, which in turn modifies the first term of the distribution due to $q(t)$.

More generally, if the heat input is of type

$$q(t)=\sum_i q_i(t), \tag{4.17}$$

assumed enthalpy profiles are of the form

$$h(x, t)=h_0+(1/\eta_w)\sum_i q_i(t)\delta_i(t)\,\mathrm{e}^{-x/\delta_i(t)} \tag{4.18}$$

with

$$\delta_i(t)=[\eta_w/q_i(t)\int_0^t q_i(\tau)\,\mathrm{d}\tau]^{1/2} \tag{4.19}$$

Now all heat inputs (even those which vanish at a given time) can be approximated by a series of Heaviside functions

$$q(t)=\sum_i q_i u(t-\tau_i) \tag{4.20}$$

for which we get from (4.18)–(4.19)

$$\delta_i^s(t)=(\eta_w(t-\tau_i))^{1/2} \tag{4.21}$$

and

$$h(x, t)=h_0+(1/\eta_w)\sum_i q_i u(t-\tau_i)[\eta_w(t-\tau_i)]^{1/2} \times \mathrm{e}^{-x/[\eta_w(t-\tau_i)]^{1/2}}. \tag{4.22}$$

We can thus define an apparent penetration depth as

$$\delta_s(t)=\max_i\,(\eta_w(t-\tau_i))^{1/2}=(\eta_w t)^{1/2}. \tag{4.23}$$

If the heat inputs are approximated by a series of ramp functions (piecewise linear variations)

$$q(t)=\sum_i q_i\cdot(t-\tau_i)u(t-\tau_i) \tag{4.24}$$

for which

$$\delta_i^r(t)=\left(\frac{\eta_w}{2}(t-\tau_i)\right)^{1/2}, \tag{4.25}$$

the effective penetration depth is defined in a similar manner as

$$\delta_r(t) = (\eta_w t/2)^{1/2} \tag{4.26}$$

The two definitions (4.23) and (4.26) imply that all amplitudes q_i in (4.20) and (4.24) are in the same range. If it were not the case and if say q_N were much larger than the others, the effective PD would be given by $\delta_N^s(t)$ or $\delta_N^r(t)$, since the transient response would be dominated by this particular component of the input.

4.2.3 Influence of the boundary condition type

If the Neumann boundary condition (4.3) is replaced by a Dirichlet condition on temperature

$$T(0, t) = T_w(t) \qquad \text{at } x = 0, \tag{4.27}$$

the transformed initial value problem (4.7) exhibits a similar condition in place of condition (4.7(b)), i.e.

$$h(0, t) = h_w(t) \qquad \text{at } x = 0. \tag{4.28}$$

An appropriate enthalpy distribution with relative PD $\delta(t)$ is then

$$h(x, t) = h_0 + (h_w(t) - h_0)\,\mathrm{e}^{-x/\delta(t)}, \tag{4.29}$$

which satisfies all initial and boundary conditions if $\delta(0) = 0$ and leads, via the integral balance

$$(\eta h_{,x})_{x=0} = \int_0^\infty \dot{h}\,\mathrm{d}x, \tag{4.30}$$

to the PD magnitude

$$\delta(t) = \left[(2/h_w^2(t)) \int_0^t \eta_w \mathrm{h}_w^2(\tau)\,\mathrm{d}\tau\right]^{1/2}. \tag{4.31}$$

This expression now depends explicitly on the history of the boundary diffusivity but is in keeping with (4.12) if we note that in the present case $q(t) = \eta_w h_w/\delta(t)$. If η_w is constant, particular expressions of the PD for Dirichlet step and ramp functions are respectively

$$\delta_s(t) = (2\eta_w t)^{1/2}, \tag{4.32}$$

$$\delta_r(t) = (2\eta_w t/3)^{1/2}. \tag{4.33}$$

The shape of the boundary condition is again preponderant, with the same conclusion as for (4.13)–(4.14): a step function involves a larger PD than a ramp one. Comparison of Neumann and Dirichlet conditions shows also that the latter exhibit larger PD than the former for a given time, which can be explained by the particular shape of heat input needed to restore the assumed Dirichlet condition.

4.3 DEFORMING FINITE ELEMENT FORMULATION

4.3.1 Finite element equations for deforming grids

If the idea of addressing the initial value problem (4.2)–(4.5) via deforming spatial grids is retained, the Galerkin spatial shape functions ψ_m will be time-dependent at every fixed point x_i:[8]

$$T(x_i, t) = \psi_m(x_i, t)u_m(t) \qquad (m = 1, 2, \ldots, N) \tag{4.34}$$

where the u_m are the N local values of the temperature acting as generalized coordinates of the problem.

As a consequence, the standard weighted residual—Galerkin weak form of the heat equation exhibits either an unsymmetrical matrix system or temperature-dependent additional loads, both of which are mesh velocity-dependent.[2,3]

An alternative formulation that can remain computationally attractive for situations without physical nonlinearities can however be set down if we note that semi-discretizations on space and time are no longer commutative in the present case. If we think of discretizing on time first and use for that purpose the so-called implicit generalized midpoint schemes (GMS),[9] the time discretization assumptions within each time step $\Delta t = t_{n+1} - t_n$ (subscripts refer to time instants) are the following:

$$\begin{aligned} \dot{T}_\alpha &= (T_{n+1} - T_n)/\Delta t \\ T_\alpha &= (1-\alpha)T_n + \alpha T_{n+1} \\ t_\alpha &= t_n + \alpha\ \Delta t \qquad \alpha \in [0, 1]. \end{aligned} \tag{4.35}$$

Introduction of Equation (4.35) into (4.2)–(4.5) leads to a spatially continuous system at t_α

$$\left.\begin{aligned} (k_\alpha T_{\alpha,x})_{,x} &= (\rho c)_\alpha (T_\alpha - T_n)/\alpha\ \Delta t \\ T_\alpha &= (T_{\rm w})_\alpha \\ \text{or} \qquad & \\ -k_\alpha T_{\alpha,x} &= q_\alpha \qquad \text{at } x = 0 \\ T_{\alpha,x} &= 0 \qquad \text{at } x = l. \end{aligned}\right\} \tag{4.36}$$

Galerkin discretizations on space are then used for T at times t_α and t_n respectively, i.e.

$$\begin{aligned} T_\alpha &= \psi_\alpha^m u_\alpha^m \\ T_n &= \psi_n^m u_n^m \end{aligned} \qquad (m = 1, 2, \ldots, N) \tag{4.37}$$

and standard weighted residual computations yield the final set of algebraic equations

$$\left(\mathbf{K}_\alpha+\frac{1}{\alpha\,\Delta t}\mathbf{C}_\alpha\right)\mathbf{u}_\alpha=\mathbf{g}_\alpha+\frac{1}{\alpha\,\Delta t}\mathbf{D}_{\alpha n}\mathbf{u}_n, \tag{4.38}$$

where typical components of **K**, **C** and **g** are similar to the classical fixed grid expressions using the shape functions and the time-dependent quantities at t_α, i.e.

$$(\mathbf{K}_\alpha)_{ij}=\int_0^l k_\alpha\psi^i_{\alpha,x}\psi^j_{\alpha,x}\,\mathrm{d}x \tag{4.39}$$

$$(\mathbf{C}_\alpha)_{ij}=\int_0^l (\rho c)_\alpha\psi^i_\alpha\psi^i_\alpha\,\mathrm{d}x \tag{4.40}$$

$$(\mathbf{g}_\alpha)_1=q_\alpha;\ (\mathbf{g}_\alpha)_i=0 \qquad (i=2,\ldots) \tag{4.41}$$

and where we define a new coupling capacitance matrix **D** accounting for mesh deformation and built up from the spatial shape functions at t_α and t_n

$$(\mathbf{D}_{\alpha n})_{ij}=\int_0^l (\rho c)_\alpha\psi^i_\alpha\psi^j_n\,\mathrm{d}x. \tag{4.42}$$

Thc deforming grid formulation (4.38) is hence very simple and does not introduce nonlinearities inherent to itself. It should therefore remain convenient in linear situations and reduces clearly to the classical equations if the grid is kept constant, sincc $\mathbf{D}_{\alpha n}=\mathbf{C}_\alpha$ in this case. It can also be shown to retain enthalpy of the model in a weak sense when passing from one grid to another.[2,3]

4.3.2 Deforming grid strategy based on PD histories

The time-dependent PD magnitude introduced in Section 4.2 yields the prescribed grid deformation law: suppose that the original (uniform or nonuniform) grid consists of N one-dimensional elements with linear shape functions (higher-order elements can also be used but will not be considered here). We then decide to include N_{PD} elements in a zone $\delta_n(t)$ corresponding to n times the PD at any time, and to dilate this zone as time proceeds according to

$$\delta_n(t)=n(\eta_{\mathrm{w}}t)^{1/2} \tag{4.43}$$

where n is an appropriate number (not necessarily integer) which accounts for the boundary condition types and shapes. The corresponding N_{PD} elements are subsequently dilated and the remaining ones accordingly contracted until they all reach their original position, i.e. their nodes are given instantaneous coordinates $x_i(t)$ such that

$$\begin{aligned} x_i(t)/x_m(t) &= x_i^0/x_m^0 \qquad i=0,\ldots,m \\ [x_i(t)-x_m(t)]/[l-x_m(t)] &= (x_i^0-x_m^0)/(l-x_m^0) \qquad i=m+1,\ldots,N \end{aligned} \tag{4.44}$$

where $m=N_{\mathrm{PD}}$, $x_m(t)=\delta_n(t)$ and the x_i^0 are the initial node positions with respect to a fixed origin.

Once the original grid is restored, the deforming grid strategy is stopped and the transient response continues with that grid.

In all cases where η_w is temperature-independent, $\delta_n(t)$ is an explicit function of time and the same holds for the deforming grid strategy. Otherwise, $\delta_n(t)$ is dependent upon the wall temperature and varies from iteration to iteration in the solution of the nonlinear system (4.38). In order to keep an explicit prediction of the mesh deformation, the surface diffusivity appearing in (4.43) is thus based on the wall temperature of the preceding time station. The free parameter n can correct possible underestimations of $\delta(t)$ due to this approximation.

The general rule for applying the deforming grid strategy is that the effective PD should definitely never be underestimated, otherwise inaccurate spatial representation of the temperature occurs in the transition zone between the estimated and the effective PD. This in turn precludes the alternative of a fixed higher-order elements grid in the vicinity of the boundary as a remedy to steep gradient modelling. Indeed as soon as the effective PD overcomes these elements, inaccurate modelling of the spatial gradients occurs.

By comparison with standard values obtained via polynomial approximations[4,5] and to comply with the different observations of Section 4.2, the values of the penetration depth factor n reported in Table 4.1 are proposed. They have to be confirmed by numerical experimentation.

Table 4.1 Penetration depth factors for deforming grids

B.C. shape / B.C. Type	Combination of step functions	Piecewise linear functions
Neumann	$n \simeq 3$	$n \simeq 2$–3
Dirichlet	$n \simeq 4$	$n \simeq 2.5$–3.5

4.4 NUMERICAL ILLUSTRATIONS

In order to demonstrate the effectiveness of the proposed method a number of model one-dimensional problems are treated in the sequel: they comprise linear and nonlinear responses of a finite carbon slab exposed to various Neumann and Dirichlet conditions, and the linear response of a composite wall. All time responses are obtained by the GMS integrator (4.35) with the particular choices $\alpha = \frac{2}{3}$ (Neumann boundary conditions) or $\alpha = 1$ (Dirichlet boundary conditions).

4.4.1 Linear responses of a finite carbon slab

The transient analyses of a finite slab of length l, initially at $T = 0$ are first realized. A fixed grid comprising 7 linear elements is chosen (Figure 4.2). The material is carbon, whose temperature-dependent thermal characteristics are also displayed on Figure 4.2. Mean values $\bar{k}$ and $\overline{\rho c}$ are used in linear situations.

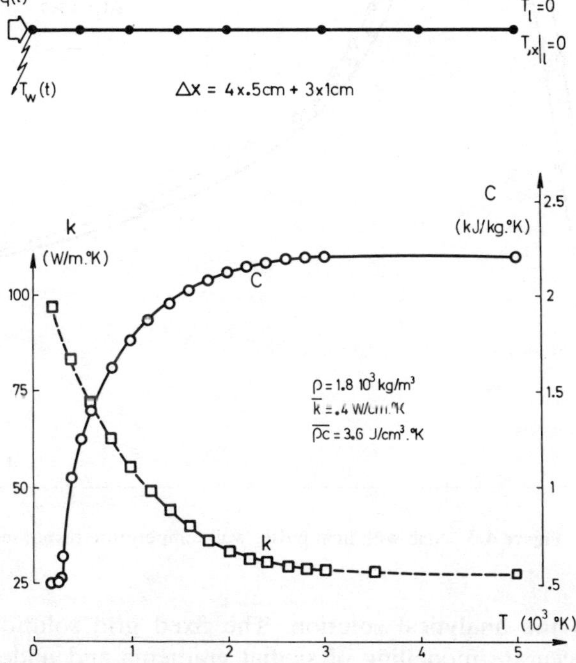

Figure 4.2 Slab discretization and material thermal data

(a) Expositions to heat input (Neumann boundary conditions)

The slab is suddenly exposed to a heat flux at $x = 0$ in $t = 0$. The face $x = l$ is kept at the initial temperature. First a heat pulse of 1 s duration is applied and the transient responses are obtained using 16 time steps of $\Delta t = 0.125$ s plus 8 steps of $\Delta t = 1$ s. Associated wall-temperature histories and spatial temperature distributions are shown in Figure 4.3 and 4.4 respectively, and

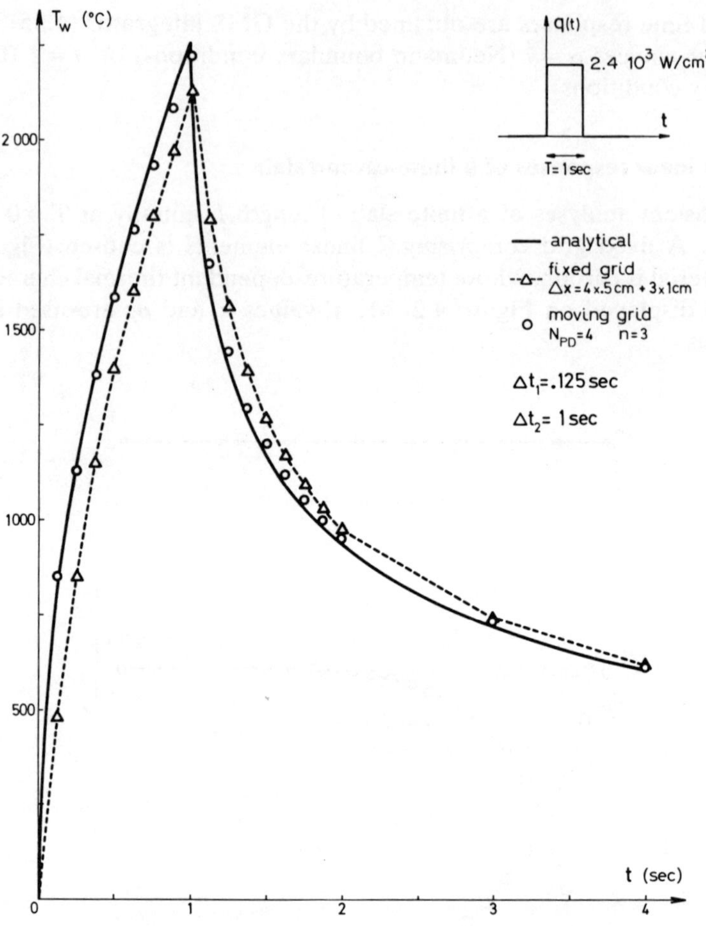

Figure 4.3 Slab with heat pulse: wall temperature responses

compared to the analytical solution. The fixed grid solution exhibits the expected inaccurate modelling of spatial gradients and underestimates the time rate of change of temperatures, producing in turn an underestimated wall temperature response during the heating phase and the reverse during the cooling phase. The spatial distributions at the time instants following the application and disappearance of the pulse illustrate the poor modelling of spatial gradients, especially at the first step.

On the other hand, the moving grid solution, with $N_{PD} = 4$ and $n = 3$, enables very satisfactory spatial responses and precise wall temperature results. As time proceeds fixed and moving grid results tend to coincide, since spatial meshes become identical.

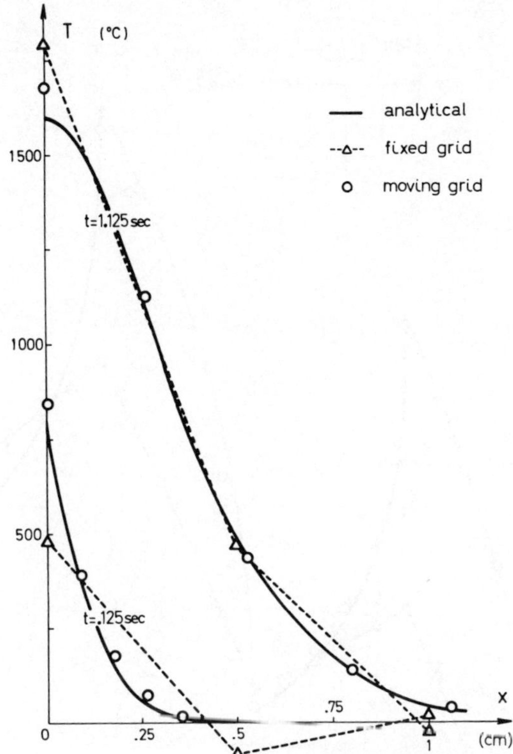

Figure 4.4 Slab with heat pulse: temperature spatial distributions

Note that the additional cost for the moving grid strategy is around 15% in this problem, as for all the linear experiments reported in the section (see Table 4.2). The ability of the procedure to deal with vanishing excitations is thus demonstrated.

Very short time responses of the same slab are then analyzed by applying a double heat pulse of duration ten times smaller, i.e. $\Delta t = 0.1$ s with the same amount of heat as in the first case (Figures 4.5 and 4.6). Transient responses with $\Delta t = 0.02$ s exhibit strong discrepancies between the fixed grid solution and both the moving grid and the analytical solutions: wall temperatures histories are crudely represented in the former case and spatial distributions present all the features due to inaccurate modelling of the gradients—appearance of negative temperatures at the interior nodes and alternate gradients to comply with global heat balance.

A triangular pulse heat flux is next applied (Figures 4.7 and 4.8) and the corresponding response analysed by the same solution techniques as before,

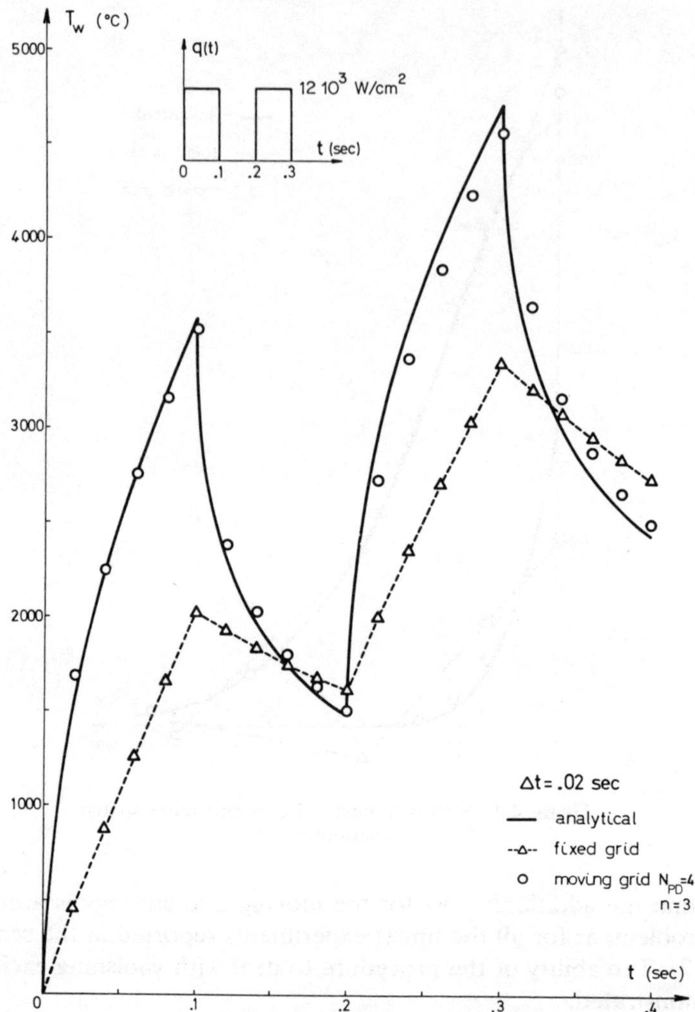

Figure 4.5 Slab with double heat pulse: wall temperature responses

in particular the penetration depth factor n is kept equal to 3 in the present case. Moving grid results are again definitely in keeping with the exact solution, contrary to the fixed grid solution.

(b) Prescribed wall temperatures (Dirichlet boundary conditions)

The wall temperature of the slab is now prescribed and the face $x = l$ is insulated.

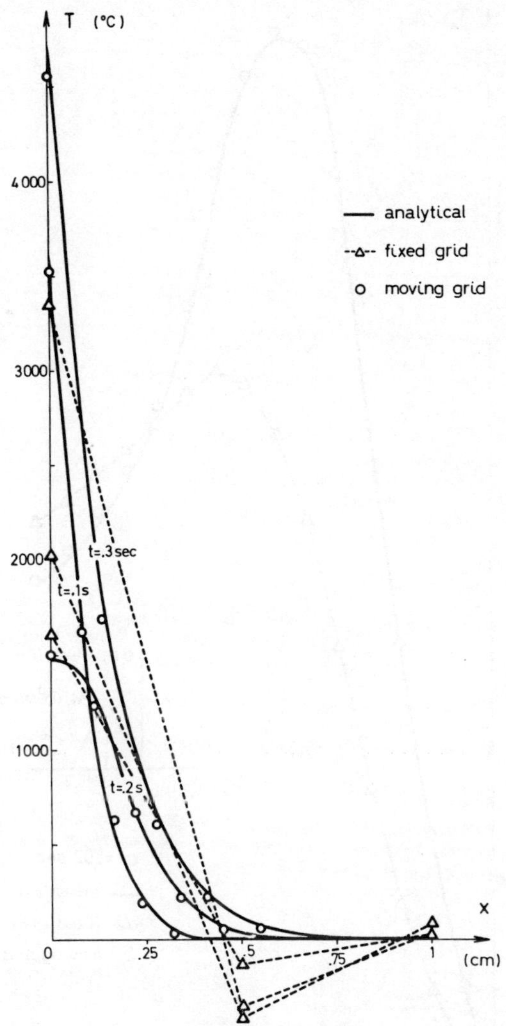

Figure 4.6 Slab with double heat pulse: temperature spatial distributions

First a double temperature pulse of 2×10^3 °C is applied during $\Delta t = 0.1$ s. Time integration is realized with steps of $\Delta t = 0.02$ s. The moving grid strategy corresponds this time to a penetration depth factor $n = 4$ (see Table 4.1) keeping $N_{PD} = 4$. Spatial distributions of temperature corresponding to time instants 0.1, 0.3 and 0.4 s are shown in Fig. 4.9, where it is clear that deforming

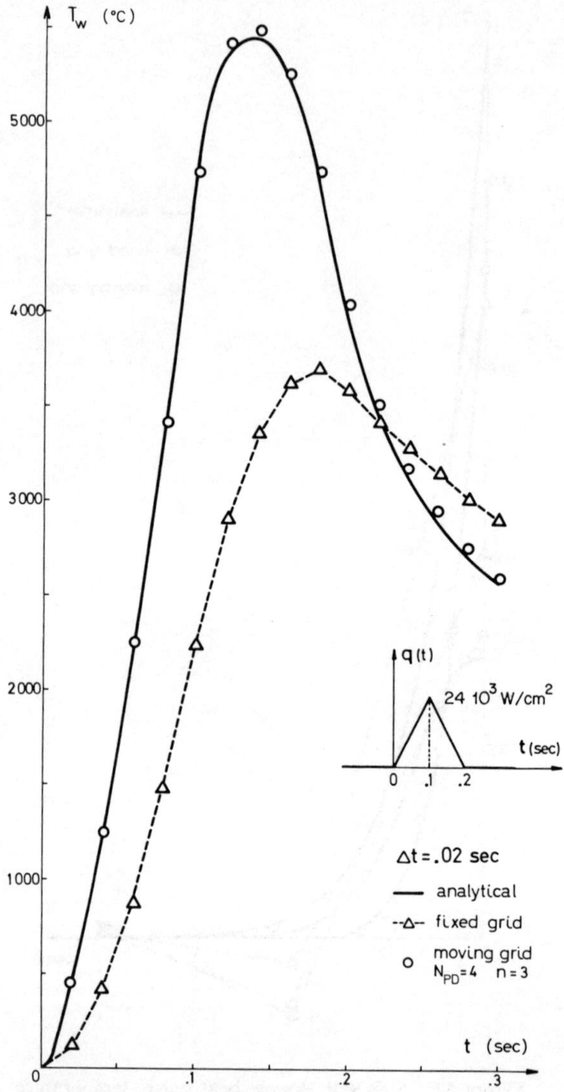

Figure 4.7 Slab with triangular heat pulse: wall temperature responses

finite elements allow for accurate modelling of the spatial gradients at no additional cost. The conclusion remains valid for the response to a triangular pulse of the same amplitude and of duration $\Delta t = 0.2$ s (Figure 4.10). Results are obtained with the same moving grid strategy as in the first case with $\Delta t = 0.04$ s.

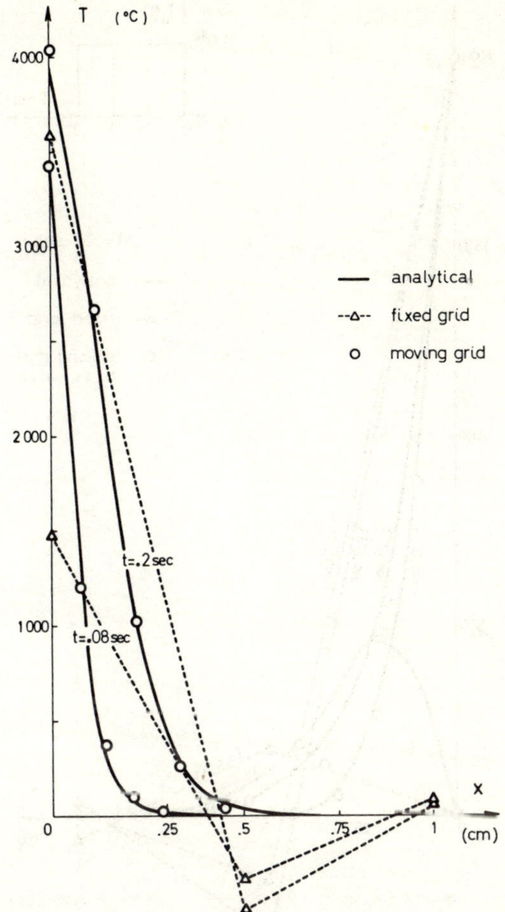

Figure 4.8 Slab with triangular heat pulse: temperature spatial distributions

4.4.2 Nonlinear responses of the slab

In order to measure the capability of treating variable surface diffusivity in the PD expressions, the same slab is analysed with effective temperature-dependent thermal characteristics (Figure 4.2). Fixed grid discretizations consist of respectively 10 and 50 equally spaced linear elements, the latter yielding converged reference results. Time steps of $\Delta t = 0.1$ s are used for expositions to respectively a step flux of magnitude 3×10^3 W/cm^2 and a sudden wall temperature of 2×10^3 °C at $x = 0$ in $t = 0$. The face $x = l$ is insulated in both cases.

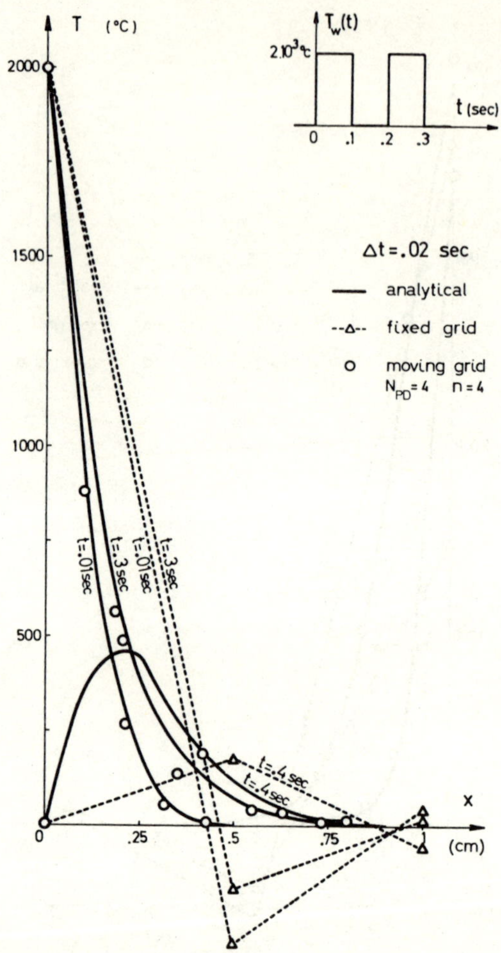

Figure 4.9 Slab with double temperature pulse: temperature spatial distributions

(a) Exposition to heat input (Figures 4.11 and 4.12)

Several moving grid tests have been conducted:

— for a given penetration depth factor ($n = 3$), include respectively 5 and 8 elements over the 10 of the spatial discretization in $\delta_3(t)$,

— a given number of elements ($N_{PD} = 5$) is located in respectively $\delta_{2.5}(t)$ and $\delta_3(t)$.

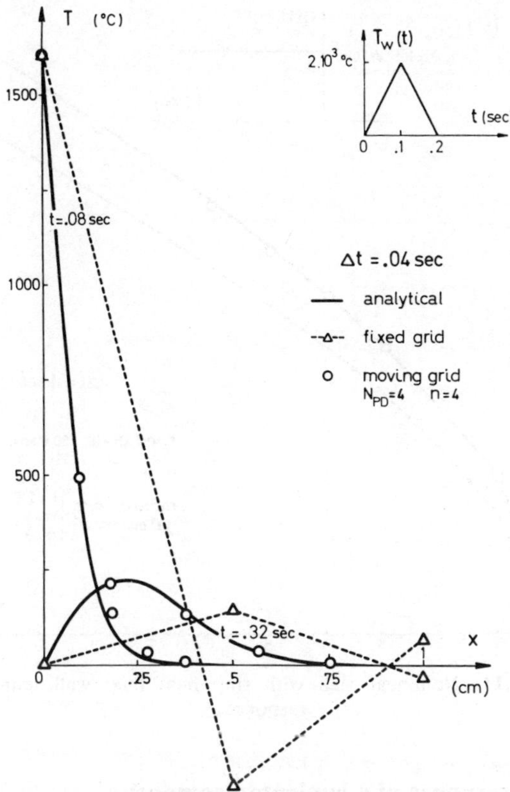

Figure 4.10 Slab with triangular temperature pulse: temperature spatial distributions

All the wall temperature responses are equivalent and in good agreement with that of the converged solution. Spatial gradients are well represented and are rather insensitive to the particular choices of the strategy parameters N_{PD} and n. This time the additional cost with respect to the fixed coarse grid solution is negligible. The latter results are still unacceptable.

(b) Prescribed wall temperature (Figure 13)

One single moving grid strategy is used with $N_{PD} = 8$ and $n = 4$, yielding excellent space distributions of temperature. Computer cost is in addition lower than the one of the fixed grid solution, which tends to confirm that moving grid strategies are particularly well suited for tracking the influence of Dirichlet boundary conditions.

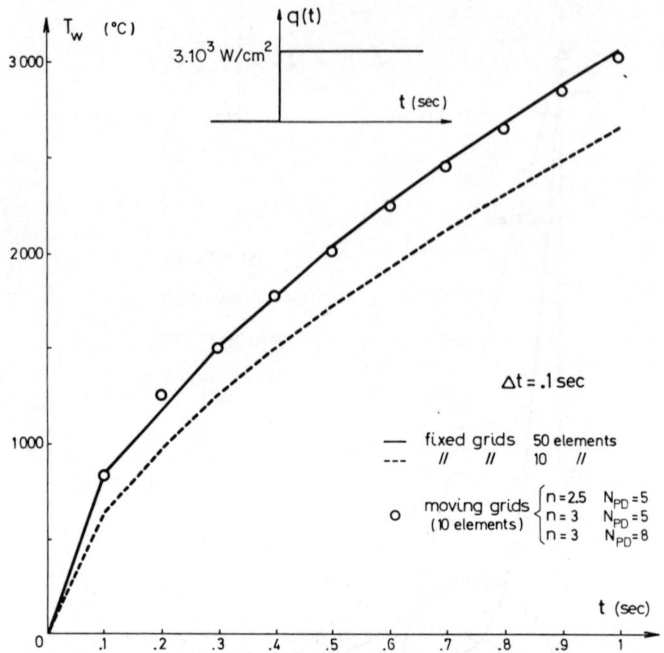

Figure 4.11 Nonlinear slab with step heat flux: wall temperature responses

4.4.3 Transient response of a laminated composite[10]

A last point to illustrate is the behaviour of the procedure in the presence of sharp spatial variations of the conductivity properties, e.g. in the case of laminated composites. For that purpose the transient response of such a composite, made of five anisotropic layers, is analysed when exposed to a heat pulse of duration $\Delta t = 2.5$ s and intensity 50 W/mm^2 at its boundary $x = 0$. The other end is insulated (Figure 4.14). Each layer has a thickness of 0.2 mm and is made of an orthotropic material with principal thermal conductivity coefficients k_L and k_T, where subscript L and T refer to the fibre orientation and to the transverse orientation respectively. Each layer has a different fibre orientation with respect to the heat penetration direction x, measured by the angle α. The heat capacity coefficient ρc, as well as the conductivity coefficients, is assumed temperature-independent. Time steps are 10 of $\Delta t = 0.5$ s and 3 of $\Delta t = 5$ s with the GMS integrator $\alpha = \frac{2}{3}$. Fixed grid solutions correspond to respectively 1 and 4 equally spaced linear elements per layer. The moving grid solution corresponds to the coarser grid with $N_{PD} = 4$ and $n = 3$. By this means, one single element can cover several layers in the moving grid procedure: effective thermal characteristics for

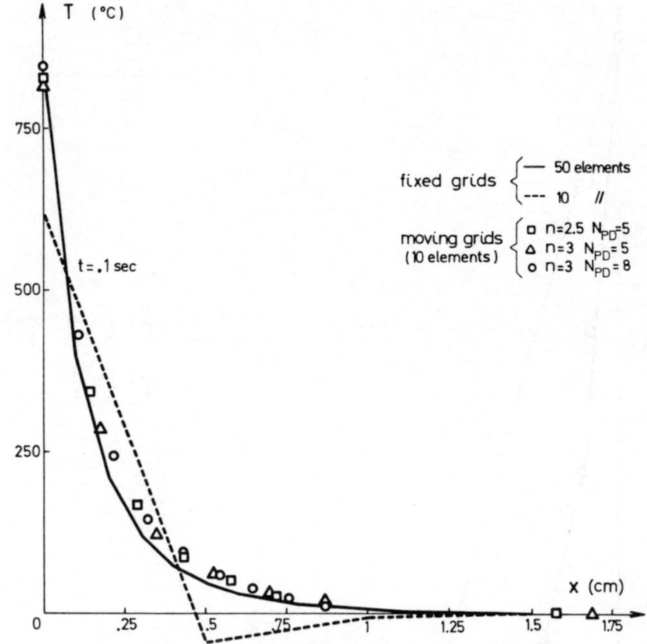

Figure 4.12 Nonlinear slab with step heat flux: temperature spatial distributions

element i in the heat penetration direction x are then assumed by taking

$$\frac{1}{k_i}=\frac{1}{l_i}\int_0^{l_i}\frac{\mathrm{d}x}{k}$$

$$(\rho c)_i=\frac{1}{l_i}\int_0^{l_i}\rho c\ \mathrm{d}x$$

where l_i denotes the length of the element.

This implies clearly an additional cost in the moving grid solution, which is more sensible for linear (temperature-independent) media. Figures 4.14 and 4.15 present the usual wall temperature history and space distributions for two particular instants. The observations raised in the preceding applications hold here again: wall temperatures are modelled adequately by the moving grid solution with marked differences with respect to the coarse fixed grid solution, until the two meshes coincide. Spatial distributions in the latter solution are poorly represented. A look at computer costs (Table 4.2) indicates however that this time the physical pseudo-nonlinearity involved in the moving grid solution (inhomogeneous elements) implies substantial additional work.

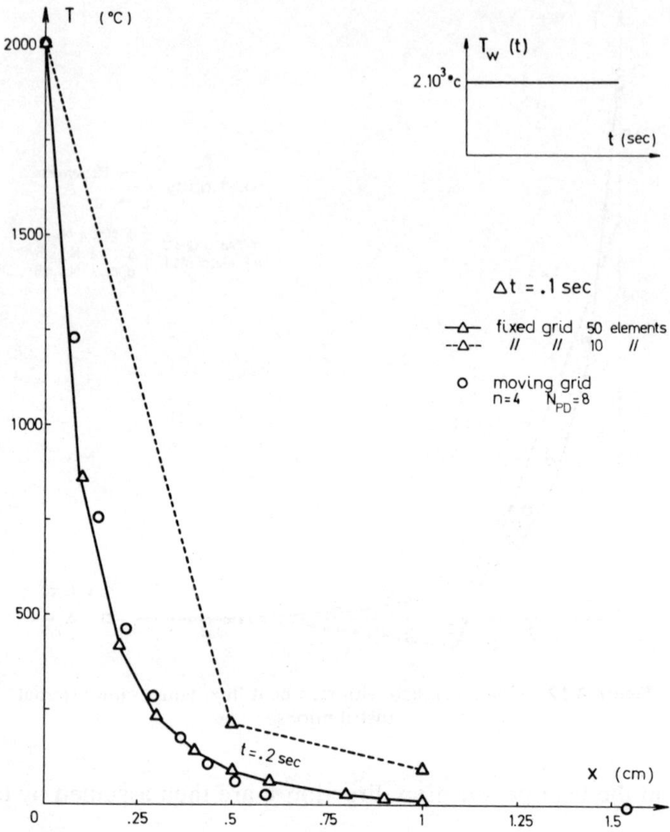

Figure 4.13 Nonlinear slab with step temperature: temperature spatial distributions

This situation could be changed if each layer were given the effective temperature-dependences of the thermal characteristics.

4.5 CONCLUSIONS AND FUTURE WORK

Moving grid finite element formulations based on a mesh coupling capacitance matrix and an explicit deformation pattern via heat penetration depth concepts have proved to allow for adequate modelling of transient steep gradients in one-dimensional situations.

Further work will be addressed to extending the techniques reported here to two-dimensional situations by considering:

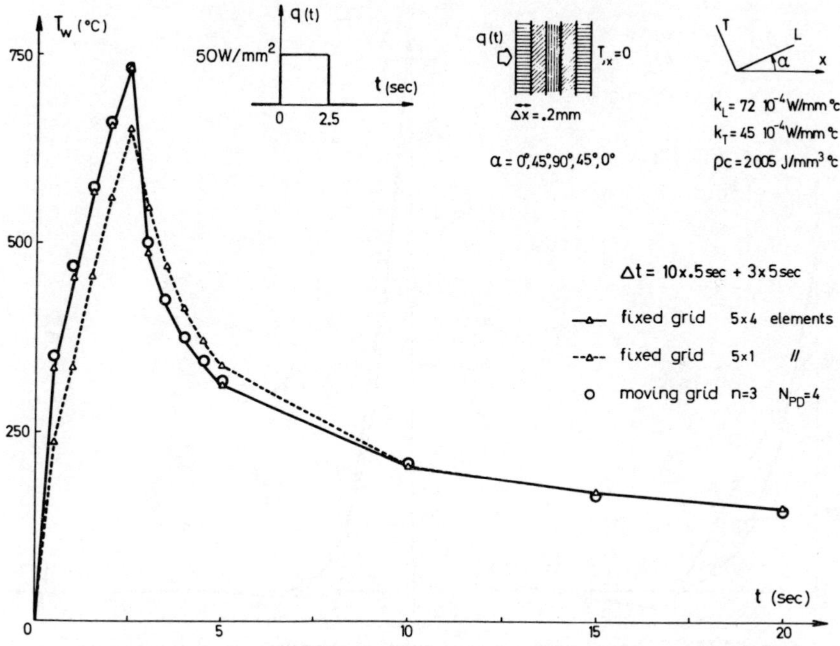

Figure 4.14 Composite wall with heat pulse: wall temperature responses

Table 4.2 Computer cost for the model problems (CPU times in seconds for IBM 370/158 computer)

Material	Problem	Nbr steps	Moving grid	Fixed grid	Converged
Linear	Heat pulse	24	88.1 (1.17)	75.0	—
	Double heat pulse	26	93.3 (1.15)	81.3	—
	Triangular heat pulse	26	95.1 (1.12)	85.2	—
	Double temperature pulse	26	95.1 (1.01)	94.5	—
	Triangular temperature pulse	16	58.4 (1.12)	52.0	—
Nonlinear	Step heat flux	10	65.0 (1.03)	63.0	173.0
	Step temperature	10	50.7 (0.83)	61.0	174.3
Linear	Composite	19	93.5 (1.58)	59.3	93.9

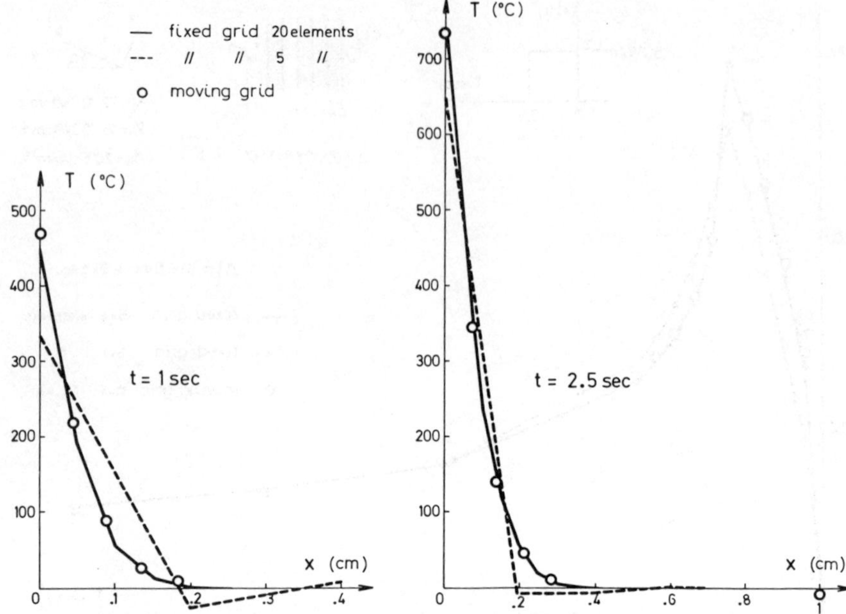

Figure 4.15 Composite wall with heat pulse: temperature spatial distributions

(a) explicit mesh deformation: the deformation pattern will be managed by defining one-dimensional discretization lines, linking the attacked boundaries to the passive ones or to symmetry planes and following the directions of heat penetration. They will follow deformation laws based on the local penetration depths. Discretization lines transverse to the latter ones will thus be moved accordingly;

(b) implicit mesh deformation: the mesh size will be adapted to distribute equally the total temperature jump from the boundary to the unattacked part of the medium between a given number of elements; steep internal gradients can also be handled by localizing first sharp spatial variations of the temperature gradients.

REFERENCES

1. O. C. Zienkiewicz, '*The Finite Element Method*', 3rd ed, McGraw-Hill, Maidenhead, (1977).
2. P. Gerrekens M. Hogge, and G. Laschet, 'Appropriate Finite Element Techniques for Heat Transfer Problems with High Gradients and Ablation', in *Numerical Methods in Thermal Problems*, Vol. II, (ed. R. W. Lewis *et al.*), Pineridge, Swansea, 168–179 (1981).

3. M. Hogge and P. Gerrekens, 'One-dimensional finite element analysis of thermal ablation with pyrolysis', *Comp. Meth. Appl. Mech. Eng.*, **33**, 609–634 (1982).
4. M. A. Biot, *Variational Principles in Heat Transfer*, Oxford Univ. Press, London (1970).
5. T. R. Goodman, 'The heat balance integral and its applications to problems involving a change of phase', *Trans. ASME*, **80**, 335–342 (1958).
6. T. F. Zien, 'Approximate calculation of transient heat conduction', *AIAA J.* **14**, 404–406 (1976).
7. M. Necati Ozisik, *Boundary Value Problems of Heat Conduction*, International Textbook Co., Scranton, Penn (1968).
8. D. R. Lynch and K. O'Neill, 'Continuously deforming finite elements for the solution of parabolic problems, with and without phase change', *Int. J. Num. Meth. Eng.*, **17**, 81–96 (1981).
9. T. J. R. Hughes, 'Unconditionally stable algorithms for non-linear heat conduction', *Comp. Meth. Appl. Mech. Eng.*, **10**, 135–139 (1977).
10. B. B. Raju *et al.*, 'Transient temperatures in laminated composite conical shells due to aerodynamic heating;, *AIAA J.* **16**, 547–548 (1978).

Numerical Methods in Heat Transfer, Volume II
Edited by R. W. Lewis, K. Morgan, and B. A. Schrefler

Chapter 5

Variational Formulation and Approximate Solutions of the Thermal Diffusion Equation

George A. Keramidas

ABSTRACT

Certain concepts of classical mechanics are utilized to derive the variational formulation for a field partial differential equation. By introducing suitable parameters, it is demonstrated that the concepts of virtual work and generalized coordinates can be extended to the general diffusion equation and this equation can be translated into Lagrange's equations of mechanics. The Lagrangian system of equations is most suitable for deriving approximate solutions and this is demonstrated by assuming a linear series expansion in terms of the generalized coordinates. Furthermore, it is shown that approximate methods, such as the finite element method, can be directly derived as a special application of the generalized approach. Examples of approximate solutions are given for some typical problems involving diffusion of heat.

5.1 INTRODUCTION

Approximate methods have been applied to the solution of partial differential equations when analytical forms of solution are not possible due to the nature of the particular problems under study. Some types of approximate solutions can be directly applied to the given equation and some require first the derivation of a variational form for this equation. The latter imposes certain restrictions on applying approximate methods since many of the variational derivations are problem dependent.

The intent of this study is to provide the formulation of a unified analysis for partial differential equations, such as the thermal diffusion equation, by methods analogous to those of classical mechanics. The implementation of certain concepts of classical mechanics opens the way to a formulation of the diffusion equation by means of generalized coordinates and leads to a

Lagrangian type of equations. This approach simplifies the application of approximate methods for obtaining solutions to many practical problems of interest.

The variational analysis presented here involves a generalization of the concept of virtual work from classical mechanics. The approach was first adapted by M. A. Biot and was applied to heat transfer phenomena and irreversible processes.[1,2] By applying the concept of virtual work to the diffusion equation a variational form of this equation can be obtained, which is problem independent, and the resulting variational equation contains terms similar to those found in Lagrangian mechanics. This approach should not be regarded as a 'new method' for deriving variational formulations. It is only an application of classical concepts and an extension of Lagrangian mechanics to derive a unified formulation. One of the advantages of such an approach is that the physical significance of each term of the differential equation can be easily identified.

The governing equation for the diffusion of heat and the introduction of a vector defined as heat displacement are given in the first section of this study. Based on the definition of the displacement vector the diffusion equation is written in a form similar to the equation of motion in mechanics. In this form the governing equation will be used in the next section to derive the variational formulation.

According to concepts of classical mechanics two types of variational formulations can be derived, the fundamental form based on variations of the displacement field and the complementary form based on variations of the stress field. The basic definitions and derivations for the variational analysis are given in the second section and the two types of variational formulations are presented for the diffusion equation. The analogy between the derived equations and those of mechanics can be justified by associating each of the terms of the variational equations with a quantity such as potential function, dissipation function and generalized body and boundary forces. The dissipation function is a generalization of the concept introduced by Rayleigh in mechanics for systems with viscous dissipation. The resulting Lagrangian equations, in terms of generalized coordinates, represent either type of variational formulation and they are not problem dependent. As a consequence the generalized equations can be used for obtaining solutions to physical problems governed by such equations as the general diffusion equation. Furthermore, many types of approximate solutions can be derived by expressing the field parameters in terms of a set of generalized coordinates. A particular example is given in the third section where a series expansion is assumed for the field parameters. This expansion expresses the field parameters as a linear combination of the generalized coordinates. Although such an approximation is restricted by the linear dependence on the generalized coordinates, it may represent such approximations as a Fourier series expansion or an expansion in terms of orthogonal functions. The derivation of the equations in terms of the above approximation is given and it is shown that the finite element method can be derived as a special case of this approximation.[3,4] The discretized system of equations for a linear finite element approximation is given as an example and four types of finite element models are derived. Two of the models correspond to the fundamental variational formulation and the other two correspond to the complementary one.

Numerical experiments and solutions to particular problems are given in the fourth section of this study. In the first part of the numerical applications the error behaviour of the finite element models is investigated for a given problem with known analytical solutions. Furthermore, the convergence of the models is investigated and the criteria for uniform convergence are discussed. In the second part of the numerical applications, examples of numerical solutions are presented for a number of physical problems. From the examples presented the efficiency of the computational models is discussed and one can observe that the unified formulation presented here can be extended to a variety of physical problems.

5.1 BASIC EQUATIONS AND DEFINITIONS

The general form of a partial differential equation describing diffusion of heat is given by

$$\frac{\partial T(x_k, t)}{\partial t} = \frac{\partial}{\partial x_i}\left(k_{ij}\frac{\partial T(x_k, t)}{\partial x_j}\right) + S, \tag{5.1}$$

where T is the temperature, x_i is the coordinate system, t is time, k_{ij} is the tensor diffusion coefficient and S is the generalized source term.

In order to describe the kinematics of a medium, reference is made to two configurations. The reference or initial configuration at time $t = t_0$ and the present configuration at time t with T_0 and T the values of the absolute temperatures respectively. A dimensionless temperature is defined in terms of T and T_0 if one considers the ratio of the change ΔT of the absolute temperatures from the reference to the present state over the absolute value of that variable at the reference state

$$\theta(x_k, t) = \frac{T - T_0}{T_0} = \frac{\Delta T}{T_0}. \tag{5.2}$$

In terms of the dimensionless variable θ, Equation (5.1) is expressed as follows:

$$\frac{\partial \theta}{\partial t} = \frac{\partial}{\partial x_i}\left(k_{ij}\frac{\partial \theta}{\partial x_j}\right) + \frac{1}{T_0} S. \tag{5.3}$$

From the physical point of view the dimensionless temperature θ can be interpreted as a generalized deformation similar to mechanical strain. Define now a vector field $H_i(x_k, t)$ as the *heat displacement vector*, which is a function of the coordinates x_k and time t of the form

$$\dot{H}_i(x_k, t) = \frac{\partial H_i(x_k, t)}{\partial t} \tag{5.4}$$

and

$$\theta(x_k, t) = \frac{\partial H_i(x_k, t)}{\partial x_i}. \tag{5.5}$$

The interpretation of θ as a deformation or dilatation (strain) is justified by Equation (5.5). Furthermore, Equation (5.5) may be considered as a constraint in the sense of classical mechanics and it must be verified by the physical solution and by the variations of δH_i and $\delta\theta$ as follows:

$$\delta\theta = \frac{\partial}{\partial x_i}(\delta H_i). \tag{5.6}$$

The variations δH_i correspond to virtual displacements with θ and H_i, being analogous to strain and displacement in mechanics. Another quantity, which is defined in mechanics and it is related to deformation, is the stress tensor. Here a similar stress can be defined as

$$\sigma = E\theta, \tag{5.7}$$

where E is a modulus similar to the bulk modulus in mechanics. Equation (5.7) may represent a constitutive relation for a particular medium and together with Equation (5.3) describes the behaviour of that medium under certain loading conditions.

The governing equation can be exrpessed in terms of the displacement field as follows

$$\frac{\partial H_i}{\partial t} = k_{ij}\frac{\partial \theta}{\partial x_j} + \dot{h}_i, \tag{5.8}$$

where h_i satisfies the equation

$$\frac{\partial h_i}{\partial x_i} = \frac{1}{T_0}\int_0^t S\,\mathrm{d}t. \tag{5.9}$$

This equation does not uniquely determine h_i but any particular field may be chosen to satisfy Equation (5.9). Such a field may be considered to be a given function of the coordinates and time. The diffusion coefficient is a tensor with six components and with the property

$$k_{ij} = k_{ji};$$

hence k_{ij} is a symmetric tensor. If λ_{ij} is the inverse of k_{ij}, then λ_{ij} is also symmetric and Equation (5.8) can be written as

$$\lambda_{ij}\frac{\partial H_i}{\partial t}-\lambda_{ij}\dot{h}_i=\frac{\partial \theta}{\partial x_j}. \tag{5.10}$$

Equations (5.5) and (5.8) provide a formulation which is defined as the displacement formulation for the diffusion equation and this formulation reduces to the conventional one, Equation (5.3), by combining Equations (5.5) and (5.8). Furthermore, Equations (5.5), (5.7) and (5.8) are respectively analogous to the kinematic relations, stress–strain relations and momentum equation in mechanics.

The concept of heat displacement has been introduced previously by this author, for describing different transport processes.[5] The introduced displacement is regarded as a generalized quantity conjugate to the generalized deformation θ. This generalized presentation of the governing equations for thermal problems has certain features which will be discussed in the next sections.

5.2 VARIATIONAL ANALYSIS

A large number of variational formulations have been introduced in the past for deriving approximate solutions. The majority of them are based on the minimization of a functional which describes a particular physical process. Such formulations are usually restricted to the conditions of the particular problem and their applicability is limited. One wishes to have a unified approach for deriving variational formulations. Such formulations should be problem independent and they should be derived from physical considerations.

Classical mechanics has provided researchers with some powerful tools for solving problems either analytically or by approximate methods. One such a tool is the principle of virtual work which has been successful in solving a variety of problems. The principle of virtual work in mechanics has been extended to such areas as thermodynamics[6] and variational equations can be derived which are not problem dependent for a variety of physical processes. Another concept largely used in mechanics is the concept of generalized coordinates which can be used to describe a physical system. Generalized coordinates can be interpreted as generalized displacements or velocities which specify the configurations of a physical system.[7]

The applications of these concepts are not restricted to mechanics or only to certain media. In this section the concept of virtual work is used to derive two variational formulations for the diffusion equation. The first is referred to as the fundamental form and is based on the displacement formulation of the diffusion equation. The second formulation is based on the conventional form of the diffusion equation and is referred to as the complementary form.

These two forms are directly analogous to the ones in classical mechanics where the fundamental form of a variational principle is given in terms of variations of the displacement field and the complementary form in terms of forces. Both of these forms are based on equilibrium or conservation laws governing the physical system.

5.2.1 Fundamental variational principle

The fundamental form is derived from the displacement formulation of the diffusion equation by implementing the principle of virtual work. Consider a variation δH_i of the displacement field H_i and the corresponding variation $\delta\theta$ given by the constraint Equation (5.6). If the medium is subjected to a virtual displacement δH_i, then the principle of virtual work requires that

$$\int_v \left[\lambda_{ij}\frac{\partial H_i}{\partial t} - \lambda_{ij}\dot{h}_i - \frac{\partial \theta}{\partial x_j}\right]\delta H_j \, dv = 0, \tag{5.11}$$

where the integration is extended over a volume v of the medium. Integration of Equation (5.11) by parts yields

$$\int_v \lambda_{ij}\frac{\partial H_i}{\partial t}\delta H_i \, dv + \int_v \theta \frac{\partial}{\partial x_i}(\delta H_i)\, dv - \int_v \lambda_{ij}\dot{h}_j \,\delta H_i \, dv = \int_s \theta n_i \,\delta H_i \, ds. \tag{5.12}$$

The surface integral is extended over the boundary surface s of volume v and n_i is the unit vector normal to the boundary pointing outward. The second term of the left-hand side is written as

$$\int_v \theta\delta\left(\frac{\partial H_i}{\partial x_i}\right) dv = \int_v \theta\delta \, dv = \delta P \tag{5.13}$$

and the scalar function P may be interpreted as a potential function, similar to the potential function of mechanics, given by

$$P = \tfrac{1}{2}\int_v \theta^2 \, dv. \tag{5.14}$$

In terms of the variations δP, Equation (5.12) is written as follows:

$$\int_v \lambda_{ij}\frac{\partial H_i}{\partial t}\delta H_i \, dv - \int_v \lambda_{ij}\dot{h}_j \,\delta H_i \, dv + \delta P = \int_s \theta n_i \,\delta H_i \, ds. \tag{5.15}$$

The variational relation given by Equation (5.15) can be considered as the variational principle for the diffusion equation. Equation (5.15) must be verified for arbitrary variations of the displacement field H_i, with θ defined as a function of H_i through the constraint, Equation (5.5). Hence the variational equations given by Equation (5.15) are nothing but a variational statement for the diffusion equation with energy conservation automatically satisfied. This variational principle must be understood in a broad sense since it corresponds to the principle of virtual work in mechanics. It represents the fundamental form since it was derived in terms of the displacement field.

5.2.2 Complementary variational principle

The analogy with mechanics suggests that it is possible to derive a complementary form of the variational principle for the diffusion equation. Since the generalized thermal deformation θ, Equation (5.5), leads to the definition of the generalized thermal stress σ, which is similar to the mechanical stress, then the complementary form is derived by varying the stress field and expressing the conservation equation, Equation (5.3), in a variational form.

In terms of the generalized stress σ, Equation (5.3) is written as

$$\frac{\partial \sigma}{\partial t}+\frac{\partial}{\partial x_i}\left(k_{ij}\frac{\partial \sigma}{\partial x_j}\right)-\frac{1}{T_0 E}S=0 \tag{5.16}$$

and for arbitrary variation $\delta\sigma$ an equivalent to Equation (5.11) is obtained

$$\int_v \left[\frac{\partial \sigma}{\partial t}-\frac{\partial}{\partial x_i}\left(k_{ij}\frac{\partial \sigma}{\partial x_j}\right)-\frac{1}{T_0 E}S\right]\delta\sigma \,\mathrm{d}v=0. \tag{5.17}$$

Integration by parts yields

$$\int_v \frac{\partial \sigma}{\partial t}\delta\sigma \,\mathrm{d}v+\int_v k_{ij}\frac{\partial \sigma}{x_j}\frac{\partial}{\partial x_i}(\delta\sigma)\,\mathrm{d}v-\int_v \frac{1}{T_0 E}S\,\delta\sigma\,\mathrm{d}v=\int_s k_{ij}n_i\frac{\partial \sigma}{\partial x_j}\,\mathrm{d}s. \tag{5.18}$$

The surface integral is extended over the boundary surface s of volume v and n_i denotes the unit vector normal to the boundary pointing outwards. Proceeding as in the previous section, Equation (5.18) is written as

$$\int_v \frac{\partial \sigma}{\partial t}\delta\sigma\,\mathrm{d}v-\int_v \frac{1}{T_0 E}S\,\delta\sigma\,\mathrm{d}v+\delta P=\int_s n_i k_{ij}\frac{\partial \sigma}{\partial x_j}\delta\sigma\,\mathrm{d}s \tag{5.19}$$

where the variation δP is given by

$$\delta P=\int_v k_{ij}\frac{\partial \sigma}{\partial x_j}\delta\left(\frac{\partial \sigma}{\partial x_i}\right)\mathrm{d}v \tag{5.20}$$

and the scalar P by

$$P=\tfrac{1}{2}\int_v k_{ij}\frac{\partial \sigma}{\partial x_j}\frac{\partial \sigma}{\partial x_i}\,\mathrm{d}v. \tag{5.21}$$

Equation (5.19) represents the complementary form of the variational principle for the diffusion equation. The different terms of the variational equation, Equation (5.19), may be evaluated either in terms of the thermal stress σ or in terms of the deformations θ. The latter case corresponds to formulating the variational principle for the diffusion equation in terms of a primitive dimensionless variable θ.

The two forms of variational principles derived here are generalized statements of conservation laws for a physical system and their existence is verified by concepts from classical mechanics. Either form can be applied to the governing equation, but it is first necessary to derive a more compact form of these variational equations which will be better suited to practical applications. Some remarks can be made regarding the form of the variational equations. A significant difference between the two forms is the fact that the complementary form involves space derivatives of the variable θ. This does not present a problem as long as the space distribution for θ is continuous. If discontinuities exist, as is the case in many practical problems, then the complementary form is not the appropriate one to be used for the solution of such a problem. In general, when using the same approximate representation for the transport variable θ, the complementary principle will yield results which are less accurate than those obtained by the fundamental form which is based on the displacement field.

5.2.3 Generalized coordinates

A physical system is usually described by a set of coordinates and the governing equations are expressed in terms of these coordinates. In many cases, a given set of coordinates is not the most convenient one to describe a physical problem. As an example, for certain problems involving the motion of a particle it is more convenient to use a moving coordinate system than a fixed one. It would be very desirable, and practical, to have a general method for describing a physical system by a set of coordinates which will provide a uniform method to represent and solve the governing equations for physical processes. The generalized coordinates of classical mechanics are such a set.

When a physical system is described by a set of generalized coordinates, it is the usual practice to designate each coordinate by q and the set of n generalized coordinates by q_k, $k=1,\ldots,n$. For a physical system of N particles a set of $3N$ generalized coordinates may be specified, and since these coordinates must have some definite set of values they are functions of the time and also of the Cartesian system of coordinates. For a physical system which is described by a set of generalized coordinates q_k, the time derivative $\dot{q}_k$ is defined as a generalized velocity associated with this coordinate. This generalized velocity, for example, can be computed in terms of Cartesian coordinates and velocities. If one assumes that q_k is a function of the Cartesian coordinates x_i and time t, then

$$q_k = q_k(x_i, t).$$

It is also possible to express the Cartesian coordinates in terms of generalized coordinates as

$$x_i = x_i(q_k, t).$$

From the definition of the generalized velocity one obtains

$$\dot{x}_i = \frac{\partial x_i}{\partial t} + \frac{\partial x_i}{\partial q_k}\dot{q}_k.$$

Other physical quantities such as energy, momentum, forces etc. can be expressed in terms of the generalized coordinates. For example the generalized momentum p_k associated with the coordinate q_k is given by

$$p_k = \frac{\partial K}{\partial \dot{q}_k}$$

when K is the kinetic energy in terms of $\dot{q}_k$.

The subject of this section is to implement the concept of generalized coordinates to the two forms of variational equations previously derived. By introducing generalized coordinates into the variational equations, they can be translated into a Lagrangian type of equations. This provides a unified approach and the diffusion equation is represented in a generalized form independent of coordinate systems or physical conditions.

Consider the displacement field H_i to be a function of a set of n generalized coordinates q_k,

$$H_i = H_i(q_k, x_j, t). \tag{5.22}$$

The variations of H_i, or small displacements δH_i, are due entirely to the variations of the generalized coordinates δq_k and they are expressed as

$$\delta H_i = \frac{\partial H_i}{\partial q_k}\delta q_k. \tag{5.23}$$

The variations δq_k can be identified as virtual displacements since they do not necessarily represent any actual motion but only any possible motion of a system. Under such a motion there is an amount of work done which is defined as the virtual work δW, and it is expressed as

$$\delta W = Q_k \delta q_k, \tag{5.24}$$

where Q_k is defined as a generalized force associated with the coordinate q_k.

The variations of the potential function P, given by Equation (5.14), are expressed as

$$\delta P = \frac{\partial P}{\partial q_k}\delta q_k \tag{5.25}$$

and the time derivative of the displacement vector is taken as

$$\dot{H}_i = \frac{\partial H_i}{\partial q_k}\dot{q}_k + \frac{\partial H_i}{\partial t}. \tag{5.26}$$

Equation (5.26) expresses the total time derivative of the vector H_i and by differentiating with respect to the generalized velocity $\dot{q}_k$ one obtains

$$\frac{\partial \dot{H}_i}{\partial \dot{q}_k} = \frac{\partial H_i}{\partial q_k}. \tag{5.27}$$

Taking into account Equations (5.23), (5.25), (5.26) and (5.27), the variational principle, Equation (5.15), can be transformed into a more compact form. The first term of Equation (5.15) is written as

$$\int_v \lambda_{ij}\frac{\partial H_i}{\partial t}\,\delta H_j\,\mathrm{d}v = \int_v \lambda_{ij}\frac{\partial H_i}{\partial t}\frac{\partial \dot{H}_j}{\partial \dot{q}_k}\,\delta q_k\,\mathrm{d}v = \frac{\partial D}{\partial \dot{q}_k}\,\delta q_k \tag{5.28}$$

where the function D is given by

$$D = \tfrac{1}{2}\int_v \lambda_{ij}\dot{H}_i\dot{H}_j\,\mathrm{d}v. \tag{5.29}$$

The vector $\dot{H}_i$ can be regarded as the diffusive local rate of heat flow and the function D represents a dissipation function. This dissipation has been previously introduced for systems with friction or viscous effects and it is known as Rayleigh's dissipation function.

In terms of the variations δP and δD, Equation (5.15) yields

$$\delta D + \delta P = \int_s \theta\eta_i\frac{\partial H_i}{\partial q_k}\,\delta q_k\,\mathrm{d}s + \delta F. \tag{5.30}$$

The term δF of the right-hand side of Equation (5.30) can be considered as a force term due to the source given by

$$\delta F = \int_v \lambda_{ij}\dot{h}_j\frac{\partial H_i}{\partial q_k}\,\delta q_k\,\mathrm{d}v \tag{5.31}$$

and

$$F = \int_v \lambda_{ij}\dot{h}_j H_i\,\mathrm{d}v. \tag{5.32}$$

The surface integral of the right-hand side of Equation (5.30) can be expressed in terms of a generalized boundary force Q_k as

$$\int_s \theta\eta_i\frac{\partial H_i}{\partial q_k}\,\delta q_k\,\mathrm{d}s = Q_k\,\delta q_k \tag{5.33}$$

where

$$Q_k = \int_s \theta \eta_i \frac{\partial H_i}{\partial q_k} \mathrm{d}s. \tag{5.34}$$

From the physical point of view, Equation (5.33) can be interpreted as the virtual work δW, done by the force Q_k associated with the virtual displacement δq_k. Comparing Equations (5.24), (5.30) and (5.33) the relation

$$\delta D + \delta P - \delta F = \delta W \tag{5.35}$$

is obtained. This can be regarded as expressing conservation, since the virtual work done by the generalized boundary forces is equal to energy gain or loss for a physical system.

For arbitrary variations of the generalized coordinates Equation (5.35) is written as

$$\frac{\partial D}{\partial \dot{q}_k} + \frac{\partial P}{\partial q_k} = Q_k + \frac{\partial F}{\partial q_k}. \tag{5.36}$$

For the set of n generalized coordinates Equation (5.36) constitute a system of n differential equations for the unknowns q_k. These equations are of the same form as the equations of Lagrangian mechanics for the slow motion of a dissipative system with negligible inertia forces. For the present case of thermal diffusion the potential function is regarded as a thermal potential and the function D as the heat dissipation function. The function F may be due to internal heat generation and/or heat transfer. Furthermore, the generalized forces Q_k represent thermal forces due to the temperature distribution at the boundary and they are defined by Equation (5.33) as the work done by Q_k under the displacement δq_k.

This generalized derivation in terms of the coordinates q_k can be extended to the complementary form of the variational principle. The derivation is identical and it yields a set of equations similar to Equations (5.36), the difference being that they are derived in terms of the temperature θ. If one assumes that θ is a function of a set of n generalized coordinates $\bar{q}_k$, then

$$\theta = \theta(\bar{q}_k, x_i, t). \tag{5.37}$$

Following the same steps as for the previous derivations the Lagrangian type of equations are

$$\frac{\partial \bar{D}}{\partial \dot{\bar{q}}_k} + \frac{\partial \bar{P}}{\partial \bar{q}_k} = \bar{Q}_k + \frac{\partial \bar{F}}{\partial \bar{q}_k}, \tag{5.38}$$

where

$$\bar{D} = \tfrac{1}{2} \int_v \dot{\theta}\dot{\theta} \, dv \tag{5.39}$$

$$\bar{P} = \tfrac{1}{2} \int_v k_{ij} \frac{\partial \theta}{\partial x_i} \frac{\partial \theta}{\partial x_j} \, dv \tag{5.40}$$

$$\bar{F} = \int_v \frac{1}{T_0 E} S\theta \, dv \tag{5.41}$$

and

$$\bar{Q}_k = \int_s k_{ij} \eta_i \frac{\partial \theta}{\partial x_j} \frac{\partial \theta}{\partial \bar{q}_k} \, ds. \tag{5.42}$$

As one may observe from Equations (5.39)–(5.42) the complementary variational principle translates to similar types of equations as the fundamental form, but there is an important difference between the two formulations. The latter form involves spatial derivatives of the temperature θ in the expressions for the function $\bar{P}$ and the generalized forces $\bar{Q}_i$. This is a disadvantage of the complementary form, as was mentioned earlier, especially for applications to problems with discontinuities on the distribution of the temperature fields. Although each of the formulations was derived in terms of a different set of generalized coordinates, both can be represented by the same type of generalized equations. This is verified by the unified approach used to derive the Lagrangian type of equations and also by the fact that the temperature field θ and the displacement field are conjugate variables related by the constraint Equation (5.5).

The concept of generalized coordinates provides the basis for deriving the Lagrangian formulation of the diffusion equation. From the physical point of view the generalized coordinates can describe a system completely and the Lagrangian equations can provide an accurate formulation of the physical problem. A particular advantage of the formulation is that it was derived independently of a particular representation of the field variables. Furthermore, the above analysis can be extended to many physical systems governed by equations other than the thermal diffusion equation.

The Lagrangians types of equations derived in this section do not represent a new mathematical theory but only a different way of expressing a conservation law. The main feature of these equations in the manner in which they were derived. It is evident that these equations have the same form in any system of generalized coordinates and from classical mechanics one can verify that the functions involved in Lagrange's equations have the same value for any coordinate system.

5.3 APPROXIMATE METHODS

The partial differential equation describing thermal diffusion was transformed into a set of ordinary differential equations expressed in terms of a set of generalized coordinates. The equations obtained can be solved analytically for a number of physical problems. For many of the problems encountered in practical applications, analytical solutions are not easy to obtain and approximate or numerical solutions are sought.

The generalized nature of the previously obtained equations is most appropriate for applying approximate solutions. Since these equations are given in terms of generalized coordinates, the field variables describing the physical process are assumed to be functions of these coordinates. Many different expressions or combinations of the generalized coordinates can be used to form the functional representing the field variable. In order to demonstrate the procedure one may assume a linear dependence of the field variables on the generalized coordinates. This specific representation by no means restricts the previously derived formulations to only linear functional approximation. The main reason for such a choice is the fact that a variety of numerical solutions are based on such approximations and, as will be shown, the finite element method can be derived as a special case of the generalized formulations.

5.3.1 Linear functionals

The linear dependence of the displacement field H_i on the generalized coordinates is expressed as

$$H_i(q_k, x_j, t) = f_{ik}(x_j)q_k(t). \tag{5.43}$$

The generalized coordinates q_k may be functions of time and they are treated as the coordinates describing the physical system. From the physical point of view they represent degrees of freedom and the functions $f_{ik}(x_j)$ specify the extent to which the q_k participate in the functions $H_i(x_j, t)$. The approximation of the displacement field given by Equation (5.43) is quite general since no specific expressions are given for the functions f_{ik}. Furthermore, Equation (5.43) may represent approximations such as Fourier series or series expansions in terms of orthogonal functions. Similar approximations can be used for the temperature field θ.

In the following the fundamental form of the variational principle will be used to derive the approximate system of equations for obtaining numerical solutions. For the complementary form the derivation will not be given since it is identical to that of the fundamental form.

The Lagrangian type of equation given by Equation (5.36) can be translated to a system of ordinary differential equations in terms of the generalized coordinates q_k by using the approximation, Equation (5.43), for the displacement field and by first defining the time and spatial derivatives as

$$\dot{H}_i = \dot{q}_k f_{ki} \tag{5.44}$$

and

$$\theta = \frac{\partial H_i}{\partial x_i} = q_k f_{ki,i}. \tag{5.45}$$

In terms of these definitions the dissipation function D is evaluated as

$$D = \tfrac{1}{2} \int_v \lambda_{ij} \dot{q}_k f_{ki} \dot{q}_l f_{lj} \, \mathrm{d}v$$

or

$$D = \tfrac{1}{2} d_{ij} \dot{q}_i \dot{q}_j, \tag{5.46}$$

where

$$d_{ij} = \int_v \lambda_{kl} f_{ki} f_{lj} \, \mathrm{d}v \tag{5.47}$$

Similarly the potential function P is evaluated as

$$P = \tfrac{1}{2} p_{ij} q_i q_j, \tag{5.48}$$

where

$$p_{ij} = \int_v f_{ik,k} f_{jl,l} \, \mathrm{d}v. \tag{5.49}$$

The function F transforms to the form

$$F = S_i q_i \tag{5.50}$$

where

$$S_i = \int_v \lambda_{ki} \dot{h}_l f_{kl} \, \mathrm{d}v. \tag{5.51}$$

Finally the boundary forces Q_i are expressed as

$$Q_i = \int_s \theta \eta_j f_{ij} \, \mathrm{d}s. \tag{5.52}$$

Substituting the expressions for D, P, F, and Q_i from the above equations into Equation (5.43) yields

$$d_{ij} \dot{q}_i + P_{ij} q_j = Q_i + S_i. \tag{5.53}$$

This matrix equation represents a system of ordinary differential equations for the unknown field parameters q_i, which may represent displacements. The generalized coordinates describe the physical system by means of a large but finite number of variables and their number determines the size of the above system of equations. The solution of Equation (5.53) yields the approximate solution for the diffusion equation. The order of approximation of the numerical solution is determined by the functions $f_{ij}(x_k)$ and depends on the accuracy of the solution required and usually on the complexity of the physical problem.

For the evaluation of the matrix coefficients of Equation (5.53) the integration domain is considered to be taken over the volume v in the n-space of the coordinates x_i, where

$$\mathrm{d}v = \mathrm{d}x_1, \mathrm{d}x_2 \cdots \mathrm{d}x_n.$$

In a more general sense the integration domain can be extended over all $n+1$ coordinates x_i and t instead of only the n coordinates x_i, where

$$\mathrm{d}\tau = \mathrm{d}x_1, \mathrm{d}x_2 \cdots \mathrm{d}x_n\, \mathrm{d}t.$$

This extension of the integration domain yields the same Lagrangian-type equations since the unknown physical quantities can be expressed in any functional form in terms of generalized coordinates. For this particular case the approximation given by Equation (5.43) will be replaced by

$$H_i(x_k, t) = f_{ij}(x_k) g_n(t) q_j^n. \tag{5.54}$$

It should be noted that the approximation given by Equation (5.54) is more restrictive in nature than the one given by Equation (5.43) since one assumes a known dependence of the variables on time through the function $g_n(t)$. The generalized coordinates, in this case q_i^n, represent values at the ith point in space and at the nth point in time. The respective time and spatial derivatives of Equation (5.54) yield

$$\dot{H}_i = f_{ij}(x_k) \dot{g}_n(t) q_j^n \tag{5.55a}$$

and

$$H_{i,i} = f_{ij,i}(x_k) g_n(t) q_j^n. \tag{5.55b}$$

A system of equations in terms of the generalized coordinates q_i^n can be obtained by substituting the approximation given by Equation (5.54) into the Lagrangian equations. The expressions for $\boldsymbol{D}$, $\boldsymbol{P}$, $\boldsymbol{F}$ and $\boldsymbol{Q}_i$ are evaluated as for the previous approximations and they are

$$\begin{aligned} D &= \tfrac{1}{2} d_{ijkl} q_i^k q_j^l, \\ P &= \tfrac{1}{2} P_{ijkl} q_i^k q_j^l, \\ F &= S_{ik} q_i^k, \end{aligned} \tag{5.56}$$

where

$$d_{ijkl} = \int_\tau \lambda_{mn} f_{im} f_{jn} \dot{g}_k \dot{g}_l \, d\tau,$$

$$p_{ijkl} = \int_\tau f_{im,m} f_{jn,n} g_k g_l \, d\tau, \tag{5.57}$$

$$S_{ik} = \int_\tau \lambda_{mn} \dot{h}_n f_{mi} g_k \, d\tau.$$

Substituting the above expressions into Equation (5.43) yields

$$[d_{ijkl} + p_{ijkl}] q_i^k = Q_j^l + S_{jl} \qquad i, j = 1, n, \ldots, k, l = 1, m, \tag{5.58}$$

where the boundary forces Q_j^l are given by

$$Q_j^l = \int_s \eta_i f_{km,m} g_n f_{ij} g_l q_k^n \, ds. \tag{5.59}$$

The matrix Equation (5.58) represents a system of $n+m$ algebraic equations for the unknown field parameters, generalized coordinates q_i^k, and the domain of integration τ of the matrix coefficients is extended over the $n+1$ coordinates x_i and t.

The Lagrangian type of equations and the two systems of differential equations derived in this section represent two examples of approximate methods which can be applied to solve problems involving thermal diffusion. The application of the derived formulation to a physical system can be carried out by dividing the domain of the system into a finite number of subdomains connected by common boundaries. Each of these domains constitutes a subsystem that can be analysed separately by the system of equations derived in terms of generalized coordinates. For each subsystem s the two systems of equations may be written in a general form as follows

$$A_{ij}^{(s)} q_j = Q_i^{(s)}, \tag{5.60}$$

where q_j are the generalized coordinates for the subsystem s, $Q_i^{(s)}$ are the associated generalized boundary forces, and $A_{ij}^{(s)}$ are the matrix coefficients of the system of equations which were derived from the Lagrangian equation for the domain s. Taking the sum of Equation (5.54) over all the domains of the subsystems, the equation for the total system is written as follows:

$$A_{ij} q_j = Q_i \tag{5.61}$$

where A_{ij} is now the matrix coefficient for the system of equations corresponding to the total domain and Q_i are the forces only at the boundary of the total system. This result is justified by the fact that the summation extends to the variables not only at the interfaces of the subsystems but also to the variables at the boundary of the total system. However, at the interfaces the forces are in pairs that cancel out. For example, for two subsystems s and s' with a common boundary the generalized coordinates at the interface are q_i and the corresponding forces are Q_i and Q_i'. From the definition of the boundary force, the outward normal vectors for the domains with common boundary are in opposite directions, hence

$$Q_i = -Q_i'$$

or

$$Q_i + Q_i' = 0. \tag{5.62}$$

This leads to the cancellation of all the forces at the interfaces and the derived Equation (5.61) include only the forces at the boundary of the total system.

The procedure of dividing the domain of a system into subspaces is commonly used in approximate methods and a typical example is the identical procedure used for the analysis of a physical system by the finite element method.

5.3.2 Finite element method

A variety of approaches can be found under the name of the finite element method. Such approaches include the well-known Galerkin methods, the Rayleigh–Ritz methods and many others which are usually modifications or derivatives of the ones mentioned. The formulation and derivation of the Galerkin method are well known and the method has been successfully applied to a variety of problems for obtaining numerical solutions. The Galerkin approach is usually referred to as the conventional finite element method in contrast to those which are based on modification of the Galerkin or nonconventional approaches. The intention here is to show that the conventional finite element method is a special application of the present generalized formulation. Consider the following approximation for the displacement field H_i

$$H_i(x, t) = a_0 + a_1 x + a_2 x^2 + a_3 x^3 + \cdots. \tag{5.63}$$

Then the temperature θ is given by

$$\theta = a_1 + 2a_2 x + 3a_3 x^2 + \cdots, \tag{5.64}$$

where the a_i $(i = 0, 1, \ldots, n)$ are time-dependent coefficients to be determined. Although the approximation given by Equation (5.63) is restricted to one-dimensional space, it can be easily extended to two or three dimensions since the derived formulations for approximate methods are of general form. In order to demonstrate the derivation of the matrix equation for the finite element method only a first-order approximation will be considered here. Higher-order approximations have been considered in previous studies[3,4] and their derivation is carried out in a similar way as the first-order one. For a first-order approximation the derived finite element models correspond to linear element models with two degrees of freedom. Four such linear elements will be considered here. The first two are linear displacement models denoted by DFEM and DFET and correspond for the approximations given by Equations (5.43) and (5.54). The last two are conventional linear models in terms of the primitive variable θ and are denoted by CFEM and CFET. The displacement models are derived from the fundamental form of the variational formulation and the conventional models from the complementary form.

5.3.2.1 *Displacement models*

(a) DFEM. From Equation (5.63) the approximation of the displacement field $H(x, t)$ for the linear element of length Δx is given by

$$H(x, t) = \left(1 - \frac{x}{\Delta x}\right) H_1(t) + \frac{x}{\Delta x} H_2(t), \tag{5.65}$$

where H_1 and H_2 are the nodal values of the displacement. They can be identified as the generalized coordinates q_1 and q_2 with the corresponding basis functions f_{x1} and f_{x2} given by

$$f_{x1} = \left(1 - \frac{x}{\Delta x}\right) \quad \text{and} \quad f_{x2} = \frac{x}{\Delta x}. \tag{5.66}$$

Then, according to Equation (5.43), the approximation Equation (5.65) is

$$H(x, t) = f_{xi}(x) H_i(t). \tag{5.67}$$

For the deformation θ the approximation is evaluated from Equation (5.66) as

$$\theta(x, t) = \frac{1}{\Delta x}(H_2(t) - H_1(t)). \tag{5.68}$$

Note that within each element θ varies only with time for the linear approximation of the displacement field. This is analogous to a constant strain element in mechanics. The matrix coefficients of Equation (5.53) are evaluated in terms of Equations (5.65) and (5.68) as

$$[d_{ij}] = \frac{A\,\Delta x}{6k}\begin{bmatrix} 2 & 1 \\ 1 & 2 \end{bmatrix}, [p_{ij}] = \frac{A}{\Delta x}\begin{bmatrix} 1 & -1 \\ -1 & 1 \end{bmatrix} \tag{5.69}$$

and

$$Q_i|_{x=0}=Q_1=-A\theta,\ Q_i|_{x=\Delta x}=Q_2=A\theta,$$

where A is the cross-sectional area of the element and k is the diffusion coefficient.

If one also assumes a linear approximation for the source term, then S_i from Equation (5.51) is evaluated as

$$\{S_i\}=\frac{A\,\Delta x}{6k}\begin{bmatrix}2 & 1\\ 1 & 2\end{bmatrix}\begin{Bmatrix}\dot{h}_1\\ \dot{h}_2\end{Bmatrix}=\frac{A\,\Delta x}{6k}\begin{Bmatrix}S_1\\ S_2\end{Bmatrix}, \tag{5.70}$$

where $S_1=2\dot{h}_1+\dot{h}_2$ and $S_2=\dot{h}_1+2\dot{h}_2$. The matrix equation for the DFEM model is then assembled as

$$\begin{bmatrix}2 & 1\\ 1 & 2\end{bmatrix}\begin{Bmatrix}\dot{H}_1\\ \dot{H}_2\end{Bmatrix}+\frac{6}{W_0^2}\begin{bmatrix}1 & -1\\ -1 & 1\end{bmatrix}\begin{Bmatrix}H_1\\ H_2\end{Bmatrix}=\frac{6}{W_0}\begin{Bmatrix}-\theta\\ \theta\end{Bmatrix}+\begin{Bmatrix}S_1\\ S_2\end{Bmatrix}. \tag{5.71}$$

The matrix Equation (5.70) has been derived in dimensionless form according to the transformations

$$\begin{gathered}\bar{x}=\frac{x}{L},\quad \bar{t}=\frac{k}{L^2}t,\quad \theta=\frac{T-T_0}{T_1-T_0},\quad \bar{H}=\frac{T_0}{T-T_0}\frac{1}{L}H\\ \bar{\dot{H}}=\frac{T_0}{T-T_0}\frac{1}{L}\dot{H},\quad W_0=\frac{\Delta x}{L},\quad W_1=\frac{L}{K}\frac{T_0}{T_1-T_0}w_1.\end{gathered} \tag{5.72}$$

Here L is a characteristic length and T_1 a constant reference value for T. The has been eliminated from Equation (5.71) for simplicity.

(b) DFET model. If one now considers the approximation given by Equation (5.54), then the linear approximation for $H(x, t)$ and $\theta(x, t)$ is given as follows:

$$\begin{aligned}H(x,t)=&\left(1-\frac{x}{\Delta x}\right)\left(1-\frac{t}{\Delta t}\right)H_1^1+\left(1-\frac{x}{\Delta x}\right)\frac{t}{\Delta t}H_1^2\\ &+\frac{x}{\Delta x}\left(1-\frac{t}{\Delta t}\right)H_2^1+\frac{x}{\Delta x}\frac{t}{\Delta t}H_2^2\end{aligned} \tag{5.73}$$

and

$$\begin{aligned}\theta(x,t)&=\frac{1}{\Delta x}\left[\left(1-\frac{t}{\Delta t}\right)(-H_1^1+H_2^1)+\frac{t}{\Delta t}(-H_1^2+H_2^2)\right]\\ &=\left(1-\frac{t}{\Delta t}\right)\theta^1+\frac{t}{\Delta t}\theta^2.\end{aligned} \tag{5.74}$$

The generalized coordinates q_i^k and the corresponding function f_{xi} and g_k are identified as

$$q_1^1=H_1^1,\quad q_1^2=H_1^2,\quad q_2^1=H_2^1,\quad q_2^2=H_2^2$$

$$f_{x1}=1-\frac{x}{\Delta x}, f_{x2}=\frac{x}{\Delta x} \tag{5.75}$$

$$g_1=\left(1-\frac{t}{\Delta t}\right),\qquad g_2=\frac{t}{\Delta t}$$

with

$$\theta^1=\frac{1}{\Delta x}(-H_1^1+H_2^1)\qquad\text{and}\qquad \theta^2=(-H_1^2+H_2^2) \tag{5.76}$$

The matrix coefficients are evaluated from Equation (5.57) and the matrix equation for the DFEMT model is assembled as

$$\left[\begin{bmatrix}-2 & -1 & 2 & 1\\ -1 & -2 & 1 & 2\\ -2 & -2 & 2 & 1\\ -1 & -2 & 1 & 2\end{bmatrix}+2\frac{\Delta t}{\Delta x^2}\begin{bmatrix}2 & -2 & 1 & -1\\ -2 & 2 & -1 & 1\\ 1 & -1 & 2 & -2\\ -1 & 1 & -2 & 2\end{bmatrix}\right]$$

$$\times\begin{Bmatrix}H_1^1\\ H_2^1\\ H_1^2\\ H_2^2\end{Bmatrix}=2\frac{\Delta t}{\Delta x}\begin{Bmatrix}-(2\theta^1+\theta_1^2)\\ (2\theta^1+\theta^2)\\ -(\theta^1+2\theta_1^2)\\ (\theta^1+2\theta_2^2)\end{Bmatrix} \tag{5.77}$$

The two sets of equations given by Equations (5.71) and (5.77) represent the displacement finite element models for obtaining numerical solutions of the transport equation. Both models were derived from the fundamental form of the variational formulation as examples of approximate methods.

(c) CFEM model. The approximation for the temperature θ for the linear element of length Δx is given by

$$\theta(x,t)=\left(1-\frac{x}{\Delta x}\right)\theta_1+\frac{x}{\Delta x}\theta_2 \tag{5.78}$$

where θ_1 and θ_2 are the nodal values of the temperature. They can be identified as the generalized coordinates q_1 and q_2 with the corresponding functions f_{x1} and f_{x2} given by Equation (5.66).

The matrix coefficients and the expressions for $\bar{D}$, $\bar{P}$, $\bar{F}$, and $\bar{Q}_1$ are evaluated from Equations (5.39)–(5.42) and the resulting matrix equation is given by

$$\begin{bmatrix} 2 & 1 \\ 1 & 2 \end{bmatrix} \begin{Bmatrix} \dot{\theta}_1 \\ \dot{\theta}_2 \end{Bmatrix} + \frac{6}{W_0^2} \begin{bmatrix} 1 & -1 \\ -1 & 1 \end{bmatrix} \begin{Bmatrix} \theta_1 \\ \theta_2 \end{Bmatrix} = \frac{6}{w_0^2} \begin{Bmatrix} -(\theta_2 - \theta_1) \\ (\theta_2 - \theta_1) \end{Bmatrix} + \begin{Bmatrix} S_1 \\ S_2 \end{Bmatrix}. \tag{5.79}$$

The above equation is derived from the complementary form of the variational formulation and is identical to the finite element equations obtained by the conventional Galerkin method. Furthermore the matrix coefficients are identical to the ones given by Equation (5.69).

(d) CFET model. The linear approximation for the temperature θ is given by

$$\theta(x, t) = \left(1 - \frac{x}{\Delta x}\right)\left(1 - \frac{t}{\Delta t}\right) \theta_1^1 + \left(1 - \frac{x}{\Delta x}\right) \frac{t}{\Delta t} \theta_1^2 + \frac{x}{\Delta x}\left(1 - \frac{t}{\Delta k}\right) \times \theta_2^1 + \frac{x}{\Delta x} \frac{t}{\Delta t} \theta_2^2. \tag{5.80}$$

As for the previous approximations the generalized coordinates can be identified as

$$q_i^j = \theta_i^j, \qquad i = 1, 2, \quad j = 1, 2, \tag{5.81}$$

and the corresponding functions f_{xi} and g_j are given by Equations (5.75) and (5.76). The matrix coefficients are evaluated in terms of Equation (5.80) and the matrix equation for the CFET model can be derived from Equation (5.77) by replacing the nodal unknowns H_i^j by θ_i^j.

The four finite element elements presented in this section can be grouped into two types of equations which correspond to the general approximation given by Equations (5.43) and (5.54). The first type can be represented by the following generalized equation

$$d_{ij}\dot{q}_j + p_{ij}q_j = Q_i + S_i \tag{5.82}$$

where the matrix coefficients are given by Equation (5.69) and the generalized coordinates are identified as

$$\begin{aligned} &\text{DFEM model:} \quad q_i = H_i \qquad i = 1, 2 \\ &\text{CFEM model:} \quad q_i = \theta_i \qquad i = 1, 2. \end{aligned} \tag{5.83}$$

For the DFEM model Equation (5.82) should be used together with the constraint

$$\theta_i = \frac{1}{\Delta x}(H_{i+1} - H_i). \tag{5.84}$$

The second type of model can be represented by

$$[d_{ijkl}+p_{ijkl}]q_j^l=Q_i^k+S_i^k, \tag{5.85}$$

where the matrices are given by Equations (5.77) and the generalized coordinates are identified as

$$\begin{aligned} &\text{DFET model:} \quad q_i^j=H_i^j \qquad i=1,2, \quad j=1,2 \\ &\text{CFET model:} \quad q_i^j=\theta_i^j \qquad i=1,2, \quad j=1,2. \end{aligned} \tag{5.86}$$

For the DFET model Equation (5.82) should be used together with the constraint given by Equation (5.84). In Table 5.1 the four models are presented in summary.

Table 5.1

Approximation	Model	Generalized coordinates	Type of matrix equations
Only for spatial coordinates S-type	DFEM	$q_i=H_i$	Ordinary differential equation (5.82)
	CFEM	$q_i=\theta_i$	
For both spatial and time coordinates S-T-type	DFET	$q_i^j=H_i^j$	Algebraic equation (5.85)
	CFET	$q_i^j=\theta_i^j$	Equation (5.85)

Both types of models will be used to solve thermal diffusion problems and examples will be given later in this study. From these examples a realistic comparison between models and also an evaluation of the models is possible.

The generalized variational formulations and the derived approximate methods presented here represent a unified approach for describing thermal phenomena. Some of the advantages of this unified approach are as follows:

- The variational formulation is problem independent.
- The presentation of the duffusion equation by generalized coordinates simplifies the complexity of the equation.
- The Lagrangian type of equations are most suitable for obtaining approximate solutions.
- The finite element method can be derived directly as a special application for approximate solutions.
- The derived finite element models can be represented by one generalized model.
- The displacement models are most suitable for solving problems with discontinuities.

In the following a number of numerical solutions will be presented for specific boundary value problems and the error behaviour of these numerical solutions will be investigated.

5.4 NUMERICAL APPLICATIONS

The thermal diffusion equation provides a good test case for numerical solutions since a variety of boundary conditions can be considered. Before numerical solutions to specific problems are presented it is appropriate to investigate the stability and the error behaviour of the four element models. Since it is necessary to use a numerical scheme to approximate the time derivatives of the S-type models, a finite difference scheme is chosen for that purpose. The scheme is known as the backward finite difference approximation. It is an implicit unconditionally stable scheme and the general approximation for the first derivative is given by the following series expansion:[8]

$$\frac{\mathrm{d}y(t_i)}{\mathrm{d}t}=\frac{1}{\Delta t}\sum_{j=0}^{n} a_j y(t_{i-j})+O(\Delta t^n) \tag{5.87}$$

where a_j are constant coefficients and their values are given for up to a fourth-order approximation in Table 5.2.

Table 5.2

Order of approximation		a_0	a_1	a_2	a_3	a_4	a_5
First	$n=1$	1.0	−1.0	0	0	0	0
Second	$n=2$	1.5	−2.0	0.5	0	0	0
Third	$n=3$	11./6.	−3.0	1.5	−1./3.	0	0
Fourth	$n=4$	25./12.	−4.0	3.0	−4./3.	0.25	0

The error analysis of the numerical solution of a particular problem will be presented in the following section and some criteria for convergence and error estimates will be discussed.

5.4.1 Problem definition

The governing equation for the one-dimensional case of thermal diffusion is given by

$$\frac{\partial\theta}{\partial t}-D_0\frac{\partial^2\theta}{\partial x^2}=0 \tag{5.88}$$

where θ is the dimensionless temperature, and D_0 a constant diffusivity. It has been also assumed that there are no heat sources or sinks present in the medium. The boundary and initial conditions are as follows:

$$\begin{aligned} \theta(x,0) &= 0.0 \quad 0 \leqslant x \leqslant L \\ \theta(0,t) &= 1.0 \quad 0 \leqslant t \\ \theta(L,t) &= 0.0 \quad 0 \leqslant t \end{aligned} \tag{5.89}$$

where L is a finite characteristic length approximating infinity. The analytical solution to the above problem is given by ($D_0 = 1.0$)

$$\theta(x,t) = \operatorname{erfc}\left(\frac{x}{2\sqrt{t}}\right). \tag{5.90}$$

The solution of the diffusion equation in the semi-infinite space is a typical example for testing numerical solutions and studying error behaviour. Its numerical solution is challenging, especially when discontinuities in the temperature field are present.

5.4.2 Error analysis

Numerical solutions of the above problem are obtained by the four finite element models previously derived and these numerical solutions are compared to the analytical one, Equation (5.90), in order to investigate the error behaviour. For the S-type models (DFEM and CFEM) a first-order backward finite difference scheme is chosen to approximate the time derivative.

The error of the numerical solution is the average error E_η over the entire length L at a particular time level and it is evaluated by the following relation

$$E_\eta = \frac{1}{N} \sum_{i=1}^{N} |\theta - \theta_i| \tag{5.91}$$

where N is the number of elements used for the space discretization and $|\theta - \theta_i|$ is the absolute local error. For illustrating numerical results, the characteristic length $L = 1$ is divided into a uniform mesh with NE elements and $NP = NE + 1$ nodal points (degrees of freedom), and the dimensionless element length Δx then equals to $1/NE$.

Four sets of results will be presented for all four models. The first set of results is obtained for a constant value of Δx and variable Δt (Table 5.3) and the second set is obtained for constant Δt and variable Δx (Table 5.4). The third for constant values of the ratio $\Delta t/\Delta x$ (Table 5.5) and the fourth set for constant values of the ratio $\Delta t/\Delta x^2$ (Table 5.6).

In all of the figures the average error is presented at three time levels. This is necessary in order to investigate the error behaviour throughout the time domain. The first set, Figures 5.1(a), 5.1(b), the value of NE is constant

Table 5.3

NE	Δx	Δt	$M_1 = \Delta t/\Delta x$	$M_2 = \Delta t/\Delta x^2$
		0.0001	0.002	0.04
		0.00025	0.005	0.10
		0.0005	0.010	0.20
		0.00125	0.025	0.50
20	1/20.0	0.0050	0.100	2.00
		0.0100	0.200	4.00
		0.0125	0.250	5.00
		0.0200	0.400	8.00
		0.0250	0.500	10.00
		0.0500	1.000	20.00

Table 5.4

Δt	NE	Δx	$M_1 = \Delta t/\Delta x$	$M_2 = \Delta t/\Delta x^2$
	5	1/5.0	0.025	0.125
	10	1/10.0	0.050	0.500
	20	1/20.0	0.100	2.000
	30	1/30.0	0.150	4.500
0.005	40	1/40.0	0.200	8.000
	50	1/50.0	0.250	12.500
	60	1/60.0	0.300	18.000
	80	1/80.0	0.400	32.000
	100	1/100.0	0.500	50.000
	200	1/200.0	1.000	200.000

Table 5.5

NE	Δx	Δt	$M_1 = \Delta t/\Delta x$	$M_2 = \Delta t/\Delta x^2$
10	0.10000	0.02500		2.50
20	0.05000	0.01250		5.00
40	0.02500	0.00625		10.00
50	0.02000	0.00500	0.25	12.50
80	0.01250	0.00312		20.00
100	0.0100	0.00250		25.00

Table 5.6

NE	Δx	Δt	$M_1 = \Delta t/\Delta x$	$M_2 = \Delta t/\Delta x^2$
10	0.1	0.0250	0.250	
20	0.050	0.00625	0.125	
40	0.0250	0.00156	0.063	2.5
50	0.0200	0.00100	0.050	
80	0.0125	0.00040	0.031	
100	0.0100	0,00025	0.025	

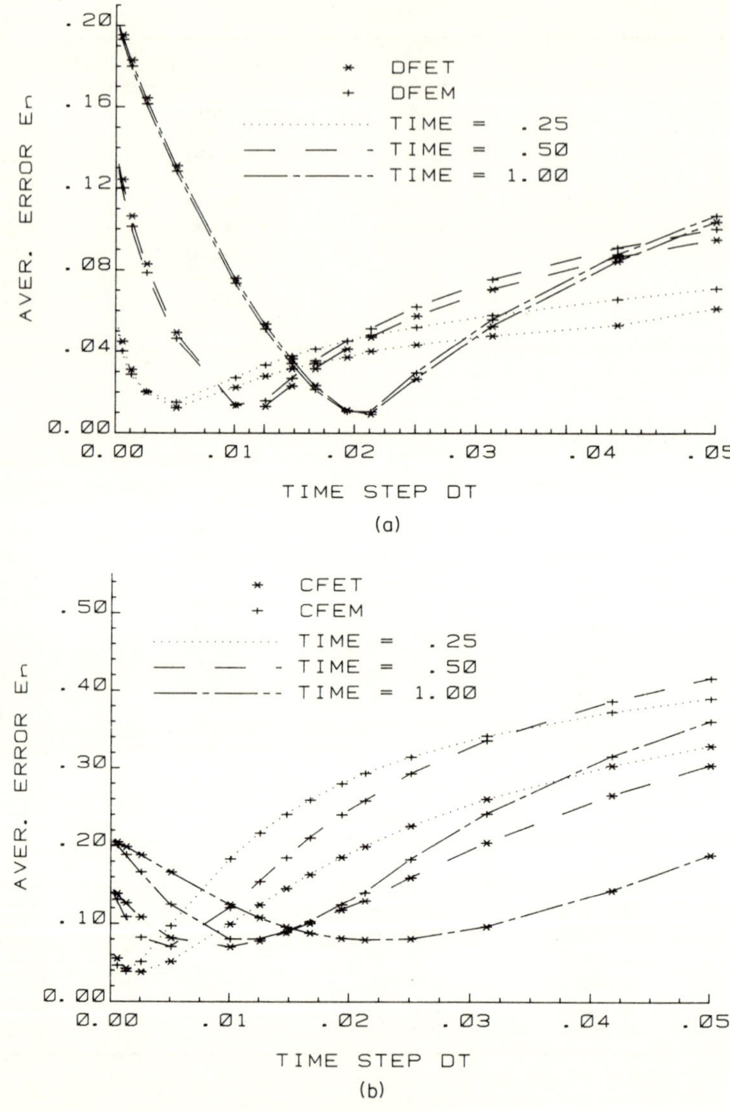

Figure 5.1 Average error for constant Δx and variable Δt: (a) displacement models; (b) conventional models

($NE = 20$) and the time step Δt was given values $0.0001 \leq \Delta t \leq 0.05$. The corresponding values for $\Delta t/\Delta x$ and $\Delta t/\Delta x^2$ are given in Table 5.3. The second set of results are given in Figures 5.2(a), 5.2(b), where Δt is constant ($\Delta t = 0.005$) and NE was given values $5 \leq NE \leq 200$. Table 5.4 contains the values for Δx and the corresponding values of $\Delta x/\Delta x$ and $\Delta t/\Delta x^2$.

5.4.2.1 Displacement models (DFET, DFEM)

In Figures 5.1(a) and 5.2(a) results for the average error E_n are given for the two displacement models. The error behaviour in both figures follows a similar pattern. As Δt or NE increases there is initially a reduction in the error of the numerical solutions, and a minimum error is reached. With further

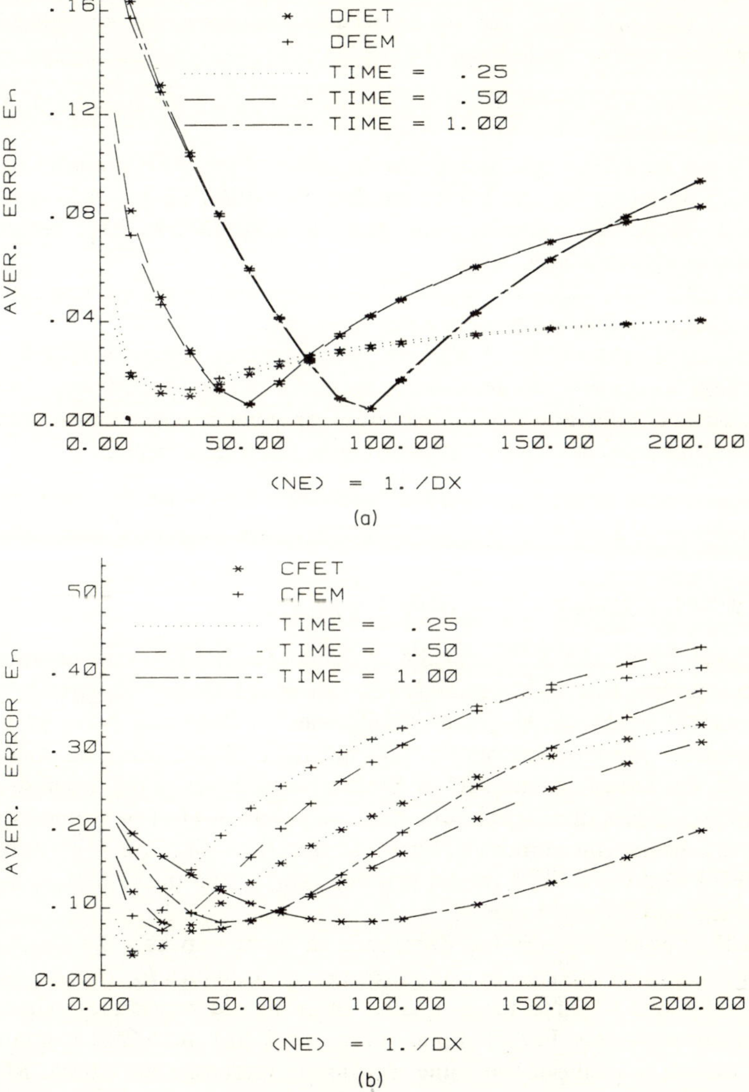

Figure 5.2 Average error for constant Δt and variable Δx: (a) displacement models; (b) conventional models

increases of Δt or *NE* the error starts to increase but the rate of change is smaller than before the minimum was reached. The difference between the values of the error for the two models is very small for small values of Δt but it becomes noticeable as Δt increases with the DFET model being the more accurate one. Similar observations can be made for the error behaviour with respect to *NE*. The error behaviour in both figures shows a nonuniform convergence as Δt or Δx decreases. This pattern is not unusual since only Δt or Δx is changed while Δx or Δt remains constant respectively. Certain observations can be made from the results presented in these figures.

- There is an optimum combination of Δt and Δx for minimum error for both models.
- To the left of the optimum point the error of the DFEM model is smaller than the error for the DFET model. The order of accuracy is reversed just before the optimum point and the difference in error between the two models becomes larger.
- As time progresses the value of the minimum error decreases and the optimum points are translated to the right.
- For large values of Δt or *NE* the error increases monotonically with time.
- There seems to be an area where the error curves for different time levels intersect, indicating that a constant error value can be achieved for certain combinations of Δt and Δx throughout the time domain.

The similarities observed in error behaviour may lead to some optimum combination of Δt and Δx for error control of the numerical solutions.

5.4.2.2 Conventional models (CFET, CFEM)

In Figures 5.1(b) and 5.2(b) results are given for the two conventional finite element models and for the same set of values as before. The error behaviour with respect to Δt or Δx shows similarities to the error behaviour of the displacement models but there is no uniformity in the error behaviour. For example, for either constant Δt or Δx the error values reach a minimum, but as time progresses these minimum error values do not follow a regular pattern as for the displacement models. Furthermore, the difference in error between the CFET and the CFEM model is much larger than the difference between the DFET and DFEM models.

Another observation is the difference in error between the displacement and conventional models. In all figures the error values for the displacement models are two to three times smaller than the corresponding values of the conventional models. For example, in Figures 5.2(a) and 5.2(b) at small values of Δt the error is about the same, but as Δt increases ($\Delta t > 0.01$, $M_1 > 0.2$) the error for the conventional models, Figure 5.2(b), becomes much larger. This is due mainly to the error involved in applying the boundary conditions (at $x = L$), and also due to the absence of the boundary forces. In the case of the conventional models the boundary forces are zero for prescribed temperature at the boundary. Furthermore, as was mentioned earlier in this study, the presence of space derivatives in the complementary formulation is the main drawback of the conventional models and they contribute to the larger error of these models. In the following discussion certain examples will be used to demonstrate this deficiency.

The previously made observations on the patterns of the error behaviour suggest that optimum combinations of Δt and Δx can be defined for error control of the numerical solution. These combinations can be based on optimum values of the ratio $M_1 = \Delta t/\Delta x$ or the ratio $M_2 = \Delta t/\Delta x^2$. Typically for a second order equation, such as the diffusion equation, the ratio M_2 is the parameter for error and stability control. The value of this parameter, when it is kept within certain limits, guarantees accuracy and stability of the numerical solution.[9]

A careful examination of the results for the error, shown in Figures 5.1(a) and 5.2(a), and the values of M_2 from Tables 5.3 and 5.4 which correspond to minimum error, shows that there is an inconsistency. For example, at time $t = 0.5$ Table 5.3 and Figure 5.1(a) show a minimum error at $M_2 = 5.00$ ($\Delta t = 0.0125$), but from Table 5.4 and Figure 5.3(a) the minimum error is at $M_2 = 12.5$ (*NE* = 50). What appears to be an inconsistency in the behaviour of the displacement models, with respect to the parameter M_2, leads to the observation that M_2 might not be the right parameter for error control of the numerical solution. In contrast an examination of the error with respect to M_1 shows that at time $t = 0.5$ the minimum values of the error, Figures 5.1(a) and 5.3(a), correspond to the same value of the parameter M_1. In fact for each value of the parameter M_1 the corresponding error values are about the same in both figures.

In order to study the dependence of the numerical solution error on the two parameters M_1 and M_2 further analysis is needed. The error of the numerical solution now is investigated for some constant values of the two parameters. Results are given for all four models and it is shown that uniform convergence can be achieved for either constant M_1 or constant M_2. Furthermore, it is shown that uniform convergence does not always correspond to minimum error.

5.4.3 Convergence of the numerical solution

Numerical experiments on the error of the solutions in the previous section show that nonuniform convergence is obtained when either Δt or Δx alone is the dependent variable. It was also observed that minimum error could be achieved for some optimum combinations of the two variables Δt and Δx. It is a well-known fact that in order to achieve uniform convergence both Δt

and Δx should be the dependent variables with the condition that the ratio $\Delta t/\Delta x$ or $\Delta t/\Delta x^2$ is kept constant. The results for the average error E_n in this section will show that uniform convergence is achieved for both the displacement and conventional models when M_1 or M_2 is kept constant. Some representative values of M_1 and M_2 are given in Tables 5.7 and 5.8, with the corresponding values of Δt and Δx, for the four models. Numerical results are presented in Figures 5.3–5.5, for all models and all cases as before, and for two values of each of the parameters M_1 and M_2.

Table 5.7

M_1	M_2	Δx	Δt
	2.50	0.10000	0.02500
	5.00	0.0500	0.01250
0.25	10.00	0.0250	0.00625
	12.50	0.0200	0.00500
	20.00	0.0125	0.00312
	25.00	0.0100	0.00250
	5.00	0.1000	0.05000
	10.00	0.0500	0.02500
0.50	20.00	0.0250	0.01250
	25.00	0.0200	0.01000
	40.00	0.0125	0.00625
	50.00	0.0100	0.00500

Table 5.8

M_1	M_2	Δx	Δt
0.250		0.1000	0.02500
0.125		0.0500	0.00625
0.063	2.5	0.0250	0.00156
0.050		0.0200	0.00100
0.031		0.0125	0.00039
0.025		0.0100	0.00025
0.500		0.1000	0.05000
0.250		0.0500	0.01250
0.125	5.0	0.0250	0.00312
0.100		0.0200	0.00200
0.063		0.0125	0.00078
0.050		0.0100	0.00050

Results are given in Figures 5.3(a) and 5.3(b), for the two models for constant $M_1 = 0.25$ and in Figures 5.4(a) and 5.4(b), results are given for $M_2 = 2.5$. The uniform convergence of the numerical solution is obvious for either parameters M_1 or M_2. Examining each pair of figures (Figures 5.3(a) and 5.4(a), 5.3(b) and 5.4(b)), which correspond to the displacement and conventional models respectively, one can observe the effect of M_1 and M_2 in the error of the solutions obtained by these models. For both parameters M_1 and M_2 the error of the displacement models reaches constant values, but these constant error values are smaller for the case of constant M_1 than for the case of constant M_2. For example, in Figures 5.3(a) and 5.4(a), for each value of Δx the corresponding error value is much larger for constant M_2 than for constant M_1. Furthermore, the corresponding value of Δt for each Δx, Table 5.7, is less for the case of $M_2 = 2.5$ than for $M_1 = 0.25$. This behaviour of error is consistent with the error behaviour of the previous section where it was shown that a smaller value of Δt for a particular value of Δx does not always mean smaller error.

Since in any numerical solution we are concerned not only with the convergence but also with the amount of error involved and the economy with which the solution is obtained, it is obvious that in the case of the displacement models the parameter for error control should be M_1 and not M_2. In Figures

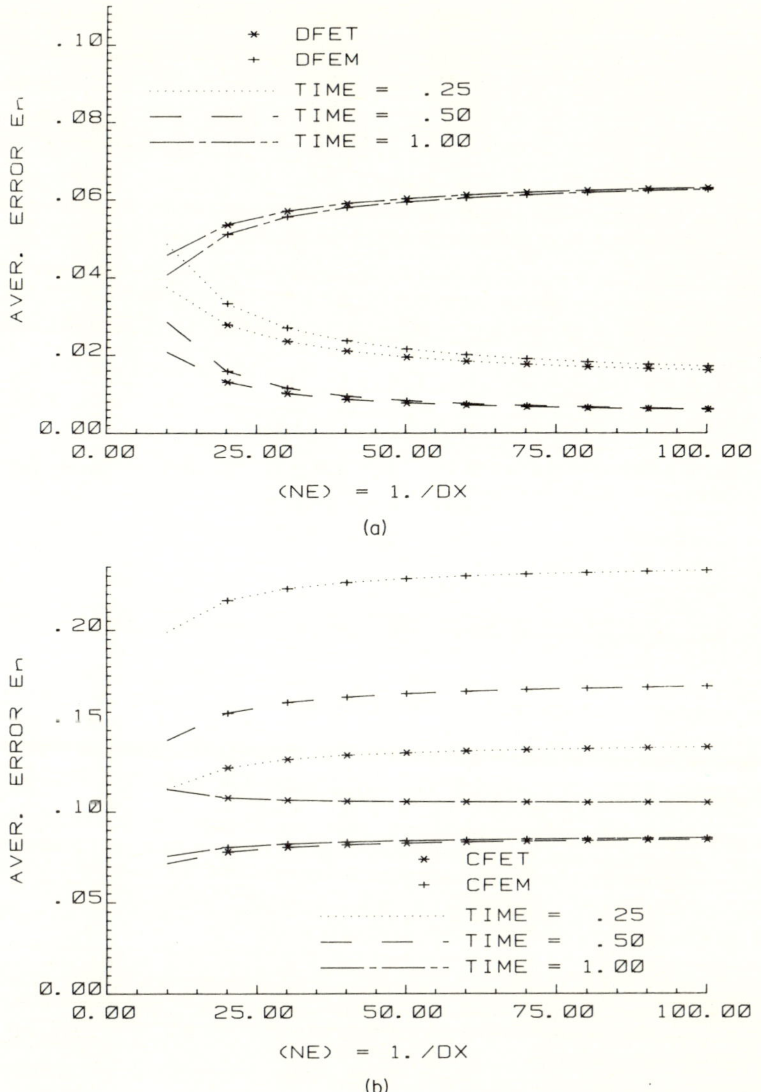

Figure 5.3 Average error for constant $M_1 = 0.25$: (a) displacement models; (b) conventional models

5.3(b) and 5.4(b) results are given for the conventional models and for the same cases as above. Again uniform convergence is obtained for both parameters M_1 and M_2 but the error values for constant M_1 are now larger than for constant M_2. This is the opposite from what was the case for the displacement models. Furthermore, the difference in error between the two

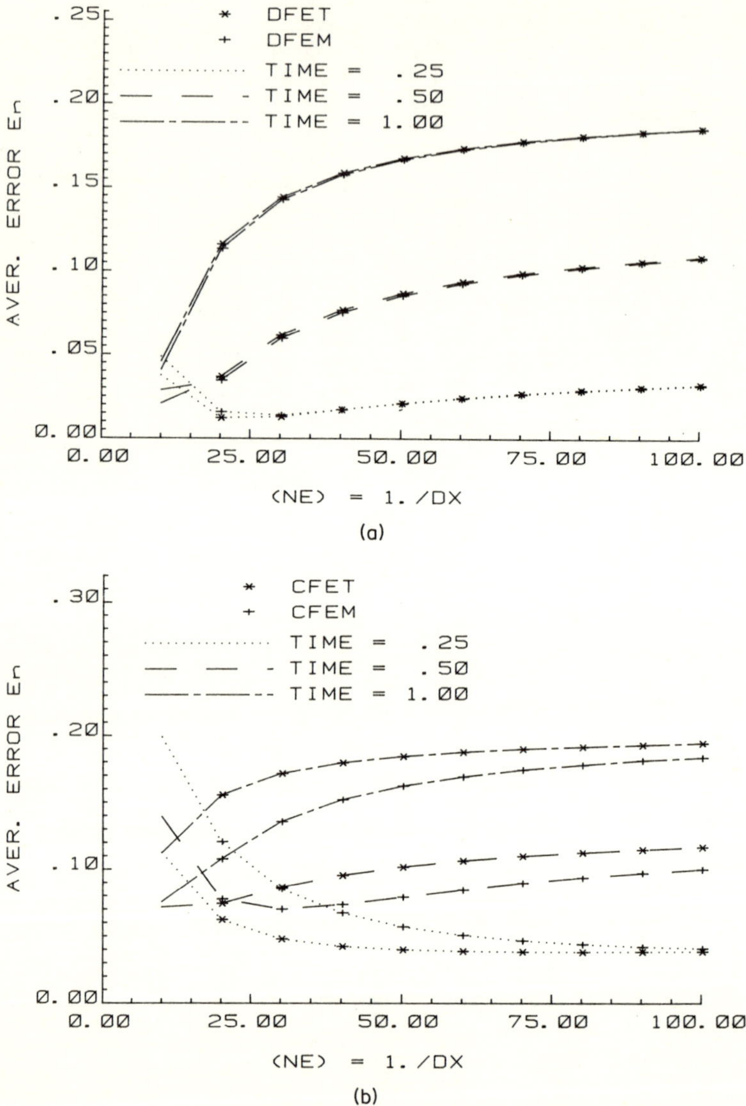

Figure 5.4 Average error for constant $M_2 = 2.5$: (a) displacement models; (b) conventional models

conventional models, CMET and CFEM, is much larger for $M_1 = 0.25$ than for $M_2 = 2.5$. It is obvious from the two figures that for the conventional models the dominant parameter is M_2 and not M_1. Also for the larger values of Δt ($M_1 = 0.25$) the numerical solution is more sensitive to the type of time integration scheme used.

It appears that for the displacement models the correct parameter for error control is the ratio M_1 and for the conventional models the correct parameter is the ratio M_2. This can be justified first by the type of equation involved in each of the formulations and second by the results obtained for the average error of the numerical solutions. The equation for which numerical solutions are sought through the displacement models, Equation (5.8), does not contain second-order space derivatives of the temperature and only the ratio $\Delta t/\Delta x$ appears in the approximations of this equation. In contrast the equation for which numerical solutions are sought through the conventional models, Equation (5.3), does contain second-order derivatives, hence the ratio $\Delta t/\Delta x^2$ appears in the approximation of this equation. The results presented in this section are in full agreement with this reasoning.

In the last part of this section two figures (5.5(a) and 5.5(b)) are given for different values of the parameters M_1 and M_2. These figures, together with Figures 5.3(a) and 5.4(a) and Figures 5.3(b) and 5.4(b), can be used to evaluate and compare the displacement and conventional models. For the displacement models, Figures 5.3(a) 5.4(a) and 5.5(a), the error increases as the value of M_1 increases from 0.25 to 0.5. A similar behaviour is observed for the conventional models as M_2 increases from 2.5 to 5.0. Comparing now Figures 5.3(a)–5.5(a) to Figures 5.3(b)–5.5(b), one can observe that the displacement models consistently result in smaller error values than the conventional ones, especially for large values of Δx (small *NE*). Furthermore, the smaller error values are obtained by the displacement models with a larger time step. For a given value of Δx, say $\Delta x = 0.025$, the values of Δt which correspond to $M_1 = 0.25$ (DFET and DFEM) and $M_2 = 2.5$ (CFET and CFEM) are $\Delta t_1 = 0.00625$ and $\Delta t = 0.00156$ respectively. This can be easily observed from Tables 5.7 and 5.8 and from Figures 5.3–5.5.

The following general observations can be made regarding the error behaviour of the numerical models:

- Uniform convergence is obtained for all models only when the ratio M_1 or the ratio M_2 is kept constant.
- Uniform convergence does not automatically guarantee minimum error.
- For the DFET and DFEM models the minimum error is obtained when the control parameter is M_1.
- For the CFET and CFEM models minimum error is obtained when the control parameter is M_2.
- Of the two time integration schemes used, the finite element in time is consistently more accurate for all cases.
- The displacement models are computationally more efficient overall since they can achieve smaller error than the conventional ones by using larger time steps.

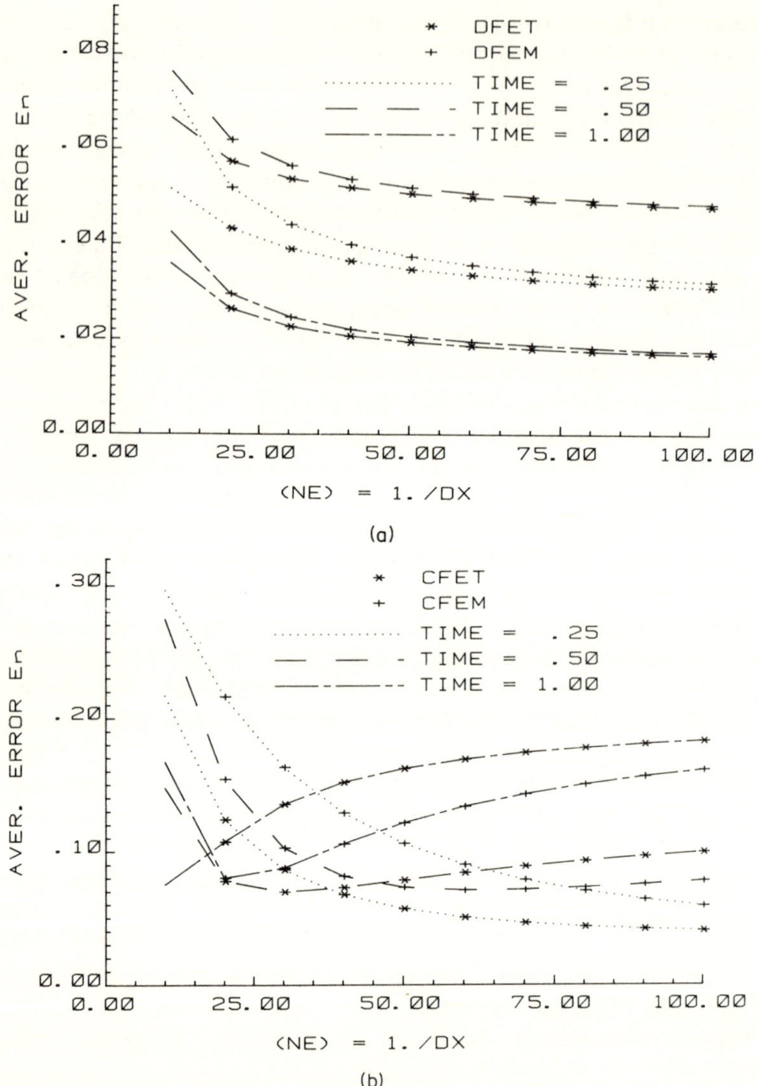

Figure 5.5 Average error: (a) constant $M_1 = 0.5$, displacement models; (b) constant $M_2 = 5.0$, conventional models

— For all of the models the error decreases significantly up to $NE = 40$ and an increase of NE beyond this value has only small effect on the numerical solution.

Numerical experimentations based on the average error result in a good understanding of the overall behaviour of the numerical models and how the numerical solutions obtained by these models are affected by the parameters M_1 or M_2. From the results presented in the last two sections and from previous studies some optimum values for M_1 and M_2 can be determined. These optimum values should correspond to small error but at the same time should hold the efficiency of the computational model at a high level. The range of values for M_1 is between 0.2 and 0.4 and for M_2 is between 2.0 and 4.00. The exact choice of values usually depends on the conditions of the specific problems under study.

The experience gained from studying the error behaviour can be used for obtaining solutions to specific problems. From such solutions one can evaluate the efficiency of a computational model in a more detailed fashion, and with respect to other parameters such as boundary conditions. In the next section several problems will be presented for that purpose.

5.4.4 Numerical examples

The computational models developed in this study will be applied to solve the thermal diffusion equation for a number of boundary conditions. Numerical solutions of this equation with the associated boundary conditions are discussed in this section, and results obtained from the displacement (DFET) and conventional (CFET) models are compared to available analytical solutions. Results for all case studies were obtained by using $NE = 50$ ($\Delta x = 0.02$) and $M_1 = 0.25$ for the DFET model and $M_2 = 2.5$ for the CFET one. The corresponding time steps are indicated on the figures.

The governing equation for the diffusion of heat is given by Equation (5.88) and the dimensionless temperature $\theta(x, t)$ is related to the heat displacement by

$$\theta(x, t) = \frac{\partial H(x, t)}{\partial x}. \tag{5.92}$$

If a heat source term is included in the diffusion equation, then Equation (5.89) is written as

$$\frac{\partial \theta}{\partial t} - D_0 \frac{\partial^2 \theta}{\partial x^2} = K_0 S, \tag{5.93}$$

where K_0 is the coefficient associated with the heat source term $S(K_0 = 0$ or $1)$.

In terms of the heat displacement field $H(x, t)$ the diffusion equation is written as

$$\frac{\partial H}{\partial t} - D_0 \frac{\partial^2 H}{\partial x^2} = K_0 \dot{h}. \tag{5.94}$$

Initial and boundary conditions associated with this equation are numerous. Here some typical boundary conditions are considered and numerical solutions will be obtained for both computational models. In Table 5.9 the initial and boundary conditions are given for each of the problems considered.

Table 5.9

Problem	D_0	K_0	Initial boundary conditions	Conditions	Figure
1	1.0	0.0	$\theta(x, 0)=0.0$	$\theta(0, t)=1.0$ $\theta(L, t)=0.0$	6
2	1.0	1.0	$\theta(x, 0)=0.0$	$\theta(0, t)=0.0$ $\theta(L, t)=0.0$	7
3	1.0	1.0	$\theta(x, 0)=0.0$	$\theta(-L, t)=0.0$ $\theta(L, t)=0.0$	8
4	1.0	0.0	$\theta(x, 0)=0.0$	$\theta(0, t)=1.0$ $\frac{\partial\theta}{\partial x}(1, t)=0.0$	9

In the first three problems the right-hand side boundary condition ($x=L$) is assumed to be applied at infinity. For the heat source term the value K_0 is 1.0 at $x=0.5$ and 0.0 elsewhere with $\dot{h}(0.5, t)=1.0$. This represents a point heat source at $x=0.5$ with unit heat generation per unit time. Numerical results for the problems described in Table 5.9 are given in Figures 5.6–5.9.

The first problem is the typical case for the diffusion equation with constant temperature applied on the left-hand boundary. The physical boundary to the right is at inifinity and the computational boundary is at $x=1.0$. Since the boundary condition $\theta(L, t)=0.0$, is at infinity, then it is physically incorrect to apply this boundary condition at $x=1.0$. The temperature at $x=1.0$ is zero for small times ($t<1, 0$), but it is not zero for larger times. Therefore no boundary condition is imposet $x=1.0$ and the computational model should be able to simulate an outflow boundary at that point.

In Figure 5.6(a) results for the DFET model are given for four time intervals and the numerical solution is compared to the analytical one (solid lines). The agreement between the two is good and the largest error occurs close to the right-hand side boundary. Similar results are given in Figure 5.6(b) for the CFET model for the temperature distribution. The error of the CFET model is larger than the DFET one throughout the computational domain. The larger error occurs mainly because of the error introduced by the boundary conditions. The boundary forces resulting from the variational formulation are given by

$$Q_i=\int_s \eta \frac{\partial\theta}{\partial x}\frac{\partial\theta}{\partial q_i}\,\mathrm{d}s \tag{5.95}$$

for the CFET model, and

$$Q_i = \int_s \eta\theta \frac{\partial H}{\partial q_i} \, \mathrm{d}s \tag{5.96}$$

for the DFET model.

If a boundary condition is described for the temperature at $x = 0.0$ ($\theta(0, t) = 1.0$), then

$$\delta\theta = 0 \rightarrow \frac{\partial \theta}{\partial q_i} = 0,$$

which means that the boundary force at $x = 0.0$ is zero for the CFET model. This is not the case for the DFET model since δH_i is not zero. At the other boundary ($x = 1.0$) the same is true for both models if the temperature is described at that boundary. As was mentioned before, no boundary condition is imposed at $x = 1.0$, which means that the boundary forces are not zero and should be included in the calculation for both models. By retaining the boundary forces the results for the DFET model, Figure 5.6(a), show good agreement with the analytical solution and the boundary force accounts for the neglected portion of the physical domain ($x > 1$). The inclusion of the boundary force in the computation for the DFET model is also in agreement with the theory since neither $\theta(1, t)$ nor δH_i is zero.

Although theory and results are in agreement for the displacement formulation the same is not true for the conventional one. If one retains the boundary force in the computations, the results, given in Figure 5.6(c), show very large error close to the boundary ($x = 1.0$) but if the boundary force is set equal to zero, even though θ or $\delta\theta$ are not zero, then the results, given in Figure 5.6(b), show much less error and the temperature distribution is similar to the DFET model. This discrepancy in the results obtained by the CFET model exists in all cases of solving problems in the semi-infinite or infinite space. There is a deficiency of the conventional formulation in terms of properly simulated outflow conditions. For all these types of problems, even if the boundary conditions are not explicitly imposed, the boundary forces should be zero for the CFET model. This is in agreement with the Galerkin approach and the term 'consistent model' will be used for the CFET model with no boundary forces.

The next example, Problem 2, is the case when a point heat source is present at $x = 0.5$. Results for the DFET model are given in Figure 5.8(a) and for the CFET model in Figures 5.7(b) and 5.7(c). In the last two figures results are given for the cases of omission and inclusion of the boundary force ($x = 1.0$), respectively, as previously done for Problem 1. As one can observe from Figure 5.7(b) (no boundary force), the results for the consistent model show lower temperature distribution than the ones of the DFET model, Figure

5.7(a). But when the boundary force is included, Figure 5.8(c), the solution results in much larger values for the temperature close to the boundary ($x = 1.0$). This behaviour of the CFET model is the same as previously in Problem 1. If one now considers a infinite domain and that the temperature is zero at $-L$ and $+L$ ($L \to \infty$), the heat generated by a point heat source will be diffused in both directions according to the governing equation and the

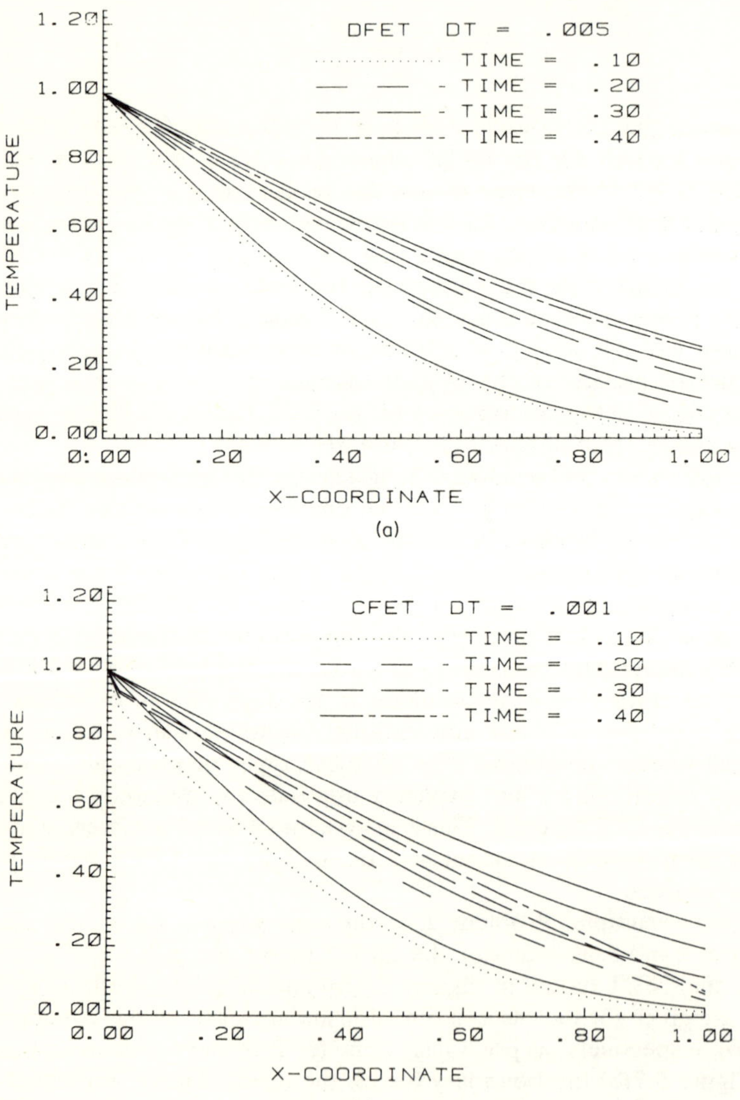

(a)

(b)

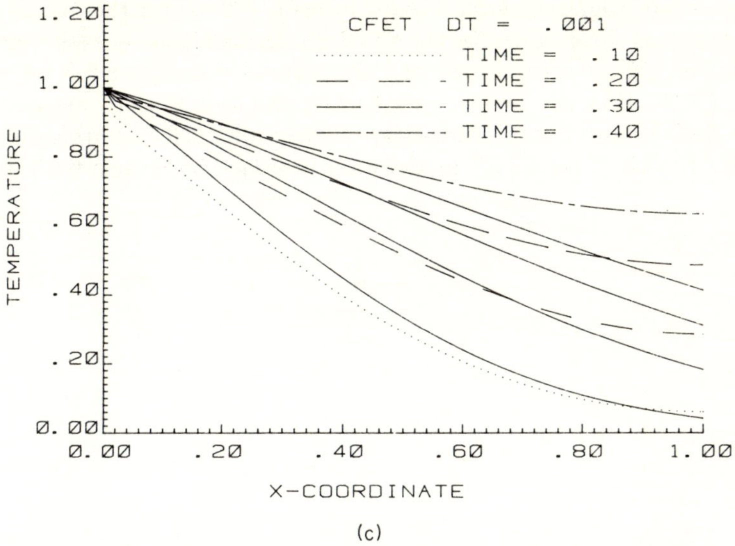

Figure 5.6 Temperature distribution, Problem 1: (a) displacement model; (b) conventional model without boundary forces; (c) conventional model with boundary forces

temperature distribution should be symmetric with respect to the point heat source. This is true for the results of the DFET model, Figure 5.8(a), and the results of the CFET model when the boundary forces are included, Figure 5.8(c), but it is not true for the results obtained by the consistent CFET model.

Throughout the results presented in the previous and present sections an important difference exists between the two models. The implementation of the DFET model to solve problems with unbounded domains results in computational models where no inconsistencies exist between the physical and mathematical aspects of the problem. Although this is the case for the DFET model, it is not true for the CFET model. In that case the absence of the boundary forces results in a better numerical solution, but the physical and mathematical aspects of the problem contradict each other. The inability of the CFET model to simulate correctly outflow conditions is mainly because the conventional model cannot account for the presence of the physical space beyond the computational one.

In the last example, Problem 4, the diffusion of heat in a finite domain is considered, a slab of thickness one, and the boundary conditions for this case is given in Table 5.9. In this example, the numerical results of the two models are of about the same accuracy. For the finite domain problem, the boundary conditions are applied on both boundaries. The boundary forces are zero for the CFET model but for the DFET model the boundary forces are zero only if the specified boundary temperature is zero. Otherwise the values of the boundary forces depend on the values of the temperature at each boundary.

The numerical results of Problem 4 are presented in Figure 5.9 for the two models. Comparing the results obtained by the two models one can easily observe that the numerical solutions are almost identical. The only difference is that the results of the CFET model were obtained by using a time step size

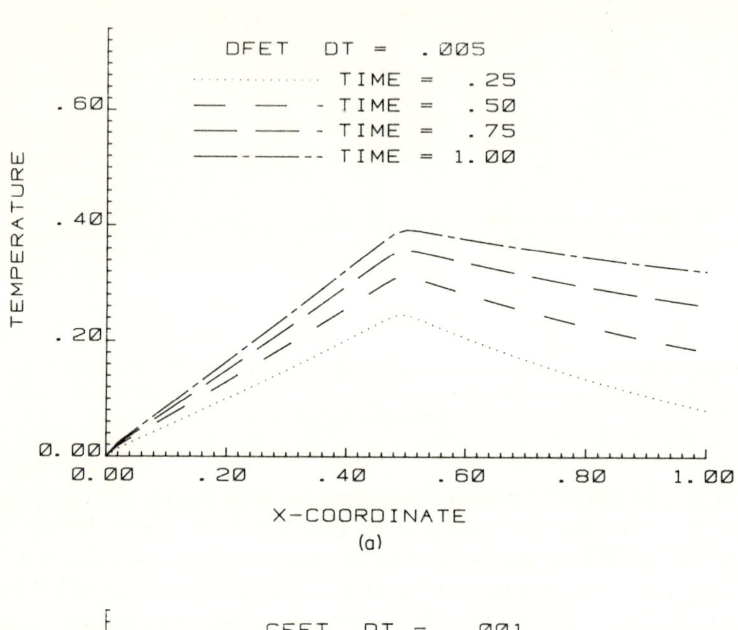

(a)

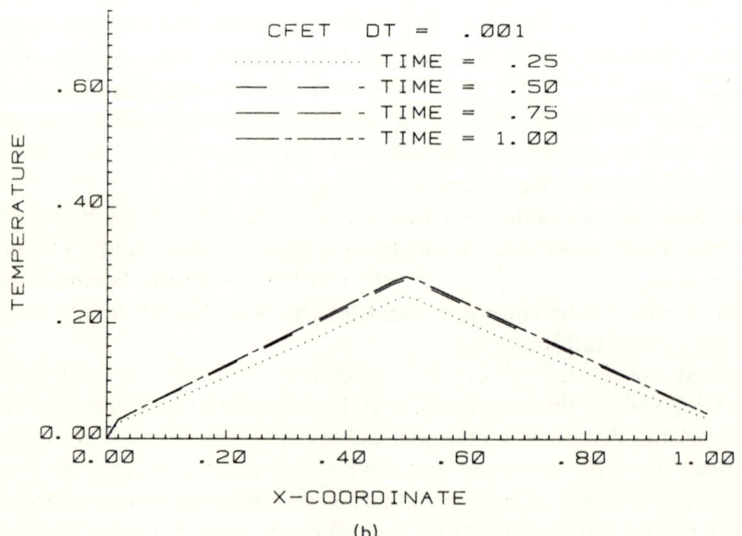

(b)

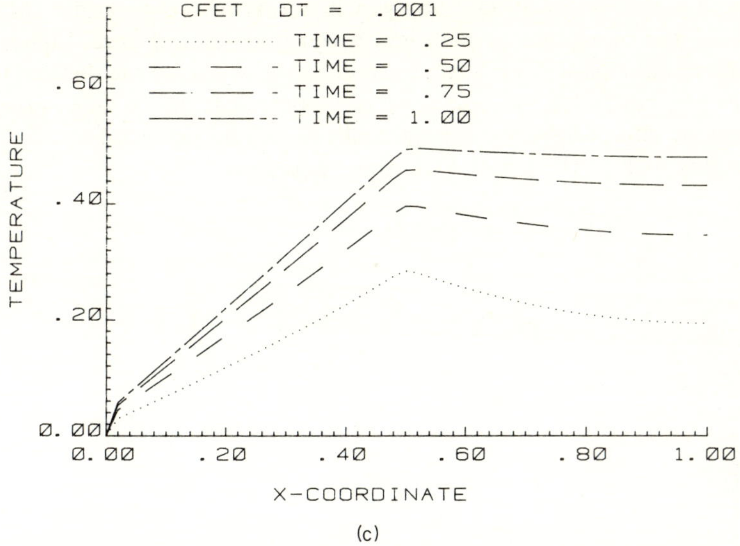

(c)

Figure 5.7 Temperature distribution, Problem 2: (a) displacement model; (b) conventional model without boundary forces; (c) conventional model with boundary forces

five times smaller than the one used for the DFET model. Therefore from the computational point of view the displacement formulation is more efficient since it requires less time to solve the same problem. If the problem to be solved does not have outflow boundary conditions, both models produce numerical solutions of about the same accuracy.

Throughout the examples presented here one can observe that the displacement formulation, the DFET model, is more consistent than the conventional one with respect to handling different physical problems. This is justified not only from the numerical solutions obtained but also from the formulation on which the displacement model is based.

5.5 SUMMARY AND CONCLUSIONS

A variational formulation for the thermal diffusion equation has been presented in this study, and, based on this formulation, a generalized approach to approximate solutions has been developed. As a special application to approximate solutions, four computational models were derived for obtaining numerical solutions to thermal diffusion problems. The models are based on first-order finite element approximations for both the spatial and time coordinates. Two of the models are derived from the fundamental variational analysis in terms of heat displacement and the other two are derived from the complementary variational analysis in terms of temperature. Although only linear and one-dimensional models were considered here, higher order and multi-dimensional models can be easily obtained since the derived formulation is not restricted to one type of approximation or one dimension. Furthermore, the present variational analysis can be extended to other types of equations such as the general transport equation.

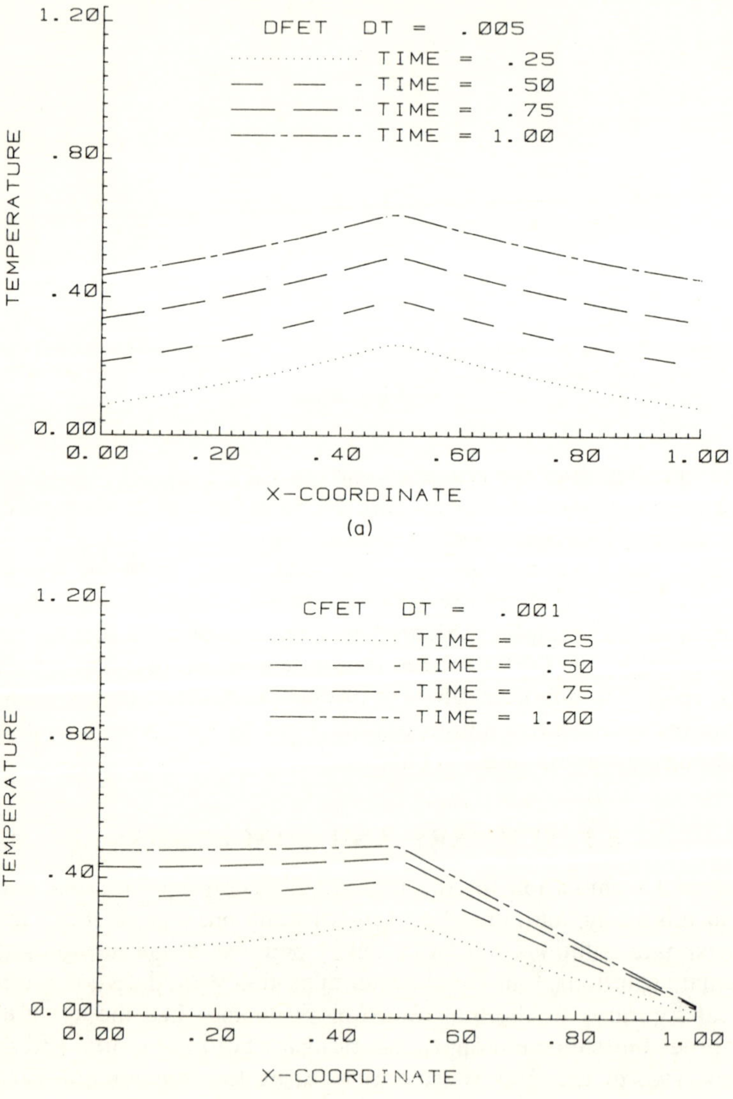

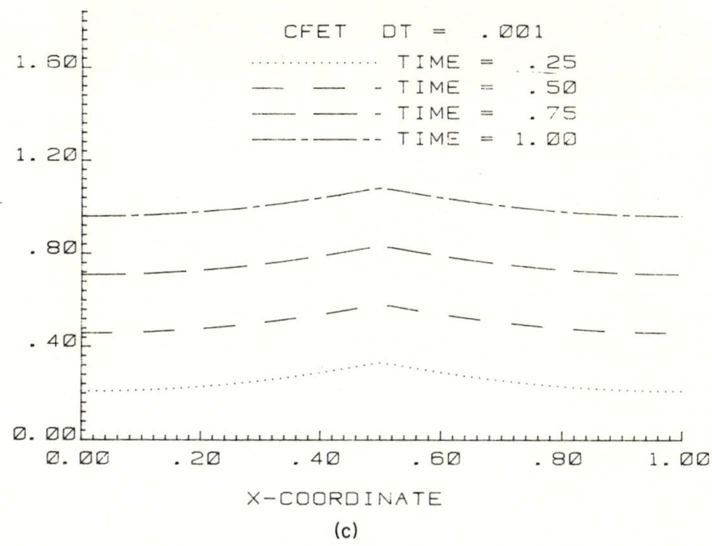

Figure 5.8 Temperature distribution, Problem 3: (a) displacement model; (b) conventional model without boundary forces; (c) conventional model with boundary forces

An evaluation of the four computational models with respect to accuracy and efficiency was carried out, first, by numerical experimentation on the error behaviour and, second, by obtaining numerical solutions to specific boundary value problems.

The investigation of the error of the numerical solution shows that the displacement models not only are more consistent in the error behaviour than the conventional ones, but also that the overall error of the numerical solutions, obtained by the displacement models, is always smaller than the error of the solutions obtained by the conventional ones. The larger error of the conventional models is partly because they are derived from the complementary variational principle, which according to the theory is less accurate than the corresponding fundamental variational principle. Numerical solutions to specific thermal diffusion problems obtained here also justify this argument. Furthermore, these solutions showed the inability of the conventional models to correctly simulate outflow boundary conditions.

Although the two types of computational models have some common characteristics in behaviour, the conventional ones are less efficient from both accuracy and computational effort points of view. Displacement models have been used widely in structures, but as it has been shown these models can be extended to other types of problems as well. Furthermore, the fundamental form of the variational analysis can be derived, based on concepts of classical mechanics presented here, for many governing equations.

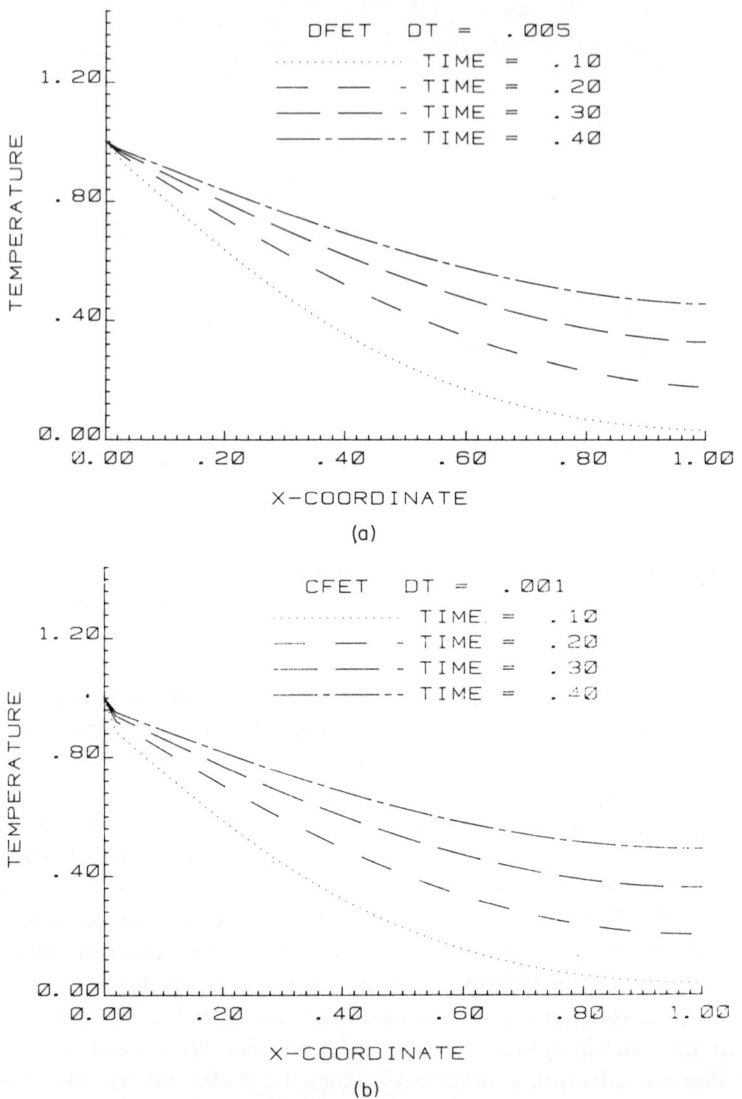

Figure 5.9 Temperature distribution, Problem 4: (a) displacement model; (b) conventional model

In conclusion, the generalized variational analysis and the unified approach in obtaining approximate solutions should be emphasized as well as the applicability of the analysis to other physical problems. It should be also noted that the finite element method, which was presented in this study, is only one restricted application of the generalized formulation.

ACKNOWLEDGEMENTS

This work was partially supported by the Office of Naval Research. The author wishes to thank R. A. Skop and O. Griffin for their valued suggestions.

REFERENCES

1. M. A. Biot, 'Thermodynamics and heat flow analysis by Lagrangian methods', *Proc. 7th Anglo-American Aeronautical Conference*, Inst. of Aero-Sciences, N.Y. (1959).
2. M. A. Biot, 'New methods in heat flow analysis with applications to flight structures', *J. Aero. Sci.* **24** (22) (1959).
3. G. A. Keramidas, 'Finite element modeling for convection–diffusion problems', *NRL Memorandum Report 4225* (1980).
4. G. A. Keramidas, 'Variational formulation and approximate solutions for transport phenomena', *NRL Memorandum Report* (1982).
5. G. A. Keramidas, 'Finite element modeling of the momentum and diffusion equations', *NRL Memorandum Report 4310* (1980).
6. M. A. Biot, 'Variational principles in irreversible thermodynamics with applications to viscoelasticity', *Phys. Rev.* **97**, 1463–1469 (1955).
7. V. D. Barger, and M. G. Olsson, 'Classical Mechanics, A Modern Perspective', McGraw-Hill (1973).
8. B. Carnahan, H. A. Luther, and J. O. Wilke, *Applied Numerical Methods*, Clarendon Press, Oxford (1959).
9. B. Gay, and P. T. Cameron, 'The efficiency of numerical solutions of the heat conduction equation', *ASME* (1967).

Numerical Methods in Heat Transfer, Volume II
Edited by R. W. Lewis, K. Morgan, and B. A. Schrefler

Chapter 6

Penalty-finite Element Methods in Conduction and Convection Heat Transfer

J. N. Reddy

6.1 INTRODUCTION

The coupled conduction and convection phenomenon due to the buoyancy-driven flow inside a closed enclosure (i.e. internal flows) subjected to differential side heating is commonly encountered in many practically important engineering problems. These include: thermal insulation of buildings, Batchelor;[1] heat transfer through double-glazed windows, Elder[2] and Gill;[3] cooling of electronic equipment, Pedersen *et al.*[4]; general circulation of planetary atmospheres, Hart;[5] crystal growth from melts, Carruthers;[6] sterilization of canned foods, Hiddink *et al*;[7] cooling fluids in channels surrounding a nuclear reactor core, Petuklov;[8] and convectively cooled underground electric cable systems, Chato and Abdulhadi;[9] and others. Although most of these flows are fully three-dimensional, the limitations imposed by experimental as well as solution techniques have forced researchers to analyse only those fluid motions that are believed to render themselves to approximation by two-dimensional models. The Navier–Stokes equations (i.e. momentum equations) and the energy equation governing the phenomenon are highly nonlinear, and the coupling between the boundary layer and the core region in the internal flows makes it difficult to obtain analytical solutions.

The development and use of computational methods for numerical solution of the Navier–Stokes equations have been under way for more than five decades, but it is only in the last decade that notable contributions in the numerical simulation of fluid flows have been made; among these, the most important one is the basic understanding of the relationship between the physics of a problem and the numerical procedure employed for its solution. With the advent of modern electronic computers, several finite difference and finite element formulations of the Navier–Stokes equations have been investigated to date (see, for example, references 10–36). All of these formulations can be classified into two major categories: (a) the primitive variables (i.e. velocities and pressure) formulation, and (b) vector potential and vorticity formulation. The temperature is an additional variable in both of the formulations when heat transfer is accounted for. Difficulties in imposing the pressure boundary conditions and satisfying the continuity equation are encountered in the primitive variables formulation. Chorin[10] and Williams[11] investigated the primitive variables approach. Aziz and Hellums[12] investigated the solution of Navier–Stokes equations in terms of the vorticity components and a vector potential. Mallinson and de Vahl Davis[13–15] developed the so-called method of false transient to obtain the steady-state solution. The method of 'upwind differencing' was employed in reference 14 to approximate the convective terms (i.e. the nonlinear terms in the momentum equations) in an effort to overcome the stability restrictions imposed by the use of central differencing. The study concludes that the assumption of two dimensionality is questionable at high values of the Reynolds number ($R_E > 500$). In reference 15, the solution of the steady-state Navier–Stokes equations in three dimensions has been obtained for the problem of buoyancy-driven flow in a rectangular cavity subjected to a thermal gradient across the sides of the cavity using the method described in reference 13. Roscoe[16] extended the exponential type approximation method of Allen and Southwell[17] to three-dimensional Navier–Stokes equations (also, see reference 18). Dennis *et al.*[19] employed the Dennis and Hudson scheme,[20] which is based on a formulation similar to that of Aziz and Hellums[12], to solve the three-dimensional Navier–Stokes equations for the wall-driven cavity problem (for $R_E = 100$). The difference between the formulation in reference 19 and that of Aziz and Hellums[12] is that in reference 19 the three equations relating the velocity and vorticity components are employed instead of a vector potential. The Dennis and Hudson[20] scheme is second-order accurate and suffers from large artificial diffusion at high Reynolds numbers. In the numerical solution of the Navier–Stokes equations by finite difference methods one encounters convective instability at high Rayleigh/Reynolds numbers. This difficulty is often overcome by employing first-order upwind differencing (see Nallasamy and Prasad[21]) or a hybrid of first-order upwind and central differencing (see Spalding[22]) for the convective terms, and central differencing for the diffusion terms. These difference operators introduce some artificial diffusion, at the expense of accuracy, into the problem. Recently, Agarwal[23,24] presented a third-order accurate upwind scheme which was shown to be accurate and stable at high Reynolds numbers. The scheme was also used in the solution of steady Navier–Stokes equations in three dimensions. The basic equations were expressed in terms of the three velocities and three vorticity components. The convective terms in the transport equations for the vorticity components were differenced using a third-order accurate upwind scheme. In another study, Ghosh *et al.*[25] developed an implicit locally one-dimensional scheme for the solution of incompressible three-dimensional Navier–Stokes equations in primitive variables.

One of the major difficulties associated with the solution of Navier–Stokes equations for fluid flows over nonrectangular boundaries is the application of the boundary conditions. Although attempt has been made to rectify this problem by constructing curvilinear meshes in finite difference methods,[25] the finite element method has a definite advantage in that any complicated geometry can be suitably represented and appropriate boundary conditions of the model can be imposed in a natural way. Natural convection heat transfer in two-dimensional enclosures has been investigated by Tabarrok and Lin[26] using the vector potential and vorticity formulation, by Gartling[27] and Upson *et al.*[28] using the so-called *mixed* formulation (which contains all of the primitive variables), and by Heinrich *et al.*[29,30] and Reddy and his colleagues[31–33] using penalty function methods.

Feasibility studies for the solution of three-dimensional Navier–Stokes equations by finite element methods have been reported recently by Gresho *et al.*[34] and by the author.[35,36] The latter approach is based on the so-called penalty finite element model, whereas the approach taken by Gresho *et al.*[34] is based on the Galerkin finite element model of the primitive variable formulation. In reference 36 the author presented numerical results for the wall-driven cavity problem and the window-cavity problem, both of which serve as model problems in establishing the reliability and accuracy of new computational techniques.

In spite of all these developments, many of the complex internal as well as external flow problems remained unresolved. For practical engineering problems, only feasibility studies can be attempted at present. Most of the previous studies were either confined to two dimensional models, or, when the three-dimensional model was used, to a model problem that was believed to typify certain physical flow features. The latter is an acceptable approach provided the reliability and accuracy of the computational scheme is established. Increasing efforts are being made to devise and test the reliability and accuracy of new computational schemes that are computationally efficient. Thus, despite the progress made to date in the numerical simulation of flows by the finite-difference and finite-element methods, much further work is needed, especially in three-dimensional flows, even to solve some of the basic problems of computational fluid dynamics.

The present paper reviews the penalty finite element method for natural convection in two- and three-dimensional enclosures subjected to differential side heating, and presents numerical results for a number of convective heat transfer problems.

6.2. THE PENALTY FUNCTION METHOD

The basis of the penalty function method comes from a method suggested by R. Courant[37] in 1941 for the minimization of a quadratic functional subjected to an equality constraint (see Reddy[38] for a review of the history of the method). The first use of the penalty function method in the finite-element analysis of a constrained minimization problem was apparently due to Babuska[39] who proved the existence and uniqueness of the finite-element solution to the penalty function formulation of the Dirichlet problem for the Poisson equation. Since the essential (or Dirichlet) boundary conditions cannot be included in the variational problem associated with the Poisson equation, Babuska[39] employed the penalty function concept of Courant to include, in an approximate sense, the essential boundary conditions in the variational problem. The variational problem consists of finding the solution u to the Poisson equation, $-\nabla^2 u = f$ in Ω, and $u = 0$ on Γ (where Ω is a two-dimensional domain and Γ is its boundary) by minimizing the modified functional

$$I_p(u) = \int_\Omega [\tfrac{1}{2}\nabla u \cdot \nabla u - fu]\,\mathrm{d}x\,\mathrm{d}y + \tfrac{1}{2}\int_\Gamma \gamma u^2\,\mathrm{d}s \tag{6.1}$$

where γ is an arbitrary positive parameter (a function of position or a constant) called the penalty parameter. The penalty parameter can be viewed as a weight function whose form and magnitude dictates the degree to which the 'constraint', $u = 0$ on Γ, is satisfied. The Euler equations of the modified functional are

$$-\nabla^2 u + f = 0 \quad \text{in } \Omega, \qquad \frac{\partial u}{\partial n} + \gamma u = 0 \quad \text{on } \Gamma. \tag{6.2}$$

Note that as γ is increased to a large value, $\gamma^{-1}(\partial u/\partial n) + u = 0$. Thus, for a large value of γ the variational problem in (6.1) corresponds to the Poisson's equation for the Dirichlet problem.

In the context of the present problem, namely, viscous incompressible flows, the penalty function method has a natural application. We illustrate the application with the Stokes problem in two dimensions: find the velocity field (u, v) and pressure P that satisfy the equations

$$\left.\begin{aligned} 2\mu\frac{\partial^2 u}{\partial x^2} + \mu\frac{\partial}{\partial y}\left(\frac{\partial u}{\partial y} + \frac{\partial v}{\partial x}\right) - \frac{\partial P}{\partial x} &= f_x \\ \mu\frac{\partial}{\partial x}\left(\frac{\partial u}{\partial y} + \frac{\partial v}{\partial x}\right) + 2\mu\frac{\partial^2 v}{\partial y^2} - \frac{\partial P}{\partial y} &= f_y \end{aligned}\right\} \quad \text{in } \Omega \tag{6.3}$$

$$\frac{\partial u}{\partial x} + \frac{\partial v}{\partial y} = 0 \qquad \text{in } \Omega. \tag{6.4}$$

where μ is the viscosity, and f_x and f_y are the body force components. The velocity field must also satisfy certain boundary conditions of the problem. For the sake of discussion, we assume that $u = v = 0$ on the boundary Γ.

The above problem can be formulated variationally. Let H be the product space $H = H_0^1(\Omega) \times H_0^1(\Omega) \times \hat{H}(\Omega)$, where $H_0^1(\Omega)$ is the Hilbert space of order 1 with compact support

$$H_0^1(\Omega) = \left\{ u: u, \frac{\partial u}{\partial x}, \frac{\partial u}{\partial y} \text{ are square integrable on } \Omega, \text{ and } u = 0 \text{ on } \Gamma \right\}, \quad (6.5)$$

and $\hat{H}(\Omega)$ is the space of functions that are orthogonal to the set of constants:

$$\hat{H}(\Omega) = \left\{ P: P \text{ is square integrable, and } \int_\Omega P \, dx \, dy = 0 \right\}. \quad (6.6)$$

The variational problem associated with (6.3) and (6.4) (with $u = v = 0$ on Γ) consists of finding (u, v, P) from the function space H such that

$$I(u, v, P) = \int_\Omega \left\{ \mu \left[\left(\frac{\partial u}{\partial x}\right)^2 + \left(\frac{\partial v}{\partial y}\right)^2 + \frac{1}{2}\left(\frac{\partial u}{\partial y} + \frac{\partial v}{\partial x}\right)^2 \right] + f_x u + f_y u \right\} dx \, dy$$
$$- \int_\Omega P\left(\frac{\partial u}{\partial x} + \frac{\partial v}{\partial y}\right) dx \, dy \quad (6.7)$$

is a minimum on H.

An alternative variational problem involves finding (u, v) from the space S,

$$S = \left\{ (u, v): u, v \text{ are from } H_0^1(\Omega), \text{ and } \frac{\partial u}{\partial x} + \frac{\partial v}{\partial y} = 0 \right\} \quad (6.8)$$

such that

$$I_0(u, v) = \int_\Omega \left\{ \mu \left[\left(\frac{\partial u}{\partial x}\right)^2 + \left(\frac{\partial v}{\partial y}\right)^2 + \frac{1}{2}\left(\frac{\partial u}{\partial y} + \frac{\partial v}{\partial x}\right)^2 + f_x u + f_y v \right\} dx \, dy \quad (6.9)$$

is a minimum on S. Since (u, v) is sought in the space S, we have

$$\frac{\partial u}{\partial x} + \frac{\partial v}{\partial y} = 0,$$

and hence the last term in (6.7) drops out, giving $I_0(u, v)$.

The last variational problem is a constrained minimization problem: minimize the functional I_0 subjected to the incompressibility constraint in (6.4). Therefore, the penalty function method can be employed to include the constraint in the variational problem: find (u_n, v_n) in $H_0^1(\Omega) \times H_0^1(\Omega)$ such that

$$I_p(u_n, v_n) = I_0(u_n, v_n) + \frac{\gamma_n}{2} \int_\Omega \left(\frac{\partial u_n}{\partial x} + \frac{\partial v_n}{\partial y}\right)^2 dx \, dy \quad (6.10)$$

is a minimum on $H_0^1(\Omega)\times H_0^1(\Omega)$ for any fixed (preassigned) parameter γ_n. In other words, the constraint is satisfied in a least squares sense. The degree to which the constraint is satisfied is directly proportional to the value of the penalty parameter γ_n. In fact one can show[38–44] that the solution (u_n, v_n) to the penalty problem converges to the solution (u, v) of the original problem as γ_n goes to infinity. The error between the two solutions is given by

$$(\|u_n - u\|_1^2 + \|v_n - v\|_1^2)^{1/2} \leqslant \frac{c}{\gamma_n}\|P\|_0 \tag{6.11}$$

where $\|\cdot\|_1$ and $\|\cdot\|_0$ denote the H^1- and L_2-norms, respectively, and c is a constant.

The Euler equations of the functional I_p are

$$2\mu\frac{\partial^2 u_n}{\partial x^2} + \mu\frac{\partial}{\partial x}\left(\frac{\partial u_n}{\partial y} + \frac{\partial v_n}{\partial x}\right) + \gamma_n\frac{\partial}{\partial x}\left(\frac{\partial u_n}{\partial x} + \frac{\partial v_n}{\partial y}\right) = f_x$$

$$\mu\frac{\partial}{\partial x}\left(\frac{\partial u_n}{\partial y} + \frac{\partial v_n}{\partial x}\right) + 2\mu\frac{\partial^2 v_n}{\partial y^2} + \gamma_n\frac{\partial}{\partial y}\left(\frac{\partial u_n}{\partial x} + \frac{\partial v_n}{\partial y}\right) = f_y \tag{6.12}$$

Comparing equations (6.12) with (6.3), it follows that (the Lagrange multiplier can be post-computed according to)

$$P_n = -\gamma_n\left(\frac{\partial u_n}{\partial x} + \frac{\partial v_n}{\partial y}\right). \tag{6.13}$$

The above discussion can easily be extended to the Navier–Stokes equations; the convective terms do not alter the form of the penalty terms.

Despite the wide use of the penalty function method in mathematical programming and optimization, the penalty function method was not regarded, until recently, as a powerful computational device because of two drawbacks:

(a) the technique was used in connection with the approximate solution of variational problems by Rayleigh–Ritz type methods, which were themselves never regarded as competitive when compared to the traditional finite difference methods;

(b) in the practical application of the penalty function method, the penalty terms 'misbehave' when proper selection of the approximation functions or integration rule is not used to evaluate the penalty terms in the coefficient matrix.

These two drawbacks were obviated by the finite element method (and *reduced integration* techniques). In 1973, the penalty function method was introduced into the finite element analysis of fluid flow problems by Zienkiewicz.[45]

However, the second shortcoming was not overcome until Zienkiewicz *et al.*[46] devised, rather ingeniously, the so-called reduced integration technique, which was later used by Zienkiewicz and his colleagues[30,47] in the numerical integration of the penalty terms (e.g., the second term on the right-hand side of Equations (6.1) and (6.10)).

In the next section we describe an application of the penalty function method to the equations governing convection heat transfer in three dimensions. In the present problem, the incompressibility condition (or divergence-free condition on the velocity field) is treated as the constraint.

6.3 GOVERNING EQUATIONS

Consider a three-dimensional domain Ω filled with an incompressible viscous heat-conducting Newtonian fluid. We assume that the flow conditions are such that the familiar Boussinesq approximation holds, and that viscous dissipation is negligible. Under these conditions, the equations of momentum and energy can be written as

$$-(\nabla\times\mathbf{u})\times\mathbf{u}-\frac{1}{\rho_0}\nabla P-\beta(T-T_0)\mathbf{g}+\nu\,\nabla^2\mathbf{u}=\mathbf{0} \tag{6.14}$$

$$\nabla\cdot\mathbf{u}=0 \tag{6.15}$$

$$-\mathbf{u}\cdot(\nabla T)+\kappa\,\nabla^2 T=0 \tag{6.16}$$

where $\mathbf{u}=(u_1, u_2, u_3)$ is the velocity vector, P is the total pressure, T is the temperature, ν is the kinematic viscosity, κ is the thermal diffusivity, $\mathbf{g}$ is the gravitational vector, ρ is the density, β is the volumetric expansion coefficient, t is the time, and the subscript zero on ρ and T denotes some reference state.

Using L, L^2/κ, κ/L, and $\rho_0\kappa^2/L^2$ as scale factors for length, time, velocity, and pressure, respectively, and introducing $\theta=(T-T_0)/(T_1-T_0)$, Equations (6.14)–(6.16) can be expressed in the nondimensional form,

$$-(\nabla\times\mathbf{u})\times\mathbf{u}-\nabla P-R_A P_R\theta\hat{\mathbf{g}}+P_R\,\nabla^2\mathbf{u}=0 \tag{6.17}$$

$$-\mathbf{u}\cdot(\nabla\theta)+\nabla^2\theta=0 \tag{6.18}$$

wherein all the variables $(\mathbf{u}, P, \theta)$ are nondimensionalized and $\mathbf{g}=-\hat{\mathbf{k}}g$. Here L denotes a dimension of the cavity, T_0 and T_1 are fixed reference temperatures, and R_A and P_R are the Rayleigh and Prandtl numbers, respectively:

$$R_A=g\beta(T_1-T_0)L^3/\kappa\nu,\ P_R=\nu/\kappa \tag{6.19}$$

Equations (6.15), (6.17) and (6.18) must be solved in conjunction with appropriate boundary conditions. For the window cavity problem considered here, we have the following boundary conditions (see Fig. 6.1):

$$
\begin{aligned}
&x=0 \quad \text{(plane of symmetry)}: && \frac{\partial\theta}{\partial x}=0,\ u_1=0\\
&x=1: && \frac{\partial\theta}{\partial x}=0,\ u_1=u_2=u_3=0\\
&y=0,1: && \theta=y-0.5,\ u_1=u_2=u_3=0\\
&z=0,1: && \frac{\partial\theta}{\partial z}=0,\ u_1=u_2=u_3=0.
\end{aligned}
\tag{6.20}
$$

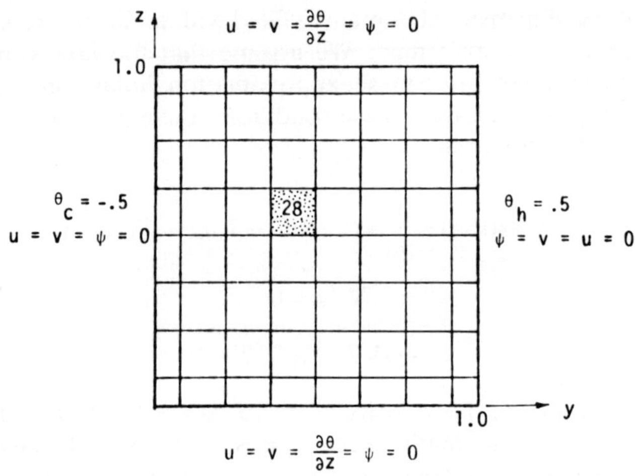

Figure 6.1 Finite element mesh and boundary conditions

The velocity potential $\boldsymbol{\psi}$ (or solenoidal vector potential for **u**) vector can be obtained by solving the Poisson's equation,

$$-\nabla^2\boldsymbol{\psi}=\nabla\times\mathbf{u} \tag{6.21}$$

subject to the boundary condition that the vector potential $\boldsymbol{\psi}$ is normal to an impermeable surface and that the normal derivative of the vector potential on such surfaces is zero. For example, on the surface $y=0$, we specify

$$\frac{\partial\psi_2}{\partial y}=0, \qquad \psi_1=\psi_3=0, \tag{6.22}$$

where $\boldsymbol{\psi}=(\psi_1,\psi_2,\psi_3)$.

6.4 PENALTY FUNCTION FORMULATION

Here we describe the penalty function formulation that avoids the pressure variable as a primary dependent unknown, and thus avoids difficulties associated with the proper imposition of pressure boundary conditions. Equations (6.17) and (6.18) can be viewed as those relating the velocity vector **u** and temperature θ, and Equation (6.15) as a constraint condition on the velocity field. That is, we wish to solve Equations (6.17) and (6.18) subject to the constraint in Equation (6.15). This viewpoint enables us to employ the penalty function method to formulate the problem variationally, and subsequently by the finite element method. The penalty function method transforms a given constrained variational problem into an (actually, a sequence of) unconstrained variational problem(s) by the introduction of a penalty on the infringement of constraints.

To fix the ideas, suppose that we wish to find the generalized solution $\mathbf{\Lambda}$ to the variational problem (nonlinear, in general),

$$\delta I(\Lambda, \delta\Lambda) = 0 \qquad \text{in } \Omega \tag{6.23}$$

subject to the constraint

$$G(\Lambda) = 0 \qquad \text{in } \Omega \tag{6.24}$$

Here δ denotes the variational symbol, and Ω is the domain of the problem. Clearly, the solution space is restricted to a subspace S of a linear vector space H on which $\delta I(\cdot\,\cdot)$ is defined; the elements of the subspace S must satisfy the additional conditions in Equation (3.10(b)). In the penalty function method, instead of seeking the solution in the subspace S (which is difficult to construct in the finite element approximation), we seek the solution in the whole space H by solving the *modified* (or perturbed) variational problem

$$\delta I(\mathbf{\Lambda}, \delta\mathbf{\Lambda}) + \delta\left\{\frac{\gamma}{2}\int_{\Omega} [G(\mathbf{\Lambda})]^2 \,\mathrm{d}x\,\mathrm{d}y\,\mathrm{d}z\right\} = 0 \tag{6.25}$$

wherein γ is the *penalty parameter*, which is a preselected (positive for extremum problems) function or constant whose value, generally speaking, has an effect on the accuracy of the solution. The quadratic functional in the square brackets is referred to as the *penalty functional.* For sufficiently large values of γ, the variational problem in Equation (6.25) is equivalent to the variational problem in Equations (6.23) and (6.24). It can also be shown that the Lagrange multiplier associated with the constraint in Equation (6.24) is given by the formula

$$\lambda_\gamma = \gamma G(\mathbf{\Lambda}_\gamma) \tag{6.26}$$

where Λ_γ is the solution of the modified problem (6.25).

Returning to the problem at hand, we apply the penalty function concept described above to the variational problem associated with Equations (6.17) and (6.18):

$$\delta I(\mathbf{\Lambda}, \delta\mathbf{\Lambda}) + \delta I_p(\mathbf{\Lambda}, \delta\mathbf{\Lambda}) = 0 \tag{6.27}$$

where $\mathbf{\Lambda} = (\mathbf{u}, \theta)$, $\delta\mathbf{\Lambda} = (\delta\mathbf{u}, \delta\theta)$, and

$$\delta I(\mathbf{\Lambda}, \delta\mathbf{\Lambda}) = \int_\Omega \{(\mathbf{u}\cdot\nabla\mathbf{u})\cdot\delta\mathbf{u} + P_R(\nabla\mathbf{u}\colon\nabla\delta\mathbf{u}) + R_A P_R \theta\mathbf{g}\cdot\delta\mathbf{u}$$
$$+[\mathbf{u}\cdot(\nabla\theta)]\delta\theta + \nabla\theta\cdot\nabla\delta\theta\}\,\mathrm{d}x\,\mathrm{d}y\,\mathrm{d}z - \int_\Gamma [\mathbf{t}\cdot\delta\mathbf{u} + q\,\delta\theta]\,\mathrm{d}s \tag{6.28}$$

$$I_P(\mathbf{\Lambda}) = \frac{\gamma}{2}\int_R (\nabla\cdot\mathbf{u})^2\,\mathrm{d}x\,\mathrm{d}y\,\mathrm{d}z. \tag{6.29}$$

In Equation (6.28), $\mathbf{t}$ denotes the specified surface stress (i.e. traction) vector, q denotes specified heat flux on the boundary Γ of Ω, the single dot between vectors denotes the scalar product, and the double dot between gradients of vectors denotes the double scalar product (e.g. $\boldsymbol{\sigma} : \boldsymbol{\varepsilon} = \sigma_{ij}\varepsilon_{ji}$). The Lagrange multiplier, which is equal to the negative of P, is given by

$$P_\gamma = -\gamma(\nabla\cdot\mathbf{u}_\gamma). \tag{6.30}$$

6.5 FINITE ELEMENT MODEL

The finite element formulation of Equation (6.27) is straightforward. Over a typical element Ω^e (assumed to be the eight-node, three-dimensional isoparametric element) of the finite element mesh Ω_h of Ω, a generic dependent variable v is interpolated by

$$v = \sum_i^8 v_i\phi_i \tag{6.31}$$

where ϕ_i are the trilinear interpolation functions associated with the three-dimensional, linear, isoparametric element, and v_i is the nodal value of v at node $i\,(i = 1, 2, \ldots, 8)$. Substitution of Equation (6.31) into Equation (6.27) results in the following pair of element equations (corresponding to $\{\delta\mathbf{u}\}$ and $\{\delta\theta\}$ coefficients)

$$\begin{bmatrix} [K^{11}] & [K^{12}] & [K^{13}] \\ [K^{12}]^{\mathrm{T}} & [K^{22}] & [K^{23}] \\ [K^{13}]^{\mathrm{T}} & [K^{23}]^{\mathrm{T}} & [K^{33}] \end{bmatrix} \begin{Bmatrix} \{u_1\} \\ \{u_2\} \\ \{u_3\} \end{Bmatrix} = \begin{Bmatrix} \{F^1\} \\ \{F^2\} \\ \{F^3\} \end{Bmatrix} \tag{6.32}$$

$$[C]\{\{\theta\} = \{Q\} \tag{6.33}$$

where the coefficient matrices $K_{ij}^{\alpha\beta}$, C_{ij}, F_j^{α} and $Q_j (I, J = 1, 2, 3;\ i, j = 1, 2, \ldots, 8)$ are given by

$$K_{ij}^{II} = \int_{\Omega^e} \left\{ \sum_{N=1}^{3} \phi_i U_N \frac{\partial \phi_j}{\partial x_N} + P_R\left(\frac{\partial \phi_i}{\partial x_I}\frac{\partial \phi_j}{\partial x_I} + \frac{\partial \phi_i}{\partial x_J}\frac{\partial \phi_i}{\partial x_J} + \frac{\partial \phi_i}{\partial x_K}\frac{\partial \phi_i}{\partial x_K} \right) + \gamma\left(\frac{\partial \phi_i}{\partial x_I}\frac{\partial \phi_j}{\partial x_I} \right) \right\} \mathrm{d}x\, \mathrm{d}y\, \mathrm{d}z,$$

($I \neq J \neq K$, and I, J and K permute in a natural order)

$$K_{ij}^{IJ} = \int_{\Omega^e} \left(\frac{\partial \phi_i}{\partial x_J}\frac{\partial \phi_j}{\partial x_I} + \gamma \frac{\partial \phi_i}{\partial x_I}\frac{\partial \phi_j}{\partial x_J} \right) \mathrm{d}x\, \mathrm{d}y\, \mathrm{d}z \qquad (I \neq J, x_1 = x, x_2 = y, x_3 = z)$$

$$C_{ij} = \int_{\Omega^e} \left\{ \sum_{N=1}^{3} \left(\phi_i U_N \frac{\partial \phi_j}{\partial x_N} + \frac{\partial \phi_i}{\partial x_N}\frac{\partial \phi_j}{\partial x_N} \right) \right\} \mathrm{d}x\, \mathrm{d}y\, \mathrm{d}z$$

$$F_j^I = -R_A P_R \int_{\Omega^e} \theta q_I^e \phi_j \, \mathrm{d}x\, \mathrm{d}y\, \mathrm{d}z + \int_{I^e} t_I^e \phi_j \, \mathrm{d}s$$

$$Q_j = \int_{\Gamma^e} q^e \phi_j \, \mathrm{d}s. \tag{6.34}$$

It should be pointed out that the coefficient matrices are *not* symmetric; the symmetry is spoiled by the presence of convective terms. Further note that K_{ij}^{IJ} and C_{ij} depend on the velocity components $u_i (i = 1, 2, 3)$ which are not known *a priori*. Also, the right-hand column $\{F\}$ depends on the temperature θ. An iteration technique is required to solve the equations. The following procedure was employed in the present study. At the beginning of the first iteration, the velocity field is set to zero, and the coefficient matrices $[C^e]$ and $[K^e]$ are evaluated (and assembled), and the solution of the assembled equation (6.33) is obtained after imposing the specified temperature boundary conditions of the problem. This solution (for the temperature field) is used to evaluate the column matrices $\{F^e\}$ and the assembled equation (6.32) is solved for the velocity field. The velocity field is then used (for the same R_A and P_R) in the second iteration to evaluate the convective terms of K_{ij}^{IJ} and C_{ij} and the procedure outlined above is repeated until the velocities and temperature computed in two consecutive iterations differ (in r.m.s. value) by, say, one percent. Once convergence is achieved for given values of R_A and R_R, the values are then increased to the next pair of values and the iterative procedure is repeated with the initial guess for the velocity field being the converged solution of the previous set of R_A and R_R values. To accelerate the convergence, a weighted sum of the last and last but one solutions was used:

$$\{\Delta\} = \varepsilon \{\Delta\}^r + (1 - \varepsilon)\{\Delta\}^{r-1} \tag{6.35}$$

where r denotes the iteration number, and ε, $0 \leqslant \varepsilon \leqslant 1$ is an acceleration parameter. A value of $\varepsilon = \frac{2}{3}$ proved to give faster convergence.

One can transfer the nonlinear (convective) terms to the right-hand side of the equation and assume that it is known from the previous iteration. This gives a constant coefficient matrix and saves computational time in recomputing the coefficient matrices during each iteration. However, this procedure is found to result in divergent solutions even for moderately high Reynolds/Rayleigh numbers.

The algebraic complexity and the nonlinear nature of the matrices in equation (6.32) forces one to use numerical integration to evaluate various matrix coefficients. Another reason for the need to use numerical integration is the 'reduced integration' required by the penalty method. The need for the use of reduced integration can be explained in terms of the conditions for the existence and uniqueness of solutions to the discretized penalty problem. The continuity and coercivity requirements (of the existence theorem) on the variational problem are satisfied by the penalty-finite-element model only if the parameter appearing in the coercivity condition is independent of the mesh size; that is, the finite element chosen for the penalty method should be such that this parameter does not depend on the mesh size. It is found that (see Oden[48]) numerical integration of matrix coefficients associated with the penalty functional with one less number of Gaussian points (in each direction) will ensure that the parameter is independent of the mesh size. For additional details on this, the reader is referred to references 29–33.

An alternate explanation of the need for reduced integration is also given here. The penalty finite element equation (6.32) has the form

$$([K_1] + \gamma[K_2])\{\Delta\} = \{F\}. \tag{6.36}$$

As γ is increased to a large value (in an attempt to satisfy the constraint more closely), the magnitude of $[K_1]$ in comparison to $\gamma[K_2]$ becomes negligible in the computer, and we have

$$\gamma[K_2]\{\Delta\} = \{\mathrm{F}\} \quad \text{or} \quad [K_2]\{\Delta\} = \frac{1}{\gamma}\{F\}. \tag{6.37}$$

This implies that as γ is made larger and larger, equations resulting solely from the constraint condition are left and the contributions of conservation of momentum and energy are lost. If $[K_2]$ is invertible, then the solution to equations associated with the constraint condition is obtained. Since $\{F\}/\gamma$ is about zero for a large value of γ, the resulting solution is almost zero. On the other hand, if γ is small, the constraint is not satisfied adequately and the resulting solution is in error. To circumvent this difficulty two things must be done. First, the magnitude of γ must be such that the matrix $[K_1]$ is not negligibly small compared to $[K_2]$. Second, the matrix $[K_2]$ must be singular so that the number of constraint equations is less than the number of unknowns (hence $[K_2]$ cannot be inverted to solve (6.37)). This can be achieved by using reduced integration on the elements of $[K_2]$. In other words, the standard 2×2 Gauss rule can be used to evaluate the elements of $[K_1]$, whereas the 1×1 Gauss rule must be used to evaluate the elements of $[K_2]$. Concerning the value of γ, experience shows that $R_A\times 10^4 \leq \gamma \leq 10^{13}$ gives good accuracy in all problems. The upper bound is a function of the word length in the machine (i.e. computer) being used. For IBM 370/158, in double precision, the upper limit is 10^{13}. The lower bound on γ can be determined and applied at the element level using the element (or cell) Rayleigh number. In this case the penalty parameter is assumed to be piecewise constant.

6.6 NUMERICAL RESULTS

All the results discussed in this paper were obtained on an IBM 370/158 computer using double precision arithmetic. The four-node isoparametric element was used. The example problems discussed herein are taken from the author's previous papers on the subject.[31–33,35,36]

To investigate the influence of the penalty parameter $\gamma_n(=10^n)$ on the solution (u_n, v_n, P_n), the problem of natural convection in a square enclosure (in the presence of a constant thermal gradient across the vertical walls, the other two horizontal walls being insulated; see Figure 6.1) is solved for $R_A = P_R = 1.0$. Table 6.1 shows the convergence of the (maximum) velocities, pressure and stream function with the penalty parameter. As can be seen, the 'accuracy' increases with increasing γ_n. Also, note that the stream function is relatively less sensitive to the penalty parameter. The solution remains unchanged for $\gamma_n = 10^6$, 10^8, 10^{10}. For $\gamma_n > 10^{10}$, the round-off errors in the computer gradually increased, and for $\gamma_n = 5\times 10^{13}$ the coefficient matrix became singular. Figure 6.2 shows a plot of $\log_{10}|u_n - u_8|$ versus $\log \gamma_n$. All slopes are equal to 1 (before $|u_n - u_8|$ becomes zero).

Table 6.1 Convergence of the solution with the penalty parameter ($x = 0$).[38]

γ_n	$u(0.22, 0.5)\times 10^3$	$-v(0.5, 0.22)\times 10^3)$	$\psi(0.5, 0.5)\times 10^3$	P (*element* 28)
1	6.1957	1.5565	1.2824	0.8781
10	4.8646	3.1060	1.2492	0.1552
10^2	4.1449	3.9074	1.2374	0.1233
10^3	4.0387	4.0243	1.2359	0.1070
10^4	4.0276	4.0365	1.2357	0.1050
10^6	4.0264	4.0379	1.2357	0.1048
10^8	4.0263	4.0379	1.2357	0.1048
10^{10}	4.0264	4.0379	1.2357	0.1048
10^{12}	4.0269	4.0387	1.2359	0.1068
10^{13}	3.8141	3.8200	1.1745	0.1523
0.5×10^{13}	* * * *	Zero appeared on the diagonal		

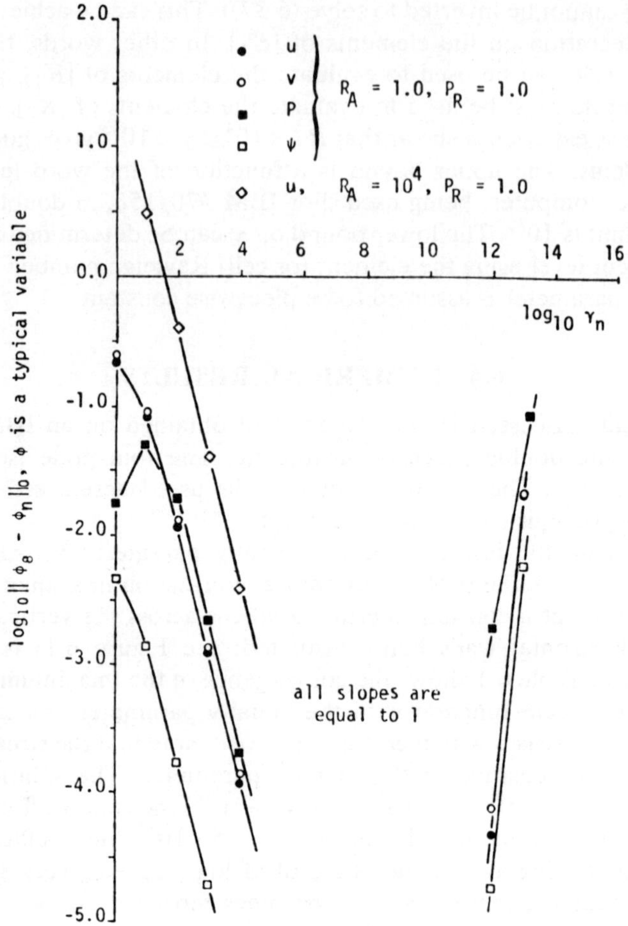

Figure 6.2 Accuracy (i.e., error estimates) of the penalty finite element solutions of natural convection in a square cavity

The effect of the Rayleigh number (for fixed Prandtl number, $P_R = 1.0$) on the velocity and temperature fields was investigated using, in the y–z plane, a 10×10 uniform mesh for $R_A = 10^3$ and 10^4, and 12×12 nonuniform mesh for $R_A = 10^5$ and 10^6. The boundary conditions are given by

$$y = 0,1: \quad \theta = y - 0.5, \quad u_2 = u_3 = 0$$

$$z = 0,1: \quad \partial\theta/\partial z = 0, \quad u_2 = u_3 = 0.$$

Figure 6.3 shows the isotherms and streamlines obtained by the penalty finite element model for $R_A = 10^4$, 10^5 and 10^6. The numerical results for the

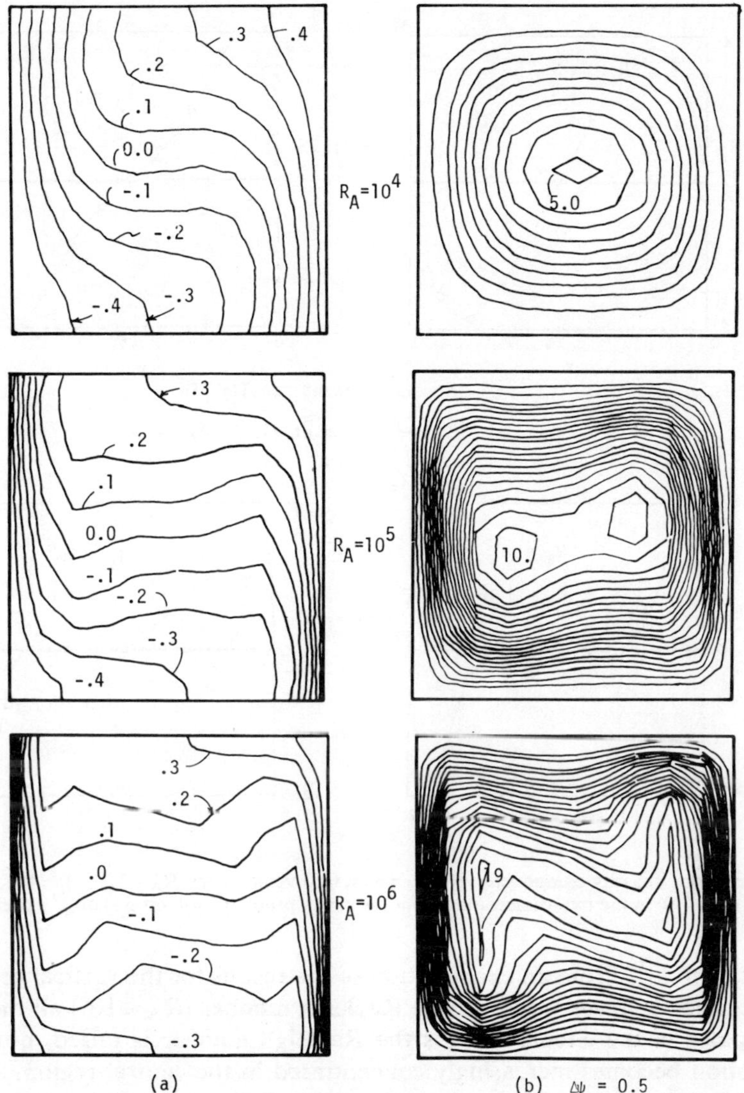

Figure 6.3 Isotherms and isostreams for $R_A = 10^6$, $P_R = 1$ by penalty formulation (mesh: 12×12): (a) isotherms; (b) isostreams

velocity component $u_3 = w(y, 0.5)$ and temperature $\theta(y, 0.5)$ for $R_A = 10^3$, 10^4, 10^5 and 10^6 ($P_R = 1.0$) are compared in Figures 6.4 and 6.5, respectively, with those obtained by Upson *et al.*,[28] who employed 168 quadratic isoparametric finite elements (with 745 nodes). Despite relatively coarse meshes used in the present study, the results compare very well with those

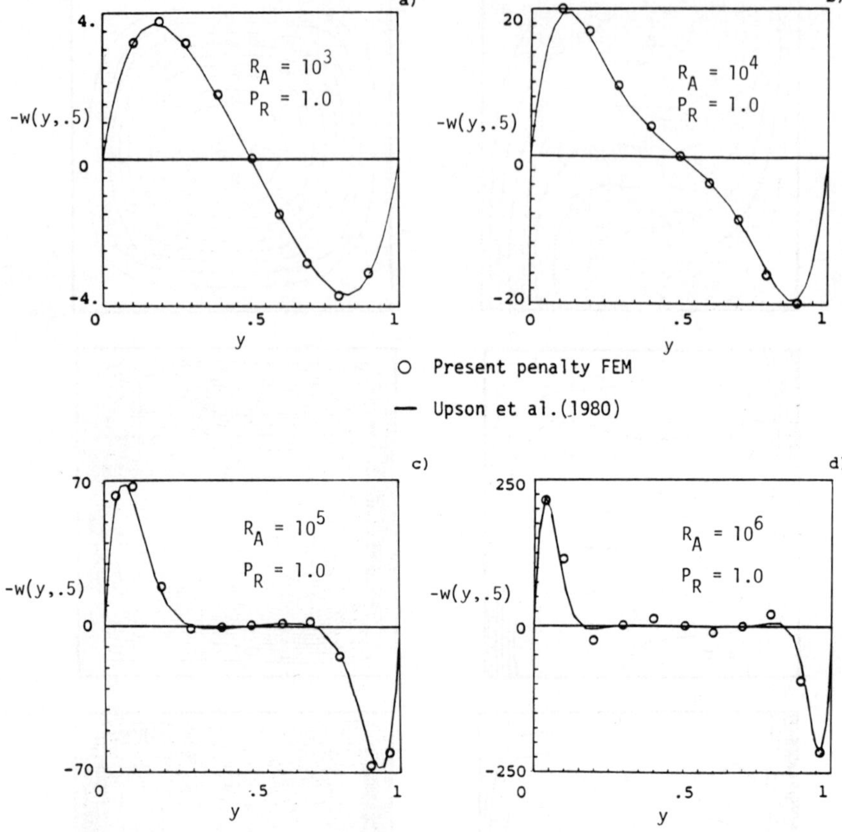

Figure 6.4 Velocity component $w(y, z) = w(y, 0.5)$ versus y for $R_A = 10^3$, 10^4, 10^5, and 10^6 ($P_R = 1.0$) in the two-dimensional window cavity problem (—Upson *et al.*,[28] ∘ present)

of Upson *et al.*[28] From an examination of the results for the vertical velocity, we note that at lower values of the Rayleigh number ($R_A \leq 10^3$) all the fluid participates in the circulation. As the Rayleigh number is raised, however, the motion becomes increasingly concentrated in the central region. A plot of the Nusselt number (a measure of heat transfer across a surface),

$$N_u = \int_0^1 \frac{\partial \theta}{\partial y} \, dz,$$

as a function of z (along the vertical wall) is shown in Figure 6.6. Again, the present results are in good agreement with the results obtained by Upson *et al.*[28] for Rayleigh numbers up to 10^6. For $R_A = 10^6$ the present results start deviating, indicating that the mesh used is too coarse to accurately predict the behaviour for $R_A \geq 10^6$.

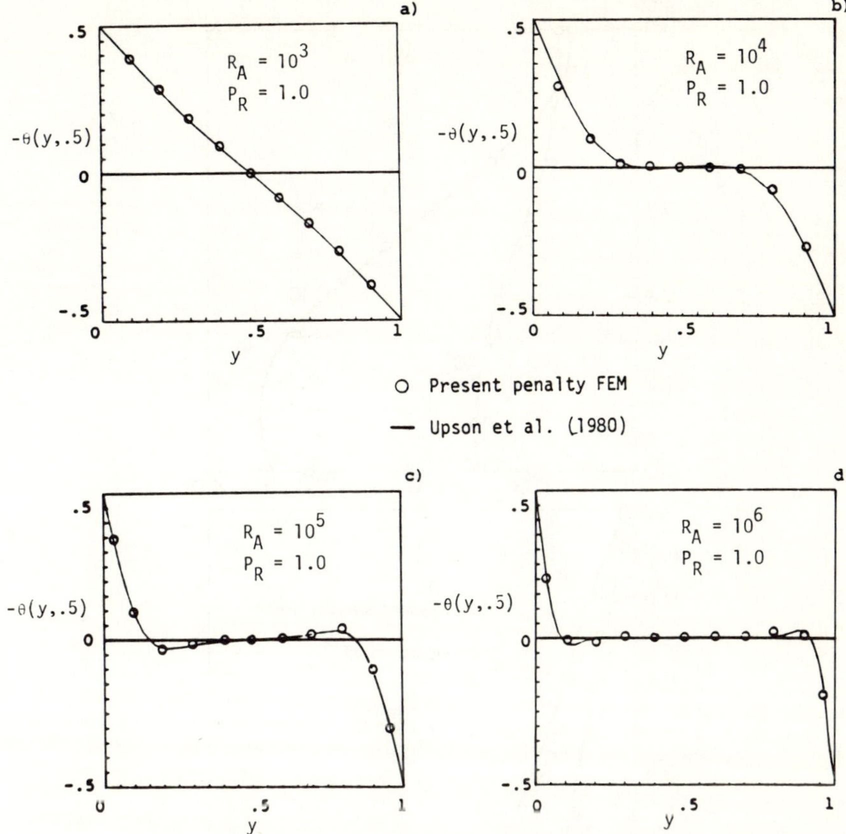

Figure 6.5 Variation of the temperature $\theta(y, z) = \theta(y, 0.5)$ for $R_A = 10^3$, 10^4, 10^5, and 10^6 ($P_R = 1.0$) in the two-dimensional window cavity problem (—Upson *et al.*[28], ◦ present FEM)

The effect of the Prandtl number (which depends only on the fluid properties) on heat transfer (across the cold wall) was also investigated for the square cavity problem. The Nusselt number, stream function (ϕ) and vorticity (ζ) values at the centre of the cavity, and the number of iterations required for convergence of the solution (convergence tolerance, 10^{-4}) are given in Table 6.2. When the Prandtl number is greater than or equal to unity, the heat transfer is virtually constant. For lower values of the Prandtl number ($P_R < 1$), heat transfer is reduced with increasing Prandtl number. This is probably because the boundary layers do not form and the flow becomes increasingly less viscous. This can also be seen from the plots of streamlines and vorticity lines shown in Figure 6.7 (for $R_A = 10^4$ and three different Prandtl numbers, $P_R = 10^2$, 1, and 10^{-2}). The effect of the Prandtl number on the computational

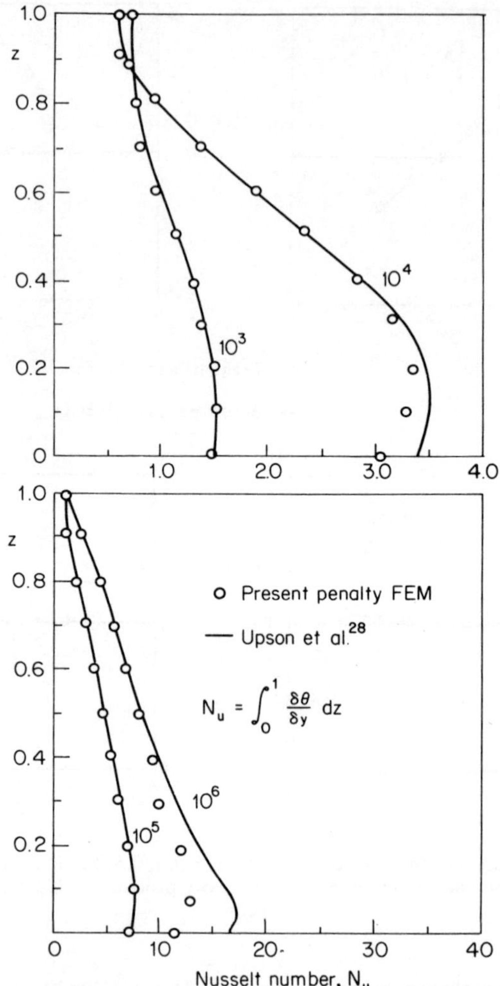

Figure 6.6 Variation of the Nusselt number along the vertical wall, z, for $R_A = 10^3$, 10^4, 10^5, and 10^6 in the two-dimensional window cavity problem ($P_R = 1.0$)

time is apparent from the results in Table 6.2. For Prandtl numbers less than unity, the number of iterations required for convergence

$$\left(\sum_i |\Delta_i^{r+1} - \Delta_i^r|^2 \leqslant 10^{-4}, \qquad r = \text{iteration number} \right)$$

increases with decreasing values of the Prandtl number.

Table 6.2 The effect of the Prandtl number on the Nusselt number, stream function and vorticity values, and number of iterations required for convergence (n) for the two-diemnsional cavity problem ($R_A = 10^3$ and 10^4)

R_A	P_R	n†	N_u	ψ	ρ
	10^{-2}	11	1.0992	1.1386	−31.26
	10^{-1}	9	1.1558	1.1561	−31.41
10^3	1	8	1.1666	1.1581	−31.31
			(1.14)†	(1.18)‡	
	10	8	1.1666	1.1581†	
	10^2	8	1.1666	1.1581	−31.32
	10^2	20	1.9993	5.0128	−96.32
	10^{-1}	20	2.0390	5.0403	−101.5
10^4	1	16	2.1318	5.1474	−103.1
			(2.49)†	(5.13)‡	
	10	17	2.1440	5.2016	−106.9
	10^2	17	2.1442	5.2070	−107.3

† convergence tolerance, 10^{-4}.
‡ values from reference 29, wherein a 4×4 mesh of 9-node rectangular elements was used.

Numerical results for aspect ratios (height to width) of the cavity other than unity were also obtained. The isotherms and streamlines for a rectangular enclosure of aspect ratio 3, with flow properties same as those used by Hellums and Churchill,[49] $R_A = 1.466\times10^4$, $P_R = 0.733$, are shown in Figure 6.8. The present results agree qualitatively with those of Hellums and Churchill. Similar results for a rectangular enclosure of aspect ratio 1.83, $R_A = 8200$, $P_R = 2450$ and with a linear temperature distribution on the horizontal walls are shown in Figure 6.9. The data of the problem is taken from the paper by Szekely and Todd,[50] who have computed experimental and finite difference solutions of the problem. The plotted values of the steady-state isotherms agree well with those of Szekely and Todd.[50] The present results for an aspect ratio of 16, $R_A = 10^5$, and $P_R = 1$ (with insulated horizontal walls) are shown in Figure 6.10. This problem was studied experimentally by Elder[51] for a slightly different Rayleigh number ($R_A = 3\times10^5$). A symmetric (about the centre) but nonuniform mesh of 24×14 linear elements was used. The isotherms and streamlines shown in Figure 6.10 are those obtained at the end of 60 iterations, with the error (in the velocity field) between the last two iterations being less than 10^{-3}. These results agree qualitatively with those of Elder.[51] Note that (for $R_A = 10^5$ and aspect ratio 16) a breakdown of the single convection cell extending full height of the slot into a number of shorter cells is produced in the core region. In each of the cells, the fluid is travelling up the hot wall and down the cold one. Due to convergence problems in the computation, it was not possible to investigate the effect of further increases of the Rayleigh number on the flow field.

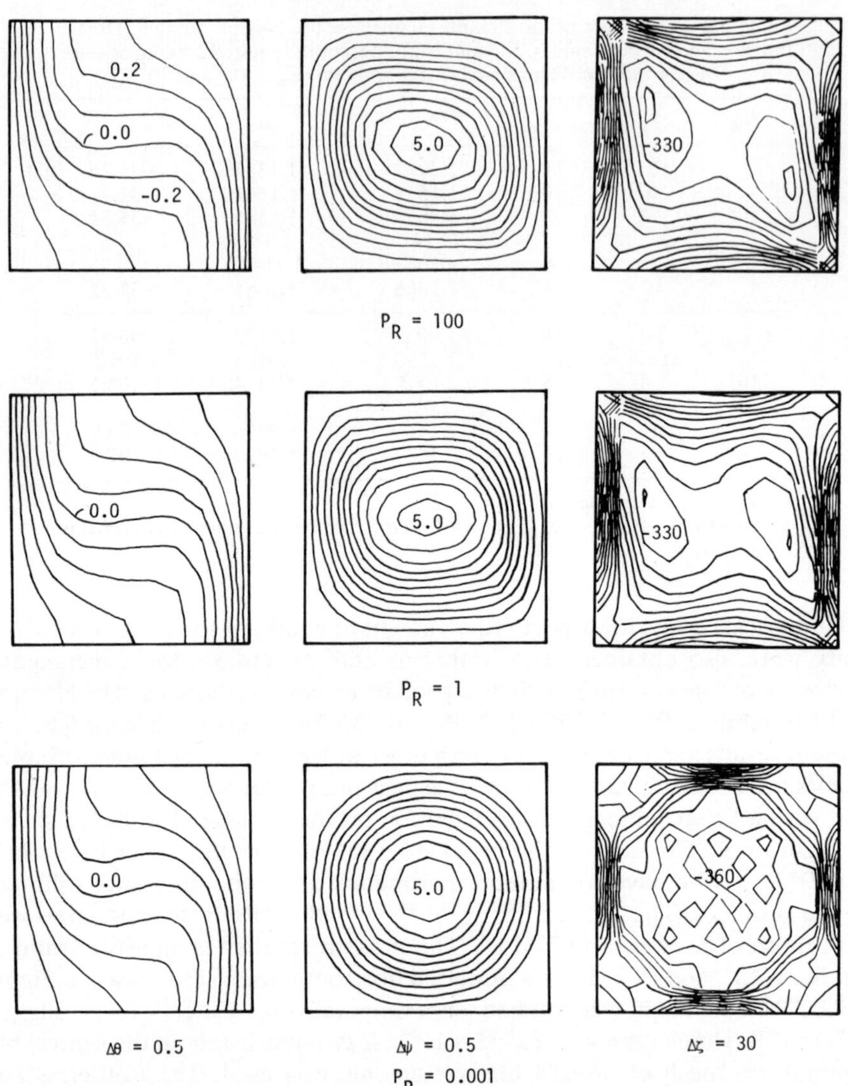

Figure 6.7 Isotherms, isostreams, and equivorticity lines for various Prandtl numbers by stream function–vorticity formulation ($R_A = 10^4$): (a) isotherms; (b) isostreams, and (c) equivorticity lines

Numerical results for some nonrectangular enclosures are presented next. Plots of the isotherms and streamlines for the natural convection ($R_A = 10^5$, $P_R = 1$) in a nonrectangular enclosure (see Figure 6.11 for the geometry, mesh, and boundary conditions) are shown in Figure 6.12. For $R_A = 10^5$, the flow is stratified parallel to the boundaries.

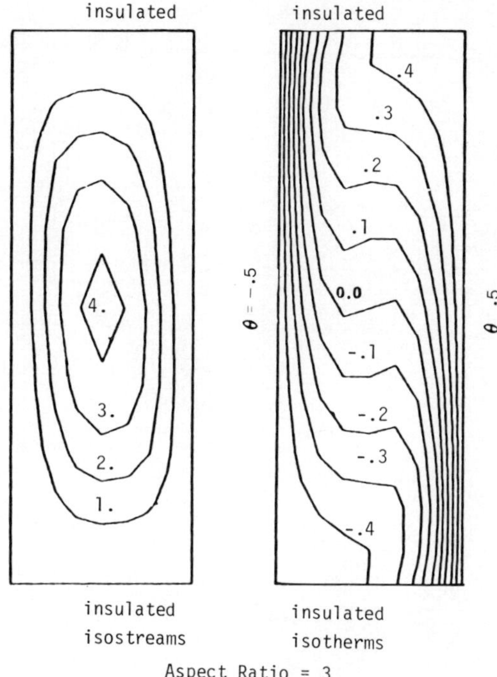

Figure 6.8 Isotherms and stream lines for $R_A = 14\,660$, $P_R = 0.733$

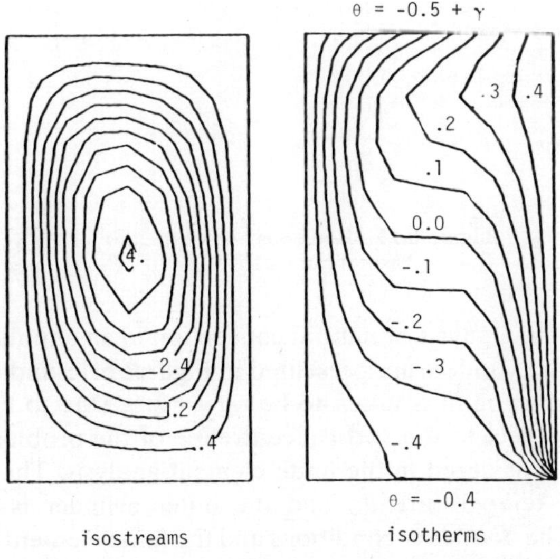

Figure 6.9 Isotherms and stream lines for $R_A = 8\,200$, $P_R = 2\,450$

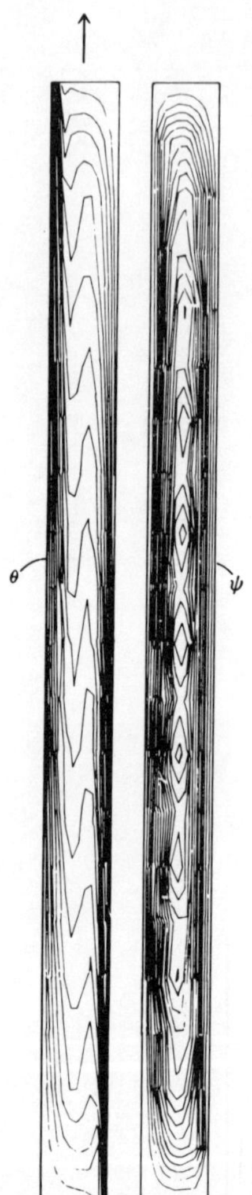

Figure 6.10 Isotherms and isostreams for the vertical slot problem (aspect ratio = 16, $R_A = 10^5$, $P_R = 1$, mesh: 24×14)

The results of the analysis of natural convection in an annulus between two concentric circular cylinders are presented in Figures 6.13 and 6.14. The ratio of the outer to inner radii is taken to be $r_o/r_i = 2.6$. Due to the existence of symmetry with respect to the vertical centreline of the problem, only half of the annulus was considered in the finite element analysis. The inner cylinder is maintained at temperature θ_h, and the outer cylinder is maintained at temperature θ_c. The boundary conditions and the finite element mesh (14×24) for the concentric cylinders are shown in Figure 6.13 (the first row of elements along the inner wall are not clear in the figure). Two Rayleigh numbers,

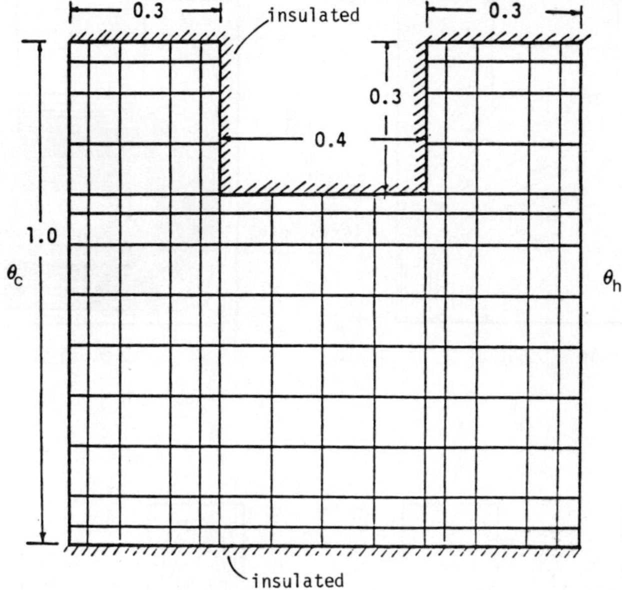

Figure 6.11 Finite element mesh for a nonrectangular cavity

$R_A = 10^3$ and 5×10^4, with $P_R = 0.706$, for which experimental results are available (see Kuehn and Goldstein[52]), were used in the present study. Due to convergence problems at $R_A = 5 \times 10^4$, which can be attributed to the coarse mesh used in the study, results are presented for $R_A = 4.7 \times 10^4$. Plots of streamlines and isotherms at Rayleigh numbers of 10^3 and 4.7×10^4 are shown in Figures 6.14(a) and 14(b), respectively. For $R_A = 10^3$, the streamlines obtained in the present study compare well visually with those of Kuehn and Goldstein; however, the isotherms show some plume behaviour compared to the essentially 'conduction' patterns of Kuehn and Goldstein.[52] For $R_A = 4.7 \times 10^4$, the streamlines obtained in the present work are somewhat different from those of Kuehn and Goldstein,[52] although the vorticity centre compares very well. The isotherms in this case have the same form as those of Kuehn and Goldstein[52] except for the rounding/sharpening of the corners.

Plots of the velocity and temperature along the horizontal cross-section through the centre of the annulus are shown in figures 6.15 and 6.16 respectively. The secondary (reverse) flow observed for $R_A = 4.7 \times 10^4$ in Figure 6.15 is apparently not seen in the streamline plot (Figure 6.14). This is because the negative velocity is very small and the streamline corresponding to the reverse flow has a magnitude smaller than the increment used in the plot.

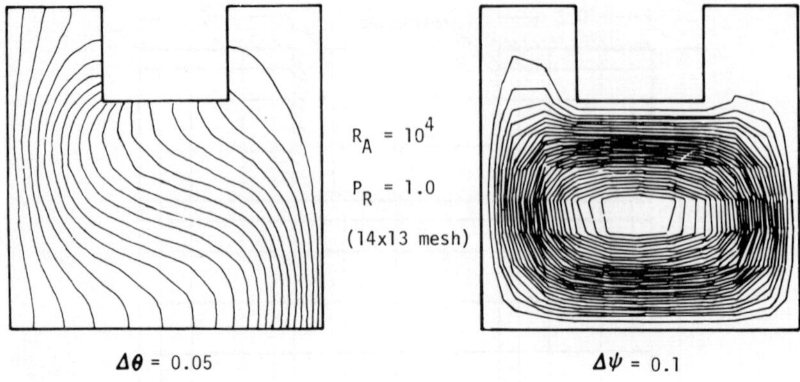

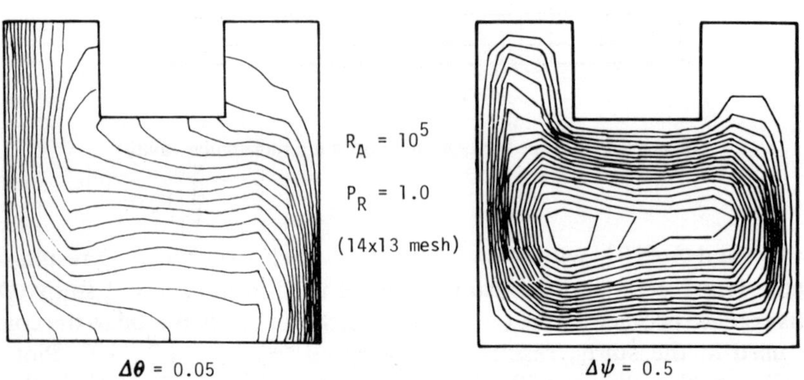

Figure 6.12 Temperautre and stream function for non-square enclosure by the penalty finite element model (a) temperature; (b) stream function

In Figure 6.17, a plot of the local equivalent conductivity, K_{eq}, versus the angular distance θ is shown (for $R_A = 10^3$ and 4.7×10^4). Here K_{eq} is given by

$$K_{eq} = N_u(\alpha) \ln (r_0/r_i), \qquad N_u(\alpha) = \frac{\partial \theta}{\partial n}(\alpha),$$

where N_u is the Nusselt number. The Nusselt number was calculated at the nodes closest to the walls. The present result is compared with the analytical solution of Jischke and Farschi[53] and the numerical solution of Kuehn and Goldstein[52]. It whould be pointed out that the solution in Reference[53] is based on an infinite Prandtl number, and is calculated on the inner wall. Kuehn and Goldstein[52] used a rather coarse finite difference mesh.

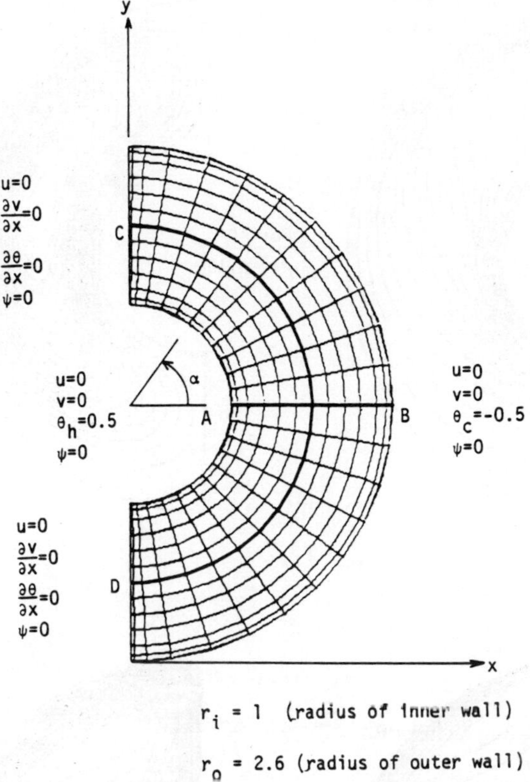

Figure 6.13 Finite elements and boundary conditions for circular annulus (375 meshes and 336 nodes)

A discussion of the results of the three-dimensional window cavity (natural convection) problem are presented next. Although it is convenient to discuss the quantitative results (i.e., in tabular and graphical form), the form of the solutions (or flow features) is of primary interest. The quantitative results, although reliable, are of declining quality with increasing Rayleigh numbers. Due to its demand on computer storage and time, the exploration of the effects of further mesh refinement on the truncation errors was not feasible.

The variations of velocity components $\bar{w} = (w \times 10^3/R_A)$ as a function of y and $\bar{v} = (v \times 10^3/R_A)$ as a function of z are shown in Figure 6.18 for $R_A = 10^3$ and 10^4 ($P_R = 1.0$). A comparison of the results of the two-dimensional model (see Figure 6.3) with those of the three-dimensional model (see Figure 6.18) indicates that the dimensionality (i.e., fixed wall at $x = 1.0$) has the effect of reducing the strength of the flow. In other words, the angular

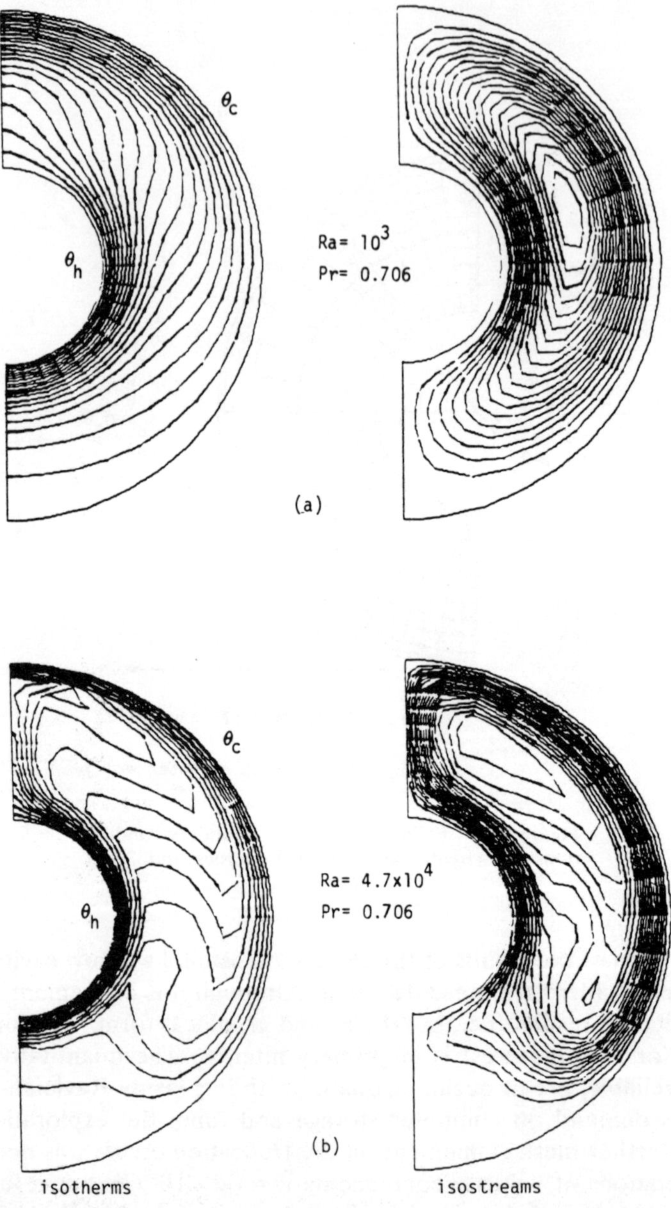

Figure 6.14 Isotherms and isostreams for (a) $R_A = 10^3$ and (b) $R_A = 4.7 \times 10^4$

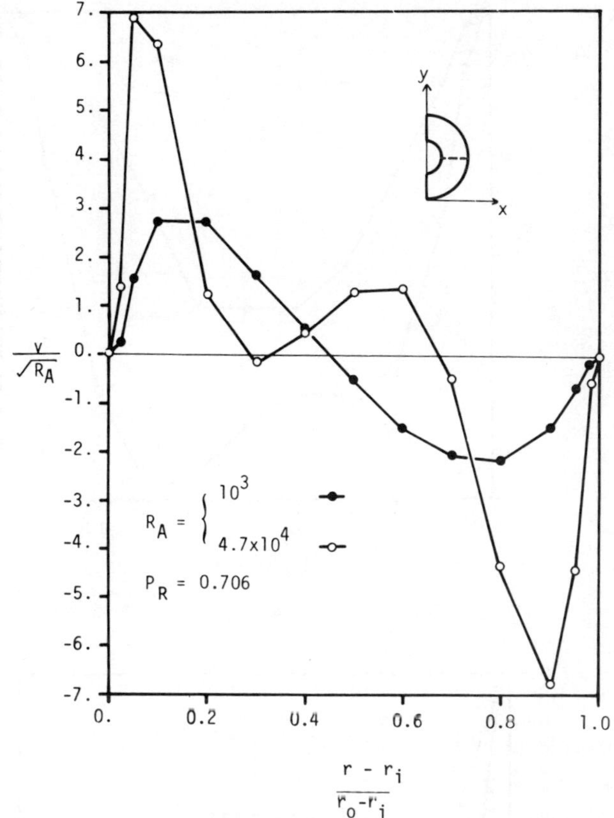

Figure 6.15 Vertical velocity distribution along horizontal cross section

velocity of the 'fluid cylinder' is reduced by the kinematical interaction of the rotating fluid (about an axis parallel to the x-axis) with the stationary walls.

The plots of the temperature variation as a function of y are shown in Figure 6.19. Near the $x = 0$ and $x = 1$ boundaries, the fluid moves more slowly than it does far from the boundary, and the convection of heat is reduced, which in turn produces temperature gradients in the x-direction.

In Figure 6.20 plots of the pressure (computed from the Poisson's equation for pressure) as a function of y and z are shown for $R_A = 10^4$ and $P_R = 1.0$. The pressure was assumed to be zero along $y = 0.5$ and $z = 0$ (boundary conditions on the pressure). The maximum pressure is obtained at $y = 0.5$ and $z = 0.5$ on the plane of symmetry.

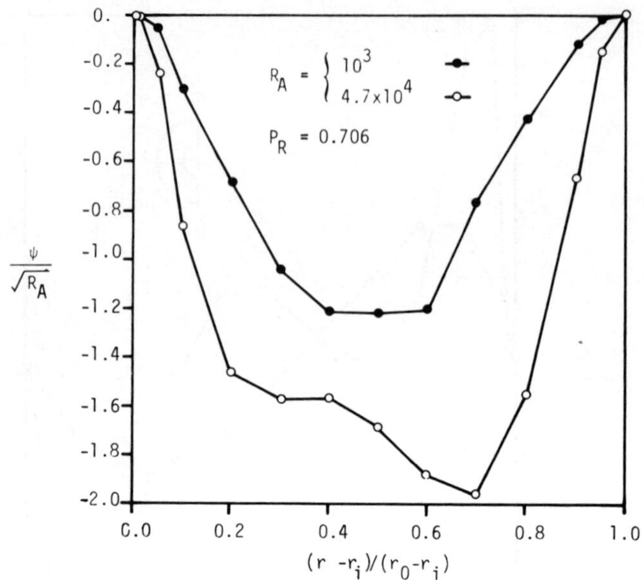

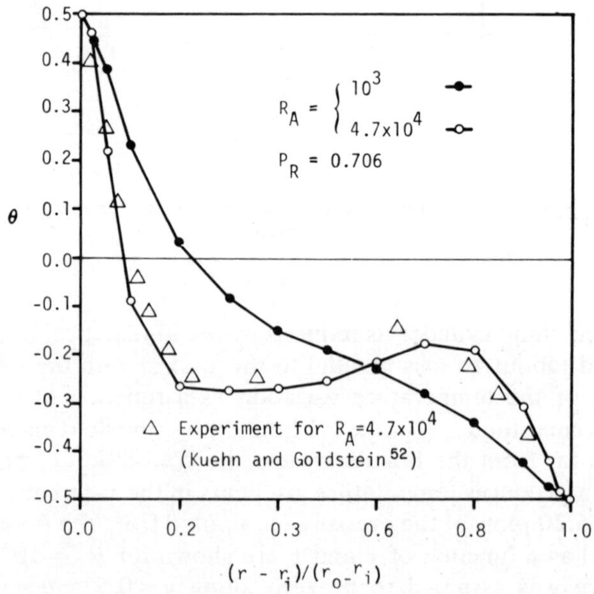

Figure 6.16 (a) Stream function along horizontal cross section; (b) temperature distribution along horizontal cross section

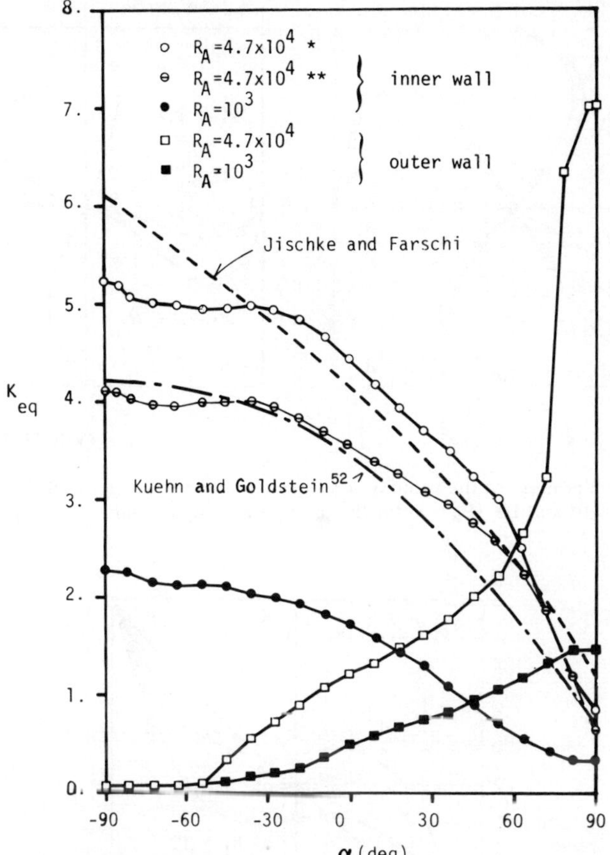

Figure 6.17 Heat transfer along inner and outer walls
* Heat transfer is calculated using temperature at $r = 1.008$
** Heat transfer is calculated using temperature at $r = 1.16$

$$K_{eq} = N_u(\theta) \ln (r_O/r_i)$$

$$N_u(\theta) = \left.\frac{\partial T}{\partial n}\right|_{\text{at}\,\alpha} \quad \text{(normal gradient)}$$

6.7 SUMMARY AND CONCLUSIONS

A finite element analysis based on the penalty function method is presented for the solution of the Navier–Stokes equations and the energy equation governing steady laminar motion of a viscous incompressible fluid. Numerical results are presented for the two-dimensional window cavity ($R_A = 10^3$, 10^4, 10^5, 10^6) and three-dimensional window cavity ($R_A = 10^3$ and 10^4) problems

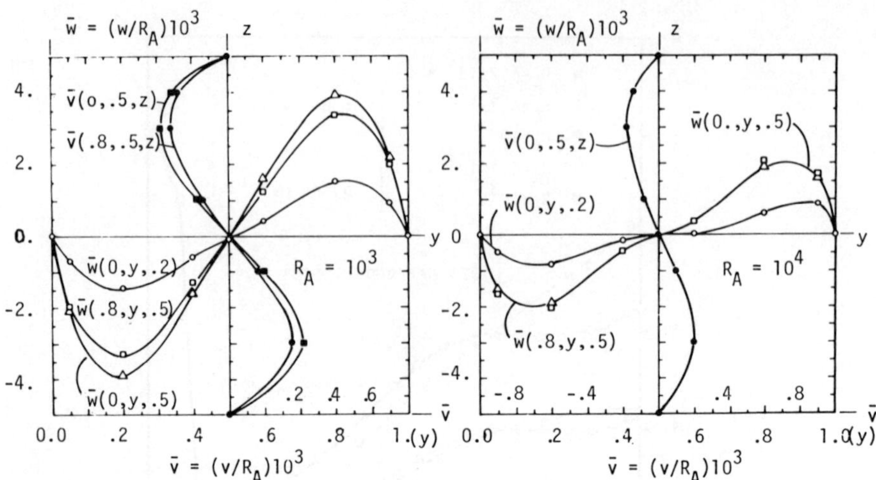

Figure 6.18 Variation of the velocity components w and v along y and z, respectively, for $R_A = 10^3$ and 10^4 ($P_R = 1$) for the three-dimensional window cavity problem

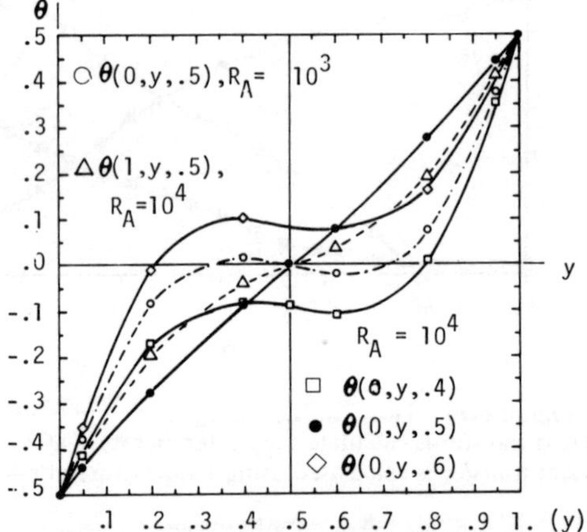

Figure 6.19 Variation of the temperature at $x = 0$ and $x = 1$ for various values of z, with y for $R_A = 10^3$ and 10^4 ($P_R = 1$) for three-dimensional window cavity problem

and a concentric cylindrical annulus problem. From the three-dimensional simulation of the flows, it is concluded that the fixed wall in the three-dimensional cavity has the (kinematic as well as thermal) effect of reducing the strength of the flow field. Due to the lack of quantitative results on the three-dimensional window cavity problem in the open literature, comparison of the results to assess the quality is not possible. However, it is noted that

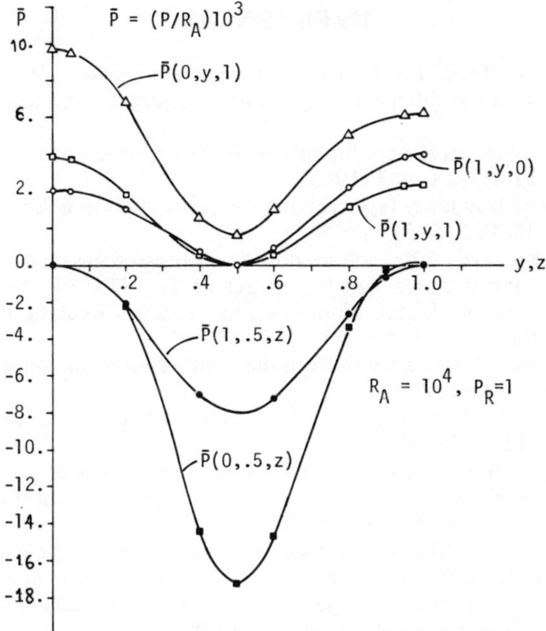

Figure 6.20 Variation of the pressure with respect to y and z for $R_A = 10^4$ ($P_R = 1$) for three-dimensional window cavity problem

the global behaviour of solutions predicted by the penalty finite element model is very accurate despite the use of coarse meshes.

In order to analyse higher Rayleigh number flows, mesh refinements must be made. Toward this end it is more practical to use out-of-core solution methods to meet the storage demands. Solution of practically important convection flow problems (especially with complicated geometry) requires large amounts of storage and time, and without the power of a CRAY-1 (or equivalent) it is not possible for researchers in academic institutions (which do not have computational facilities that match CRAY-1) to perform any detailed investigations such as those mentioned above.

ACKNOWLEDGEMENTS

Many of the results of two-dimensional analyses presented herein are taken from papers co-authored by the author with Dr. Akio Satake and Mr. D. R. Mamidi. The author is thankful to both of them for the computational assistance. It is also a pleasure to acknowledge the help of Mr. Richard Tam in proof-reading the manuscript and Mrs. Vanessa McCoy for typing it.

REFERENCES

1. G. K. Batchelor, 'Heat transfer by free convection across a closed cavity between vertical boundaries at different temperature', *Quart. J. App. Math.*, **12**, 209–233 (1954).
2. J. W. Elder, 'Numerical experiments with free convection in a vertical slot', *J. Fluid Mech.*, **24**(4), 823–843 (1965).
3. A. E. Gill, 'The boundary layer regime for convection in a rectangular cavity', *J. Fluid Mech.*, **26**(3), 515–536 (1966).
4. B. D. Pedersen, *et al.*, 'Development of a compressed gas-insulated transmission line', *I.E.E.E. Winter Power Meeting*, paper 71 TP 193 PWR (1971).
5. J. E. Hart, 'Stability of Thin Non-Rotating Hadley Circulation', *J. Atmos. Sci.*, **29**, 687 (1972).
6. J. P. Carruthers, 'Crystal growth from the melt', *Treatise on Solid State Chemistry*, Vol. 5, Plenum Press, 325 (1975).
7. J. Hiddink, *et al.*, 'Natural convection heating of liquids in closed containers', *App. Sci. res.*, **32**, 217 (1976).
8. B. S. Petuklov, 'Actual problems of heat transfer in nuclear power engineering', *Int. Seminar on Future Energy Production*, Hemisphere Publishing Corporation, Washington, D.C., pp. 151–163 (1976).
9. J. C. Chato and R. S. Abdulhadi, 'Flow and Heat Transfer in Convectively Cooled Underground Electric Cable Systems: Part 1—Velocity Distributions and Pressure Drop Correlations; Part 2—Temperature Distributions and heat Transfer Correlations', *J. Heat Transfer*, 30–40 (February 1978).
10. A. J. Chorin, 'Numerical solutions of the Navier–Stokes equations', *Mathematics of Computation*, **22**, 745 (1968).
11. G. P. Williams, 'Numerical integration of the three-dimensional Navier–Stokes equations for incompressible flows', *J. Fluid Mech.* **10**, 727 (1969).
12. K. Aziz, and J. D. Hellums, 'Numerical solution of the three-dimensional equations of motion for laminar natural convection', *The Physics of Fluids*, **10**(2), 314–324 (1967).
13. G. D. Mallinson and G. de Vahl Davis, 'The method of the false transient for the solution of coupled elliptic equations', *J. Comp. Phys.*, **12**, 435–461 (1973).
14. G. de Vahl Davis and G. D. Mallinson, 'An evaluation of upwind and central difference approximations by a study of recirculating flow', *Computers and Fluids*, **4**, 29–43 (1976).
15. G. D. Mallinson and G. de Vahl Davis, 'Three-dimensional natural convection in a box: a numerical study', *J. Fluid Mech.*, **83**(1), 1–31 (1977).
16. D. F. Roscoe, 'The solution of the three-dimensional Navier–Stokes equations using a new finite difference approach', *Int. J. Num. Meth. Eng.*, **10**, 1299–1308 (1976).
17. D. N. de G. Allen and R. V. Southwell, 'Relaxation methods applied to determine the motion, in two dimensions, of viscous fluid: Part a—Fixed Cylinder', *Quart. J. Mech. and App. Math.*, **8**, 128 (1955).
18. S. M. Richardson and A. R. H. Cornish, 'Solution of three-dimensional incompressible flow problems', *J. Fluid Mech.*, **83**, 309 (1977).
19. S. C. R. Dennis, D. B. Ingham and R. N. Cook, 'Finite difference methods for calculating steady incompressible flows in three dimensions', *J. Comp. Phys.*, **33**, 325–339 (1979).
20. S. C. R. Dennis and J. D. Hudson 'A difference method for solving the Navier–Stokes equations', *Proc. Int. Conf. Num. Meth. Laminar and Turbulent Flows*, Swansea, U.K., Pentech Press, London (1978).

21. M. Nallasamy and K. K. Prasad, 'Numerical studies on quasilinear and linear elliptic equations', *J. Comp. Phys.*, **15**, 429 (1974).
22. D. B. Spalding, 'A novel finite-difference formulation for differential expressions involving both first and second derivatives', *Int. J. Num. Meth. Eng.*, **4**, 551 (1972).
23. R. K. Agarwal, 'A third-order-accurate upwind scheme for Navier–Stokes solutions at high Reynolds numbers', *AIAA Paper No. 81-0112* (1981).
24. R. K. Agarwal, 'A third-order-accurate upwind scheme for Navier–Stokes solutions in three dimensions', McDonnell Douglas Research Laboratories, Report No. MDRL 81–20; *ASME Winter Annual Meeting*, Washington, D.C. (1981).
25. A. Ghosh, C. W. Mastin, and J. F. Thompson, 'Locally one-dimensional scheme for the solution of three-dimensional Navier–Stokes equations', in *Computers in Flow Predictions and Fluid Dynamics Experiments*, (ed. K. N. Ghia, *et al.*) American Society of Mechanical Engineers, New York, pp. 3–10 (1981).
26. B. Tabarrok and R. C. Lin, 'Finite element analysis of free convection flows', *Int. J. Heat Mass Transfer*, **20**, 945 (1977).
27. D. K. Gartling, 'Convective heat transfer analysis by the finite element method', *Computer Meth. Appl. Mech. Eng.*, **12**, 365–382 (1977).
28. C. D. Upson, P. M. Gresho, and R. L. Lee, 'Finite element simulations of thermally induced convection in a closed cavity', *Informal Report*, UCID-18602, Lawrence Livermore National Laboratory, Livermore, CA (1979).
29. J. C. Heinrich, R. S. Marshall, and O. C. Zienkiewicz, 'Penalty function solution of coupled convective and conductive heat transfer', *Int. Conf. Numer. meth. in Laminar and Turbulent Flow*, Swansea (1978).
30. R. S. Marshall, J. C. Heinrich and O. C. Zienkiewicz, 'Natural convection in a square enclosure by a finite-element penalty function method using primitive fluid variables', *Num. Heat Transfer*, **1**, 315–330 (1978).
31. J. N. Reddy and D. R. Mamidi, 'Penalty velocity-stream function finite element models for free convection heat transfer problems', in *Recent Advances in Engineering Science*, (ed. R. L. Sierakowski) University of Florida, Gainesville, pp. 381–386 (1978).
32. J. N. Reddy, 'Penalty finite element methods for the solution of advection and free convection flows', *Finite Element Methods in Engineering*, (ed. A. P. Kabaila and V. A. Pulmano). The University of New South Wales, Sydney, Australia, pp. 583–598 (1979).
33. J. N. Reddy and A. Satake, 'A comparison of various finite element models of natural convection in enclosures', *J. Heat Transfer*, **102**, 659–666 (1980).
34. P. M. Gresho, S. T. K. Chan, R. L. Lee and C. D. Upson, 'Solution of the time-dependent, three-dimensional incompressible Navier–Stokes equations via FEM', *Proc. Int. Conf. Num. Meth. for Laminar and Turbulent Flow*, (ed. C. Taylor and B. A. Shreffler) Pineridge Press, Swansea, UK, pp. 27–39 (1981).
35. J. N. Reddy, 'Finite-element simulation of natural convection in three-dimensional enclosures', *AIAA/ASME 3rd Joint Thermophysics, Fluids, Plasma and Heat Transfer Conference*, St. Louis, Mo. (1982).
36. J. N. Reddy, 'Penalty-finite-element analysis of 3-D Navier–Stokes equations', *Computer Meth. App. Mech. and Eng.* **35**, 87–106 (1982).
37. R. Courant, *Calculus of Variations and Supplementary Notes and Exercise*, revised and amended by J. Moser, New York University, New York (1956).
38. J. N. Reddy, 'On penalty function methods in the finite-element analysis of flow problems', *Int. J. Num. Meth. Fluids*, **2**, 151–171 (1982).
39. I. Babuska, 'The finite element method with penalty', *Technical Note BN-710*, The Institute for Fluid Dynamics and Applied Mathematics, University of Maryland (August 1971).

40. B. T. Polya, 'The convergence rate of the penalty function method', *Zh. vychisl. Mat. mat. fiz.*, **11**, 3–11 (1971); English translation: *U.S.S.R. Computational Mathematics and Mathematical Physics*, **11**, 1–12 (1971).
41. J. N. Reddy, 'On the accuracy and existence of solutions to primitive variable models of viscous incompressible fluids', *Int. J. Eng. Sci.*, **16**, 921–929 (1978).
42. M. Bercovier, 'Perturbations of mixed variational problems, applications to mixed finite element methods', *R.A.I.R.O., Analyze Numerique/Numerical Analysis*, **12**, 211–236 (1978).
43. J. N. Reddy, 'On the finite element method with penalty for incompressible fluid flow problems', in *The Mathematics of Finite Elements and Applications III*, (ed. J. R. Whiteman) Academic Press, New York, pp. 227–235 (1979).
44. J. N. Reddy, 'On the mathematical theory of penalty-finite elements for Navier–Stokes equations', *Proc. Third Int. Conf. Finite Elements in Flow Problems*, June 10–13, Banff, Alberta, Canada (1980).
45. O. C. Zienkiewicz, ''Constrained variational principles and penalty analysis function methods in finite elements', *Lecture Notes on Mathematics*, Springer-Verlag, Berlin (1973).
46. O. C. Zienkiewicz, R. L. Taylor and J. M. Too, 'Reduced integration techniques in general analysis of plates and shells', *Int. J. Num. Meth. Eng.*, **3**, 275–290 (1971).
47. O. C. Zienkiewicz and P. N. Godbole, 'Viscous, incompressible flow with special reference to non-Newtonian plastic fluids', in *Finite Elements in Fluids*, Vol. 1, (ed. R. H. Gallagher, J. T. Oden, C. Taylor and O. C. Zienkiewicz) Wiley–Interscience, London, pp. 25–55 (1975).
48. J. T. Oden, 'Penalty methods and selective reduced integration for Stokesian flows', *Thirs International Conference on Finite Elements in Flow Problems*, Banff Center, Alberta, June 10–13 (1980).
49. J. D. Hellums and S. W. Churchilll, 'Transient and steady state, free and natural convection, numerical solutions', *A.I.Ch.E. J.*, **8**, 690–695 (1962).
50. J. Szekely and M. R. Todd, 'Natural convection in a rectangular cavity: transient behavior and two phase systems in laminar flow', *Int. J. Heat Mass Transfer*, **14**, 467–482 (1971).
51. J. W. Elder, 'Laminar free convection in a vertical slot', *J. Fluid Mech.*, **23**, 77–98 (1965).
52. T. H. Kuehn and R. J. Goldstein, 'An experimental and theoretical study of natural convection in the annulus between horizontal concentric cylinders', *J. Fluid Mech.*, **74**(4), 695–719 (1976).
53. M. C. Jischke and M. Farschi, 'Boundary layer regime for laminar free convection between horizontal circular cylinders', *J. Heat Transfer*, **102**, 228–235 (1980).

Numerical Methods in Heat Transfer, Volume II
Edited by R. W. Lewis, K. Morgan, and B. A. Schrefler

Chapter 7

Predictions of Laminar Natural Convection in Heated Cavities

Keith H. Winters

ABSTRACT

This chapter discusses, with illustrative calculations, several examples of laminar, natural convection in heated cavities. These include convection in the following types of cavity: a square cavity at high Rayleigh number; a narrow cavity at moderate aspect ratio; a rectangular cavity heated from below; a trapezoidal cavity; a rectangular cavity containing a conducting obstruction. The steady equations for the velocity, pressure and temperature are solved in the Boussinesq approximation, using a standard Galerkin formulation of the finite element method.

NOMENCLATURE

A	aspect ratio of the cavity, H/D
d	thickness of the divider
D	width of the cavity
f_1, f_2, f_3 f_4, f_5	weight functions in the spaces H'_u, H'_w, H'_p, H'_T and H'_s respectively
g	acceleration due to gravity
h	height of the divider
H	height of the cavity
$H'_u(\Omega), H'_w(\Omega),$ $H'_T(\Omega), H'_s(\Omega)$	spaces of functions having square-integrable first derivatives over Ω and which are zero on the part of Γ with specified u, w, T and s respectively.
$H'_p(\Omega)$	space of functions having square-integrable first derivatives over Ω and which satisfy the boundary conditions for p.
i, j, k	indices denoting nodal values of the fields u, w, p, T and s
k_r	ratio of thermal conductivities of divider and air
n	coordinate normal to a boundary

N_u	local Nusselt number on hot or cold face, $\partial T/\partial x$ at $x=0$ or 1
$\bar{N}_u$	average value of N_u on hot or cold face, $1/A \int_0^A N_u \, dz$
p^*	pressure
p	nondimensional pressure, $p^*D^2/\rho_0\kappa^2$
$\tilde{p}$	centrosymmetric part of p, $p - R_aP_r(z-0.5)/2$
P_r	Prandtl number, μ/κ
R_a	Rayleigh number, $g\beta(T_H^* - T_c^*)D^3/\kappa\nu$
R_{a_c}	critical Rayleigh number for Bénard convection
s	stream function
s	finite element approximation to stream function
T_H^*	temperature of hot face
T_c^*	temperature of cold face
T^*	temperature
T	nondimensional temperature, $(T^* - T_c^*)/(T_H^* - T_c^*)$
T	finite element approximation to temperature
u^*	velocity in x direction
u	nondimensional velocity, u^*D/κ
u	finite element approximation to u velocity
w^*	velocity in z direction
w	nondimensional velocity, w^*D/κ
w	finite element approximation to w velocity
x^*, z^*	planar Cartesian coordinates
x, z	nondimensional coordinates $(x^*, z^*)/D$
β	coefficient of volumetric expansion
γ	stratification, $(\tau_z R_a/4)^{1/4}$
Γ	boundary of the cavity
$\Gamma_1, \Gamma_2, \Gamma_4$	part of Γ for which Dirichlet boundary conditions are not specified for u, w and T respectively
κ	thermal diffusivity
ν	kinematic viscosity
τ_x	horizontal temperature gradient at cavity centre, $\left.\dfrac{\partial T}{\partial x}\right\|_{x=1/2, z=A/2}$
τ_z	vertical temperature gradient at cavity centre, $\left.\dfrac{\partial T}{\partial z}\right\|_{x=1/2, z=A/2}$
ψ	shape function of quadratic variation
ϕ	shape function of linear variation

ρ_0	reference density at temperature T_c^*
Ω	the cavity domain
Ω_F	the fluid domain
Ω_s	the solid domain.

7.1 INTRODUCTION

A wide range of problems in physics and engineering are concerned with the phenomenon of natural convection. Not surprisingly, it is attracting a large and expanding research effort both experimental and theoretical. For example, the 1980 review of the heat transfer literature[1] lists 72 papers on confined natural convection alone, almost as many as were cited on channel flow. The purpose of this chapter is to discuss, with illustrative calculations, a few of the many interesting problems that have arisen recently in the prediction of convection in heated cavities.

The scope of this study is limited to steady-state, two dimensional, laminar convection, and predictions are made within the framework of the finite element approximation. However, one point which will emerge is the possibility of predicting the onset of unsteady behaviour in a time-independent calculation such as this.

The organization of the chapter is as follows. The next section presents the mathematical description of natural convection in terms of velocities, pressure and temperature. The Boussinesq approximation is applied to the equations, and their finite element discretization is given. Subsequent sections present the results of specific, illustrative examples. Section 7.3 is concerned with the problem of convection in a square cavity with differentially heated sidewalls, at a Rayleigh number of 10^7. Section 7.4 discusses the extension to convection in a cavity with a large height-to-width ratio, for which there are convergence problems, possibly related to limit points, at sufficiently high Rayleigh number. Section 7.5 considers the case of a square cavity heated from below. This is the well-known Rayleigh–Bénard problem where the prediction of the bifurcation to non-unique solutions of the field equations is of interest. Section 7.6 is concerned with convection in a particular case of a nonrectangular cavity showing a strong sensitivity to the temperature boundary condition. Finally in Section 7.7 the effect of conducting inserts on the flow and heat transfer in a confined region is studied.

All the results are obtained using the program ENTWIFE, which is derived from the library of finite element subroutines known as TGSL and developed by the Theory of Fluids Group at Harwell. The variety of the problems that have been considered demonstrates the robustness of the code for the prediction of laminar natural convection in heated cavities. The results of Sections 7.3–5 are based on work reported previously.[2]

7.2 THEORY

Two-dimensional, steady, laminar natural convection is described by the conservation equations for momentum, mass and energy. In the present work these are expressed in the Boussinesq approximation. Adopting the non-dimensionalization of Mallinson and de Vahl Davis,[3] the field equations for the two velocity components, the pressure and the temperature become:

$$u\frac{\partial u}{\partial x}+w\frac{\partial u}{\partial z}+\frac{\partial p}{\partial x}-P_r\nabla^2 u=0, \tag{7.1}$$

$$u\frac{\partial w}{\partial x}+w\frac{\partial w}{\partial z}+\frac{\partial p}{\partial z}-P_r\nabla^2 w-R_aP_rT=0, \tag{7.2}$$

$$\frac{\partial u}{\partial x}+\frac{\partial w}{\partial z}=0, \tag{7.3}$$

and

$$u\frac{\partial T}{\partial x}+w\frac{\partial T}{\partial z}-\nabla^2 T=0, \tag{7.4}$$

where the direction of gravity is along $-\mathbf{z}$. The Rayleigh number R_a in the above equations is based on the width of the cavity.

As an aid to flow visualization, it is also convenient to obtain the stream function s, from the equation

$$\nabla^2 s=\frac{\partial u}{\partial z}-\frac{\partial w}{\partial x}. \tag{7.5}$$

The equations are now re-expressed in a weak form which incorporates their boundary conditions. For example, Equation (7.1) is multiplied by an arbitrary test function f_1 in the space $H'_u(\Omega)$, and the $\nabla^2 u$ term is integrated by parts, to give

$$\int_\Omega f_1\left(u\frac{\partial u}{\partial x}+w\frac{\partial u}{\partial z}+\frac{\partial p}{\partial x}\right)+P_r\int_\Omega \nabla f_1\cdot\nabla u-P_r\int_{\Gamma_1} f_1\frac{\partial u}{\partial n}=0. \tag{7.6}$$

Note that the choice of test function restricts the integration of the boundary term to that part of the boundary Γ_1 where Dirichlet conditions are not specified for u.

In the same way, the weak forms of Equations (7.2)–(7.5) are obtained:

$$\int_\Omega f_2\left(u\frac{\partial w}{\partial x}+w\frac{\partial w}{\partial z}+\frac{\partial p}{\partial z}-R_aP_rT\right)+P_r\int_\Omega \nabla f_2\cdot\nabla w-P_r\int_{\Gamma_2} f_2\frac{\partial w}{\partial n}=0, \tag{7.7}$$

$$\int_\Omega f_3\left(\frac{\partial u}{\partial x}+\frac{\partial w}{\partial z}\right)=0, \tag{7.8}$$

$$\int_\Omega f_4\left(u\frac{\partial T}{\partial x}+w\frac{\partial T}{\partial z}\right)+\int_\Omega \nabla f_4\cdot\nabla T-\int_{\Gamma_4} f_4\frac{\partial T}{\partial n}=0, \tag{7.9}$$

and

$$\int_{\Omega} \nabla f_5 \cdot \nabla s + \int_{\Omega} f_5 \left(\frac{\partial u}{\partial z} - \frac{\partial w}{\partial x} \right) = 0. \tag{7.10}$$

These equations are now discretized by dividing the domain Ω into a number of regions, the finite elements, each of which has a number of nodes. Throughout this study straight-sided triangles with nodes at the vertices and mid-points are used. The fields u, w, p, T and s are now approximated by the following expansions in sets of piecewise polynomial basis functions:

$$u = \sum_{i=1}^{N} u_i\psi_i, \quad w = \sum_{i=1}^{N} w_i\psi_i, \quad p = \sum_{i=1}^{M} p_i\phi_i, \quad T = \sum_{i=1}^{N} T_i\psi_i, \quad \text{and} \quad s = \sum_{i=1}^{N} s_i\psi_i. \tag{7.11}$$

The set ψ consists of functions ψ_i which are unity at node i, zero at every other node and have quadratic variation. The set ϕ consists of functions ϕ_i that are unity at every vertex node i, zero at every other vertex node and have linear variation.

This is the so-called 'mixed interpolation' which ensures that the pressure p is uniquely determined[4]. The Galerkin formulation is now applied, whereby the test functions for each equation are taken to be the set of basis functions which are used to represent the variable corresponding to that equation. For the problems considered here, the boundary conditions are such that whenever a Dirichlet condition is not appropriate, then the normal derivative of the variable is zero on the boundary. As a result, all the remaining boundary integrals in Equations (7.6)–(7.10) vanish. Finally, therefore, we obtain the equations

$$\int_{\Omega} \psi_j \left(u \frac{\partial u}{\partial x} + w \frac{\partial u}{\partial z} + \frac{\partial p}{\partial x} \right) + P_r \int_{\Omega} \nabla\psi_j \cdot \nabla u = 0, \tag{7.12}$$

$$\int_{\Omega} \psi_j \left(u \frac{\partial w}{\partial x} + w \frac{\partial w}{\partial z} + \frac{\partial p}{\partial z} - R_a P_r T \right) + P_r \int_{\Omega} \nabla\psi_j \cdot \nabla w = 0, \tag{7.13}$$

$$\int_{\Omega} \phi_k \left(\frac{\partial u}{\partial x} + \frac{\partial w}{\partial z} \right) = 0, \tag{7.14}$$

$$\int_{\Omega} \psi_j \left(u \frac{\partial T}{\partial x} + w \frac{\partial T}{\partial z} \right) + \int_{\Omega} \nabla\psi_j \cdot \nabla T = 0, \tag{7.15}$$

and

$$\int_{\Omega} \nabla\psi_j \cdot \nabla s + \int_{\Omega} \psi_j \left(\frac{\partial u}{\partial z} - \frac{\partial w}{\partial x} \right) = 0, \tag{7.16}$$

where the index j runs over the set of all nodes interior to Ω, and the index k runs over the set of all vertex nodes, excluding that at which the pressure is specified. All the integrals in these equations are evaluated numerically by a 7-point Gauss formula.

The original field equations have now been approximated by a set of coupled, non-linear, algebraic equations in the nodal values u_i, w_i, T_i, p_i and s_i. These are solved by the Newton–Raphson algorithm, and the linear system which results at each iterate is solved by the frontal method.[5,6] The Newton–Raphson linearization requires an initial guess for the finite element fields. In general this is taken to be the solution at a lower Rayleigh number (and possibly on a coarser grid). At sufficiently low R_a an initial guess of zero is adequate. In this way good convergence of the Newton–Raphson procedure is obtained, usually after about four iterations. This is in fact a rather crude way of stepping in Rayleigh number, the disadvantage being that the choice of step length is made by the user. An improved technique based on Davidenko continuation which automates the stepping has recently been introduced by Jackson and Winters.[7]

7.3 THE SQUARE CAVITY WITH HEATED SIDEWALLS

The first example that I wish to consider consists of convection in a square, air-filled cavity with differentially heated sidewalls. This is the so-called 'double-glazing' problem which was proposed as a comparison exercise by de Vahl Davis *et al.*[8,9] It has been the subject, therefore, of intensive study for the suggested range of Rayleigh numbers from 10^3 to 10^6. The contributions to the comparison exercise have been collated by Jones and Thompson[10] and a preliminary assessment made by de Vahl Davis and Jones.[11] From this it is clear that differences between methods are magnified as the Rayleigh number increases. However, the highest value of 10^6 was insufficient to distinguish the best results for some of the predicted quantities, particularly the average Nusselt number. The benchmark values themselves are only accurate to about 1% at 10^6 and a significant number of contributors obtained that level of precision for their prediction of $\bar{N}_u$.

An obvious extension of the comparison problem is therefore to the case of a higher Rayleigh number of 10^7, since the computation becomes rapidly more difficult above $R_a = 10^6$. The geometry for the problem is shown in Figure 7.1. The boundary conditions are as follows:

$$\textit{on ab:} \qquad u = w = 0, \qquad T = 1 \tag{7.17}$$

$$\textit{on bc and da:} \qquad u = w = 0, \qquad \frac{\partial T}{\partial z} = 0 \tag{7.18}$$

$$\textit{on cd:} \qquad u = w = 0, \qquad T = 0 \tag{7.19}$$

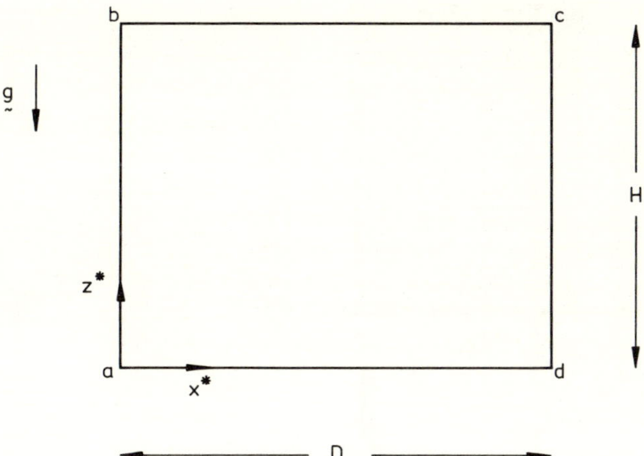

Figure 7.1 Geometry of rectangular enclosure

In addition the pressure is specified as zero at the centre of the cavity. The aspect ratio A is unity.

The results were obtained on a grid of 37×37 nodes with irregular spacing to give a higher density of freedoms near the walls. This should be sufficient to place two or three nodes within the vertical velocity boundary layers, as defined by Quon,[12] but no attempt was made to optimize the irregularity of the spacing. The grid and contours of stream function, temperature and centrosymmetric pressure $\bar{p}$ are displayed in Figure 7.2. These are in broad agreement with the results of previous studies of Quon,[12] Upson *et al.*[13] and Marshall *et al.*[14] The streamlines show the strong flow past the vertical boundaries and the weak recirculating flow in the core where the temperature is highly stratified. There is a pronounced reversal of the temperature gradient in the x direction, outside the vertical boundary layers.

Figure 7.3(a) presents profiles of temperature along the horizontal, for various heights, and complements the isotherms of Figure 7.2(c). Figures 7.3(b) and (c) show profiles of the w and u velocity along the horizontal and vertical mid-planes. The extremely intense and narrow vertical boundary layer, generated by buoyancy effects, is apparent in the w velocity profile.

Figure 7.3(d) shows the variation of the local Nusselt number N_u along the hot wall. This was calculated as the gradient of the computed, finite element temperature field. It is therefore linear and discontinuous across element boundaries, and the results are presented without smoothing, as in the case of R_a from 10^3 to 10^6.[15] Because of the linear variation, the maximum and minimum values of N_u are attained at vertex nodes, and their magnitudes are taken to be the average of the discontinuous values.

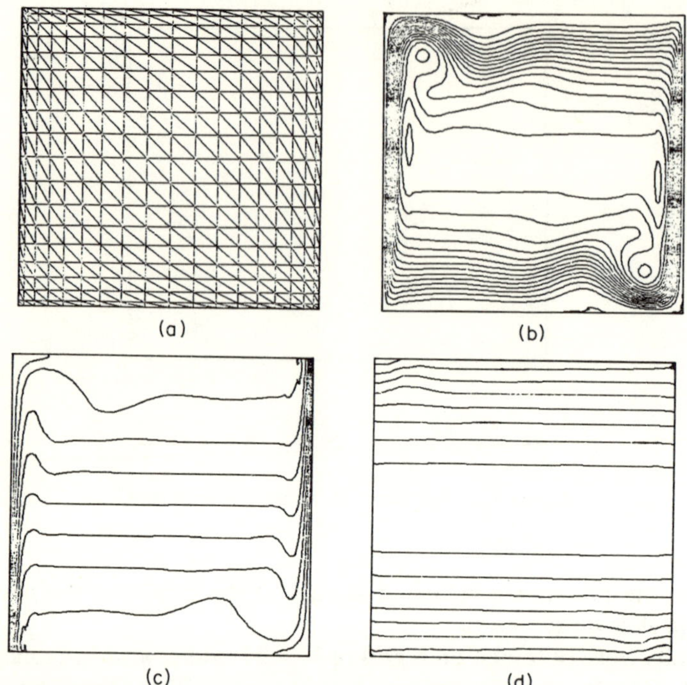

Figure 7.2 Natural convection of air in a square cavity at $Ra = 10^7$: (a) the 37×37 non-uniform grid; (b) streamfunction contours at intervals of 2.0; (c) temperature contour at intervals of 0.1; (d) contours of centrosymmetric pressure

This method of calculating the Nusselt number is rather crude, although it does give a useful guide to the discretization error which is reflected in the degree of discontinuity. I have also calculated a Nusselt number N_{u_c} which is constrained to satisfy global energy conservation (unlike N_u, where conservation is a fortuitous result of the symmetry of this particular problem). In the terminology of Upson *et al.*[13] and Gresho and Lee[16] this gives a consistent flux Nusselt number. However, the present computation differs from theirs, and is based on a new method of imposing Dirichlet boundary conditions in the finite element method, due to Cliffe and Jackson.[17] It should be noted that it is not yet clear whether the consistent flux method gives any greater accuracy for the Nusselt number, for general problems, than other methods would give on the same size grid. One problem which I experienced was that the consistent method imposed wiggles between nodal values of the local Nusselt number. This affected the accuracy of the prediction of $N_{u_{max}}$. Of course, the wiggles are reduced by appropriate mesh refinement, but then so are the discontinuities in N_u obtained from the gradient of the finite element field T.

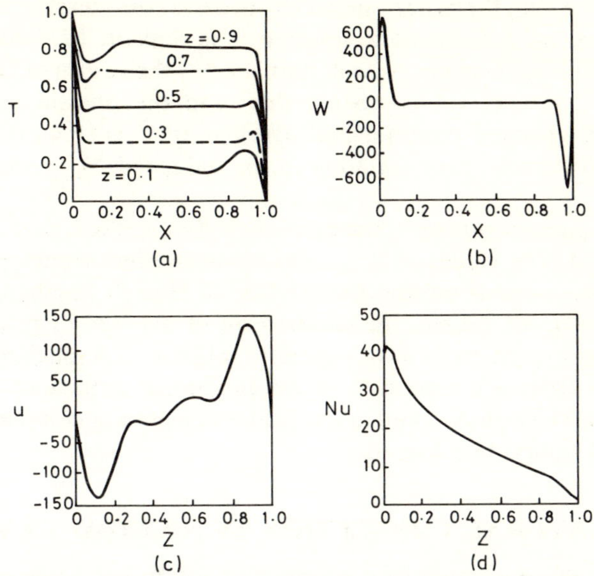

Figure 7.3 Natural convection of air in a square cavity at $Ra = 10^7$: (a) horizontal temperature profiles, at various heights; (b) w, velocity along horizontal mid-plane; (c) u, velocity along vertical mid-plane; (d) local Nusselt number along hot face

Table 7.1 gives values of the average, maximum and minimum Nusselt numbers at $R_a = 10^7$ for both methods, and also includes the results of other work.

The authors represented in the table (apart from Marshall *et al.*[14]) contributed to the comparison exercise of de Vahl Davis *et al.*[8,9] and for Rayleigh numbers in the range 10^3–10^6 their predictions were counted amongst the

Table 7.1 Nusselt numbers for the natural convection in an air-filled square cavity at $R_a = 10^7$

	$\bar{N}_u$	N_{umax}	N_{umin}
Present	17.2	41.8	1.14
Present—consistent flux	16.5	41.2	1.23
Kessler and Oertel†	16.22	38.93	2.405
Upson *et al.*[13]†	18.18	39.18	1.430
Upson *et al.*[13]†—consistent flux	16.34	36.50	1.416
Quon[12]	16.61	39.7	1.35
Gartling†	18.07	41.4	1.48
Marshall *et al.*[14]	17.88	—	—
Marshall *et al.*[14]—consistent flux	15.74	~35	—

† Quoted in Jones and Thompson.[10]

best. It is interesting therefore that at 10^7 there is considerable variation, and in the absence of a benchmark solution it is difficult to draw meaningful conclusions. For the average Nusselt number $\bar{N}_u$, the values of Quon, Kessler and Oertel, Upson *et al.* (consistent flux) and the present consistent flux results are all grouped together and average at $\bar{N}_u = 16.4 \pm 0.2$. The same authors predict much more variation in the values of $N_{u_{max}}$ and particularly $N_{u_{min}}$.

In the comparison problem, the consistent flux method gave good predictions of $\bar{N}_u$, but poor values of $N_{u_{max}}$, and a similar effect seems apparent here. Further work is needed on the computation of Nusselt numbers in the finite element method. Of course, the comparison of average or peak values can be misleading, a point made most cogently by Quon.[12] A much more accurate picture arises from a comparison of the functional behaviour of quantities such as the Nusselt number, and in any future benchmark comparison exercise this approach should be adopted.

7.4 NATURAL CONVECTION IN NARROW CAVITIES

The problem of convection in a rectangular cavity has been much studied, and a comprehensive review is given by Schinkel.[18] Despite this effort, a clear overall picture of the processes involved, as a function of the Prandtl number, Rayleigh number and aspect ratio, has yet to emerge. The importance of the Prandtl number in determining the type of flow was established by Elder[19] for high P_r fluids, such as oils, and more recently through the predictions of Jones [20] for the low Prandtl numbers typical of liquid metals. However, the effect of the aspect ratio on the convection is less clear. Measurements have been made at very high aspect ratios, for example up to $A = 78$, by Yin *et al.*[21] However, computational difficulties have limited predictions to more modest aspect ratios, although Raithby and Wong[22] have considered A up to 80. As a result there is no generally accepted figure for the dependence of the Nusselt number on the aspect ratio.

The main features of natural convection in narrow cavities can be interpreted from the behaviour of the horizontal and vertical temperature gradients at the cavity centre, τ_x and τ_z, and the stratification parameter $\gamma = (\frac{1}{4}\tau_z R_a)^{1/4}$. These quantities play an important role, both in the classification of the flow[23] and its analysis by stability theory[19,24]. However, they are often ignored in reports of simulations of natural convection, despite being easily obtainable from the computation.

The functional dependence of τ_x and τ_z on the Rayleigh number and aspect ratio is interesting. For example, τ_x correlates well with the ratio R_a/A, which has led to some attempts to quantify the flow regimes of Eckert and Carlson,[23] but with little agreement. Defining the transition region to be that for which the horizontal temperature gradient τ_x is in the range $-0.9 \leq \tau_x \leq 0$, then Gilly *et al.*[25] and Raithby and Wong[22] find it to be bounded in the range

$$250 \leqslant \frac{R_a}{A} \leqslant 2500 \qquad \text{for } A \leqslant 4.$$

Duxbury,[26] on the other hand, obtains the range

$$300 \leqslant \frac{R_a}{A} \leqslant 4000$$

from his measurements. This work also shows clearly the further subdivision of the boundary layer region (when τ_x is positive) into two types.

Regarding τ_z, Elder,[19] Schinkel,[18] and Raithby and Wong[22] show that $A\tau_z$ correlates well with R_a/A, while Duxbury[26] and Thomas and de Vahl Davis[27] correlate $A\tau_z$ with $R_a/A^{1.25}$. In all cases, $A\tau_z$ rises from zero in the conduction region, attains a maximum value in the transition region and decreases to an approximate constant value for higher R_a/A in the boundary layer region.

I have carried out a simulation of natural convection in an air-filled cavity of aspect ratio 10, for the geometry of Figure 7.1 and the boundary conditions of the previous section. A Prandtl number of 0.71 is assumed and a regular grid of 21×61 nodes used. For this case there are detailed measurements of Duxbury[26,28] and finite difference calculations of Jones[29] with which to compare.

The streamlines and isotherms obtained at a Rayleigh number of 1.71×10^4 are in good agreement with the contour plots presented by Duxbury[28] and Jones,[29] and are not shown. Instead, values of τ_x, τ_z, γ and $\bar{N}_u$ are given in Table 7.2.

Table 7.2 The horizontal and vertical temperature gradients at the cavity centre, τ_x and τ_z, the stratification γ and the average Nusselt number $\bar{N}_u$ for convection in a cavity of aspect ratio 10, at various Rayleigh numbers

Ra	τ_x	τ_z	γ	$\bar{N}_u$
1.71×10^4	−0.067	0.094	4.48	2.01
5×10^4	0.208	0.059	5.22	2.74
10^5	0.295	0.061	6.24	3.28
1.2×10^5	0.299	0.063	6.61	3.47
1.5×10^5	0.291	0.067	7.07	3.70

At $R_a = 1.71 \times 10^4$, the value of τ_x is small and negative, which shows that the flow is at the upper end of the transition region. A smaller value of −0.05 is derived from the measurements of Duxbury[26] at this Rayleigh number. The stratification of 4.48 is also consistent with a flow in the transition region; Bergholz[24] gives a value of $\gamma = 4.8$ as the upper limit of the transition region.

The Nusselt number of 2.01 is higher than the value of 1.88 predicted by Jones[29] and compares with measured values of 2.2 and 1.7 for the hot and cold faces respectively (Duxbury[28]).

As the Rayleigh number increases, the flow becomes increasingly complex; the sign change in τ_x shows that the boundary layer region is entered between 1.71×10^4 and 5×10^4. Both τ_x and τ_z show the characteristic behaviour discussed by Duxbury[26] and Schinkel.[18] At 10^5, the predicted τ_z and $\bar{N}_u$ are in fair agreement with the recent work of Raithby and Wong.[22] The streamlines and isotherms for $R_a = 10^5$ and 1.5×10^5 are shown in Figure 7.4. These show an interesting transition in the flow. In both cases the central core flow is rather weak, but in passing from 10^5 to 1.5×10^5 additional weak eddies are developing above the central core recirculation.

Finally, it was found to be impossible to obtain converged solutions above 2×10^5 on the 21×61 grid used for the calculation, indicating a limit point in the solution of the discrete equations. While it is usual for a given grid to fail at sufficiently high Rayleigh number, there are reasons to believe that there may be a limit point in the exact steady equations at about this Rayleigh number. Both computations and experiment[18,26,29], show that the flow becomes unsteady at about $R_a = 2 \times 10^5$ for an aspect ratio of 10. Furthermore, plotting the stratification γ from Table 7.2 on the R_a–γ stability diagram of Bergholz[24] shows that the 2×10^5 value lies within the region in which travelling-wave instabilities are predicted. Schinkel[18] tentatively suggests that the instability he observes is indeed of this type. Raithby and Wong[22] also report convergence problems for cases where the analysis of Bergholz[24] predicts instabilities. Further work is clearly needed to establish whether the physical instability manifests itself as a limit point in the steady equations (the techniques of Keller[30] are relevant here). There are well-known counter-examples where no such effect is seen (e.g. vortex-shedding by a cylinder [31]).

7.5 BÉNARD CONVECTION IN A SQUARE CAVITY

Bénard convection is the motion that results when a fluid is heated from below and only when the Rayleigh number is above a certain critical value R_{a_c}. A considerable number of studies of this problem have been made (see Azouni[32] for a recent review). Its interest lies not only in its applications in many branches of physics and engineering, but also in its being a simple example of a bifurcation phenomenon.

Many previous studies have used linear stability theory to predict the critical Rayleigh number, but only a small number of different boundary conditions can be solved in this way. Numerical solution is necessary for the case of Bénard convection confined to a cavity with rigid walls, for example. Luijkx and Platten[33] studied this particular problem for a wide range of aspect ratios, by a Galerkin solution of the linearized equations.

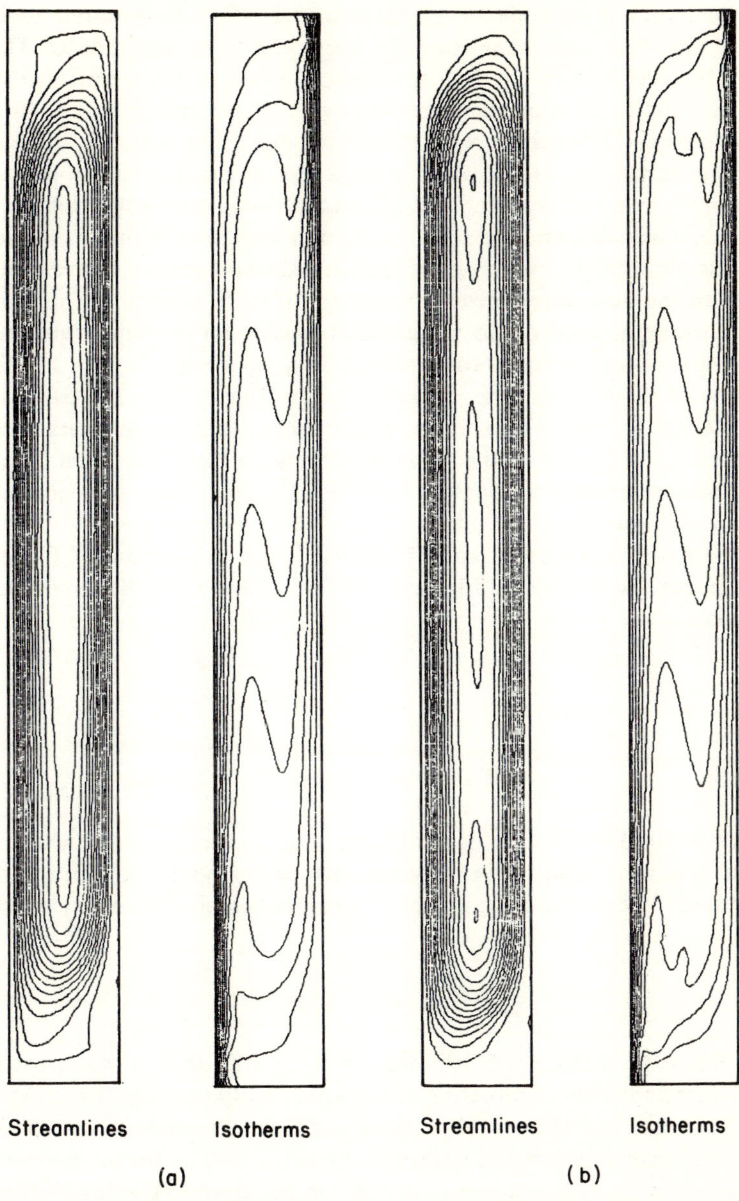

Figure 7.4 Natural convection of air in a cavity of aspect ratio 10: $Ra = 10^5$: streamlines at intervals of 4.457, isotherms at intervals of 0.1; (b) $Ra = 1.5 \times 10^5$: streamlines at intervals of 5.333, isotherms at intervals of 0.1

In general a full solution of Equations (7.1)–(7.4) is required to obtain details of the flow and heat transfer, including their magnitude. There is considerable scope here for the application of the many elliptic solvers in current use. Finite element packages in particular can handle irregularities in the geometry and boundary conditions with ease. The solution of the non-Boussinesq equations is a straightforward extension; the effect of variable fluid properties on the critical Rayleigh number would be a worthwhile study in itself (see Leonardi and Reizes[34] for an application to inclined cavities).

Previous numerical simulations of the Boussinesq equations have paid little attention to the non-uniqueness inherent in them. In particular the ability of the finite element method to generate solutions on chosen branches of a bifurcation problem remains to be demonstrated. Upson *et al.*[13] considered a rigid box of aspect ratio one, and obtained a flow at $R_a = 10^6$ consisting of a single clockwise eddy. This solution is plainly non-unique, since its reflection about the line $x = \frac{1}{2}$ satisfies the same equations. In fact there must be at least three solutions, since the trivial one $u = w = 0$ satisfies the steady equations at *all* Rayleigh numbers.

I have applied the present code ENTWIFE to the problem of Figure 7.1, for an aspect ratio of 1, Prandtl number of 0.71, and the following boundary conditions:

$$\textit{on bc}: \qquad u = w = 0, \qquad T = 0$$

$$\textit{on da}: \qquad u = w = 0, \qquad T = 1$$

$$\textit{on ab and cd}: \qquad u = w = 0, \qquad \frac{\partial T}{\partial x} = 0$$

These correspond to the case of rigid horizontal surfaces and side walls. Some computations were also done for free surface boundary conditions on the horizontal surfaces bc and da, in which case the velocity boundary conditions become

$$\frac{\partial u}{\partial z} = 0 \qquad \text{and} \qquad w = 0.$$

At a Rayleigh number of 10^4 and on a 13×13 grid it was possible to generate three solutions, consisting of a primary eddy rotating either clockwise or anticlockwise, and the trivial no-flow solution. Streamlines and isotherms for the clockwise solution are shown in Figure 7.5. The initial guess for the nontrivial solution was taken to be that of the double-glazing problem of Section 7.3, at the same Rayleigh number, with the appropriate sense of rotation. A zero initial guess converged to the trivial solution as expected.

At higher Rayleigh numbers there are successive bifurcations from the no-flow branch to solutions consisting of 2,3, and higher numbers of eddies. Attempts to trigger these solutions from an *ad hoc* initial guess failed, however.

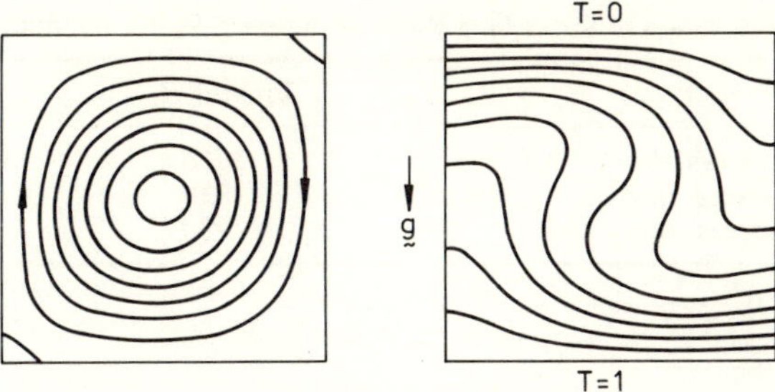

Figure 7.5 Bénard convection of air in a square cavity at $Ra = 10^4$: (a) streamfunction contours; (b) temperature contours

Clearly, a more systematic approach is required, and progress in this direction has been made by Jackson and Winters.[7]

A number of computations with the 13×13 grid were carried out to demonstrate the ability of the present code to predict the critical Rayleigh number. The amplitude of the flow varies as $(R_a - R_{a_c})^{1/2}$ in the linear region above R_{a_c}. By computing the square of the maximum value of the streamfunction ψ^2_{max}, for example, at various R_a the value of R_{a_c} is obtained by simple extrapolation to $\psi_{max} = 0$. A similar method has been used by Van Steeg and Wesseling[35] and Gresho *et al.*[36] In this way I have obtained values of ψ^2_{max} for the above geometry and boundary conditions, and also for the case of horizontal free surfaces and $A = 1$ and 0.5. Figure 7.6 shows a plot of ψ^2_{max} against R_a for these three cases, and it is apparent that the flow remains linear well above the critical value.

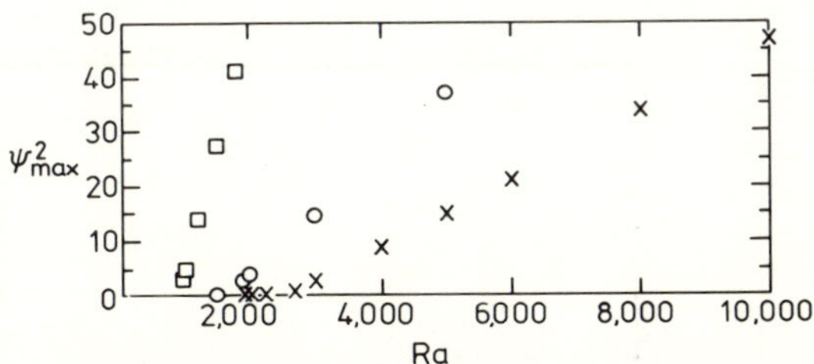

Figure 7.6 The dependence of the square of the maximum streamfunction value on the Rayleigh number, for Bénard convection of air in a square cavity: ×, aspect ratio 1, rigid walls; ○, aspect ratio 1, free horizontal surfaces; □, aspect ratio 1/2, free horizontal surfaces

Extrapolating these results gives the values of R_{a_c} (Table 7.3), which are in excellent agreement with those of other work.

The disadvantage of this method of obtaining the critical Rayleigh number is that it requires the flow on one of the bifurcating branches to be triggered

Table 7.3 Predicted values of the critical Rayleigh number R_{ac} for confined Bénard convection

	Present R_{ac}	Other work
Rigid horizontal surfaces, $A = 1$	2610	2585.03†
Free horizontal surfaces, $A = 1$	1830	—
Free horizontal surfaces, $A = \frac{1}{2}$	920	885.21‡

† Luijkx and Platten[33]
‡ Hall and Walton[37]

in some way. This may be difficult in practice, as we have already seen in the case of bifurcation to higher modes from the no-flow branch. An improved technique based on monitoring the Jacobian matrix has been introduced by Jackson and Winters[7] and overcomes this difficulty.

7.6 CONVECTION IN NON-RECTANGULAR GEOMETRIES

The problems considered so far have been for cavities of rectangular cross-section. Much less work seems to have been done on convection in non-rectangular geometries, where many important and varied effects are possible. This section describes convection of water in a trapezoidal cavity, where the sloping floor causes a drastic sensitivity to the prescribed wall temperature.

The geometry of the enclosure is shown in Figure 7.7, with the height-to-width ratio H/D equal to one. A constant Prandtl number of 3.5 was assumed.

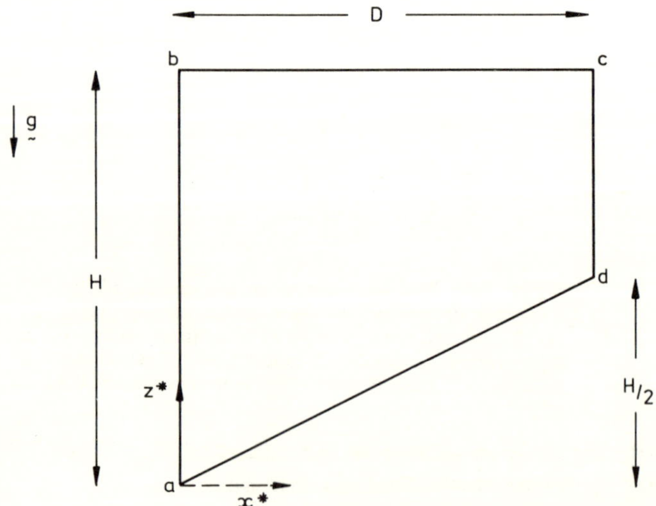

Figure 7.7 The geometry of the trapezoidal cavity

The boundary conditions are:

$$\textit{on bc}: \quad u = w = 0, \qquad T = 1$$

$$\textit{on cd}: \quad u = w = 0, \qquad \frac{\partial T}{\partial x} = 0$$

$$\textit{on da}: \quad u = w = 0, \qquad T = 0.$$

This particular example was a model of a physical problem for which the boundary condition on ab was uncertain. Two choices are considered:

$$\text{(a)}\ u = w = 0, \quad T = \begin{cases} 2z, & 0 \leqslant z \leqslant \frac{1}{2} \\ 1, & \frac{1}{2} \leqslant z \leqslant 1 \end{cases} \tag{7.20}$$

so that the temperature increases linearly from zero at the base of the wall, to a value of one at mid-height at which it remains over the upper half, and

$$(b)\ u = w = 0, \quad T = \begin{cases} 0, & 0 \leqslant z \leqslant \frac{1}{2} \\ 2(z - 1/2), & \frac{1}{2} \leqslant z \leqslant 1 \end{cases} \tag{7.21}$$

where the temperature is zero over the lower half of the wall and then increases linearly to a value of one at the top.

The streamlines and isotherms for case (a) are shown in Figures 7.8 and 7.9 for Rayleigh numbers of 10^3 and 10^8 respectively. At 10^3 the flow consists of a single, clockwise eddy up past the vertical wall. This develops into the

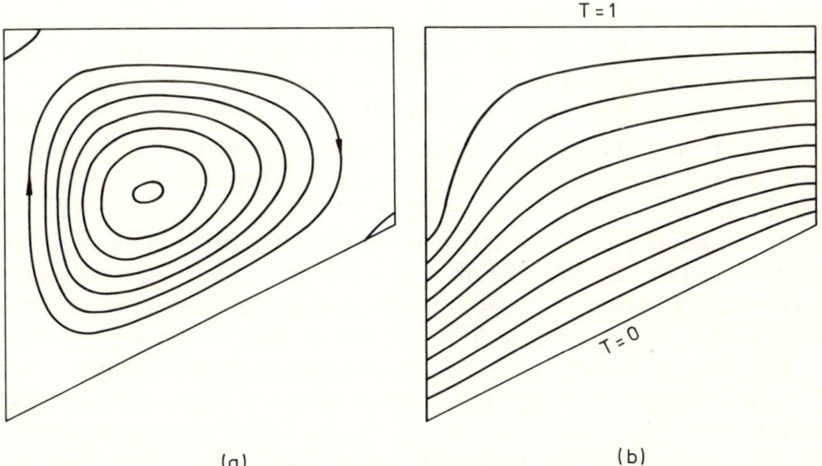

Figure 7.8 Natural convection of water in a trapezoidal cavity at $Ra = 10^3$, for boundary condition (a) of equation 20: (a) streamlines; (b) isotherms

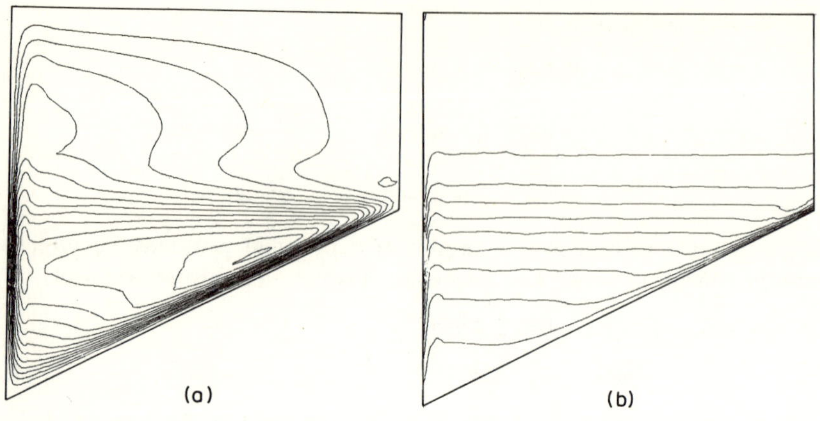

Figure 7.9 As for figure 8, but at $Ra = 10^8$

complex flow at $R_a = 10^8$ shown in Figure 7.9(a). Although there is still a single, clockwise primary recirculation, the core now consists of two eddies rotating in the same sense. There is strong, boundary-layer flow down the sloping flow and up the longer vertical wall, but the flow in much of the upper section of the cavity is weak. There is pronounced stratification of the isotherms in much of the cavity, with the uppermost part being close to isothermal. There is an interesting region of reversed horizontal temperature gradient outside the vertical boundary layer, which is associated presumably with the two-eddy structure of the core flow.

The streamlines and isotherms for boundary condition (b) are shown on Figures 7.10 and 7.11 for the same two Rayleigh numbers. In complete contrast to the previous case, the convection at $R_a = 10^3$ consists of two counter-rotating eddies, with the flow directed *down* past the longer vertical wall. As the Rayleigh number increases the eddies move apart and approach the vertical wall and sloping floor, where boundary layers develop, particularly along the upper part of the floor. At $R_a = 10^8$ the flow is confined to the lower triangular section. The region above is relatively stagnant, and there is a third, very weak region of recirculation under the roof. There is a considerable degree of stratification in the upper section, and the lower triangular part is almost isothermal, in contrast to the previous case. It is apparent that the heat flux through the left vertical wall is directed *out* of the cavity.

The twin counter-rotating eddy structure which is supported by boundary condition (b) is readily understood from Figure 7.10(b). Because of the $T = 0$ boundary condition on the walls of the lower triangular section of the cavity, the isotherms must dip in this region even if they were perfectly stratified for $z = 0.5$. It is this dip in the temperature contours which generates the two-eddy

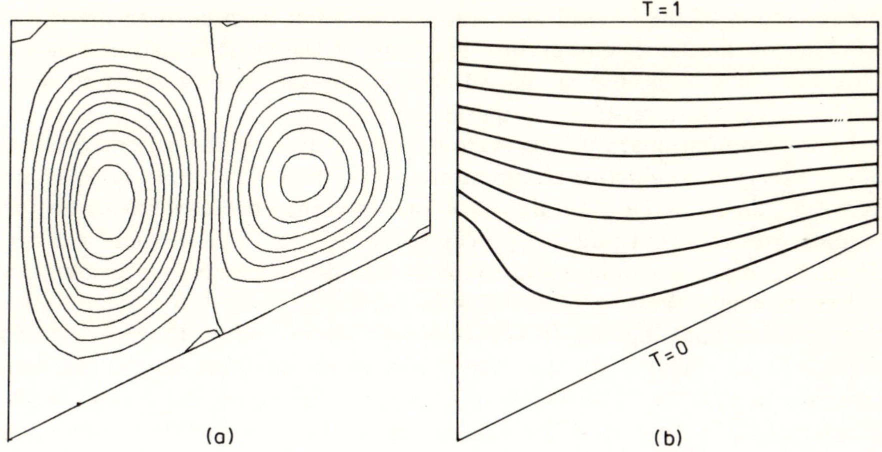

Figure 7.10 Natural convection of water in a trapezoidal cavity at $Ra = 10^3$, for boundary condition (b) of equation 21: (a) streamlines; (b) isotherms

structure. It is removed by ensuring that the lower part of the left vertical wall is always at a higher temperature than the sloping floor.

7.7 CONVECTION IN A CAVITY WITH A CONDUCTING INSERT

A few experiments have been done recently on the effects of conducting inserts on natural convection in heated cavities.[26,38,39] The interest here is on

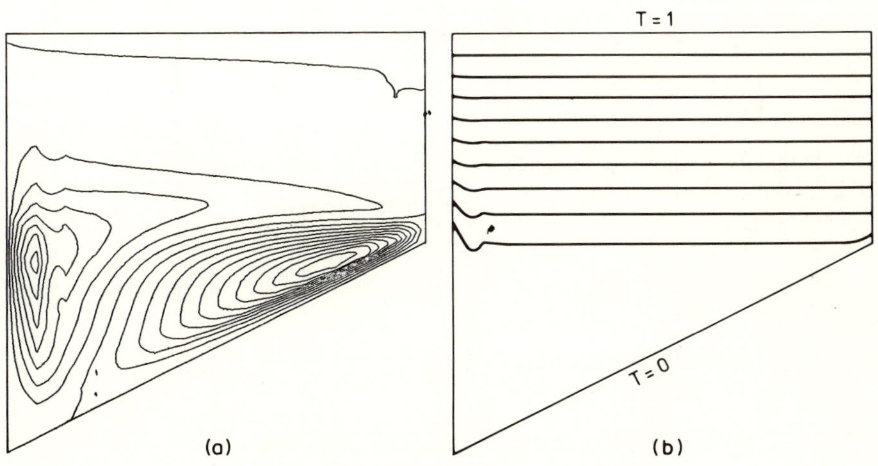

Figure 7.11 As for figure 10, but at $Ra = 10^8$

the reduction of the overall heat transfer resulting from such an insert, but its effect on the local temperature gradients at the walls is also of concern. Duxbury[26] has suggested the use of appropriate inserts in a cavity to shape the local Nusselt number to a required distribution.

Little computational work seems to have been done on the general problem of the effects of conducting obstructions on confined convection. Jackson and Winters[40] developed a general facility for modelling problems having distinct regions with different physical processes, within the finite element method. There are many possible applications of this development, for example to problems where there is Darcy flow in porous regions, and Navier–Stokes flow in other fluid regions. It was used by Winters[41] to study the effect of a partial, brass division on the convection in an air-filled cavity, allowing comparison with the experiments of Duxbury.[26] However, the divider in that problem is very thin, and the full power of the new facility is not really exploited.

In this section I should like to illustrate its application to the case of a closed air-filled cavity containing a thick obstruction attached to the floor. The geometry is shown in Figure 7.12. The aspect ratio of the cavity is $\frac{5}{8}$, the nondimensional thickness of the divider, d/D, is $\frac{1}{2}$, its height h is half that of the cavity and it is centrally located. Note that the corresponding problem of the same obstruction attached to the roof has a related solution, due to the symmetry of the Boussinesq equations[41]. Two extreme cases for the conductivity of the divider are considered. First it is assumed to have a conductivity

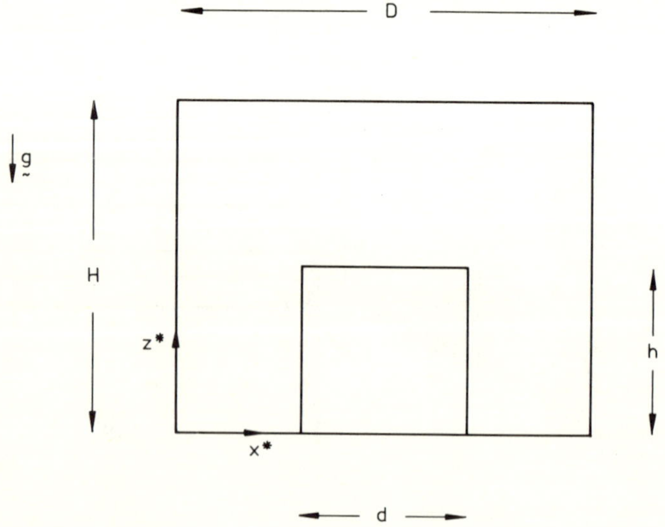

Figure 7.12 Geometry of the rectangular cavity with a partial obstruction

k_r relative to air of 4000, corresponding to a brass obstruction. Second, it is assumed to be a fictitious, highly insulating material with a relative conductivity of 0.1.

Equations (7.1)–(7.5) are to be solved in the fluid region Ω_F, together with

$$k_r \nabla^2 T = 0 \tag{7.22}$$

in the solid region Ω_s. The relative conductivity must be retained in the above equation to ensure continuity of heat flux, in the discrete equations, across the solid/fluid interface. The equations for the two sets of fields in regions Ω_f and Ω_s are then solved simultaneously, with boundary conditions (7.17–19).

Figures 7.13 and 7.14 show streamlines and isotherms for the case of a conducting and insulating obstruction respectively, at a Rayleigh number of

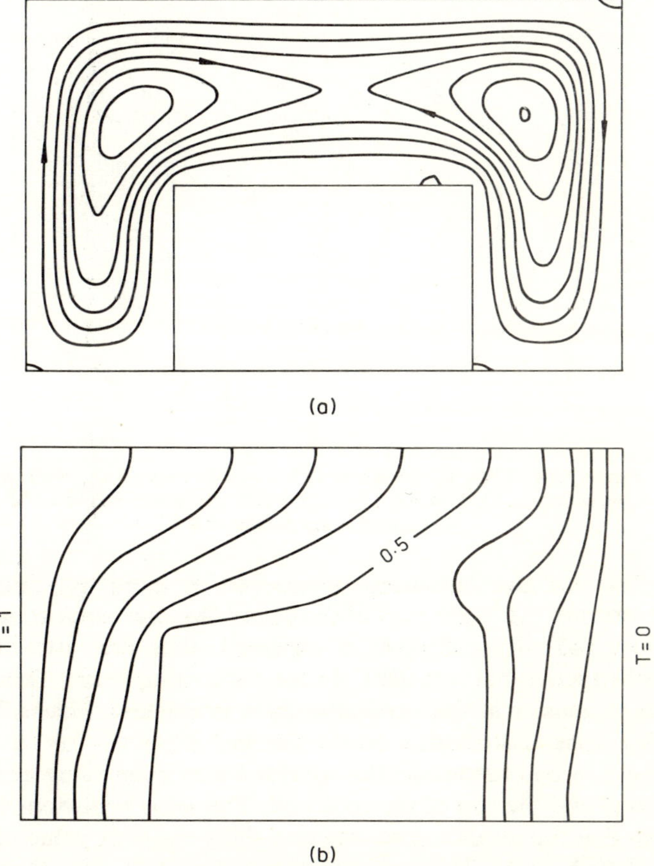

Figure 7.13 Natural convection of air in a cavity with a conducting brass obstruction ($k_r = 4000$) at $Ra = 10^5$: (a) streamlines at intervals of 0.3846; (b) isotherms at intervals of 0.1

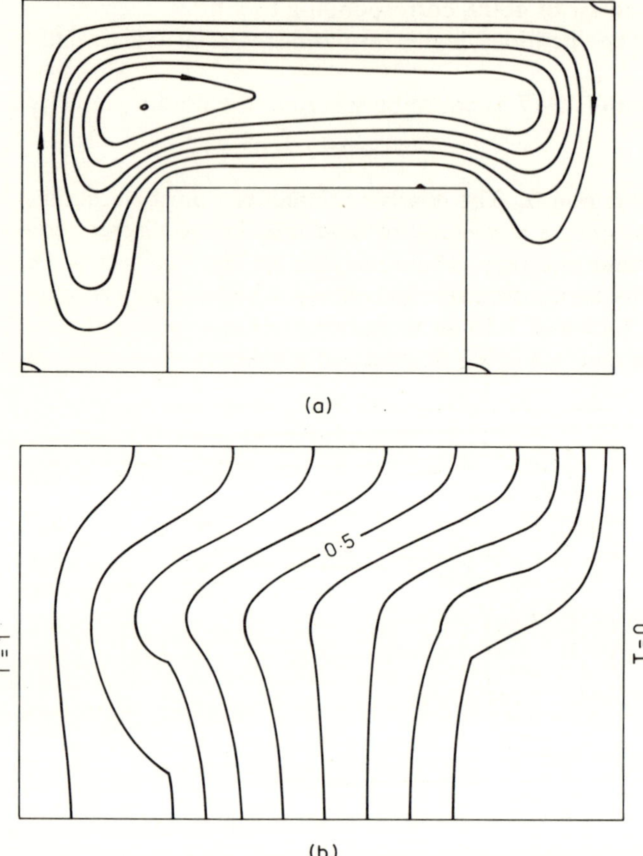

Figure 7.14 Natural convection of air in a cavity with an insulating obstruction ($k_r = 0.1$) at $Ra = 10^5$: (a) streamlines at intervals of 0.3729; (b) isotherms at intervals of 0.1

10^5. The flow for the insulating obstruction is more asymmetrical and penetrates less into the lower part of the cavity than the conducting case. The isotherms are radically different, as expected; they are 'attracted' to the insulating obstruction but 'repelled' by the conducting insert. Their effect on the local heat transfer at the vertical walls is interesting. Figure 7.15 shows the Nusselt number distribution on the hot and cold walls for the two cases. In the absence of an obstruction this attains a maximum value near the base of the hot wall and the top of the cold wall. The same behaviour is apparent here, except that the insulating obstruction shifts the peak value of N_u to the mid-point of the hot wall. The overall heat transfer is higher for the conducting obstruction, as expected.

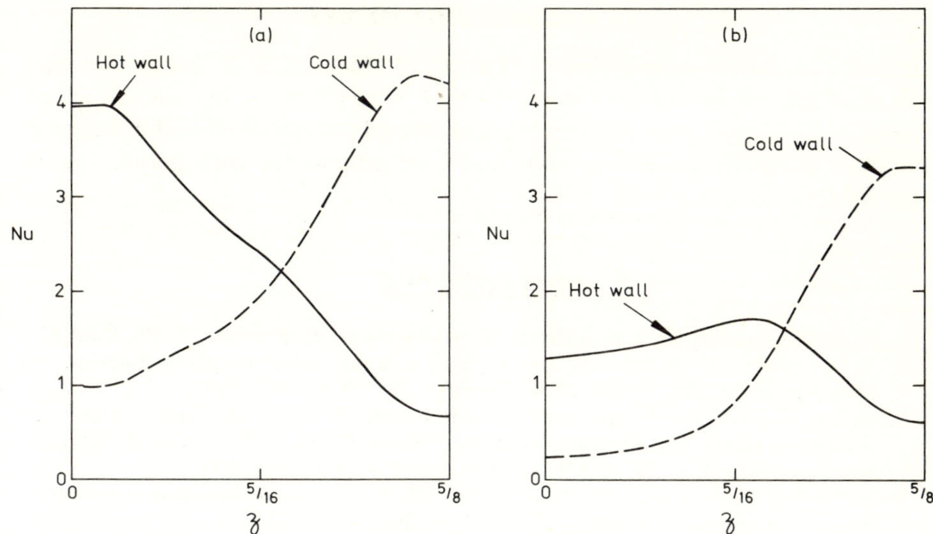

Figure 7.15 The variation of the local Nusselt number on the hot and cold vertical walls at $Ra = 10^5$ for an (a) conducting and (b) insulating obstruction

There are clearly many important simulations to be made on the effect of conducting inserts, in particular, and on problems having distinct regions with different physics, in general. It is hoped that more work will be done in this area.

7.8 CONCLUSIONS

This chapter has described some of the work done on the prediction of natural convection in heated cavities, using the Harwell code ENTWIFE. The examples chosen have demonstrated some of the power and flexibility of the finite element package. In particular, they point the way to further work in many areas, of which the more obvious are: the clarification of the aspect ratio dependence of heat transfer in rectangular cavities; the study of limit points and bifurcations in the steady equations for natural convection; the simulation of problems having distinct regions with different physics; the effect of the conductivity, size and position of inserts on heat transfer in cavities; the study of geometrical effects on natural convection. All such investigations can be carried out with the present code. Its development to allow other important areas to be studied, such as turbulent and three-dimensional convection, is in progress.

ACKNOWLEDGEMENTS

It is a pleasure to thank Drs. J. Rae, C. P. Jackson, I. P. Jones and Mr. K. A. Cliffe of Harwell for many helpful discussions on the simulation of natural convection. Part of this work was carried out on behalf of the National Nuclear Corporation (Risley) Ltd., and I am grateful for their permission to publish.

REFERENCES

1. E. R. G. Eckert, R. J. Goldstein, S. V. Patankar, E. Pfender, J. W. Ramsey, T. W. Simon, E. M. Sparrow and K. Y. Teichman, 'Heat transfer—a review of 1980 literature', *Int. J. Heat Mass Transfer*, **24**, 1863–1902 (1981).
2. K. H. Winters, 'The finite-element simulation of buoyancy-driven flows', in *Numerical Methods in Thermal Problems, Volume II*, eds. R. W. Lewis, K. Morgan and B. A. Schrefler, pp. 920–931, Pineridge Press, Swansea (1981).
3. G. D. Mallinson, and G. de Vahl Davis, 'Three-dimensional natural convection in a box; a numerical study', *J. Fluid Mech.*, **83**, 1–31 (1977).
4. C. P. Jackson and K. A. Cliffe, 'Mixed interpolation in primitive variable finite element formulations for incompressible flow', *Int. J. Num. Meth. Eng.*, **17**, 1659–1688 (1981).
5. B. M. Irons, 'A frontal solution program for finite-element analysis', *Int. J. Num. Meth. Eng.*, **2**, 5–32 (1970).
6. P. Hood, 'Frontal solution program for unsymmetric matrices', *Int. J. Num. Meth. Eng.*, **10**, 379–399 (1975).
7. C. P. Jackson and K. H. Winters, 'A finite-element study of the Bénard problem using parameter-stepping and bifurcation search', *Harwell Report AERE-TP.936*, AERE Harwell, U.K. and *Int. J. Num. Meth. Fluids* (submitted) (1982).
8. G. de Vahl Davis, I. P. Jones, and P. J. Roache, *Computers and Fluids*, **7**, 315–316 (1979).
9. G. de Vahl Davis, I. P. Jones and P. J. Roache, *J. Fluid Mech.* **95**(4), inside back cover (1979).
10. I. P. Jones and C. P. Thompson, 'Numerical solutions for a comparison problem on natural convection in an enclosed cavity', *Harwell Report AERE-R9955 and R9955 Supplement I,* AERE Harwell, England (1981).
11. G. de Vahl Davis and I. P. Jones, 'Natural convection in a square cavity—a comparison exercise', in *Numerical Methods in Thermal Problems, Volume II*, ed. R. W. Lewis, K. Morgan and B. A. Schrefler, pp. 552–572, Pineridge Press, Swansea (1981).
12. C. Quon, 'Effects of grid distribution on the computation of high Rayleigh number convection in a differentially heated cavity', *2nd National Symposium on Numerical Methods in Heat Transfer*, Univ. of Maryland, U.S.A., and Preprint (1981).
13. C. D. Upson, P. M. Gresho and R. L. Lee, 'Finite-element simulation of thermally induced convection in an enclosed cavity', *Report UCID-18602,* Lawrence Livermore Laboratory, California, U.S.A. (1980).
14. R. S. Marshall, J. C. Heinrich and O. C. Zienkiewicz, 'Natural convection in a square enclosure by a finite-element, penalty function method using primitive fluid variables', *Num. Heat Transfer*, **1**, 315–330 (1978).
15. K. H. Winters, 'A numerical study of natural convection in a square cavity', *Harwell Report AERE-R9747,* AERE Harwell, England (1980).

16. P. M. Gresho and R. L. Lee, 'The consistent method for computing derived boundary quantities when the Galerkin FEM is used to solve thermal and/or fluid problems', in *Numerical Methods in Thermal Problems, Volume II*, ed. R. W. Lewis, K. Morgan and B. A. Schrefler, pp. 663–675, Pineridge Press, Swansea (1981).
17. K. A. Cliffe and C. P. Jackson, 'On the implementation of Dirichlet boundary conditions in the finite-element method' (in preparation).
18. W. M. M. Schinkel, 'Natural convection in inclined air-filled enclosures', *Ph.D. Thesis*, Delft Univ. of Technology (1980).
19. J. W. Elder, 'Laminar free convection in a vertical slot', *J. Fluid Mech.*, **23**, 77–98 (1965).
20. I. P. Jones, 'Low Prandtl number free convection in a vertical slot', *Harwell Report AERE-R10416*, AERE Harwell, England (1982).
21. S. H. Yin, T. Y. Wung and K. Chen, 'Natural convection in an air layer enclosed within rectangular cavities', *Int. J. Heat Mass Transfer*, **21**, 307–315 (1978).
22. G. D. Raithby and H. H. Wong, 'Heat transfer by natural convection across vertical air layers', *Num. Heat Transfer*, **4**, 447–457 (1981).
23. E. R. G. Eckert and W. O. Carlson, 'Natural convection in an air layer enclosed between two vertical plates with different temperatures', *Int. J. Heat Mass Transfer*. **2**, 106–120 (1961).
24. R. F. Bergholz, 'Instability of steady natural convection in a vertical fluid layer', *J. Fluid Mech.*, **84**, 743–768 (1978).
25. B. Gilly, P. Bontoux and B. Roux, 'Influence des conditions thermiques de paroi sur la convection naturelle dans une cavité rectangulaire verticale, différentiellement chauffée, *Int. J. Heat Mass Transfer*, **24**, 829–841 (1981).
26. D. Duxbury, 'An interferometric study of natural convection in enclosed plane air layers with complete and partial central vertical divisions', *Ph.D. Thesis*, University of Salford (1979).
27. R. W. Thomas and G. de Vahl Davis, 'Natural convection in annular and rectangular cavities', *Proc. 4th Int. Heat Transfer Conference*, Vol **4**, paper N.C. 2.4, Versailles (1970).
28. D. Duxbury, 'An interferometric study of natural convection in enclosed plane air layers, *M.Sc. Thesis*, Univ. of Salford, England (1972).
29. I. P. Jones, 'A numerical study of natural convection in an air-filled cavity: comparison with experiment', *Num. Heat Transfer*, **2**, 193–213 (1979).
30. H. B. Keller, 'Numerical solution of bifurcation and non-linear eigenvalue problems', in *Applications of Bifurcation Theory* (ed. P. H. Rabinowitz) pp. 359–384, Academic Press, London (1977).
31. B. Fornberg, 'A numerical study of steady viscous flow past a circular cylinder' *J. Fluid Mech.* **98**, 819–855 (1980).
32. M. A. Azouni, 'Survey of thermoconvective instabilities of confined fluids', *J. Non-Equilib. Thermodyn.*, **4**, 321–348 (1979).
33. J. M. Luijkx and J. K. Platten, 'On the onset of free convection in a rectangular channel', *J. Non-Equilib. Thermodyn.*, **6**, 141–158 (1981).
34. E. Leonardi and J. A. Reizes, 'Convective flows in closed cavities with variable fluid properties' in *Numerical Methods in Heat Transfer*, ed. R. W. Lewis, K. Morgan, and O. C. Zienkiewicz, pp. 387–412, Wiley, London (1981).
35. J. G. Van Steeg and P. Wesseling, 'Solution of the Boussinesq equations by means of the finite-element method', *Computers and Fluids*, **6**, 93–101 (1978).
36. P. M. Gresho, R. L. Lee, S. T. Chan and J. M. Leone, 'A new finite element for incompressible or Boussinesq fluids', *Third Int. Conf. on Finite Elements in Flow Problems*, Banff, Canada (1980).

37. P. Hall and I. C. Walton, 'The smooth transition to a convective régime in a two-dimensional box', *Proc. Roy. Soc. A*, **358**, 199–221 (1978).
38. H. E. Janikowski, J. Ward, and S. D. Probert, 'Free convection in vertical air-filled rectangular cavities fitted with baffles', *6th Int. Heat Transfer Conf.* Vol. 2, pp. 257–262, Toronto, Canada (1978).
39. M. W. Nansteel and R. Greif, 'Natural convection in undivided and partially divided rectangular enclosures', *J. Heat Transfer*, **103**, 623–629 (1981).
40. C. P. Jackson and K. H. Winters, 'An extension of the finite-element package TGSL to problems of many regions each with different physics, *Harwell Report AERE M.3186,* AERE Harwell, England (1981).
41. K. H. Winters, 'The effect of conducting divisions on the natural convection of air in a rectangular cavity with heated sidewalls', *AIAA/ASME 3rd Joint Thermophysics, Fluids, Plasma and Heat Transfer Conference*, Paper ASME-82-HT 69, St. Louis, U.S.A. (1982).

Numerical Methods in Heat Transfer, Volume II
Edited by R. W. Lewis, K. Morgan, and B. A. Schrefler

Chapter 8

Influence of Thermal Wall Conditions on the Natural Convection in Heated Cavities

B. Gilly, B. Roux, and P. Bontoux

SUMMARY

The natural convection in differentially heated cavities is usually studied for ideal conditions: isothermal warm and cold, active walls and adiabatic passive lateral walls. In practice, the thermal conditions on the active walls may not be uniform. Two models of non-uniform temperature distribution are studied by the method of numerical simulation. Modifications are proposed for the correlation laws given in the ideal case between the Nusselt number and the Rayleigh number. The influence of conducting lateral walls is also investigated.

8.1 INTRODUCTION

The study of natural convection in differentially heated cavities has many practical applications: solar cells, thermal insulation, crystal growth in vapour transport process, etc. A large number of theoretical, numerical and experimental works have been published on this phenomena. A review paper has been written by Catton.[1] These works have been developed for the case of the 'ideal' configuration, which applies to a rectangular cavity when the hot and the cold walls (called 'active' walls) are supposed isothermal and when the lateral walls (called 'passive' walls) are generally supposed adiabatic.

In practice, active walls are generally not isothermal. For a solar absorber, for instance, pipes in thermal contact with the water heater plate, introduce a temperature gradient on the absorber plate. Likewise, experimental work shows that the temperature of the cold wall (glass cover plate) is not uniform: the temperature in the upper part of the glass may be twice as much as that in the lower part.[2] Moreover, passive walls are never perfectly adiabatic, but in an intermediate state between perfectly adiabatic and perfectly conducting walls. The influence of changes in the conductivity of the passive walls has been studied in the case of high Prandtl number values[3] and of square air cavities.[4]

The aim of this paper is to propose modifications to the evaluation of the Nusselt number when ideal conditions are not satisfied.

The physical model to be studied is shown in Figure 8.1. The rectangular cavity is vertical; the two active walls, separated by a distance H, are maintained at different temperatures $\bar{T}_a$ and $\bar{T}_v$ with $\bar{T}_a > \bar{T}_v$. This temperature difference causes a convective flow which is supposed laminar, steady, and two-dimensional. The principal parameters of the problem are the aspect ratio $l = L/H$, the Rayleigh number (R_a) based on the temperature difference between the hot and cold active walls, and the Prandtl number (P_r) characterizing the fluid inside the cavity.

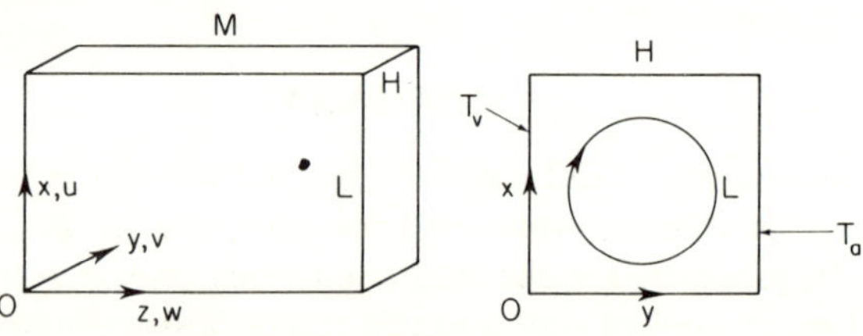

Figure 8.1 Geometry of the cavity

8.2 MATHEMATICAL MODEL

The equations governing the motion of a fluid by natural convection are derived from the conservation of mass, momentum and energy. When density variations resulting from temperature differences are small ($\Delta\rho/\rho < 10\%$), the simplified Boussinesq approximation is valid (constant density except in the buoyancy term, negligible viscous dissipation, constant fluid properties). With the stream function Ψ, the vorticity ζ, and the temperature T as dependent dimensionless variables, the unsteady governing equations may be written in the following dimensionless form:[4]

$$\frac{D\zeta}{Dt} = P_r \nabla^2 \zeta - R_a P_r \frac{\partial T}{\partial y}, \tag{8.1}$$

$$\frac{DT}{Dt} = \nabla^2 T, \tag{8.2}$$

$$\nabla^2 \Psi = -\zeta, \tag{8.3}$$

where

$$u=\frac{\partial\Psi}{\partial y}, \qquad v=-\frac{\partial\Psi}{\partial x},$$

$$\frac{\mathrm{D}}{\mathrm{D}t}=\frac{\partial}{\partial t}+u\frac{\partial}{\partial x}+v\frac{\partial}{\partial y}, \qquad \nabla^2=\frac{\partial^2}{\partial x^2}+\frac{\partial^2}{\partial y^2},$$

with (note that, for example, t is a nondimensional variable derived from the time $\bar{t}$ and likewise for the other variables):

$$t=\frac{\bar{t}\chi}{H^2} \qquad \text{(nondimensional time)}$$

$$x=\frac{\bar{x}}{H}, \quad y=\frac{\bar{y}}{H} \qquad \text{(dimensionless coordinates)}$$

$$u=\frac{\bar{u}H}{\chi}, \quad v=\frac{\bar{v}H}{\chi} \qquad \text{(dimensionless velocities in the } x \text{ and } y \text{ directions)}$$

$$\Psi=\frac{\bar{\Psi}}{\chi}, \quad \zeta=\frac{\bar{\zeta}H^2}{\chi}, \quad T=\frac{\bar{T}-(\bar{T}_a+\bar{T}_v)/2}{\bar{T}_a-\bar{T}_v}$$

$$R_a=\frac{g\bar{\beta}(\bar{T}_a-\bar{T}_v)H^3}{\nu\chi}=\text{Rayleigh number},$$

$$P_r=\frac{\nu}{\chi}=\text{Prandtl number},$$

$$\chi=\frac{k}{\rho C_p}=\text{thermal diffusivity},$$

g is the acceleration due to gravity, $\bar{\beta}$ is the thermal expansion coefficient, ν is the kinematic viscosity, ρ is the density, k is the thermal conductivity and C_p is the specific heat at constant pressure.

The dynamical boundary conditions for impermeable no-slip walls are

$$\Psi(0,y)=\Psi(l,y)=\Psi(x,0)=\Psi(x,1)=0, \tag{8.4}$$

$$\frac{\partial\Psi}{\partial x}(0,y)=\frac{\partial\Psi}{\partial x}(l,y)=\frac{\partial\Psi}{\partial y}(x,0)=\frac{\partial\Psi}{\partial y}(x,1)=0. \tag{8.5}$$

In the ideal case, the thermal boundary conditions on active walls are

$$T(x,0)=-\tfrac{1}{2} \qquad \text{and} \qquad T(x,1)=\tfrac{1}{2}. \tag{8.6}$$

Non-uniform temperature distributions have been simulated by linear and sinusoidal models:

$$T(x,0)=-\frac{1}{2}+\alpha\left(\frac{x}{l}-\frac{1}{2}\right) \tag{8.7(a)}$$

and

$$T(x, 1)=\frac{1}{2}+\beta\left(\frac{x}{l}-\frac{1}{2}\right), \qquad (8.7(b))$$

$$T(x, 0)=-\frac{1}{2}+\frac{\alpha}{2}\sin\left[\pi\left(\frac{x}{l}-\frac{1}{2}\right)\right] \qquad (8.8(a))$$

and

$$T(x, 1)=\frac{1}{2}+\frac{\beta}{2}\sin\left[\pi\left(\frac{x}{l}-\frac{1}{2}\right)\right]. \qquad (8.8(b))$$

The thermal boundary conditions on passive walls are written, in the case of adiabatic walls,

$$\frac{\partial T}{\partial x}(0, y)=\frac{\partial T}{\partial x}(l, y)=0 \qquad (8.9)$$

and in the case of perfectly conducting walls,

$$\frac{\partial T}{\partial y}(0, y)=\frac{\partial T}{\partial y}(l, y)=1. \qquad (8.10)$$

8.3 NUMERICAL METHODS

Although we are interested only in the steady solution, nevertheless the governing equations (8.1)–(8.3) are solved in the pseudo-unsteady form by adding the term $\partial\Psi/\partial t$ to the right-hand side of Equation (8.3). Each of the above equations is then parabolic in the temporal direction and elliptic in the two spatial directions. Such a partial differential equation has been solved using the alternating direction implicit method (ADI)[6,7] which leads, at each step, to a second-order ordinary differential equation. This differential equation can be considered in either a conservation or a nonconservation form, and solved with first-order accuracy using an upwind differencing scheme, with second-order accuracy using a centred scheme and with fourth-order accuracy using a compact Hermitian method. For Hermitian methods which use only three nodes, first and second derivatives are also considered as unknowns as well as the undifferentiated variables.[8] In fact we used a method based on a Hermitian approximation for Equation (8.3), with an accuracy $O(h^4)$ (h is the mesh spacing), combined with a classical approximation for the vorticity transport equation (8.1), with an accuracy $O(h^2)$ or $O(h)$.[4,9,10] In the case of Dirichlet boundary conditions on temperature as given in relations (8.10) (perfectly conductive passive walls), an Hermitian approximation $O(h^4)$ for the energy equation has also been considered. The conservative formulation improves the algorithm convergence and the solution accuracy. In the case of flow configurations for which $R_a > 100{,}000$, $l \geqslant 10$, it is useful to use an upwind differencing scheme which introduces a numerical 'false diffusion', but gives an approximate solution with a relatively small number of nodes (less than 2400). Convergence to a more accurate solution can be then obtained by pursuing iterations using a centred combined method.

Vorticity boundary conditions are given by the discretization of Equation (8.3) using the Hirsch relation which, in the case of the combined methods, gives a better accuracy for the approximation of ζ at the wall:[9]

$$\zeta_w = -\left(\frac{\partial^2\Psi}{\partial n^2}\right)_{w+1} + \frac{6}{h}\left(\frac{\partial\Psi}{\partial n}\right)_{w+1} + \frac{12}{h^2}(\Psi_w - \Psi_{w+1}) \tag{8.11}$$

8.4 IDEAL CONFIGURATION

We give first some previous results obtained for a square air cavity in the ideal case of isothermal active walls and perfectly adiabatic passive walls.[4,11]

These studies have shown that the appropriate correlations between heat losses and the characteristic parameters of the cavity and of the fluid depend on the flow régimes inside the cavity, and it is important to know the flow régime limits. In the case of laminar two-dimensional flow, different régimes have been classified by Eckert and Carlson:[12] conduction, transition and boundary-layer régimes. We use the definitions given by Thomas and De Vahl Davis[13] for the upper conduction régime bound, denoted by R_{a_1}, and for the lower boundary-layer regime bound, denoted by R_{a_2}: R_{a_1} is the Rayleigh number value for which the horizontal temperature gradient at the cavity midpoint $(\partial T/\partial y)_m$ is equal to 0.9 and R_{a_2} is the Rayleigh number value for which $(\partial T/\partial y)_m$ is equal to zero. For air and for the case of a vertical cavity, the following bound values have been proposed:[4,11]

$$R_{a_1} = 750 \quad \text{and} \quad R_{a_2} = 7500 \qquad \text{for } l = 1, \tag{8.12}$$

$$R_{a_1} = 250l \quad \text{and} \quad R_{a_2} = 2500l \qquad \text{for } l \geqslant 4. \tag{8.13}$$

Convective losses, for the boundary layer regime, can be estimated by correlations between the average Nusselt number, $\bar{N}_u$, calculated on the median plane, ($\bar{N}_u = (1/l)\int_0^1 N_u\,dx$, where N_u is the ratio of the heat flux density (q) for combined conduction and convection, and the heat flux density (q_{ref}) for the case of conduction only), the Rayleigh number R_a, and the aspect ratio, l:

$$\bar{N}_u = 0.138 R_a^{0.30} \qquad \text{for } l = 1 \tag{8.14}$$

$$\bar{N}_u = 0.29 R_a^{0.25} l^{-0.25} \qquad \text{for } l \geqslant 4 \tag{8.15}$$

The mass flux inside the cavity is characterized by the value of the stream function Ψ_m at the centre of the cavity. An approximation, deduced from

Gill's theory,[14] has been proposed

$$\Psi_m \simeq C^3 l (R_a l^{-1})^{0.25} \tag{8.16}$$

where C is a constant expressed analytically by

$$C = \left[4l\left(\frac{\partial T}{\partial x}\right)_m\right]^{-1/4}$$

and $(\partial T/\partial x)_m$ is the vertical temperature gradient at the centre of the cavity, which is a characteristic parameter of thermal stratification.[15] Moreover, if we choose the values $C = 0.80$ for $l = 1$ and $C = 0.85$ for $l \geq 4$, formula (8.16) gives a good approximation of Ψ_m.[16]

8.5 INFLUENCE OF THE PASSIVE WALLS CONDITIONS

For a square air cavity, with perfectly conductive passive walls, the boundary R_a of the transition regime, $R_{a_1}^{cond}$ and $R_{a_2}^{cond}$, are increased respectively by 33 and 13% with respect to the values obtained in the ideal case (Roux *et al.*[4])

$$R_{a_1}^{cond} = 1000 \quad \text{and} \quad R_{a_2}^{cond} = 8500 \qquad \text{for } l = 1. \tag{8.17}$$

For the boundary-layer régime, an empirical correlation which fits the computed values of the average Nusselt number at the cold wall and the Rayleigh number is[4]

$$\bar{N}_u^{cond} = 0.130 R_a^{0.28}. \tag{8.18}$$

For a given R_a, this result exhibits a decrease of $\bar{N}_u$ compared with the ideal case, this decrease being about 21% at $R_a = 100{,}000$. This is not paradoxical because the considered value of the Nusselt number takes into account the energy which is lost through the cold wall of the cavity, but not the losses through the passive walls which are zero in the ideal case.

We studied the influence of the passive surfaces for rectangular cavities. We characterized it by the parameter

$$\varepsilon = 2\left|\frac{\bar{N}_u - \bar{N}_u^{cond}}{\bar{N}_u + \bar{N}_u^{cond}}\right| \tag{8.19}$$

The variation of ε with R_a, for various values of the aspect ratio l, is shown in Figure 8.2. Examination of the figure shows the following behaviour:

(a) The parameter ε is close to zero for the conduction regime ($R_a < R_{a_1}^{cond}$). It remains nearly constant for the boundary-layer regime but varies strongly with R_a when $l \leq 1$ in the transition regime.

(b) It decreases with l in the boundary-layer regime. This behaviour is apparent in Figure 8.3 where the variation of ε with l is shown for $R_a/l = 10{,}000$. The parameter ε is strongly dependent on l when $l < 5$.

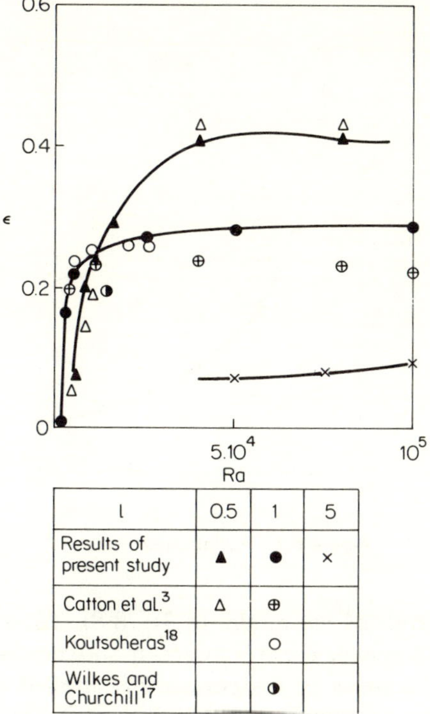

Figure 8.2 Variation of ε with R_a

For $l \geqslant 5$, the decrease is more moderate, and this demonstrates the progressive diminution of the influence of the end walls when l increases.

(c) The results obtained by Wilkes and Churchill,[17] Koutsoheras,[18] and Catton *et al.*[3] are shown in Figure 8.2. The agreement with the calculations of Koutsoheras is quite good for air with $l = 1$. Our results are very close those of Catton obtained for $P_r = \infty$ and $R_a \leqslant 10{,}000$. When $R_a >$ 10,000, some differences can be seen which probably arise from the effect of the Prandtl number.

8.6 INFLUENCE OF NON-ISOTHERMAL ACTIVE WALLS

The influence of the active walls was studied in the case of an air cavity with perfectly insulated passive walls. Linear and sinusoidal models (relations (8.7) and (8.8)) were chosen to represent the variation of the temperature on an active wall. The rates of variation, α and β, were taken between 0 and 0.6.

As shown in Reference 16, the density of heat flux, $q_{ref} = k\ (\bar{T}_a - \bar{T}_v)/H$, in the definition of $\bar{N}_u$ must be the same as in the ideal case (8.6) because of

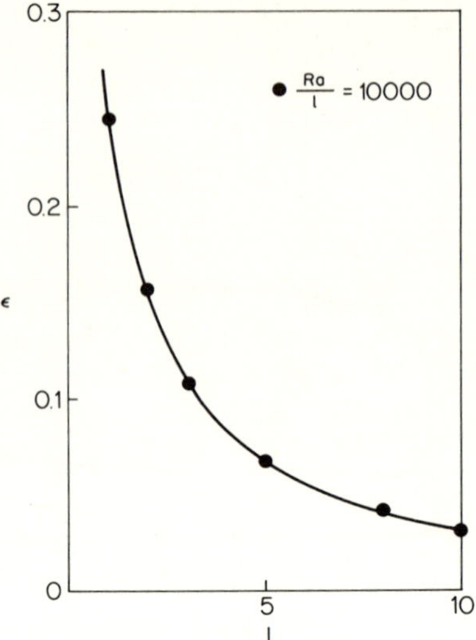

Figure 8.3 Variation of ε with l

the symmetry properties of the model (8.7), (8.8). Then the usual definition of $\bar{N}_u$ may be kept. Moreover there is a correspondence between the solutions obtained when the variation of temperature is applied at the cold wall and at the hot wall: these two solutions are such that the second can be derived from the first by a reflection with respect to the centre of the cavity for Ψ, ζ, $\partial T/\partial x$, $\partial T/\partial y$, N_u and a reflection with a change of sign for T, u, v.

8.6.1 Influence of one non-isothermal active wall

The relations (8.7(b)) or (8.8(b)) were used for $l = 1$.

8.6.1.1 Flow régimes

The horizontal gradient of temperature in the middle of the cavity, $(\partial T/\partial y)_m$, is given in terms of R_a (Figure 8.4). As β increases from 0 to 0.6, $(\partial T/\partial y)_m$ decreases with respect to its value for $\beta = 0$ if $R_a < 25{,}000$ and increases for $R_a > 25{,}000$. By using the criteria of Thomas and De Vahl Davis,[13] the boundary values, R_{a_1} and R_{a_2}, corresponding to the ideal case ((relations (8.12)):

linear case (relation (8.7(b))

$$R_{a_1} \cong (1 - 0.25\beta) R_{a_1},$$
$$R_{a_2} \cong (1 - 0.17\beta) R_{a_2}. \qquad (8.20)$$

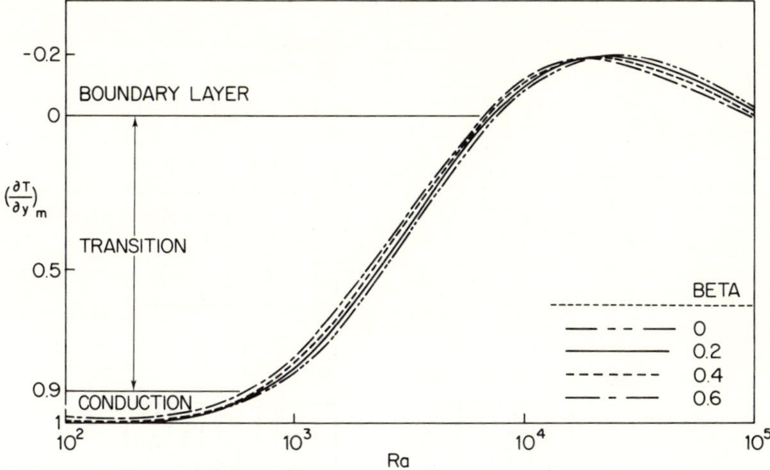

Figure 8.4 Horizontal temperature gradient at cavity midpoint (adiabatic case)

Sinusoidal case (relation (8.8(b).)

$$\begin{aligned} R_{a_1} &= (1-0.25\beta)R_{a_1}, \\ R_{a_2} &= (1-0.25\beta)R_{a_2}. \end{aligned} \tag{8.21}$$

For each of the three régimes (conduction, transition, boundary layer) the isotherm patterns exhibit distortions in the neighbourhood of the non-isothermal wall which increase as β increases (see Figure 8.5). The region close to the isothermal wall is practically unperturbed. Figure 8.5 shows that the wall gradient of temperature also substantially affects the core of the cavity for the conduction regime (Figure 8.5(a)). For the boundary-layer régime (Figure 8.5(c)), the influence is limited to the boundary layer close to the non-isothermal wall.

The streamline patterns (Figure 8.6) show a different behaviour. The perturbations appear small for the conduction and transition régimes (Figure 8.6(a)) and 8.6(b)). However, they become very important in the boundary-layer régime (Figure 8.6(c)) where the core of the cavity is directly affected. These perturbations correspond to a diminution of the mass flux when β increases. The slowing down of the fluid is shown by the velocity profiles in Figure 8.7. The mass flux, characterized by the value of stream function in the middle of the cavity (Ψ_m^β), is related to the stream function value obtained in the ideal case (relation (8.16)) as follows:

$$\Psi_m \cong (1-0.14\beta)\Psi_m, \tag{8.22}$$

RA = 100

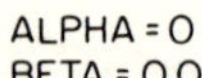

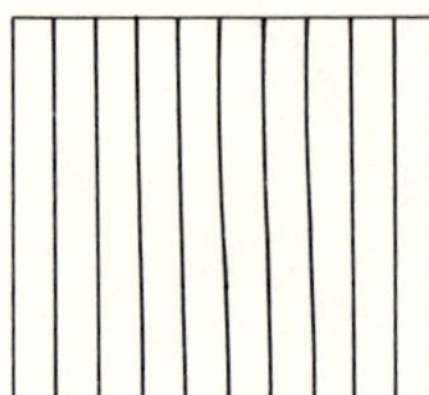

ALPHA = 0
BETA = 0.2

ALPHA = 0
BETA = 0.4

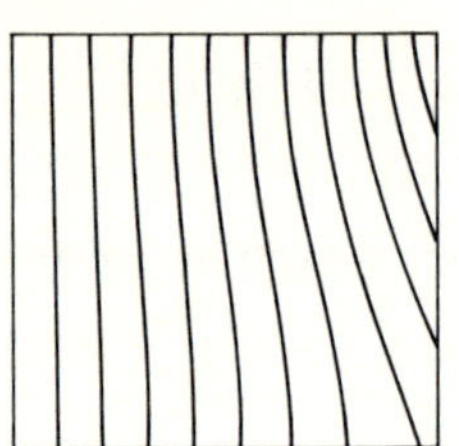

ALPHA = 0
BETA = 0.6

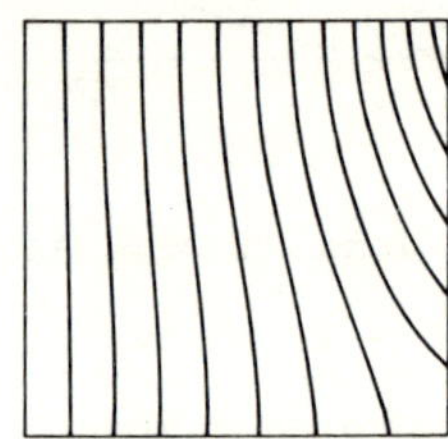

(a)

RA = 2500

ALPHA = 0
BETA = 0.0

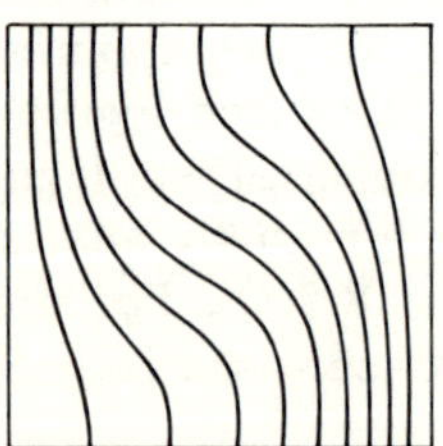

ALPHA = 0
BETA = 0.2

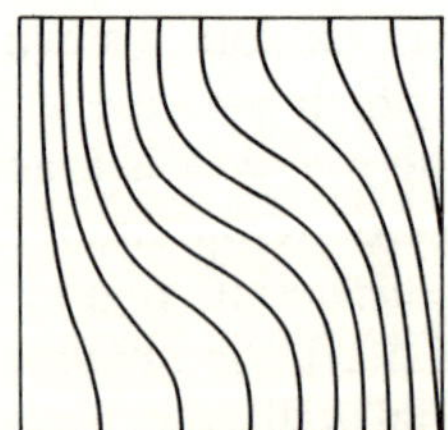

ALPHA = 0
BETA = 0.4

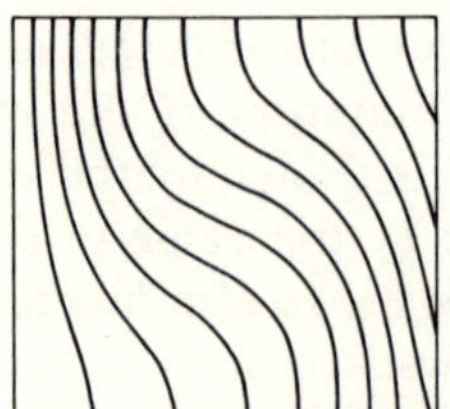

ALPHA = 0
BETA = 0.6

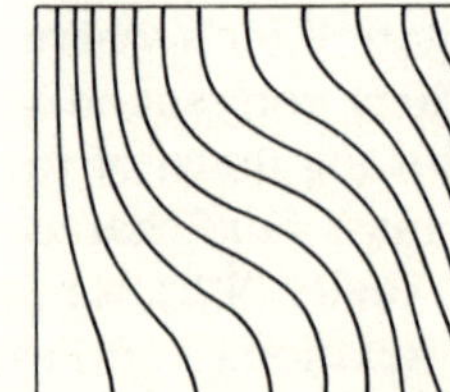

(b)

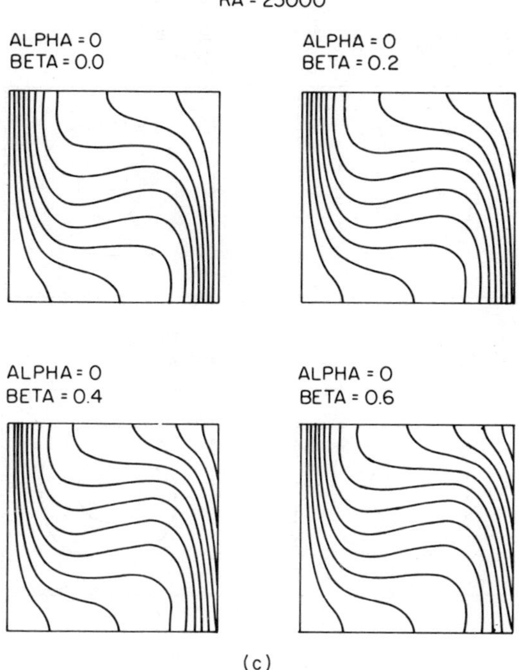

Figure 8.5 Influence of the temperature variation at the hot wall on the isothermal patterns: (a) $R_a = 100$; (b) $R_a = 2500$; (c) $R_a = 25{,}000$

where Ψ_m varies in proportion to $(\partial T/\partial x)^{-3/4}$, where $(\partial T/\partial x)_m$ is the vertical temperature gradient at the centre of the cavity. The gradient $(\partial T/\partial x)_m$ increases with β compared with the ideal case (Fig. 8.8).

The extension of the relation (8.16) to the case of a non-isothermal wall gives approximate values of Ψ_m^β close to those obtained with (8.22)

$$\Psi_m^\beta \cong \Psi_m \left[\left(\frac{\partial T}{\partial x} \right)_m \bigg/ \left(\frac{\partial T}{\partial x} \right)_m^\beta \right]^{3/4}. \tag{8.23}$$

8.6.1.2 *Heat losses by convection*

The average Nusselt number is shown in terms of the Rayleigh number in Figure 8.9 when the variation of temperature is linear (relation (8.7(b))). When β varies from 0 to 0.6, the Nusselt number increases slightly for a given R_a, which reveals an increase of the convective losses. For the boundary-layer régime, the Nusselt number may be related to its value in the ideal case

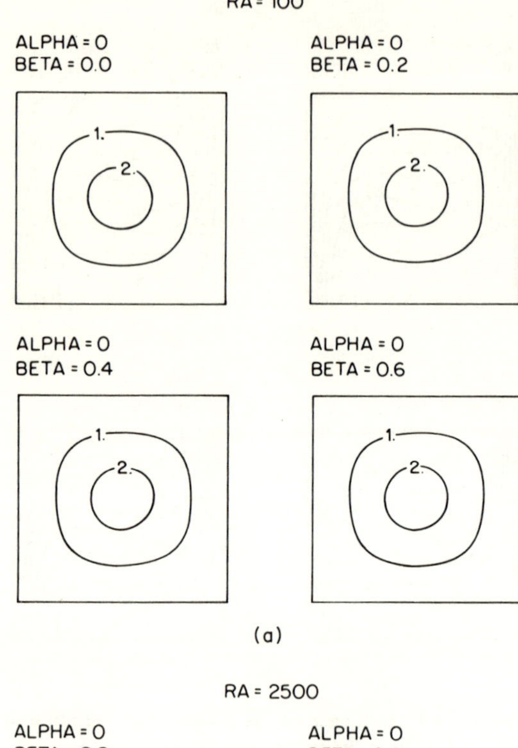
RA = 100
ALPHA = 0
BETA = 0.0
1.
2.
ALPHA = 0
BETA = 0.2
1.
2.
ALPHA = 0
BETA = 0.4
1.
2.
ALPHA = 0
BETA = 0.6
1.
2.

(a)

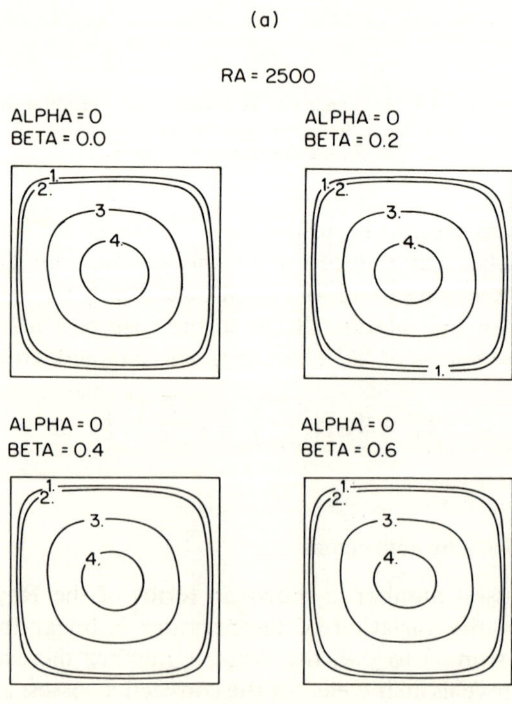
RA = 2500
ALPHA = 0
BETA = 0.0
1.
2.
3.
4.
ALPHA = 0
BETA = 0.2
1.
2.
3.
4.
1.
ALPHA = 0
BETA = 0.4
1.
2.
3.
4.
ALPHA = 0
BETA = 0.6
1.
2.
3.
4.

(b)

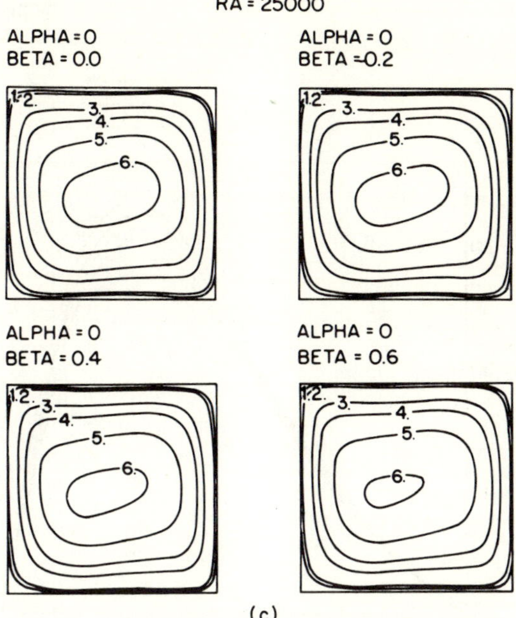

Figure 8.6 Influence of the temperature variation at the hot wall on the streamline patterns: (a) $R_a = 100$; (b) $R_a = 2500$; (c) $R_a = 25{,}000$

(relation (8.14)) by the following approximate correlation:

$$\bar{N}_u^\beta = (1 + 0.08\beta)\bar{N}_u. \qquad (8.24)$$

In the case of the sinusoidal model (relation (8.8b))), the correlation is slightly different:

$$\bar{N}_u^\beta = (1 + 0.06\beta)\bar{N}_u. \qquad (8.25)$$

The local heat losses can be characterized by $N_u(x, 1)$. The variation of $N_u(x, 1)$ with x is shown in Figure 8.10, from which we see that:

(a) the variation of the local Nusselt number is very similar to the one obtained by Catton *et al.*:[3] the slope near the end walls is nearly zero. However, our results show a small extremum near the insulated wall $x = 0$, which is not observed by Catton. An explanation may be found in the fact that Catton's results correspond to $P_r = \infty$ and $R_a = 300{,}000$.

(b) The presence of temperature gradient (relation (8.7(b)) correspond to a sensible diminution of the local heat losses in the lower part of the cavity and the maximum heat loss moves toward the centre of the cavity.

(c) In the upper part of the cavity, the losses increase when β increases. A relative minimum appears close to the upper insulated wall.

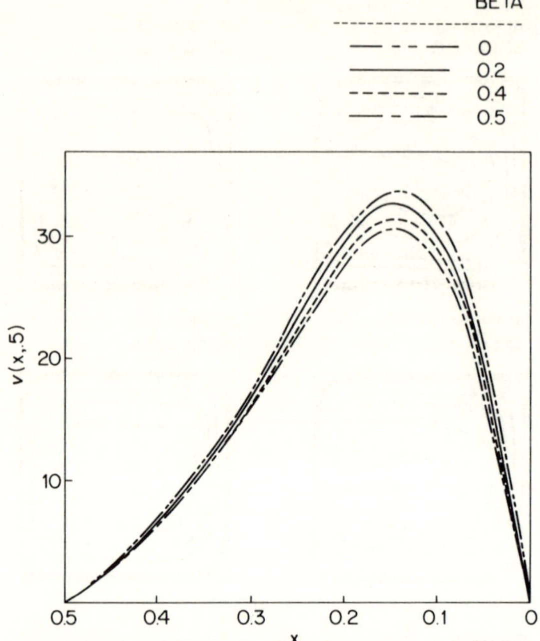

Figure 8.7 Horizontal velocity component, v, along the vertical median axis ($R_a = 100{,}000$, $\alpha = 0$)

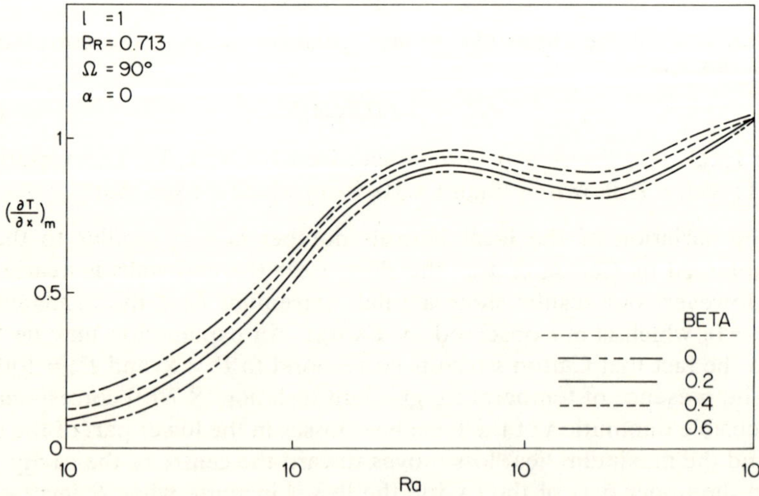

Figure 8.8 Vertical temperature gradient at cavity midpoint

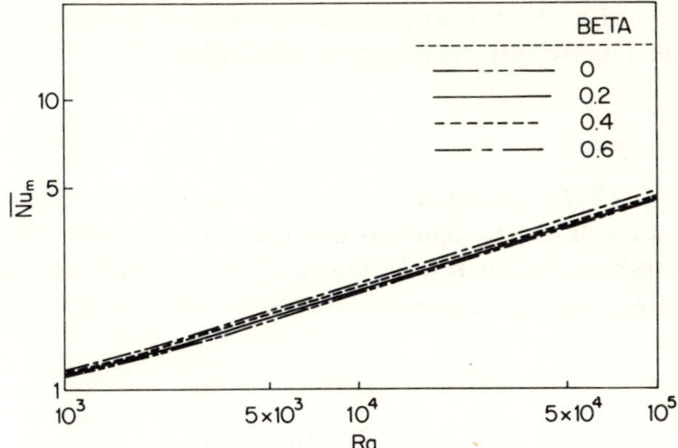

Figure 8.9 Mean Nusselt number, $\bar{N}_{um}$, versus Rayleigh number, R_a, ($\alpha = 0$)

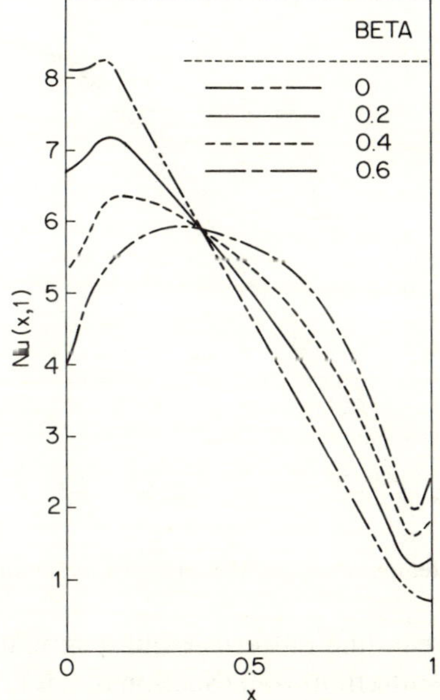

Figure 8.10 Local Nusselt number along the hot wall ($R_a = 100{,}000$, $\alpha = 0$)

8.6.2 Influence of two non-isothermal active walls

The study has been performed in the case of square ($l = 1$) and rectangular ($l = 4$, 6, and 10) cavities. The model is linear (relation (8.7), $0 < \alpha$, $\beta < 0, 6$). It was shown (8.16) that $\alpha = \beta$ restores the symmetry found in the ideal case.

The influence of a second non-isothermal wall is similar to the influence of the first one. The two effects augment each other.

8.6.2.1 Flow regimes

The variation of $(\partial T/\partial y)_m$ with R_a is shown in Figure 8.11 for $l=1$ and $\alpha=\beta$ and various values of α. The figure is qualitatively similar to Figure 8.4. The R_a values corresponding to the frontiers of the conduction, transition and boundary layer régimes are expressed as follows with respect to the ideal case values:

$$R_{a_1}^{\alpha\beta}=[1-0.25(\alpha+\beta)]R_{a_1} \tag{8.26a}$$

$$R_{a_2}^{\alpha\beta}=[1-0.17(\alpha+\beta)]R_{a_2} \tag{8.26b}$$

The relation (8.26(b)) is again valid when $l \geqslant 4$.

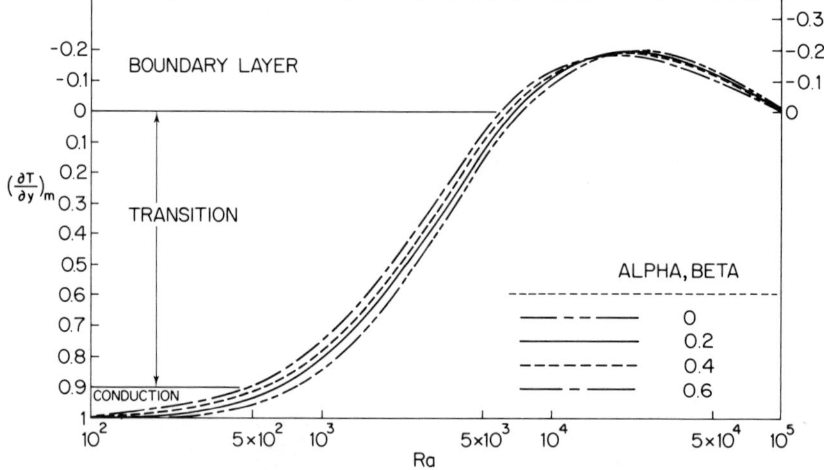

Figure 8.11 Horizontal temperature gradient at cavity midpoint, versus Rayleigh number

The isotherm and streamline patterns exhibit similar types of behaviour as in the case of one non-isothermal wall (Section 8.6.1.1) but with the symmetry properties. The mass flux diminishes in the boundary-layer régime in the following manner:

$$\Psi_m^{\alpha\beta} \cong [1-0.14(\alpha+\beta)]\Psi_m \qquad \text{for } l=1 \tag{8.27a}$$

$$\Psi_m^{\alpha\beta} \cong [1-0.25(\alpha+\beta)]\Psi_m \qquad \text{for } l \geqslant 4 \tag{8.27b}$$

We note that, as in the preceding paragraph, the only effect of the stratification parameter $(\partial T/\partial x)_m$ can account for the variation of the mass flux according to relation (8.23).

The streamline pictures for a rectangular cavity, displayed in Figure 8.12, show that the transverse instabilities observed in the ideal case remain. The values of Ψ_m on the boundary of these vortices decreases as β increases from $\beta = 0$, as described by (8.27(b)). Figure 8.13 shows the isothermals in the rectangular cavity for various values of α. It is clear from the figure that the main effects of a change in α are observable near the ends of the cavity; there is also an increase in the local temperature gradient at the centre of the cavity $(\partial T/\partial x)_m$ as β increases.

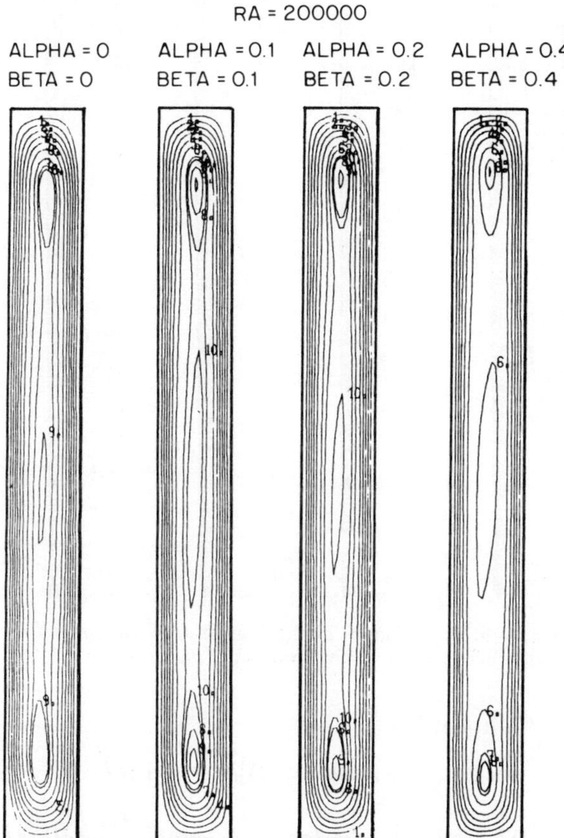

Figure 8.12 Influence of the temperature variations at the active walls on the streamline patterns ($l = 10$)

8.6.2.2 Heat losses by convection

The average Nusselt number displayed in Figure 8.14 for $l = 1$ and $0 \leqslant \alpha = \beta \leqslant 0.6$ shows that the respective effects of each wall are additive. The following correlation has been constructed for the boundary-layer regime ($R_a > 7500$)

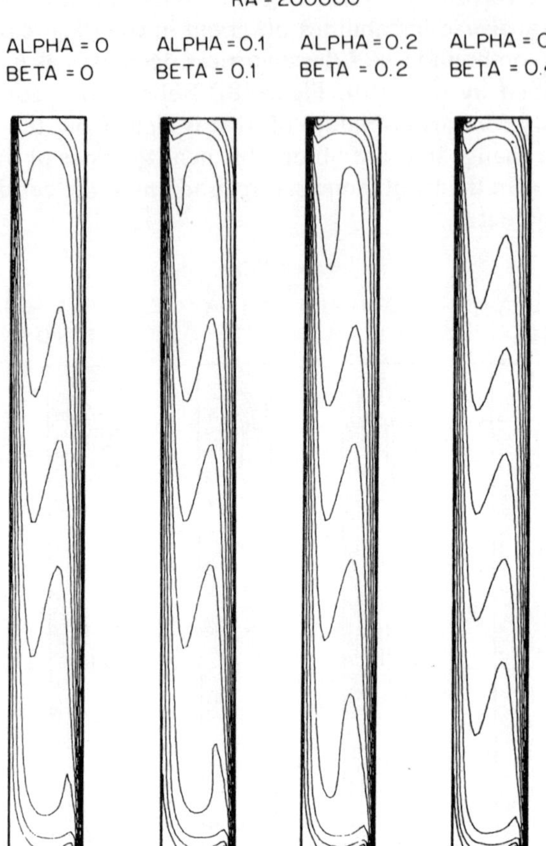

Figure 8.13 Influence of the temperature variations at the active walls on the isothermal patterns ($l=10$)

under the form of a linear law:

$$\bar{N}_u^{\alpha\beta} \cong [1+0.08(\alpha+\beta)]\bar{N}_u \qquad \text{for } l=1 \tag{8.28a}$$

$$\bar{N}_u^{\alpha\beta} \cong [1+0.07(\alpha+\beta)]\bar{N}_u \qquad \text{for } l \geqslant 4 \quad \text{and} \quad \frac{R_a}{l} \geqslant 10{,}000 \tag{8.28b}$$

The local Nusselt number, $N_u(x, 1)$, along the hot wall is shown in Fig. 8.15 for $l=10$, $R_a=30{,}000$ and 100,000. The results for $\alpha=\beta=0$ are in good agreement with the experimental results of Dulnev *et al* (Figure 6 of Reference 19) for $R_a \cong 30{,}000$. The agreement is particularly apparent for the upper region ($x \geqslant 2.5$). The variation of $N_u(x, 1)$ is also close to that predicted by Catton *et al.* (Figure 5 of Reference 3) for $\alpha=\beta=0$, $l=5$ and $R_a=300{,}000$,

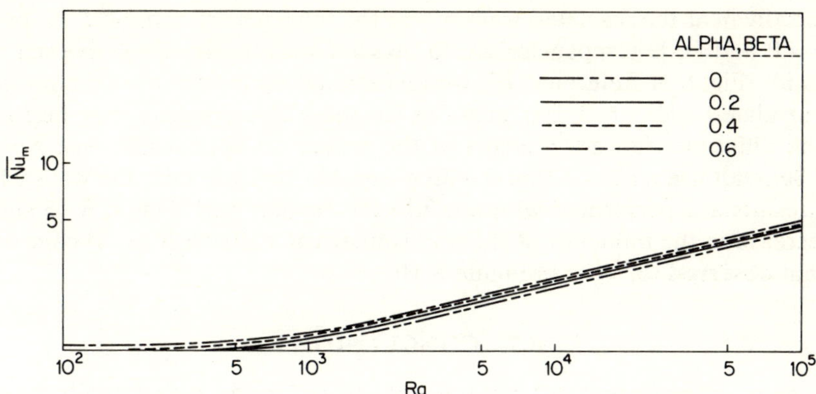

Figure 8.14 Mean Nusselt number, $\bar{N}_{u_m}$ versus Rayleigh number

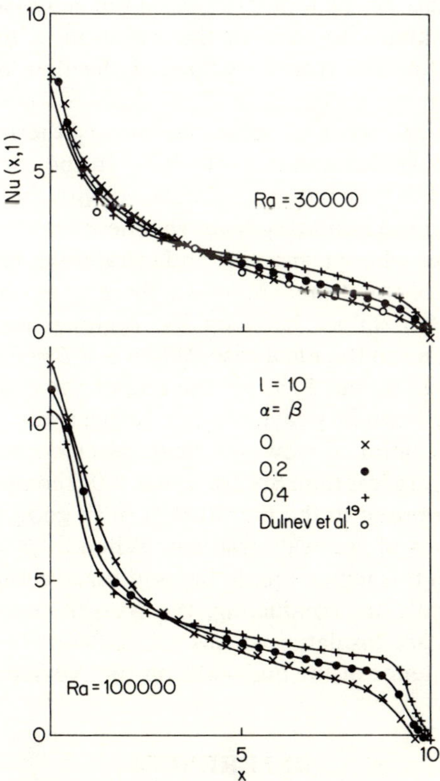

Figure 8.15 Local Nusselt number along the hot wall

especially near the insulated walls where the tangents are zero ($\partial^2 T/\partial x\ \partial y = 0$ at $x = 0$ and l). Discrepancies are to be seen with the results of Newell and Schmidt (Fig. 6 of Reference 20) whose calculations predict a maximum close the insulated wall $x = 0$. Explanations for these discrepancies may be found in the different truncation errors of the numerical approximations: Newell and Schmidt use a (25×13) grid with a variable step size near the walls while our results are performed with a (120×20) regular grid. Figure 8.15 shows a decrease in the influence of the non-isothermal walls on $N_u(x, 1)$ compared to that observed for $l = 1$ in Figure 8.10.

8.7 CONCLUSIONS

The main investigations of natural convection inside differentially heated cavities were limited to the case of isothermal active walls and insulated end walls. The contribution of this work is the study of the modifications introduced by non-isothermal active walls in vertical cavities. Two models of variation (linear and sinusoidal) for the temperature on the active walls are considered. New correlations taking into account the variation of the wall temperature are proposed to define the frontier values of the flow regime and the beat losses by convection.

From the dynamical point of view, the non-isothermal walls produce a diminution of the mass-flux and the velocities. In the case of square cavities, the modification of the flow patterns, compared with the ideal case $\alpha = \beta = 0$ is confined mainly to the boundary-layer régime.

For narrow cavities where transverse end instabilities exist in the ideal case, these instabilities do not disappear when there is a wall gradient of temperature. The frontier values between the régimes are reached for lower values of R_a compared to the ideal case. When $\alpha = \beta = 0.5$, the critical values of R_a decreased by 25% and 17% for the end of the conduction régime and the beginning of the boundary-layer régime respectively.

From the thermal point of view, the heat losses do not increase strongly. The variations of $\bar{N}_u$ are less than 8% for $\alpha = \beta = 0.5$ compared with $\alpha = \beta = 0$. The correlations proposed in the literature furnish good first approximations for the smaller values of the wall gradient. When α, $\beta \geqslant 0.50$, the modified correlations given in this paper should be used for accurate evaluation.

When the end walls are conducting, the Nusselt number is smaller than when the end walls are insulated when $l \leqslant 1$ (20% at $l = 1$, 40% at $l = 0.5$). When $l \geqslant 5$, the influence of the end walls on the Nusselt number is less than 10%.

REFERENCES

1. I. Catton, 'Convection naturelle en espace clos', *Sixième Congrès International sur le transfert de chaleur*, Toronto, Canada (1978).

2. Y. Severi, 'Etude expérimentale de la convection naturelle à l'intérieur d'une cavité simulant un convertisseur plan d'énergie solaire', *Thèse de 3ème Cycle*, Université d'Aix-Marseille (1980).
3. I. Catton, P. S. Ayyaswamy and R. M. Clever, 'Natural convection flow in a finite rectangular slot arbitrarily oriented with respect to the gravity vector', *Int. J. Heat Mass Transfer*, **17**, 173–184 (1974).
4. B. Roux, J. C. Grondin, P. Bontoux and B. Gilly, 'On a high-order accurate method for the numerical study of natural convection in a vertical square cavity', *Numerical Heat Transfer*, **1**, 331–349 (1978).
5. D. D. Joseph, *Stability of Fluid Motions II*, Springer Tracts in Natural Philosophy 28, Springer-Verlag, New York (1976).
6. D. W. Peaceman and H. H. Rachford, 'The numerical solution of parabolic and elliptic differential equations', *J. Soc. Ind. Appl. Math.*, **3**(1), 28–41 (1955).
7. G. D. Mallinson and G. de Vahl Davis, 'The method of the false transient for the solution of coupled elliptic equations', *J. Comp. Phys.*, **12**, 435–461 (1973).
8. R. S. Hirsh, 'Higher order accurate difference solutions of fluid mechanics problems by a compact differencing technique', *J. Comput. Phys.*, **19**, 90–109 (1975).
9. B. Roux, P. Bontoux, Ta Phuoc Loc and O. Daube, Optimisation of Hermitian methods for Navier–Stokes equations in the vorticity and stream function formulation', *Lecture Notes in Mathematics 771,* Springer-Verlag, 450–468 (1979).
10. P. Bontoux, B. Gilly and B. Roux, 'Natural convection in cavities for high RAYLEIGH numbers', *Notes on Numerical Fluid Mechanics*, Vieweg-Verlag, 22–35 (1979).
11. J. C. Grondin and B. Roux, 'Recherche de corrélations simples exprimant les pertes convectives dans une cavité bidimensionnelle, inclinée, chauffée différentiellement', *Revue de Physique Appliquée*, **14**, 49–56 (1979).
12. E. R. G. Eckert and W. O. Carlson, Natural convection in an air layer enclosed between two vertical plates with different temperatures, *Int. J. Heat Mass Transfer*, **2**, 106–120 (1961).
13. R. W. Thomas and G. de Vahl Davis, Natural convection in annular and rectangular cavities. Proceedings of the 4th *Int. Heat Transfer Conf.*, Versailles, Paper No. 2.4. (1970).
14. A. E. Gill, 'The boundary layer regime for convection in a rectangular cavity', *J. Fluid Mech.*, **26**(3), 515–536 (1966).
15. J. W. Elder, 'Laminar free convection in a vertical slot', *J. Fluid Mech.*, **23**(1), 77–98 (1965).
16. B. Gilly, 'Modélisation et simulation numérique de la convection naturelle dans une cavité parallélépipédique inclinée', *Thèse de 3ème Cycle*, Université d'Aix-Marseille (1979).
17. J. O. Wilkes and S. W. Churchill, 'The finite difference computation of natural convection in a rectangular enclosure', *A.I.Ch.E. J.*, **12**(1), 161–166 (1966).
18. W. Koutsoheras, 'Natural convection phenomena in inclined cells with finite side-walls—A numerical solution', *M. Eng. Sc. Thesis*, Melbourne Univ. (1976).
19. G. N. Dulnev, Yu. P. Zarichnyak and A. V. Sharkov, 'Free convection at a vertical plate and in closed interlayer at different gas pressures', *Int. J. Heat Mass Transfer*, **18**, 213–218 (1975).
20. M. E. Newell and F. W. Schmidt, 'Heat transfer by laminar natural convection within rectangular enclosures', *J. Heat Transfer*, **92C**, 159–168 (1970).

Numerical Methods in Heat Transfer, Volume II
Edited by R. W. Lewis, K. Morgan, and B. A. Schrefler

Chapter 9

Numerical Solution of Coupled Conduction–convection Problems using Lumped-parameter Methods

C. R. Gane, A. J. Oliver, D. R. Soulsby, and P. L. Stephenson

9.1 INTRODUCTION

9.1.1 The problem considered

The problem considered here is the calculation of the transient or steady-state temperature distribution within a solid that is cooled by fluid flowing through a duct or system of ducts within it. There are many practical situations which involve the solution of this type of problem. Examples encountered by the present authors include the cooling of the stators and rotors of large motors and turbogenerators, the cooling of large transformers and of buried electrical cables and auxiliaries, and the flow of residual fuel oil in a buried pipeline.

This type of problem differs from the more commonly encountered heat conduction problems because the temperature of the fluid cooling the solid is not known *a priori*. Instead, the fluid temperatures must be calculated at the same time as the solid temperatures.

The present chapter is largely concerned with describing a variety of numerical methods and computer programs developed for the solution of coupled heat conduction and convection problems, and the application of these methods to a range of problems.

Although the term coupled heat conduction and convection is used here, it is worth noting that other authors have described this type of problem as combined or conjugate heat transfer or simply heat conduction–convection.

9.1.2 Previous work

There are many situations that concern actual engineering plant and involve coupled heat conduction and convection, and therefore many papers have been published in this field. The following survey of the literature does not claim to be exhaustive, but it is intended to give a summary of the available information.

Two fundamentally different approaches to coupled heat conduction and convection have been used in the past. The first involves the calculation of the detailed temperature distribution in the fluid as well as the solid. The second approach aims to solve for the detailed temperature in the solid and only the bulk temperature variation in the fluid. The first approach requires knowledge of the detailed velocity distribution in the fluid, and this can be obtained either from separate calculations made before the temperature calculations or at the same time. No values of heat transfer coefficients have to be specified as part of the input data. This approach is clearly the more rigorous of the two, but it is limited to comparatively simple geometries for which the fluid velocities can be calculated. Consequently, solutions for practical problems have generally been obtained from the second approach which requires values of the fluid mass flow rates and fluid/solid heat transfer coefficients as part of the input data to the problem.

Studies using the first approach have either assumed a velocity profile (for example, slug flow or fully developed laminar pipe flow) or have calculated the velocity distribution as well as the temperature. As mentioned above, these studies have been limited to comparatively simple geometries, such as flow in a pipe or between parallel plates.[1–9] Other studies have been made for flow over a cylinder[10–13] or over a small sphere.[14] Boundary-layer flow over a flat plate has also been considered.[15–19] For laminar flow over a flat plate, a variant of the second approach has sometimes been used. This involves using an analytical expression relating the heat transfer coefficient to the temperature distribution along the plate surface.[20–24]

A general study of coupled conduction–convection problems, plus some examples, has been given by Chida and Katto.[25,26]

A related problem that has been solved using the first approach is that of the temperature distribution in two fluids separated by a dividing wall. These studies have often neglected conduction in the wall in the flow direction, and have sometimes assumed zero temperature drop across the wall.[27–29]

Some of the earliest studies using the second approach (calculating only the bulk fluid temperature) were concerned with simplified models of parallel-flow heat exchangers. The related problem of flow in a pipe with heat generation in the wall was also studied. Various simplifications were made, such as neglecting axial conduction, or the heat capacity of the fluid, and analytical solutions were obtained using Laplace transforms.[30–34] The resulting expressions were difficult to evaluate, and could not easily be extended to more realistic situations. An early attempt at an approximate solution involved the use of a lumped-parameter electromechanical analogue.[35] Also various workers started developing digital numerical techniques (for example, Dusinberre[36,37]), and this has led to the various large computer programs that are in use today. More recent studies of the pipe flow problems using numerical methods have been reported by Siegel and Keshock,[38] and Landram[39]. A further analytical study has been made by Kutznetov and Belousov.[40]

Various numerical methods have been employed in solving coupled heat conduction and convection problems. One frequently used is the lumped-parameter method, and numerical solution methods for it have been developed by Gane and Stephenson.[41,42] The lumped-parameter approach was also used at first by Thornton and Wieting and co-workers. However, these workers have frequently required temperature distributions for input into stress calculation programs. As the stress programs used the finite element technique, they have more recently developed finite element methods for calculating the temperatures.[43–51] MacNeal[52] and Winslow[53] describe two different types of non-uniform triangle meshes for use with finite differences. Bender[54] has studied a numerical method based on finite differences. A general method for solving coupled conduction–convection problems using the first method has been reported by Patankar.[55] A comparison of the two approaches mentioned above has been given by Thornton and Dechaumphai.[56] A survey of various numerical methods for solving conduction and convection problems is given by Roache,[57] and Hogge[58] is concerned with schemes for nonlinear conduction equations. An informative survey of sparse matrix iterative methods is given by Jacobs[59] for elliptic problems, to which Roache[60] has applied general elliptic marching schemes.

The use of the boundary element method to solve coupled heat conduction and convection problems has been studied by Ramkrishna and Amundson.[61,62]

Studies of coupled heat conduction and convection using the second approach have often entailed calculating the mass flow rates from a flow network. There are several ways in which the flow network can be linked to the temperature calculations. The simplest is for when the flows are independent of temperature. In this case, the flow network is solved once only before the temperature calculations. If the mass flow rates depend on the temperatures (for example, because of property variations), it is necessary to alternate between flow and temperature calculations (for example, Oliver,[63] and Imre *et al.*[64]). A situation between these two extremes is where the mass flow rates are specified but it is necessary to calculate how the pressure drop varies with temperature (perhaps because of viscosity variations).

Studies have been reported of coupled heat conduction and convection, for situations relating to a range of engineering plant. These include the cooling of the rotors and stators of large turbogenerators.[65–69] The cooling of transformers has been studied by Oliver[63] and Imre *et al.*[70] Other electrical plant studied include motors[71] and brushgear.[72] Buried cables and their auxiliaries have been studied by Gane and Soulsby[73,74] and Gane *et al.*[75] Other studies have related to hypersonic aircraft and to space vehicles.[76–78] Studies have also been reported for flow and heat transfer in a buried pipeline carrying residual fuel oil,[79] and for a long gas pipeline[80] and a steam pipe.[81] A solar collector has been studied by Phillips.[82]

Details have been published of various computer programs capable of solving coupled heat conduction and convection problems. These include the Central Electricity Generating Board program ANDUIN[83] which uses a lumped-parameter method. Two finite element programs, TAP1 and TAP2, have been reported by Thornton.[84,85] A lumped-parameter program THETA, suitable for a desk-top computer, has been described by Moir and Stoker.[86] Kennedy[87] has described a program for certain coupled conduction–convection problems.

9.1.3 The present contribution

The computer program ANDUIN has been used to solve a range of coupled conduction–convection problems that apply to engineering plant operated by the C.E.G.B. These problems have involved both steady and transient calculations, and have required a number of numerical methods for their solution. In addition, various mesh generation programs have been required. The purpose of this chapter, therefore, is to describe the application of the program to a number of problems in order to demonstrate the range of situations that can be handled. However, before this is done, the basic lumped-parameter equations solved by the program are outlined, followed by a description of the various numerical methods that have been employed.

9.2 DERIVATION OF LUMPED-PARAMETER EQUATIONS

This section outlines the derivation of the lumped-parameter equations used in the present work. The numerical solution of these equations is described in Section 9.3. The method for handling solid regions is described first and this enables the basic principles to be described. After this, the equations for fluid regions are derived from a heat balance. The section closes with a discussion of how the various boundary conditions are handled, and an outline is given of various methods for calculating the thermal admittances.

9.2.1 Application to solid regions

The basic approach of the lumped-parameter method is that the solid region is divided into a number of elements, and one node is placed in each element (see Figure 9.1). The node is often, but not always, at the centre of the element. The total thermal capacity of the element and all heat generation within it are assumed to occur at the node. Finally, the nodes in adjacent elements are joined by thermal admittances.

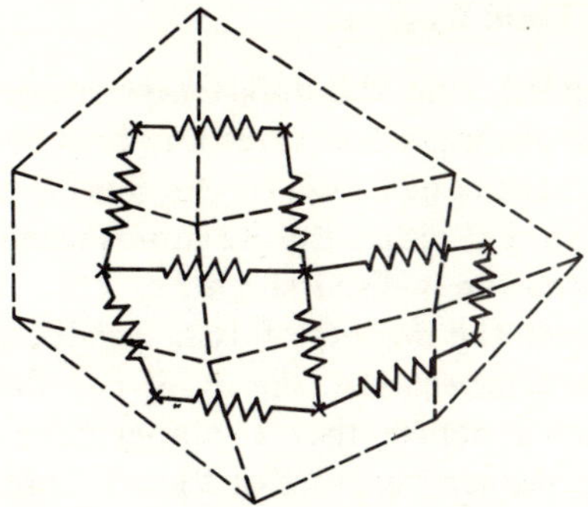

Figure 9.1 Typical part of lumped-parameter mesh: --- element boundary; × node; -ww- thermal admittance

The equation for the temperature of node i within element i is derived simply by writing a heat balance equation for that element, as follows:

$$C_i \frac{\partial T_i}{\partial t} = \sum_{\substack{j \neq i \\ j}} y_{i,j}(T_j - T_i) + Q_i, \tag{9.1}$$

where

$$Q_i = a_i + b_i T_i.$$

Here, C_i and Q_i are the thermal capacity and heat generation rate for element i, T_i is the temperature of solid node i, j is any node connected to node i, and $y_{i,j}$ is the thermal admittance between nodes i and j. Also, the heat generation term a_i is usually independent of temperature, and b_i is usually temperature dependent.

One equation of type (9.1) is obtained for every node whose temperature is to be calculated. To solve the resulting set of simultaneous equations, it is clearly necessary to know the values of C_i, Q_i and $y_{i,j}$. The first two are easily calculated, as follows

$$C_i = \rho c_p V_i \tag{9.2}$$

and

$$Q_i = q_i V_i. \tag{9.3}$$

Here, q_i and V_i are the volumetric heat generation rate and total volume respectively for element i. The calculation of the $y_{i,j}$ values is more complicated and is outlined in Section 9.2.4.

9.2.2 Application to Fluid Regions

As discussed in Section 9.1, only the bulk temperature variation is calculated in the fluid, and the various mass flow rates and heat transfer coefficients have to be specified. Axial heat conduction in the fluid is neglected because it is likely to be important only if either the Reynolds number or Prandtl number is small, and neither situation is discussed here.

Two points follow from the neglect of axial conduction and the decision to calculate only the bulk temperature. Firstly, if two fluid streams at different temperatures merge at a junction, then a temperature discontinuity occurs at the junction. Secondly, the temperature of a given fluid node does not depend on the temperature of any node downstream of it, so that a marching integration scheme can be used.

Two methods of handling the fluid region are used in the present work, and will now be derived from heat balance considerations. The equations given here can also be derived from a finite difference approximation of the governing equations.[42]

The first method of handling the fluid regions is upwind differencing, and this is illustrated by the geometry shown in Figure 9.2. The temperature of each fluid element is assumed to be constant in the flow direction, and fluid is assumed to flow out of an element and into the next at the temperature corresponding to the element it is leaving.

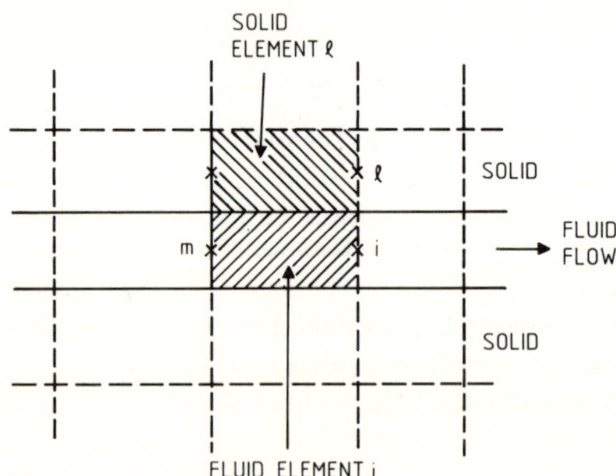

Figure 9.2 Details of upwind differencing method; × node

The heat balance for fluid element i yields the following:

$$C_i \frac{\partial T_i}{\partial t} = wc_p(T_m - T_i) + \sum_l y_{i,l}(T_l - T_i) + Q_i, \tag{9.4}$$

where w and c_p are the fluid mass flow rate and specific heat respectively.

Also, m is the fluid node upstream of fluid node i, and l is any solid element adjacent to fluid element i. The thermal admittance y in general combines the effects of heat transfer between fluid and solid surface, and heat conduction within the solid in a direction perpendicular to the bulk flow. Provided that fluid-to-fluid admittances are defined as

$$y_{i,m} = wc_{\mathrm{p}} \qquad \text{and} \qquad y_{m,i} = 0, \tag{9.5}$$

then Equation (9.4) has the same form as Equation (9.1). It must be noted that fluid-to-fluid admittances are not symmetric (that is, $y_{i,m} \neq y_{m,i}$). In this respect, they differ from all other admittances.

A more detailed consideration of Equation (9.4) shows that the fluid temperature actually calculated is that at the downstream edge of the fluid element, and that the solid temperature also corresponds to that position. Therefore both the fluid and solid nodes should be placed at the downstream edge of their respective elements (see Figure 9.2). It is difficult to define the downstream edge of a solid element when fluid flows in ducts in more than one direction and therefore upwind differencing cannot be used for such situations. However, it is used to obtain the results reported in Section 9.5.

The second method of handling the fluid region is called modified centre differencing. For this approach, the solid node is placed in the centre of its element and therefore its position is independent of the flow direction (see Figure 9.3). The fluid node, however, remains at the downstream edge of its

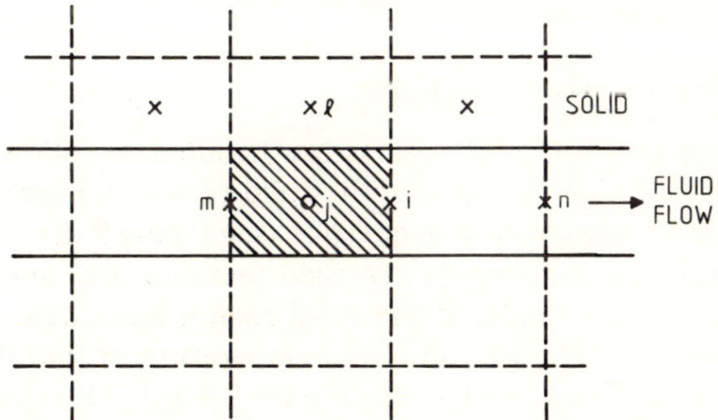

Figure 9.3 Details of modified centre difference method: ○ point used in m.c.d. method, × node

element. The fluid temperature within an element is assumed to vary linearly between fluid nodes. Thus a heat balance for fluid element i gives the following result:

$$\frac{C_i}{2}\left(\frac{\partial T_m}{\partial t}+\frac{\partial T_i}{\partial t}\right) = wc_{\mathrm{p}}(T_m - T_i) + \sum_l y_{i,l}\{T_l - \tfrac{1}{2}(T_i + T_m)\} + Q_i. \tag{9.6}$$

Here m is the fluid node upstream of fluid node i, and l is any solid element adjacent to fluid element i.

Equation (9.6) is then generalized to include pipe junctions by replacing $wc_p(T_m - T_i)$ by the sum of all of all such terms (m is then any fluid node upstream of i) and by defining a mean upstream temperature for element i as follows:

$$U_i = \frac{\sum_m w_{i,m} c_p T_m}{\sum_m w_{i,m}}. \tag{9.7}$$

Clearly, $U_i = T_m$ away from pipe junctions. In this way, Equation (9.6) becomes:

$$\frac{C_i}{2}\left(\frac{\partial U_i}{\partial t} + \frac{\partial T_i}{\partial t}\right) = \sum_m w_{i,m} c_p (U_i - T_i) + \sum_l y_{i,l}\{T_l - \tfrac{1}{2}(T_i + U_i)\} + Q_i. \tag{9.8}$$

With this approach to handling fluid junctions, the junction region itself must be described by a single fluid element, and the temperature of the corresponding fluid node becomes the bulk mixed temperature for the junction region.

The modified centre difference method, unlike the conventional centre difference method, does have a false diffusion error (Gane and Stephenson[42]). However, this error is second order in the node spacing, whereas the corresponding error for upwind differencing is first order.

9.2.3 Handling boundary conditions

Boundary conditions for solid regions are handled as follows. To specify a fixed temperature along the edge of the solid region, a number of boundary nodes are located along the edge in question (see Figure 9.4(a)). Such boundary nodes have only one property (a specified temperature) and are connected to at least one interior node. If the solid region loses heat to surrounding fluid at a known temperature, then the temperature of that fluid is specified by one or more boundary nodes (see Figure 9.4(b)). The simplest approach is to connect these boundary nodes directly to interior nodes with thermal admittances that combine the effects of heat conduction within the solid and heat transfer between the solid surface and the fluid. Alternatively a surface node (with zero thermal capacity and heat generation rate) can be located on the solid surface, and two separate thermal admittances used (see Figure 9.4(c)). With this second approach the temperature of the solid surface is also calculated.

The sole boundary condition for fluid regions is the specification of the fluid temperature at the inlet to a pipe or duct. This is done by using a fluid inlet node.

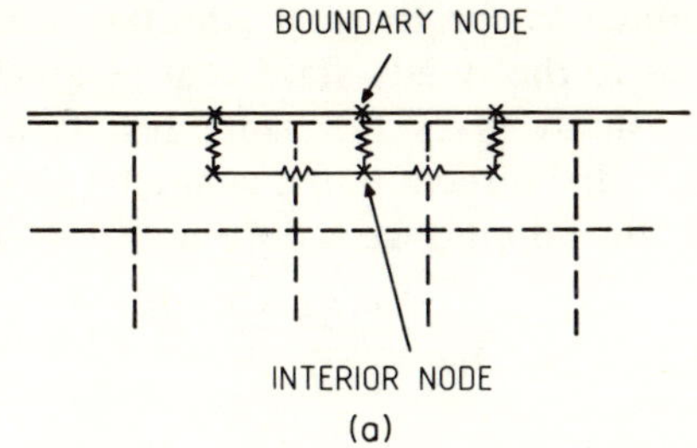

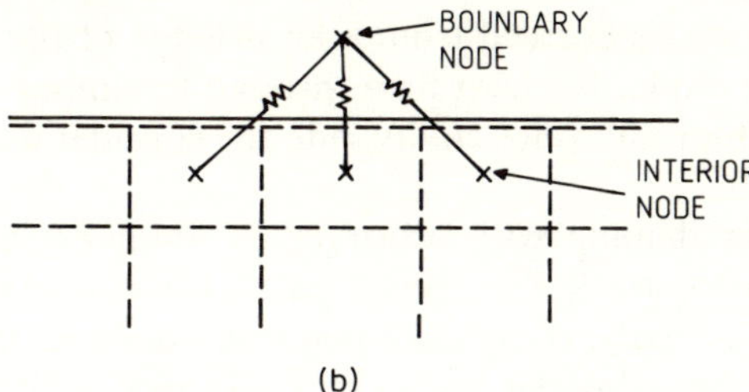

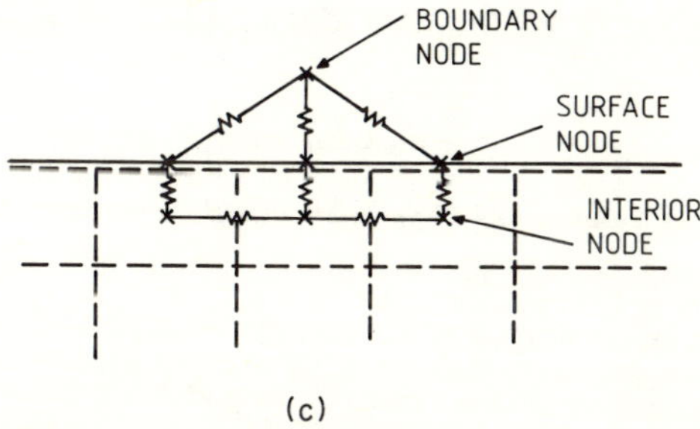

Figure 9.4 Handling of solid region boundary conditions. (a) use of boundary node for fixed temperature boundary condition; (b) use of boundary node to specify convective boundary condition; (c) use of boundary and surface nodes to specify convective boundary condition

9.2.4 Calculation of Thermal Admittances

One of the major requirements of a lumped parameter model is the calculation of the thermal admittances. A number of methods can be used, and will now be summarized.

(a) For a regular geometry, the thermal admittances can be derived from finite difference expressions that have, in turn, been derived from Taylor series expansions.

(b) For a regular geometry, the thermal admittance can be derived from analytical solutions to the steady-state heat conduction equation in one dimension. This usually gives the same results as (a). However, for conduction in the radial direction in a geometry described by cylindrical polar coordinates, the admittance is different from (a), and has the form:

$$y_{i,j} = \frac{k\ \Delta\theta\ \Delta z}{\ln\ (r_j/r_i)} \tag{9.9}$$

where r_i and r_j are the radii of the two nodes.

(c) For a simplified model of a complex geometry, approximate thermal admittances can be derived from a knowledge of the controlling resistances. For example, for heat flow through a number of regions of both high and low thermal conductivity, one can consider only the low conductivity regions.

(d) A method for dealing with arbitrary two-dimensional regions is now described. In Figure 9.5 is shown part of an interface defined by the curve $y = f(x)$. Firstly, coordinate points are defined on the interface and an equal number of nodes are to be positioned a distance Δn from the interface on both sides of it along the normals through the interface points. If (x_b, y_b) is on the boundary and (x, y) is on one side of the boundary, then

$$\begin{aligned} x &= x_b \pm \Delta n\ \sin\theta, \\ y &= y_b \mp \Delta n\ \cos\theta, \end{aligned} \tag{9.10}$$

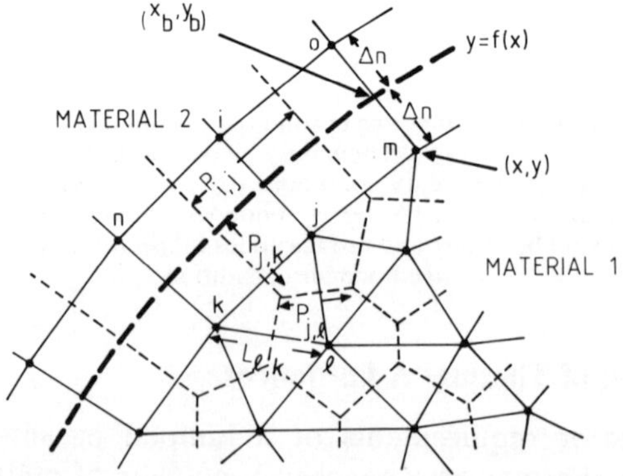

Figure 9.5 Lumped-parameter mesh arrangements for arbitrarily shaped plane regions: —— connector between adjacent nodes; – – – material interface forming part of boundaries around adjacent nodes; - - - lines forming boundary around a node (nodal boundaries)

where

$$\theta = \tan^{-1}\left\{\left(\frac{dy}{dx}\right)_{x_b, y_b}\right\}.$$

All these nodes are then interconnected both parallel to and across the interface. The regions on either side of the interface are then covered by a triangular grid with no obtuse angles[52] which may merge with rectangular networks away from the interface. The control area for a node is formed by joining the perpendicular bisectors of the lines connecting neighbouring nodes, thus forming a polygon around each node. If a sectional length Δz is considered in the third plane, then the thermal admittance between nodes j and l in material 1 (see Figure 9.5) is defined by

$$Y_{j,l} = \frac{K_1 P_{j,l}\,\Delta z}{L_{j,l}} \tag{9.11}$$

where

K_1 = thermal conductivity of material 1,

$L_{j,l}$ = distance between nodes j and l,

and

$P_{j,l}$ = length of side of polygon bisecting the line $L_{j,l}$,

with similar expressions for node pairings (k, l) and (j, k). For nodes i and j in materials 2 and 1, respectively, $2(K_1^{-1} + K_2^{-1})^{-1}$ replaces K_1 in Equation (9.11). If node i is on the interface, then $K_2^{-1} = 0$. The thermal capacitance associated with a general node j is given by

$$C_j = \rho c_p \frac{\Delta z}{4} \sum_k L_{j,k} P_{j,k} \tag{9.12}$$

where ρ and c_p are the density and specific heat of the material containing node j. An example of generating such meshes is described in Section 9.6.

9.3 NUMERICAL METHODS

This section describes the various methods that have been employed for solving the transient and steady-state equations presented in the previous section. Sections 9.3.1 and 9.3.2 are concerned with temperature solutions only, whereas Section 9.3.3 outlines a procedure for the solution of the mass flowrates as well.

9.3.1 Transient temperature equations

The following methods have been used for solving various transient problems. Their suitability for particular problems is discussed here and illustrated in later sections. It is assumed that thermal capacitances are constant, although admittances and heat generation rates can be temperature dependent.

9.3.1.1 The DUF method

The application of a method based on that of Du Fort and Frankel[88] to the solution of the solid temperature equations proceeds as follows. Firstly, at time $n\,\Delta t$ from the start of the transient, $(\partial T/\partial t)_i^{(n)}$ and $T_i^{(n)}$ on the left and right of Equation (9.1) are replaced by

$$\left\{\frac{\partial T}{\partial t}\right\}_i^{(n)} = \frac{T_i^{(n+1)} - T_i^{(n-1)}}{2\,\Delta t} \qquad \text{and} \qquad T_i^{(n)} = \frac{T_i^{(n+1)} + T_i^{(n-1)}}{2}. \tag{9.13}$$

Then combining Equations (9.1) and (9.13) and rearranging yields

$$T_i^{(n+1)} = \left(\frac{1-d_i^{(n)}}{1+d_i^{(n)}}\right) T_i^{(n-1)} + \frac{2\,\Delta t}{C_i}\left(\sum_{\substack{j\neq i\\ j}} y_{i,j}^{(n)} T_j^{(n)} + a_i^{(n)}\right), \tag{9.14}$$

where

$$d_i^{(n)} = \frac{\Delta t}{C_i}\left(\sum_{\substack{j\neq i\\ j}} y_{i,j}^{(n)} - b_i^{(n)}\right).$$

The same type of formula can be applied to the solution of the upwind difference fluid temperature equation (9.4) which has the same form as (9.1), since $y_{i,m} = wc_p$. Numerical and theoretical results obtained[41] have suggested that Equation (9.14) for solid temperatures is numerically stable if $d_i^{(n)} > 0$ for all i and n, although this has not been proved conclusively. However, for the case of a single duct with no heat generation in the fluid and fixed outside surface temperatures, it can be proved that the fluid temperature version of Equation (9.14) using upwind differencing is stable only if $0 < d_i^{(n)} \leqslant 1$, for all i and n, where here, $d_i^{(n)} = (\Delta t\, wc_p)/C_i$ is the well-known Courant number (e.g. see Gane and Stephenson[42]). This condition restricts the general use of the method for fluid nodes. However, with $d_i^{(n)} = 1$, an exact formula is obtained for the case of zero heat transfer between solid and fluid. This can be a useful property for problems in which the amount of heat transfer is small, but the changes in fluid temperatures along the duct are large (e.g. see Section 9.5).

The dominant term in the overall truncation error in using Equations (9.13) to approximate Equation (9.1) is

$$\varepsilon_i^{(n)} = -\frac{\Delta t}{2} d_i^{(n)} \left(\frac{\partial^2 T}{\partial t^2}\right)_i^{(n)}. \qquad (9.15)$$

To ensure that Equation (9.14) converges to the true solution, the $d_i^{(n)}$ values should not be allowed to increase when reducing the time step and grid spacing. This can imply a slower overall convergence than other methods such as the Crank–Nicolson approach, but like this procedure, the DUF method has $O(\Delta t^2)$ rate of convergence for a fixed grid spacing. The main advantage of the DUF method is that it is explicit, viz. a node-by-node calculation procedure.

9.3.1.2 The FIDUF method

The FIDUF method (Gane and Stephenson[41,42]) combines the DUF method for solving the solid temperature equations (Equation (9.14)) with a Fully Implicit solution of the fluid temperature equations. If the fluid temperature equation (9.8) is approximated at time $(n+1)\,\Delta t$ using backward difference replacements of time derivatives and evaluating temperature coefficients at the previous time level, then the fluid temperature $T_i^{(n+1)}$ at the downstream end of a fluid element or junction is obtained from

$$T_i^{(n+1)} = \frac{1}{(e_i^{(n)} + f_i^{(n)})}\Big[(e_i^{(n)} - f_i^{(n)})U_i^{(n+1)} + U_i^{(n)} + T_i^{(n)} + \frac{2\Delta t}{C_i}\Big(\sum_l y_{i,l}^{(n)} T_l^{(n+1)} + a_i^{(n)}\Big)\Big], \qquad (9.16)$$

where $Q_i^{(n+1)}$ has been replaced with $a_i^{(n)} + \frac{1}{2} b_i^{(n)} (U_i^{(n+1)} + T_i^{(n+1)})$,

$$e_i^{(n)} = \frac{2\,\Delta t}{C_i} \sum_m w_{i,m} c_p,$$

$$f_i^{(n)} = 1 + \frac{\Delta t}{C_i}\Big(\sum_l y_{i,l}^{(n)} - b_i^{(n)}\Big),$$

and $U_i = T_m$ and $\sum_m w_{i,m} = w_{i,m}$ away from pipe junctions. A similar procedure can be used to obtain a fully implicit replacement for the upwind difference equation (9.4) for fluid flow in a single duct. Following similar arguments to those given by Gane and Stephenson[41,42] it can be shown that Equation (9.16) is numerically stable if the following inequality is satisfied at each time level n and for all i:

$$e_i^{(n)} + f_i^{(n)} \geqslant 1 + \frac{\Delta t}{C_i}\Big(\sum_l (y_{i,l}^{(n)} - Y_{i,l})\Big(\frac{T_l^{(n)}}{T_j^{(n)}}\Big)\Big), \qquad (9.17)$$

where $T_j^{(n)} = \frac{1}{2}(U_i^{(n)} + T_i^{(n)})$ and $y_{i,l}^{(n)}$ has been expressed as

$$y_{i,l}^{(n)} = Y_{i,l} + g_{i,l}^{(n)} T_j^{(n)}$$

with $Y_{i,l}$ independent of temperature.

The computation proceeds in two stages. Initially, the simple explicit approximation of Equation (9.1) is used to advance the solid temperature solution from the starting time, t_0 to time $t_0 + \Delta t$. Then the three-level equation (9.14) is used to calculate the solid temperatures explicitly. After the solid temperatures have been calculated at a particular time level, the fully implicit equation (9.16) is used to compute the fluid temperatures $T_i^{(n+1)}$ at the downstream ends of the elements by starting next to the fluid inlet positions and working downstream from these, ensuring that at a junction, all the upstream temperatures have been calculated first. In this way, both the solid and fluid node temperatures are computed explicitly. The starting procedure is repeated if a change in time step or in one of the governing parameters is specified.

This scheme has considerable advantages on large complex grid arrangements in allowing any order of calculation of solid node temperatures. Moreover, Inequality (9.17) is a very mild condition and Equation (9.16) is normally stable for all values of Δt. Hence, the computing time per step and the computer storage required is comparatively low, although this must be measured against the restriction on time step size resulting from Equation (9.15). For very stiff problems (viz. large variations in $d_i^{(n)}$ between nodes), Equation (9.15) can be restrictive, since the maximum value of $(\sum_{j(\neq i)} y_{i,j}^{(n)} - b_i^{(n)})/C_i$ governs the time step size at level n. In practice, these quantities must be monitored as the calculation progresses. Typically, it has been found that for reasonable temporal accuracy $d_i^{(n)} < 10$.

One of the disadvantages of this method is the introduction of a false diffusion error as mentioned in Section 2.2. This can be substantial for problems involving large changes in fluid temperature gradients along a duct. In order to overcome this drawback the following method has been considered.

9.3.1.3 The IEDUF method

The IEDUF method was developed as an alternative to the DUF and FIDUF methods. All three use the DUF method, Equation (9.14), for solid temperatures. For fluid temperatures, the IEDUF method uses an Implicit formula for the term describing heat flow between fluid and solid and an Explicit formula for the convection term. This method uses the fluid temperature at the downstream end of each element in calculating the heat stored and also the heat transferred to the solid. It is thus reserved for problems without duct junctions and with unidirectional flow. The equation for fluid temperatures in a duct, based on Equation (9.4), is then given by

$$\frac{C_i(T_i^{(n+1)}-T_i^{(n)})}{\Delta t}=wc_p(T_m^{(n)}-T_i^{(n)})+\sum_l y_{i,l}^{(n)}(T_l^{(n+1)}-T_i^{(n+1)})$$
$$+a_i^{(n)}+b_i^{(n)}T_i^{(n;n+1)}, \qquad (9.18)$$

where $T_i^{(n;n+1)}=T_i^{(n)}$ if $b_i^{(n)}>0$ or $T_i^{(n+1)}$ if $b_i^{(n)}<0$. So far, this equation has been used only with $a_i=b_i=0$. It is re-arranged to give an explicit expression for $T_i^{(n+1)}$ involving previously calculated fluid and solid temperatures. An overall explicit procedure results, in the same manner as for the FIDUF scheme. This method has been used only with the Courant number, $C_r=wc_p\,\Delta t/C_i=1$, since Equation (9.18) then gives an exact solution in the case of no heat gain/loss and no heat generation in the fluid, namely $T_i^{(n+1)}=T_m^{(n)}$. Hence, for problems with comparatively small heat gain/loss and generation in the fluid, variations in $y_{i,l}$ due to fluid temperature changes can be handled by evaluating $y_{i,l}$ at the upstream node position at the previous time level, viz. using $T_m^{(n)}$. This method is more accurate than some others for such problems when there are large changes in fluid temperatures along a duct; e.g. see Section 9.5.

With $C_r=1$ and $b_i^{(n)}\leqslant 0$, it is easily shown that Equation (9.18) is numerically stable at time level n for all $\Delta t>0$, if it is assumed in the analysis that the solid temperatures, $T_l^{(n+1)}$, are constant.

9.3.1.4 The DJ method

Certain problems considered by the present authors can be represented solely by one-dimensional conduction equations which are both nonlinear and transient. Such problems can be solved using the general DUF method described in Section 3.1. However, they can also be solved more efficiently using implicit methods which employ a tridiagonal algorithm to solve the equations, while achieving comparable accuracy with a larger time-step size. Moreover, when problems involving possible thermal instability are concerned, it is desirable to use a method whose numerical stability is guaranteed. A method which possesses all these properties is the predictor–corrector scheme devised by Douglas and Jones[89].

The DJ method is applied to Equation (9.1), representing in this case a one-dimensional conduction equation, for the overall time step Δt from time t_n to time $t_n+\Delta t$, as follows. The predictor step is written as:

$$-\sum_{\substack{j\neq i\\ j}} y_{i,j}T_j^{(n+1/2)}+\left(\frac{4C_i}{\Delta t}+\sum_{\substack{j\neq i\\ j}} y_{i,j}\right)T_i^{(n+1/2)}$$
$$=\sum_{\substack{j\neq i\\ j}} y_{i,j}T_j^{(n)}+\left(\frac{4C_i}{\Delta t}-\sum_{\substack{j\neq i\\ j}} y_{i,j}\right)T_i^{(n)}+2Q_i, \qquad (9.19(a))$$

where the time-dependent terms in the y's and Q's are evaluated at time level $(n+1/2)$, and the temperature-dependent terms at time level (n). The corrector step is written as follows:

$$-\sum_{\substack{j\neq i\\ j}} y_{i,j}T_j^{(n+1)}+\left(\frac{2C_i}{\Delta t}+\sum_{\substack{j\neq i\\ j}} y_{i,j}\right)T_i^{(n+1)}$$

$$=\sum_{\substack{j\neq i\\ j}} y_{i,j}T_j^{(n)}+\left(\frac{2C_i}{\Delta t}-\sum_{\substack{j\neq i\\ j}} y_{i,j}\right)T_i^{(n)}+2Q_i, \qquad (9.19(b))$$

where both the time-dependent and temperature-dependent terms in the y's and Q's are evaluated at time level $(n+\frac{1}{2})$. Douglas and Jones have proved their method to be unconditionally numerically stable, and to be second-order accurate in time and space for uniform mesh spacing. Moreover, using sequential node numbering, Equations (9.19) lead to linear tridiagonal systems of equations, which are easily solved using the well-known algorithm based on Gaussian elimination.

9.3.2 Steady-state temperature equations

The following two iterative methods have been used to solve the nonlinear steady-state equations arising from the problems studied by the present authors. They have been mainly adopted for their ease of use, and in the case of the IMSI method, its general applicability.

9.3.2.1 The IMSI method

The Chebyshev Semi-Iterative method[90] has been used to solve Equation (9.1) with $\partial T_i/\partial t=0$, namely

$$T_i^{(n+1)}=\beta_n\left[\left(\sum_{\substack{j\neq i\\ j}} y_{i,j}^{(n)}-b_i^{(n)}\right)^{-1}a_i^{(n)}+\left(b_i^{(n)}-\sum_{\substack{j\neq i\\ j}} y_{i,j}^{(n)}\right)^{-1}\sum_{\substack{j\neq i\\ j}} T_j^{(n)}y_{i,j}^{(n)}-T_i^{(n-1)}\right]+T_i^{(n-1)}, \qquad (9.20)$$

where $\beta_1=1, \beta_2=2/(2-\mu^2)$ and $\beta_{n+1}=1/(1-\mu^2\beta_n/4)$, $n\geqslant 2$; μ is the maximum modulus eigenvalue of the associated Jacobi matrix which must be estimated *a priori* and be less than unity for the method to be convergent. If $b_i^{(n)}\leqslant 0$, then $\mu<1$. The method has been found in practice to converge for moderate values of $b_i^{(n)}>0$. However, if $b_i^{(n)}$ is very large and positive then $\mu>1$ and the iteration process will not converge. It is normally taken that $T_i^{(1)}=T_i^{(0)}$, the initial guess at the solution, for all i. One of the advantages of this particular scheme is that it is an explicit formula, irrespective of node orderings, and computer storage is kept to a minimum compared with implicit methods. These are very useful properties for large complex grid arrangements. Its main disadvantage is that for large and/or ill-conditioned sets of equations, optimum rates of convergence are very sensitive to the value of μ. From problems considered so far by the authors, the optimum value of μ has generally been within the range 0.9 to 0.995.

The fluid temperature equation (9.8), applied to ducts and junctions, with $\partial U_i/\partial t = \partial T_i/\partial t = 0$, is approximated by an Implicit formula akin to Equation (9.16). Hence $T_i^{(n+1)}$ is obtained from

$$T_i^{(n+1)} = \frac{1}{(\sum_m w_{i,m}c_p + \frac{1}{2}\sum_l y_{i,l}^{(n)} - \frac{1}{2}b_i^{(n)})} \times \left[\left(\sum_m w_{i,m}c_p - \frac{1}{2}\sum_L y_{i,l}^{(n)} + \frac{1}{2}b_i^{(n)}\right) U_i^{(n+1)} + \sum_l y_{i,l}^{(n)} T_l^{(n+1)} + a_i^{(n)}\right]. \quad (9.21)$$

The calculation procedure is the same as for the FIDUF method, namely the calculation firstly of the solid temperatures, node-by-node, using Equation (9.20), followed by the fluid temperatures using Equation (9.21) at each fluid node following the flow directions. This then gives an explicit procedure. When convergence is obtained (a maximum fractional change in temperature less than 10^{-5} between successive iterations is normally used), the overall heat balance is checked (e.g. less than a 0.1% error).

9.3.2.2 The SLR method

In the Section 3.1, the DJ method was put forward for solving nonlinear transient heat conduction problems in one space dimension. The following scheme has been adopted for solving the corresponding steady-state equations. An intermediate set of temperatures at iteration $(n+1)$ is predicted using the following Successive Line Relaxation replacement of Equation (9.1) with $\partial T_i/\partial t = 0$:

$$\sum_{\substack{j \neq i \\ j}} y_{i,j}^{(n)} T_i^* - \sum_{\substack{j \neq i \\ j}} y_{i,j}^{(n)} T_j^* = Q_i^{(n)}. \quad (9.22\text{(a)})$$

These form a linear set of tridiagonal equations which are solved using the standard algorithm. The temperatures are then relaxed by applying the following equation to each in turn:

$$T_i^{(n+1)} = \omega T_i^* + (1-\omega) T_i^{(n)}. \quad (9.22\text{(b)})$$

9.3.3 Coupled heat and flow calculations

The solution methods that have been described so far assume that the mass flow in any duct is known. In some problems the mass flow is not known *a priori*. An example is duct flows driven by natural convection in which case the flow depends on the fluid temperature in the ducts; then it is necessary to solve for the mass flow and the fluid temperature simultaneously.

This section describes how the method has been extended to handle a problem where the duct flow is initially unknown and is a function of the fluid temperature.

9.3.3.1 *The problem considered*

To illustrate the method, the steady-state example illustrated in Figure 9.6 is considered. A duct is bounded on one side by a solid material. The other side allows fluid to flow between the duct and an adjacent reservoir of fluid which has a known pressure and temperature field. For example this second wall could be porous or it could have a slot or holes in it. The fluid density is temperature dependent and the gravitational head of fluid in the duct is assumed to play a significant part in driving the flow through the duct—this provides the coupling between the heat and flow calculations. The unknown flow through the duct varies in the direction z due to the inflow or outflow of fluid through the wall.

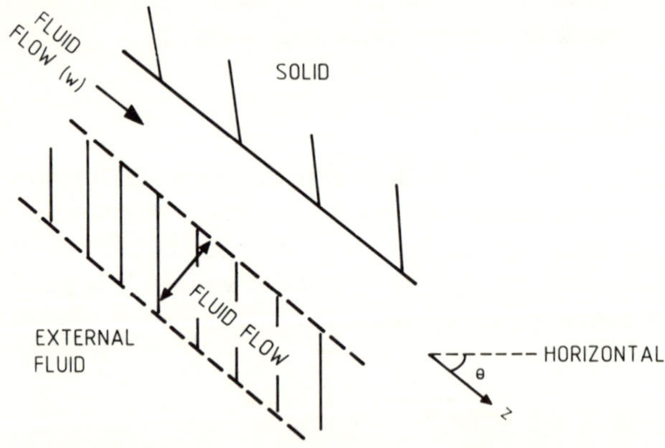

Figure 9.6 Details of a coupled heat and flow problem

The equation for temperature in the solid wall is given by Equation (9.1). The fluid temperature equation can be expressed as:

$$c_p \frac{d}{dz}(wT_f) + y_{s,f}(T_f - T_s) + y_{e,f}(T_f - T_e) - qA_d = 0. \qquad (9.23)$$

Here, $y_{e,f}$ is the thermal admittance between the duct fluid and the external fluid. It represents heat transfer between the external fluid and the insulation, conduction through the insulation and heat transfer between the insulation and duct fluid. The term (qA_d) is regarded as a heat source/sink which represents the heat transport due to the inflow or outflow of mass in the duct. This term may be expressed as

$$qA_d=\frac{1}{2}\left(\frac{dw}{dz}+\left|\frac{dw}{dz}\right|\right)c_pT_e+\frac{1}{2}\left(\frac{dw}{dz}-\left|\frac{dw}{dz}\right|\right)c_pT_f. \tag{9.24}$$

If there is a mass flow into the duct, then the second expression on the right of Equation (9.24) is automatically zero and the first expression represents the resulting heat in. If the mass flow is out, then the first expression is zero and the second expression is the heat out.

It is now necessary to provide an equation for w, the mass flow in the duct. This is given by the integral form of the momentum equation for the flow in the duct in the z-direction,

$$\rho A_d u\frac{du}{dz}=-A_d\frac{dp}{dz}-L_p\tau_w+\rho A_d g\sin\theta, \tag{9.25}$$

where L_p is the duct wetted perimeter, and τ_w is the average wall shear stress which is generally a function of u and z. The last term is the gravitational force for the duct at an angle θ to the horizontal. Also

$$w=\rho A_d u. \tag{9.26}$$

In order to eliminate the unknown duct pressure, p, in Equation (9.25), the mass balance for the duct and the pressure loss characteristics of the flow path between the external fluid and the duct are considered. These give:

$$\dot{m}_w=\frac{dw}{dz} \tag{9.27}$$

and

$$p_e-p=K_1\dot{m}_w+K_2\dot{m}_w|\dot{m}_w|, \tag{9.28}$$

where $\dot{m}$ is the mass flow into (or out of) the duct for unit length of the duct. A quadratic equation with loss coefficients K_1 and K_2 has been used for the pressure drop along the flow path. The direction of $\dot{m}_w$ (in or out) is not known, so the sign convention of $\dot{m}_w$ positive for inflow and negative for outflow has been used.

Using Equation (9.27) in Equation (9.28) and differentiating gives:

$$\frac{dp}{dz}=\frac{dp_e}{dz}-K_1\frac{d^2w}{dz^2}-2K_2\left|\frac{dw}{dz}\right|\frac{d^2w}{dz^2}. \tag{9.29}$$

Combining Equations (9.29) and (9.25) and using Equation (9.26) gives

$$w\frac{\mathrm{d}}{\mathrm{d}z}\left(\frac{w}{\rho A_{\mathrm{d}}}\right)=-A_{\mathrm{d}}\frac{\mathrm{d}p_{\mathrm{e}}}{\mathrm{d}z}+K_1A_{\mathrm{d}}\frac{\mathrm{d}^2w}{\mathrm{d}z^2}+2K_2A_{\mathrm{d}}\left|\frac{\mathrm{d}w}{\mathrm{d}z}\right|\frac{\mathrm{d}^2w}{\mathrm{d}z^2}-L_{\mathrm{p}}\tau_{\mathrm{w}}+\rho A_{\mathrm{d}}g\sin\theta. \tag{9.30}$$

Equation (9.30) is the required equation for the duct mass flow.

9.3.3.2 *Solution of equations*

The equations given in the preceding section are differential equations. Before they can be solved, they are expressed in the lumped-parameter form. The equations for the solid region have already been presented in this latter form, see Equation (9.1).

For the fluid temperature, expanding the first term of Equation (9.23) and using Equation (9.24) for (qA_{d}) gives

$$wc_{\mathrm{p}}\frac{\mathrm{d}T_{\mathrm{f}}}{\mathrm{d}z}+y_{\mathrm{s,f}}(T_{\mathrm{f}}-T_{\mathrm{s}})+y_{\mathrm{e,f}}(T_{\mathrm{f}}-T_{\mathrm{e}})+\left[\frac{1}{2}\left(\frac{\mathrm{d}w}{\mathrm{d}z}+\left|\frac{\mathrm{d}w}{\mathrm{d}z}\right|\right)c_{\mathrm{p}}(T_{\mathrm{f}}-T_{\mathrm{e}})\right]=0. \tag{9.31}$$

Using the modified centre difference method of Section 2.2 in Equation (9.31) gives

$$[\tfrac{1}{2}(w_i+w_m)]c_{\mathrm{p}}(T_m-T_i)+\sum_l y_{i,l}[T_l-\tfrac{1}{2}(T_i+T_m)]+Q_i=0 \tag{9.32}$$

with

$$Q_i=\tfrac{1}{2}c_{\mathrm{p}}[(w_i-w_m)+|w_i-w_m|][T_{\mathrm{e}}-\tfrac{1}{2}(T_i+T_m)].$$

This is chosen to be consistent with the modified centre difference method of Section 9.2.2 where $\frac{1}{2}(T_i+T_m)$ is the mean temperature of the fluid element. Equation (9.32) now corresponds to the steady-state form of Equation (9.6) with the above expression for Q_i being regarded as a pseudo heat generation term. Also because the mass flow is allowed to vary along the duct, the first term of Equation (9.31) is now written using an average mass flow for the fluid element.

The further treatment of equation (9.32) to get it into its final lumped-parameter form is as described in Section 9.2.2.

The method for handling the duct mass flow equation, Equation (9.30) will now be described. To illustrate the treatment, central differences with a uniform grid spacing are used. Using the notation of Figure 9.3, for node i with a nodal spacing of Δz,

$$\frac{w_i(w_n-w_m)}{\rho_iA_{\mathrm{d}}2\,\Delta z}=A_{\mathrm{d}}\left(\frac{\mathrm{d}p_{\mathrm{e}}}{\mathrm{d}z}\right)_i+\left[K_1A_{\mathrm{d}}+2K_2A_{\mathrm{d}}\left|\frac{w_n-w_m}{2\,\Delta z}\right|\right]\left(\frac{w_n-2w_i+w_m}{2\,\Delta z}\right)$$
$$-(L_{\mathrm{p}}\tau_{\mathrm{w}})_i+\rho_iA_{\mathrm{d}}g\sin\theta_i \tag{9.33}$$

where the subscript i indicates that the term is to be evaluated at node i. Equation (9.33) can be written as

$$a_m u_m + b_i u_i + c_n u_n = \alpha_i;$$

thus the mass flow equations form a tridiagonal set of equations.

To solve the resulting equations an iterative method has been used. Each iteration cycle consists of the following two steps:

(a) Solve the solid and fluid temperature equations (Equations (9.1) and 9.32)) using the IMSI method described in Section 3.2. Use the latest values of fluid mass flow.
(b) Solve the fluid flow equations, Equation (9.33), using a tridiagonal solution algorithm of Thomas (Varga[91]) with the latest values of fluid and solid temperatures.

In past applications it has been found necessary to under-relax the mass flow from one iteration to the next.

The resulting predictions give the fluid mass flow, fluid temperature and solid temperature at the nodes used in the problem.

9.4 THERMAL INSTABILITY IN ELECTRICAL INSULATION

Amongst the heat conduction problems involving actual engineering plant that have been studied by the present authors are two involving possible thermal instability in electrical insulation. Thermal instability occurs when the heat generation rates rise more rapidly with increasing temperature than do the heat dissipation rates, and leads to a rapid rise in temperature. The calculations were therefore highly nonlinear. However, they could be adequately represented by heat conduction models using one space dimension.

Most of the calculations reported in this section used the DJ method (Section 9.3.1) for transient calculations, and the SLR method (Section 9.3.2) for steady-state calculations. A few additional calculations have been made using explicit methods that are more easily applicable to problems in more than one space dimension.

The engineering problems studied were the prediction of possible thermal instability in the stator bars of large turbogenerators, and in 11 kV cables in a nuclear reactor following a loss of cooling accident.

9.4.1 Thermal instability in turbogenerator stator bar insulation

The stator bars (or conductors) of large turbogenerators are made of copper and are cooled by pumping water through ducts within the main insulation. The complete stator bars are surrounded by dielectric (electrical insulation) and are held in slots in the stator core. Two fault conditions can cause high dielectric temperatures; they are hot spots in the core itself, and loss of water cooling on the conductors. In either case, it is possible that the rise in dielectric losses (heat generation) with temperature could lead to thermal instability, namely a rapid rise in temperature and heat generation. The present work was aimed at a basic understanding of the problem. The heat generation rates depend on the electrical stress distribution and this was not known accurately for the actual stator bar geometry. Therefore a simple thermal model was used in which the dielectric was represented as an annulus and temperatures were assumed to vary only in a radial direction. Several dielectric materials were considered.

The power factor and permittivity of the dielectric both varied with temperature, and these directly altered the heat generation rate. In addition, most calculations allowed for the variation in permittivity with temperature and consequent redistribution of voltage across the dielectric, and this in turn altered the heat generation rates.

Calculations were made for the core hot spot condition (simulated by raising the dielectric outer surface temperature), and the loss of coolant conditions (simulated by applying a heat flux on the dielectric inner surface equal to the heat generation in the conductor). The core hot spot calculations included the effect of voltage redistribution. For all three dielectrics considered, the transient predictions tended towards high but steady values, indicating thermal stability. Similar results were obtained for the loss of coolant condition (Figure 9.7). However, one of the latter calculations was repeated in which the voltage redistribution was neglected; in this case, thermal instability occurred (Figure 9.7). This was a physically unrealistic condition but was used to demonstrate that the method used here can handle thermal instability. No such instability was predicted when voltage redistribution was included in the calculations. These results illustrate both the ability of the DJ method to handle a thermally unstable situation and also the need to include as many nonlinearities as possible.

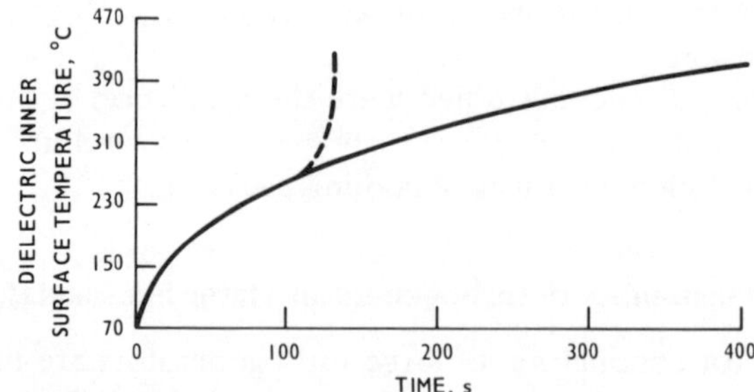

Figure 9.7 Thermal instability in stator bar insulation—, typical results for the loss of coolant condition: —— with voltage redistribution; - - - without voltage redistribution

9.4.2 Possible thermal instability in cable insulation following a LOCA

If a loss of cooling accident (LOCA) occurred at a nuclear power station with water-cooled reactors and steam as the primary heat transfer medium, the primary containment could be filled with contaminated steam and a high level of radioactivity. The dielectric of any power cables within the primary containment could, therefore, be subject to high temperatures and radiation levels. As the radiation itself causes heat generation in the cable and as both the dielectric losses and conductor losses increase with increasing temperature, there is the possibility of thermal instability. A preliminary study was therefore made of possible thermal instability in 11 kV cables subject to such conditions. The model ignored the effects of any cable supports or adjacent cables and assumed that the temperatures varied only in the radial direction. Further details of the physical problem are given by Scarisbrick and Shroff.[92]

The heat transfer from the outside of the cable to its surroundings was nonlinear as it was caused by natural convection and radiation. Standard correlations given by Bayley *et al.*[93] were used. The losses (heat generation rates) in the dielectric rose significantly with increasing temperature. Also in most calculations the losses in the conductor were taken to increase linearly with temperature. Voltage redistribution and subsequent changes in losses in the dielectric were not considered as measurements showed that the permittivity for the dielectric materials considered here did not vary significantly with temperature.

For some of the cables considered, no thermal instability was predicted. However, to illustrate the method, typical predictions for a cable design and LOCA conditions for which thermal instability was predicted are shown in Figure 9.8 and were obtained using the DJ method. Results are given for the

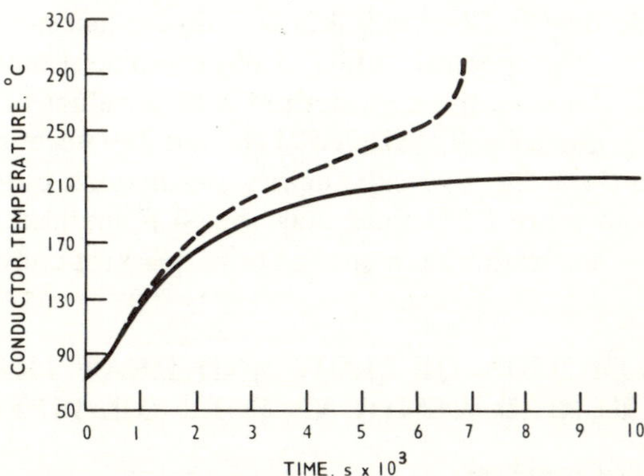

Figure 9.8 Typical results for 11 kV cables under LOCA conditions: – – – conductor losses varying with temperature, —— constant conductor losses

conductor losses increasing linearly with temperature and for the artificial problem of constant conductor losses. As only the former situation is predicted to be thermally unstable, it demonstrates again the need to include as many nonlinearities as possible into a thermal model. The time step that can be used for these calculations depends on whether it is necessary to predict the detailed temperature variation at the onset of instability, or whether it is sufficient to predict that instability will occur and, if so, the approximate time at which it will occur. The former case requires a small time step because of the rapid rise in temperature. For the latter case, the DJ method can be used with a much larger time step. For example, for time steps of 20, 100, 500 and 1000 s the conductor temperature after 6 ks is predicted to be 252.36, 252.37, 251.42 and 251.02 °C respectively, and a very large rise in temperature is correctly predicted after this point.

Predictions of steady-state temperatures for cable designs and LOCA conditions that did not produce thermal instability have been obtained satisfactorily using the SLR method. For a typical calculation, values of relaxation parameter, ω, of 0.8, 0.9, 1.0, 1.1 and 1.2 were used, and 9, 8, 6, 8 and 11 iterations respectively were required. Thus convergence was rapid and did not depend strongly on the value of ω.

Transient calculations for the thermally unstable case shown in Figure 9.8 were repeated using the DUF method (Section 9.3.1). For time steps of 5, 10 and 20 s the conductor temperatures after 6 ks were 252.38, 252.52 and 253.07 °C, and therefore were in good agreement with the DJ results. However, as the onset of thermal instability was reached, the predictions became numerically unstable. This demonstrates the superiority of the DJ method for this problem.

Steady-state calculations for the thermally stable problem mentioned earlier were also obtained using the CSI method. This is the solid node part of the IMSI method described in Section 9.3.2. It is an explicit method that entails the estimation of μ, the spectral radius of the coefficient matrix ($\mu < 1$). For the problem studied above, the CSI method with μ values of 0.9, 0.95, 0.99, 0.995, and 0.999 required 619, 457, 205, 113, and 229 iterations respectively. Thus the CSI method did give satisfactory predictions but required many more iterations and more CPU time than the SLR method. Also, μ had to be estimated more accurately than ω to get the quickest convergence.

9.5 PREDICTION OF FLOW AND HEAT TRANSFER IN A BURIED RESIDUAL FUEL OIL PIPELINE

9.5.1 The physical problem

A 450 mm nominal bore pipeline has been built to transfer residual fuel oil from Kingsnorth to Grain Power Stations over a distance of about 12 km.

The line is insulated throughout its length and, apart from short sections at each end which are above ground, it is buried underground at a minimum depth of 1 m. Heaters at the inlet to the pipe allow the oil entering the line to be heated to a maximum temperature of 70 °C but there is no additional heating along the pipeline itself. The viscosity of the oil increases with decrease in temperature until at a certain critical temperature the oil may become so viscous that it will not flow along the pipeline. The start-up procedure is therefore firstly to pump hot light fuel oil, which has a relatively low viscosity even at ambient temperature, to warm up the pipeline. The heavier residual oil is then introduced. It is desirable to minimize the amount of light fuel oil to reduce costs, but if too little light oil were to be pumped, the residual oil might cool so much that it would not be possible to maintain a flow of oil. This must be avoided, as it would not then be possible to restart the flow; this would be a major operational failure.

Pumping is normally stopped with the pipeline full of light fuel oil. However, in an emergency, pumping may cease with the line full of residual oil, and must then be restarted before the oil has cooled below a critical temperature. If this is not done the restart pressure drop is beyond the capability of the pumps and the pipeline becomes blocked.

To provide information on the operating characteristics of the pipeline, computer models were developed by Stephenson and Soulsby[79,94] for both the start-up and static cooling situations.

9.5.2 Mathematical model of the start-up procedure

Preliminary calculations showed that the pipe insulation was the dominant thermal resistance, so that temperature rises in the soil were comparatively small. Therefore, in order to reduce the problem from three to two space dimensions, the soil region was replaced by an annular region with outer radius equal to twice the burial depth of the pipeline and with a fixed temperature all round this outer radius. The steady-state thermal resistance of this annulus was therefore the same as for the actual soil geometry.[95]

The oil viscosity, and therefore the wall to oil heat transfer coefficient, was permitted to vary with temperature, as was the insulation thermal conductivity. The oil density was permitted to vary with temperature only when allowing for natural convection, which was found to increase significantly the heat transfer in laminar flow. Apart from this, all material properties were assumed to be constant, as the variation was small over the temperature range considered.

Heat transfer coefficients for turbulent flow were based on those recommended by Spiers[96] and Rosenhow and Hartnett.[97] The correlation for laminar flow was based on that derived by Oliver and Jenson.[98]

After computer runs had been made for various nodal distributions, it was found that satisfactory accuracy could be obtained with 100 divisions along the length of the pipeline, and 8 divisions in the radial direction.

The temperatures were calculated using the DUF and IEDUF methods (Section 9.3). Two features of the problem required special treatment. The first was the fact that the start-up procedure resulted in an abrupt change in temperature in the oil, and this propagated along the pipeline. The second was that there was a large (typically a factor of ten) variation in wall to oil heat transfer coefficient between the cold and hot oil.

To handle the abrupt temperature change it was decided to use numerical methods for the fluid nodes which would predict a discontinuity exactly in the absence of heat loss from oil to wall, and which were therefore assumed to be suitable for the present situation where the heat loss was small.

The variation in oil to wall heat transfer coefficient was handled by calculating the oil viscosity and heat transfer coefficient using the pipe wall and bulk oil temperatures for the corresponding upstream nodes at the start of the time step. The justification for this was that, with $C_r = 1$ and comparatively small heat loss from the oil, the upstream oil temperature at the start of a step was a good approximation to the temperature of the node itself at the end of a step.

9.5.3 Predictions for start-up

The first predictions were obtained for comparison with an analytical solution derived by Ferris (see Stephenson and Soulsby[79]) for a simplified version of the start-up model. A lumped-parameter model was therefore developed making the same assumptions, and predictions were made for both the DUF and IEDUF methods, and compared with his analytical solution.

A typical comparison is shown in Figure 9.9; the DUF method agrees well with the analytical solution except near the interface between hot and cold oil, whereas the IEDUF method is everywhere in good agreement. This gives confidence in the use of the IEDUF method for the full model outlined earlier.

Typical predictions for the full model are given in Figure 9.10; these are for pumping hot light oil followed by residual oil. The sharp change in axial temperature gradient at about 12 °C results from the flow being laminar below this temperature (for the flow rate of Figure 9.10), with a consequent drop in heat transfer coefficient. The abrupt change in temperature gradient at about 68 °C for 100% of transit time marks the residual oil front.

Typical predictions of the lowest residual oil temperature are shown in Figure 9.11 for a higher flow rate than in Figure 9.10 and for two pumping times for hot light oil. Only the longer time keeps the residual oil above its pour point (defined by BS4452[99]). As it has been decided that the pipeline will be operated only with oil temperature above the pour point, the lower

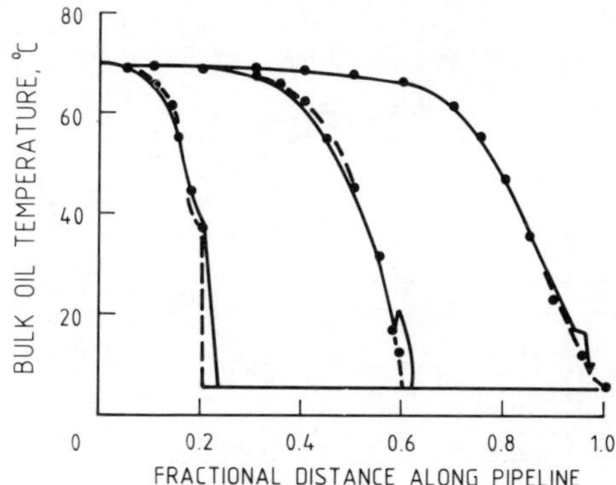

Figure 9.9 Comparison with analytical solution for pipe-line start-up (hot light fuel oil pumped into a line full of cold light oil). Results for 20, 60 and 100% of transit time: ∘ analytical solution; — DUF method; -- IEDUF method

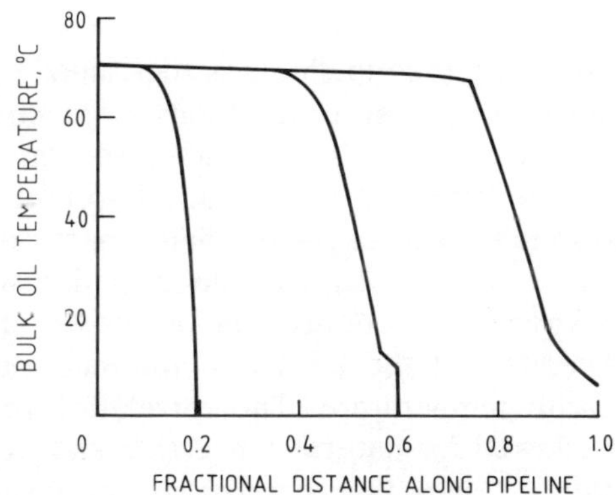

Figure 9.10 Typical predictions for start-up. Hot light oil pumped for 25% of transit time, followed by residual oil. Results for 20, 60 and 100% of transit time

light oil pumping time is an unsatisfactory start-up procedure. The abrupt drops in temperature at around 45 °C in Figure 9.11 are caused by the residual oil becoming laminar.

9.5.4 Mathematical model for static cooling

The calculations for static cooling assumed that oil had previously been pumped sufficiently long for steady temperatures to be reached. As the

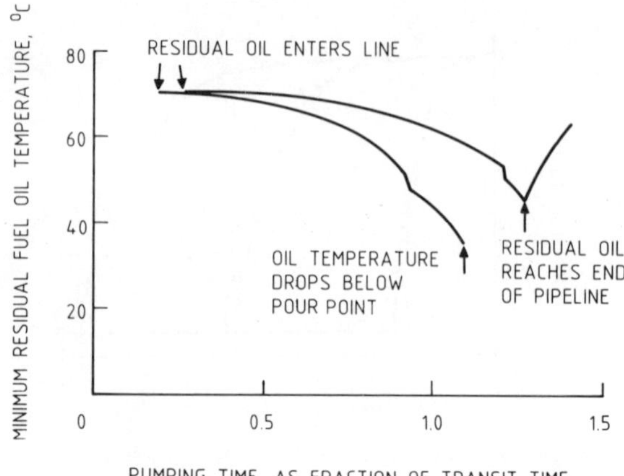

Figure 9.11 Predictions of minimum residual oil temperature during start-up. Hot light oil pumped for 19% or 25% of transit time, followed by residual oil

temperature gradient along the pipeline was then small (typically under 5 K in 12 km), only radial temperature variations were considered. Separate models were constructed for the main buried section and for the short overground sections; the latter were important despite their short length, as the oil was likely to solidify in these regions first. For the overground section, the boundary condition on the outside of the insulation was heat transfer to surrounding air at a known temperature. The oil-to-wall and insulation-to-air heat-transfer coefficients and the insulation thermal conductivity were all permitted to vary with temperature. The correlation for oil-to-wall heat-transfer coefficient allowed for natural convection and was based on that of Kuehn and Goldstein,[100] and the correlations for insulation-to-air heat-transfer coefficient were those recommended by Spiers[96] for moving air and Rose and Cooper[101] for still air.

The first part of the procedure was to calculate the steady-state radial temperature profile for a specified bulk oil temperature. This profile was then used as the initial conditions for the transient calculations.

Explicit numerical methods were first used to calculate both the steady-state and transient temperatures. The methods used were CSI and DUF (see Section 9.3; CSI is the solid node part of the IMSI method). Later studies showed that the number of iterations required for the steady-state calculations could be reduced by a factor of 4 to 5 by employing the SLR method. Also, the time step for the transient calculations could be greatly increased by using the DJ method. Hence the DJ and SLR methods required less overall CPU time than the DUF and CSI methods.

9.5.5 Predictions for static cooling

A typical cooling curve is shown in Figure 9.12, for cooling of the overground section from an initial bulk oil temperature of 70 °C. Predictions for the DJ method for time steps of 2×10^4 and 5×10^4 s are shown, and are in good agreement. For comparison, predictions are also shown for the largest time step that could be used for the DUF method, namely 500 s.

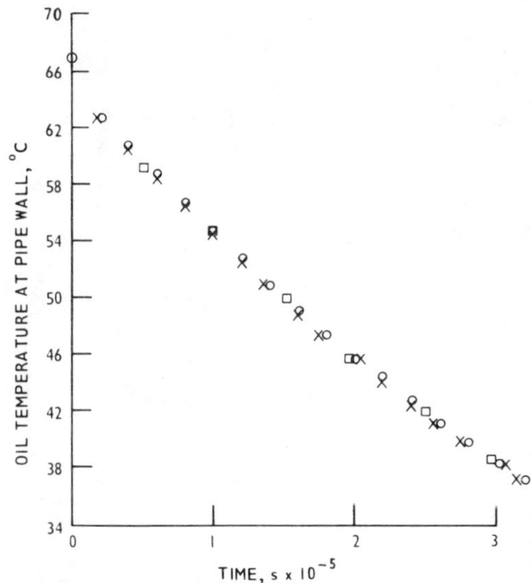

Figure 9.12 Typical predictions for cooling of overground pipeline: ⊙ DJ, $\Delta t = 2\times10^4$ s; □ DJ, $\Delta t = 5\times10^4$ s; × DUF, $\Delta t = 5\times10\,s^2$

9.6 COOLING OF E.H.V. CABLES AND THEIR ACCESSORIES

This section describes the application of two of the numerical methods outlined in Section 9.3 to the prediction of transient and steady-state temperatures in a group of forced-cooled joint assemblies linking lengths of buried, high voltage cables. The cables and joints are cooled by pumping water through an arrangement of pipes buried alongside.

9.6.1 Problem description

A plan view of a cable joint bay and various cross-sections through it are illustrated in Figure 9.13. A cable comprises copper conductor surrounded by dielectric, a metal sheath and PVC servings. When two lengths of cable

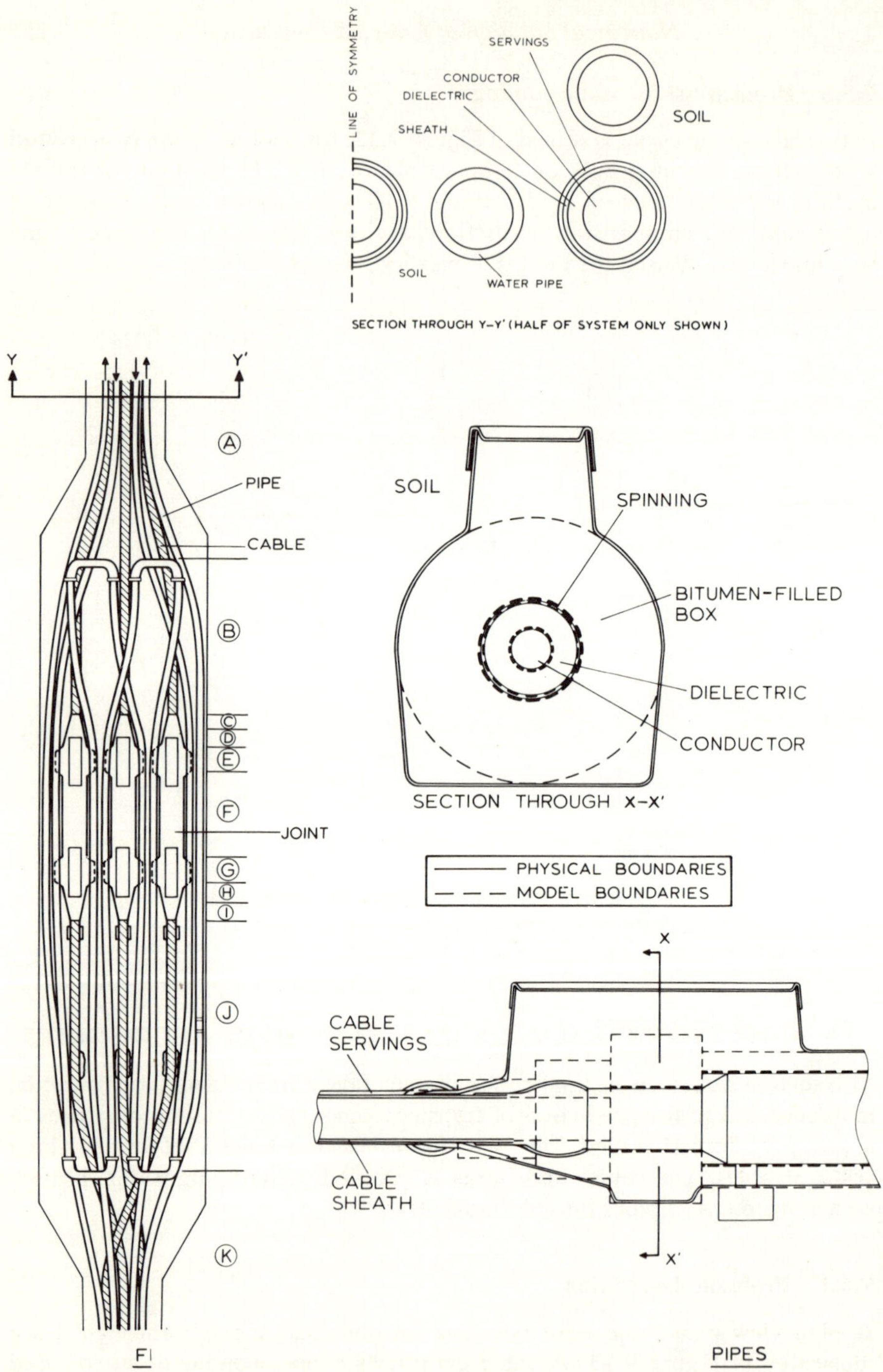

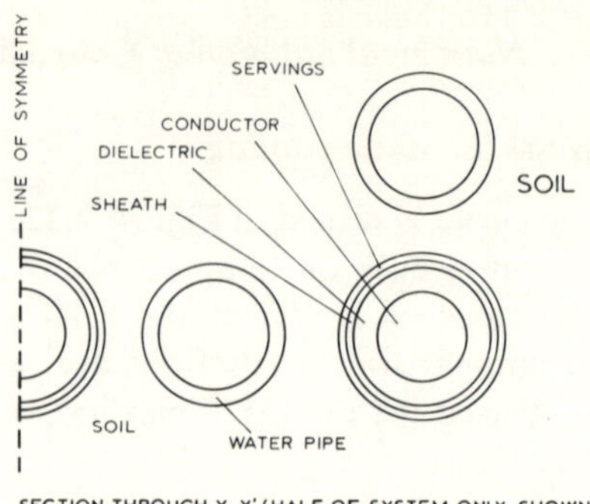

Figure 9.13 Arrangement of cables, joints and water pipes

are jointed together, the joint dielectric is considerably thicker than that of the cable. A copper cylinder (spinning) is slid over the joint dielectric and plumbed at each end to the cable sheath. The complete assembly is then encased in a bitumen-filled glass-fibre box.

For the cooling arrangement depicted in Figure 9.13, water is pumped from a cooling station at the start of section A along the 'forward' pipes on either side of the centre cable. These forward pipes are looped around the first joint bay and then continue along the cable route where further joint bays may be situated. Finally, when the main turnround position for the water pipes is reached, at the end of section K, the forward pipes return above the cables to the start of section A, where heat exchangers cool the water before it is returned to the start of the cable route.

As a result of alternating currents and voltage, electrical losses (heat generation) are produced in the conductor, dielectric and sheath of the cables, and also in the conductor and dielectric of the joint section. Ageing of the cable dielectric increases with temperature, and to achieve an economic life time, the maximum dielectric temperature (which occurs immediately adjacent to the conductor) is usually limited to 85 °C. The current loading of the cables may vary with time and a method is required to determine the corresponding transient temperature distribution in the forced-cooled cable and joint assembly.

9.6.2 Assumptions for heat transfer calculations

The main aim is to determine the maximum dielectric temperature along the cable route. In most two-way systems with identical laying conditions along the route, the joint bay with the hottest temperatures is the last one before the water pipe main turnround position. Hence the model represents this bay and the total length of cables and water pipes from inlet to turnround, but it excludes the other joint bays on the assumption that the representation of these by the equivalent length of cable does not significantly alter the water temperatures along the route.

Axial heat conduction has been ignored in the nonmetallic components and all materials are assumed to have constant thermal properties.

9.6.3 Geometrical simplifications

The joint bay has been divided into 9 main sections, B to J, and the remainder of the cable route before and after the bay are represented by sections A and K. So far, a simplified model has been developed which relies on all sections of cable joints and pipes being cylindrical and parallel to each other. The main simplifications that have been made are: (a) abrupt changes in cable spacing at the interface of sections A and B and sections J and K; (b) the pipework layout in sections B to E and G to J is the same as that in section F, with abrupt changes in pipe spacing at the interface of sections A and B and J and K; (c) the bitumen-filled turrets on top of the joint boxes have been replaced with soil and the remainder of these sections are assumed to be circular in cross-section (see Figure 9.13).

9.6.4 Three-dimensional mesh generation and calculation of thermal parameters

9.6.4.1 Two-dimensional mesh normal to cable axes

Figure 9.14 shows a typical mesh arrangement for a section through cable joints and adjacent water pipes surrounded by soil (in this case for cross-section X–X′ in Figure 9.13). Firstly, each cable (or cable joint) and pipe has an annular ring of soil of prescribed thickness associated with it. The soil region in the vicinity of the cables and pipes is then covered with a pre-defined square mesh. Any nodes which fall inside the aforementioned soil rings are moved onto the outside surfaces of the same rings and are referred to as 'soil ring nodes'. Nodes are then positioned halfway through each pipe wall at the same azimuthal positions as the adjacent soil ring nodes. At each axial plane, the pipe wall nodes are connected to a single water node in the pipe. In the case of a cable joint, the bitumen and glass-fibre box combined (referred to globally as the 'box') is divided radially into a prescribed number of equal volumes, and nodes are positioned midway between the inside and outside radii of each volume at the same azimuthal positions as the surrounding soil ring nodes. These box nodes are interconnected both radially and azimuthally. The next material radially inwards, copper spinning, is assumed circumferentially isothermal because of its high thermal conductivity. Hence, the box nodes which are radially adjacent to the spinning are all connected to a single spinning node representing one axial section of the spinning (see Figure 9.14). The dielectric between conductor and spinning, which has a low conductivity, is not divided azimuthally, but is divided radially into equal volumes with a node placed at the radial midpoint of each volume. The innermost region, copper conductor, is assumed circumferentially isothermal. The cable sections A, B, J and K in Figure 9.14 are treated in a similar manner (see Gane and Soulsby[74]).

When the inner part of the mesh has been constructed, the outer part is formed by successive triangulation (see Figure 9.15), so that the mesh is gradually coarsened away from the cables/joints and pipes. This part of the mesh is terminated at the ground surface and at boundaries sufficiently remote from the cables to be treated as adiabatic.

In this manner, each axial division A to K in Figure 9.13 has a two-dimensional geometrical mesh constructed normal to the cable axes. In the

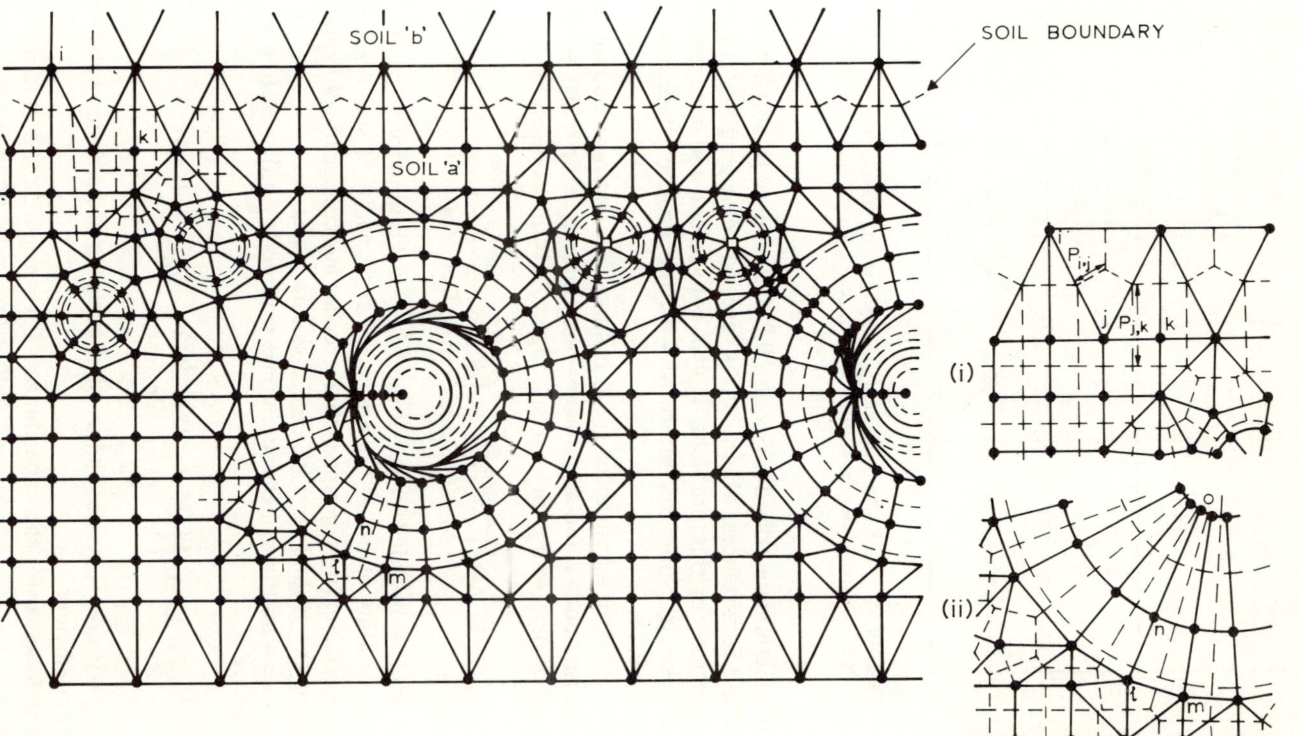

Figure 9.14 Mesh arrangement through X–X′ for cable joints, water pipes and soil: – – – material and nodal boundary; — connector between adjacent nodes; ● solid node; □ water node

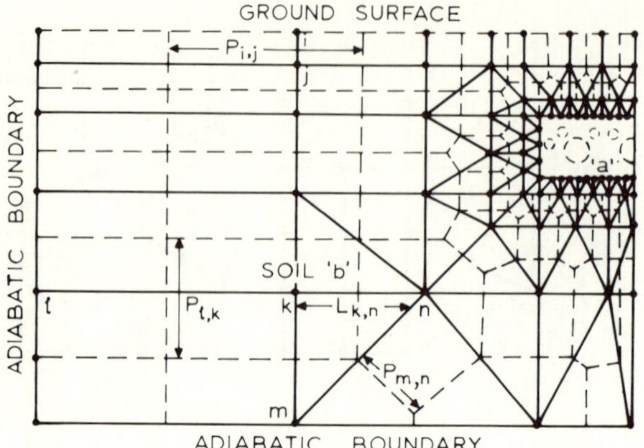

Figure 9.15 Mesh arrangement for outer soil region at section X–X′: - - - nodal boundary; — nodal connector; ● solid node

soil region the method described in Section 9.2.4 (d) is used to obtain thermal capacitances and admittances, where, in this case, Δz is the length of the axial section in question. In this problem, it was convenient and acceptable to approximate a straight-line soil interface with the broken, staggered line shown in Figure 9.14.

Two other important admittances are as follows. The composite admittance between a soil ring node l and an adjacent box node (in cable joint), servings node (in cable) or pipe wall node (e.g. see Figure 9.14(ii)) is given by

$$y_{l,n} = \Delta z \ \Delta\theta_n \left[\left(\frac{K_{s0} r_{0,n}}{\Delta r_{sr}} \right)^{-1} + \left(K \Big/ \ln \frac{2 r_{0,n}}{r_{0,n} + r_{i,n}} \right)^{-1} \right]^{-1} \tag{9.34}$$

where $\Delta\theta_n$ = angle subtended by radial lines of nodal boundary n; $r_{0,n}(r_{i,n})$ = outside (inside) radius of nodal boundary n; Δr_{sr} = width of soil ring; K = conductivity of box, servings or pipe wall. Here, use has been made of Equation (9.9) in deriving the second term inside the square brackets.

The composite admittance between a pipe wall node j adjacent to a water node l is given by

$$y_{j,l} = \Delta z \ \Delta\theta_j \left[\left(K_p \Big/ \ln \frac{r_{0,p} + r_{i,p}}{2 r_{i,p}} \right)^{-1} + (h_c r_{i,p})^{-1} \right]^{-1} \tag{9.35}$$

where $r_{i,p}(r_{0,p})$ = inside (outside) radius of the pipe and h_c is the water-side heat-transfer coefficient.

The control areas A_i for the capacitances, $C_i(=\rho c_p A_i \ \Delta z)$, of nodes in the pipes, cables and joints are contained within the nodal boundaries (e.g. see Figure 9.14(ii)). Further details are given by Gane and Soulsby.[74]

9.6.4.2 Axial mesh along cable route

With the exception of sections C, D, H, and I, the sections A to K are subdivided axially so that axial temperature gradients are predicted more accurately. Hence, if a section is subdivided n times, the two-dimensional mesh for this section is repeated n times in series along the cable axis at the specified intervals. Each metallic node is then connected to its axially adjacent nodes. Due to symmetry, the planes at the water inlet and main turnround positions are adiabatic, so that nodes next to these planes have only one axial connection, i.e. not across the adiabatic boundary.

The arrangement of axial solid and water nodes is shown in Figure 9.3 for the modified centre differencing method used here. The admittance between two axially adjacent nodes i and j in metallic materials at planes k and $k+1$, respectively, is defined by

$$y_{i,j} = 2\left[\left(\frac{K_i A_i}{\Delta z_k}\right)^{-1} + \left(\frac{K_j A_j}{\Delta z_{k+1}}\right)^{-1}\right]^{-1}. \tag{9.36}$$

9.6.4.3 Heat generation rates

For each conductor node i and sheath node j, typical expressions for the internal heat generation rates, Q_i and Q_j, at axial position k, are

$$Q_i = 0.00001\,\Delta z_k I^2 (1 + 0.0059 T_i), \qquad Q_j = f_j Q_i \tag{9.37}$$

where I = conductor current and f_j is constant. There is constant heat generation in the dielectric. Elsewhere there is no heat generation.

9.6.5 Governing equations and numerical solution procedures

The equations solved for the solid and water temperatures are (9.1) and (9.6) respectively. The water inlet temperature is specified at the beginning of axial section A (Figure 9.13) for the middle two forward pipes. A heat flux boundary condition is specified on the ground surface.

Equations (9.1) and (9.6) applied to this problem have been solved using the FIDUF method described in Section 9.3.1. The steady-state versions of these equations have been solved iteratively using the IMSI method described in Section 9.3.2.

9.6.6 Numerical results for a straight joint linking 275 kV cables

The method described above has been applied to a typical straight joint linking 275 kV cables which are externally cooled by water pipes as illustrated in Figure 9.13. In this problem, it was assumed that air and ground surface were at the same temperature, which was held constant at 10 °C. The water inlet temperature at the start of section A was fixed at 15 °C. Typical values for the heat-generation parameters were used. A single joint bay was considered which was known *a priori* to have the highest temperatures. The length of the cable route was 2647 m and that of the joint bay was 7 m. The entire soil region was assumed homogeneous with given constant properties for a stabilized backfill.

9.6.6.1 Steady current loading

A steady current of 1600 A was applied. Axial sections A–K were successively subdivided so that a total of 15, 22, and 29 axial subdivisions resulted. Equation (9.20) with $\mu = 0.995$ and Equation (9.21) were then used to predict the steady-state temperatures. Figure 9.16 shows the conductor steady-state

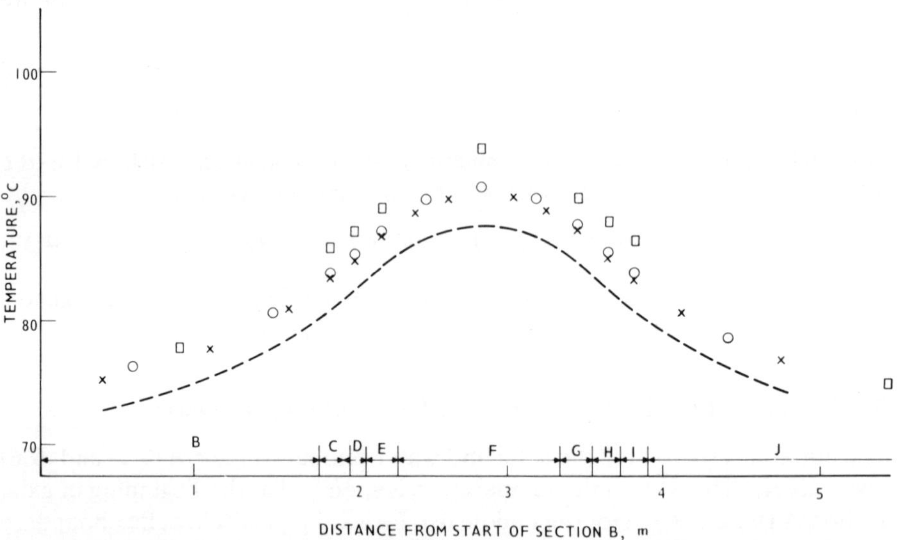

Figure 9.16 Variation of conductor steady-state temperatures in a joint bay: current 1600 A; ground temperature 10 °C; inlet water temperature 15 °C; no. of axial divisions: ⊡ 15; ⊙ 22; × 29. Results shown are for centre cable conductor. --- Results for outer cable conductor with 29 axial divisions in total

temperatures along the centre cable in sections B to J for a total of 15, 22, and 29 axial subdivisions with no radial subdivisions of the dielectric and box materials. The calculated maximum conductor temperature, which occurred in the centre of section F, was 93.9 °C, 90.8 °C, and 90.2 °C for these respective cases. It was therefore considered that 22 axial subdivisions gave sufficient accuracy and this number was then fixed.

9.6.6.2 Step change in loading

The steady-state starting temperatures were calculated for a load current of 800 A and no radial divisions of the dielectric and the box. Six values of the parameter μ were used, and the optimum value of 0.995 gave convergence in 209 iterations, taking 4.5 minutes of CPU time on an IBM 370/168 computer for a total of 11,518 nodes.

The variation of maximum conductor temperature with time following a load change from 800 A to 1600 A is shown in Figure 9.17 for 1, 5, and 10 radial divisions of each dielectric. It can be seen that 5 divisions gives sufficient temporal accuracy. At later stages in the transient, 1 division was sufficiently accurate. Similar tests showed that 4 radial divisions in the bitumen-filled box sections produced the required degree of accuracy. In all cases, the time step Δt was reduced sufficiently to eliminate temporal errors, e.g. $\Delta t = 800$ s, giving maximum $d_i = 14.5$ in Equation (9.14), was acceptable for one of the cases considered. This method is therefore appreciably faster than using the simple explicit approximation of Equation (9.1), since the latter scheme requires that maximum $d_i \leq 1$.

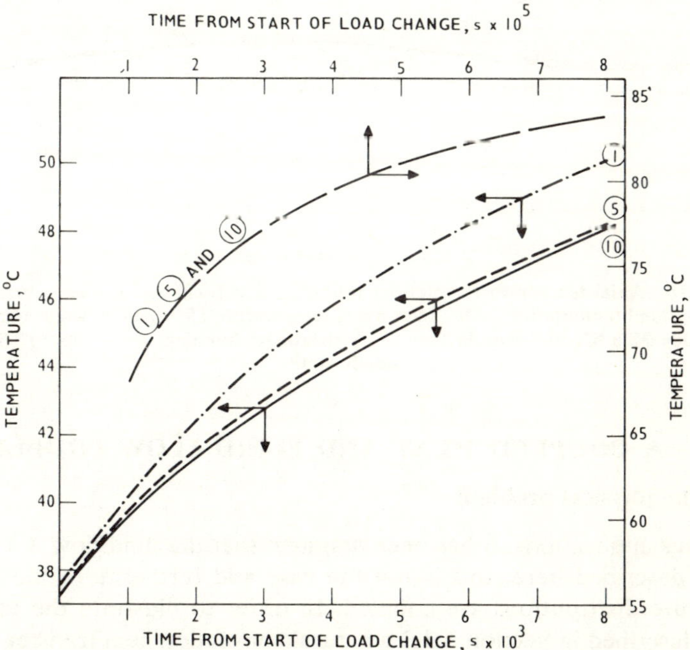

Figure 9.17 Variation of maximum conductor temperature with time, following a load change.

Number of radial dielectric nodes: – · – · – 1; – – – 5; —— 10.

— — Identical results for 1, 5 and 10 radial dielectric nodes after long time interval

Figure 9.18 shows the temperature variation with distance along the conductor and sheath/spinning for the case of 5 radial subdivisions in each dielectric and 4 in each box section. It can be seen that the maximum conductor temperature occurred at the centre of the cable joint at all times during the transient. The maximum sheath temperature occurred at a position close to the plumb between sheath and spinning.

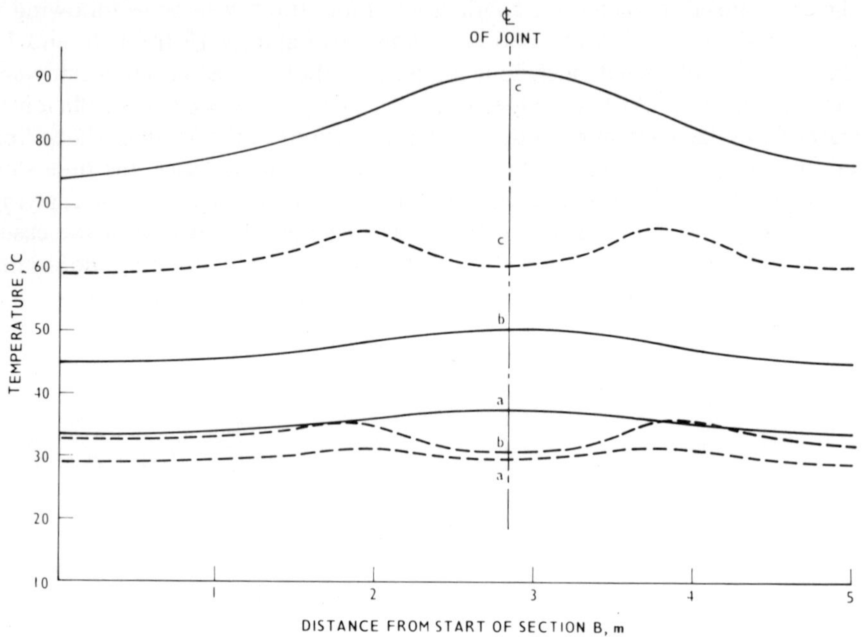

Figure 9.18 Axial temperature variations following a step change in load. Current: 800–1600 A; ground temperature 10 °C; inlet water temperature 15 °C; mean water temperature in joint bay 32.8 °C: —— conductor; ---- sheath or spinning; a = 0 s; b = 10 800 s; c = steady-state

9.7 A COUPLED HEAT AND FLUID FLOW PROBLEM

9.7.1 The physical problem

In previous applications, it has been assumed that the fluid flow is known. In the case described here, this is not the case and furthermore the flow and temperature distributions are coupled. In order to illustrate the use of the method described in Section 9.3.3 a natural convection heat transfer situation is considered. The problem is illustrated in Figure 9.19. A hot gas flows through a pipe which is insulated on both the inside and outside. The internal insulation could, for example, be used to ensure that the pipe material does not reach the fluid temperature which may be a requirement with a very hot

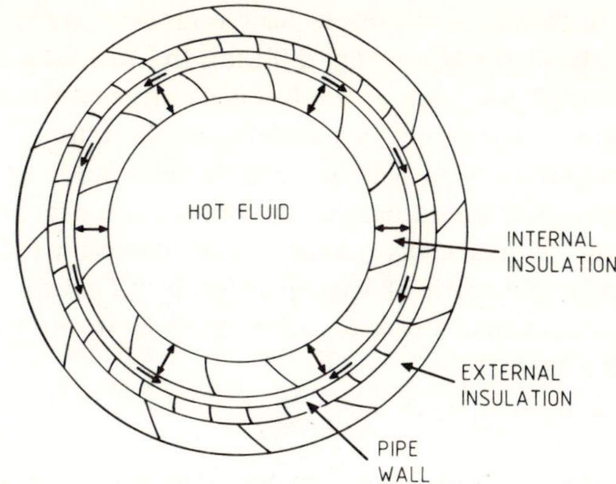

Figure 9.19 Schematic diagram of flow in a pipe with ill-fitting porous insulation: → ↔ fluid directions

fluid. Outside the external insulation is a heat sink which may, for example, be cooling water pipes or just ambient air.

Insulation is often a fibrous or cellular material and because of this, the internal insulation is taken as porous in a radial direction with a known flow resistance. Between the inner surface of the pipe wall and the internal insulation is a small annular gap which occurs because it is very difficult to make up some types of insulation such that they provide an exact fit inside a pipe.

Because of the porosity of the insulation, a thermosyphon flow loop can be set up if the conditions are suitable. This flow loop consists of hot gas flowing radially through the insulation into the gap between the insulation and the pipe. The gas in the gap is cooled by the external heat sink and flows circumferentially down through the gap. This cooled gas then flows radially back through the insulation into the main hot gas.

The problem is to calculate the details of the gas flow and temperature in the gap and to predict the resulting pipe temperatures. Clearly the equations for the gas flow, gas temperature and pipe temperature are strongly coupled. The calculation procedure must allow for a radial flow occurring over all angles. Both the magnitude of the radial flow and its direction are determined by the gas pressure difference between the gap and the main gas and this, which is not known *a priori*, varies with angle. For the purposes of this illustration, variations in the axial direction are taken as negligible.

9.7.2 Details of the solution

The solution procedure described in Section 9.3.3 was used, with the direction z referring to the azimuthal direction. An under-relaxation coefficient of 0.5

was used in the iteration of the mass flow equation. Before predictions can be obtained the physical parameters defining the problem must be supplied. Some of these are obvious, such as the hot gas temperature, sink temperature, geometry etc. Other parameters need a little explanation.

Firstly the flow parameters will be considered. These appear in Equation (9.30). It is assumed that the radial porosity of the internal insulation is known and that the radial pressure drop across the insulation for a mass flow $\dot{m}_w$ can be expressed in the form of Equation (9.28). This then gives values for K_1 and K_2. The wall shear stress (τ_w) for the flow in the gap is obtained by assuming a friction factor

$$f = aR_e^{-b}, \tag{9.38}$$

where the coefficients a and b have to be estimated. The Reynolds number (R_e) relates to flow in the gap. For sufficiently small gap sizes the flow is laminar and standard formulae for forced flow in parallel-sided ducts could be used, see for example Rosenhow and Hartnett.[97]

The external pressure gradient in the z-direction, ($\mathrm{d}p_e/\mathrm{d}z$), refers to the pressure within the main gas inside the pipe. Typically this is derived from the gravitational head of the fluid within the pipe assuming the main gas is either all at the known bulk temperature or has a known temperature distribution. This then gives

$$\frac{\mathrm{d}p_e}{\mathrm{d}z} \cong \rho_e g \sin\theta, \tag{9.39}$$

where θ is the azimuthal angle measured from the vertical and $z = r_g\theta$ where r_g is the mean radius of the gap.

The thermal parameters to be specified are those appearing in Equation (9.31), namely $y_{s,f}$ and $y_{e,f}$. They take the same form as those given in Section 9.6. The term $y_{e,f}$ represents the thermal admittance between the main fluid in the pipe and the fluid in the gap. It consists of heat transfer from the main fluid to the internal insulation, conduction through the insulation and heat transfer from the insulation to the gas in the gap. The gap heat-transfer coefficient is obtained from the following Nusselt number definition:

$$N_u = \alpha R_e^{\beta} P_r^{\gamma}, \tag{9.40}$$

where the coefficients α, β and γ must be estimated. Again for a sufficiently small gap, a suitable laminar flow correlation could be used.

The parameter $y_{s,f}$ represents the thermal admittance between the gap fluid and the external heat sink. It consists of heat transfer from the gap fluid to the pipe, conduction through the pipe wall and heat transfer from the pipe wall to the external heat sink through any medium between the pipe and the sink.

9.7.3 Predictions

A particular example has been considered consisting of a hot fluid at 370 °C with an external sink at 30 °C. The pipe is such that the radial temperature drop through the wall is very small, so its temperature at a given circumferential position is characterized by a single value. The pipe diameter is 0.85 m with a radial gap size of 5.0 mm. Typical values of insulation thickness, thermal conductivity and resistance to radial flow have been assumed.

Figure 9.20 shows the predicted flows in the gap. The predictions were obtained with 5 azimuthal nodes between 0 °C and 180 °C. Results were also obtained with 10 azimuthal nodes to provide a check on the numerical accuracy. It may be seen that inflow into the gap occurs over the top half of the gap, resulting in the increase in flow with angle. The outflow occurs over the bottom half. The inflow is a maximum at the top. Symmetry conditions mean that the flow in the gap at 0 °C and 180 °C is zero.

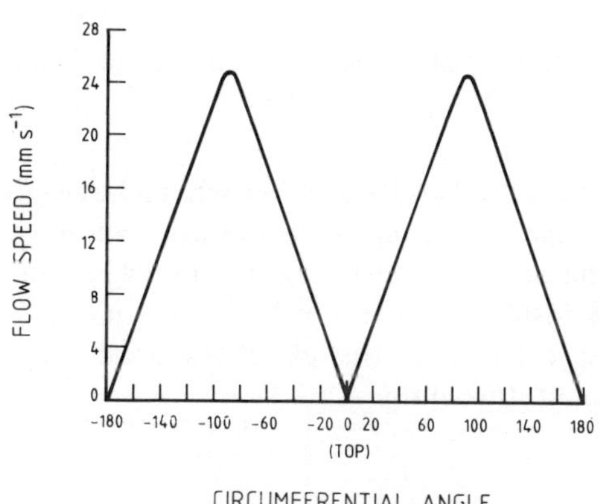

Figure 9.20 Variation of flow in the gap

The variation of predicted fluid temperatures in the gap with angle are shown in Figure 9.21. The high gas temperatures over the top ±90° are a direct result of the inflow of hot gas. Over the lower part of the pipe, the inflow ceases and outflow begins, and the heat loss from the fluid in the gap to the external sink quickly reduces the fluid temperature. Also shown on Figure 9.21 are the predicted pipe wall temperatures. These follow the fluid temperatures although circumferential conduction in the wall does have some effect. Finally, Figure 9.21 shows what the pipe temperature would be if the flow into the gap were prevented or ignored. In this case the heat transfer to the pipe is by conduction through the insulation only and there is no variation with angle.

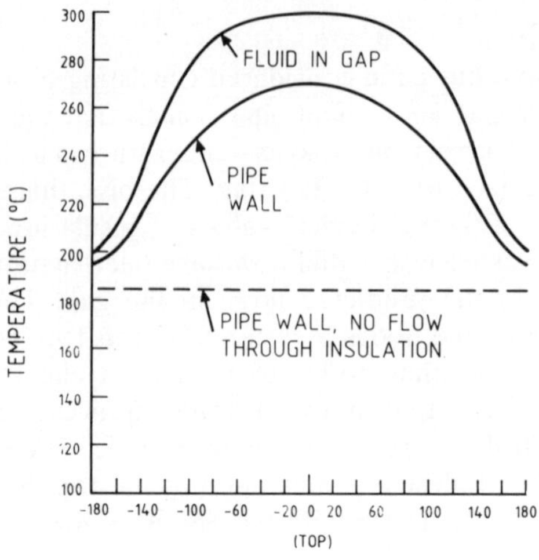

Figure 9.21 Variation of temperature with angle

The results in Figures 9.20 and 9.21 show what a large effect a comparatively low fluid flow through the internal insulation can have on the pipe wall temperature. For the case considered this fluid flow has increased the effective conductance of the insulation by an order of magnitude. This indicates how important it can be to allow for this effect, especially if the pipe wall has to be below some temperature limit.

9.8 CONCLUDING REMARKS

A general lumped-parameter method has been described for solving transient and steady-state problems involving solid materials exchanging heat with a flowing fluid. Either the fluid flows can be prescribed or they can be calculated along with the solid and fluid temperatures. Examples have been given of finite difference methods employing non-uniform grid arrangements that can be used to define the thermal admittances between nodal points. A control volume (element) surrounding each node is associated with the nodal temperature.

Two methods of handling the heat flow in the fluid have been investigated. The upwind difference scheme uses the temperature at the downstream end of a fluid element when evaluating heat transfer to the solid and heat stored in the fluid. This approach is advantageous for transient problems involving large changes in fluid temperatures along a duct, but little heat exchange with the solid, if the Courant number is unity, since a very accurate solution results. In other cases where these stipulations do not apply, a modified centre differencing scheme, which assumes linear temperature variations in the fluid flow directions, is generally more accurate.

The Du Fort–Frankel scheme combined with a locally fully implicit method for computing transient solid and fluid temperatures, respectively, can be made an explicit scheme overall and has been found suitable for large non-uniform networks. However, for one-dimensional, highly nonlinear problems, the DJ predictor–corrector scheme has been found to be more accurate and efficient to use. Similar conclusions were reached for steady-state problems, when comparing the explicit Chebyshev semi-iterative method with a successive line relaxation method of iterating.

The methods have been demonstrated by applying them to a range of engineering plant, starting with problems that have only solid regions and continuing with problems involving solid and fluid regions, with known fluid flow rates. Problems involving complicated nonlinearities and complex geometries have been solved successfully. Finally, an example has been given of a problem where the solid and fluid temperatures and the fluid flow rates have to be calculated simultaneously.

ACKNOWLEDGEMENT

This work was carried out at the Central Electricity Research Laboratories and is published by permission of the Central Electricity Generating Board.

REFERENCES

1. A. V. Luikov, V. A. Aleksashenko and A. A. Aleksashenko, 'Analytical methods of solution of conjugated problems in convective heat transfer, *Int. J. Heat Mass Transfer*, **14**, 1047 (1971).
2. E. J. Davis and W. N. Gill, 'The effects of axial conduction in the wall on heat transfer with laminar flow', *Int. J. Heat Mass Transfer*, **13**, 459 (1970).
3. M. B. Hsu and R. E. Nickell, 'Coupled convective and conductive heat transfer by finite element methods, Finite elements in flow problems' (ed. J. T. Oden *et al.*), UAH Press, Univ. Alabama in Huntsville, Alabama, U.S.A. (1974).
4. S. Mori, M. Sakakibara and A. Tanimoto, 'Steady heat transfer to laminar flow in a circular tube with conduction in the tube wall', *Heat Transfer—Japanese Research*, **3**, 37 (1974).
5. M. D. Mikhailov and B. K. Shishedjiev, 'Coupled at boundary heat and mass transfer in entrance concurrent flow', *Int. J. Heat Mass Transfer*, **19**, 553 (1976).
6. C. Taylor and A. Ijam, 'Coupled convective/conductive heat transfer including velocity field evaluation', in *The Mathematics of Finite Elements and Applications; II, MAFELAP 1975* (ed. J. R. Whiteman), Academic Press, London (1976).
7. M. Sakakibara and K. Endoh, 'Effect of conduction in wall on heat transfer with turbulent flow between parallel plates', *Int. J. Heat Mass Transfer*, **20**, 507 (1977).

8. D. K. Gartling and R. E. Nickell, 'Finite element analysis of free and forced convection', in *Finite Elements in Fluids*, Vol. 3 (ed. R. H. Gallagher, O. C. Zinkiewicz, J. T. Oden, M. Morandi Cecchi and C. Taylor), Wiley, Chichester (1978).
9. S. Mori, T. Inoue and A. Tanimoto, 'Heat transfer to laminar flow with temperature-dependent heat generation', *Can. J. Chem. Eng.*, **55**, 138 (1977).
10. B. Sunden, 'A coupled conduction–convection problem at low Reynolds number flow', *Numerical Methods in Thermal Problems*, Pineridge Press, Swansea, U.K. (1979).
11. B. Sunden, 'Conjugated heat transfer from circular cylinders in low Reynolds number flow', *Int. J. Heat Mass Transfer*, **23**, 1359 (1980).
12. B. Sunden, 'A numerical study of coupled conduction-mixed convection', in *Numerical Methods for Nonlinear Problems*, Pineridge Press, Swansea, U.K. (1980).
13. B. Sunden, 'A coupled conduction–convection study in the slip flow regime', in *Numerical Methods in Thermal Problems*, Vol. II, Pineridge Press, Swansea, U.K. (1981).
14. B. Abramzon and C. Elata, 'Numerical analysis of unsteady conjugate heat transfer between a single spherical particle and surrounding flow at intermediate Reynolds and Peclet numbers', in *Numerical Methods in Thermal Problems* (ed. R. W. Lewis, K. Morgan and B. A. Schrefler), Pineridge Press, Swansea, U.K. (1981).
15. H. S. Carslaw and J. C. Jaeger, *Conduction of Heat in Solids*, 2nd edn. Clarendon Press, Oxford, p. 391 (1959).
16. M. Sakakibara, S. Mori and A. Tanimoto, 'Effect of wall conduction on convective heat transfer with laminar boundary layer', *Heat Transfer—Japanese Research*, **2**, 94 (1973).
17. A. V. Luikov, 'Conjugate convective heat transfer problems', *Int. J. Heat Mass Transfer*, **17**, 257 (1974).
18. J. Sucec, 'Unsteady heat transfer between a fluid with time varying temperature, and a plate: an exact solution', *Int. J. Heat Mass Transfer*, **18**, 25 (1975).
19. P. Payvar, 'Convective heat transfer to laminar flow over a plate of finite thickness', *Int. J. Heat Mass Transfer*, **20**, 431 (1977).
20. M. S. Sohal and J. R. Howell, 'Determination of plate temperature in case of combined conduction, convection and radiation heat exchange', *Int. J. Heat Mass Transfer*, **16**, 2055 (1973).
21. R. Karvinen, 'Note on conjugated heat transfer in a flat plate', *Lett. Heat Mass Transfer*, **5**, 197 (1978).
22. R. Karvinen, 'Some new results for conjugated heat transfer in a flat plate', *Int. J. Heat Mass Transfer*, **21**, 1261 (1978).
23. R. Karvinen, 'Transient convective heat transfer from a wall with capacitance and resistance', *Lett. Heat Mass Transfer*, **7**, 385 (1980).
24. R. Karvinen, 'Method to solve some coupled convection and conduction problems', *Numerical Methods in Heat Transfer*, Vol. II, Pineridge Press, Swansea, U.K. (1981).
25. K. Chida and Y. Katto, 'Study on conjugate heat transfer by vectorial dimensional analysis', *Int. J. Heat Mass Transfer*, **19**, 453 (1976a).
26. K. Chida and Y. Katto, 'Conjugate heat transfer of continuously moving surfaces', *Int. J. Heat Mass Transfer*, **19**, 461 (1976b).
27. R. Viskanta and M. Abrams, 'Thermal interaction of two streams in boundary-layer flow separated by a plate', *Int. J. Heat Mass Transfer*, **14**, 1311 (1971).

28. G. S. H. Lock and R. O. Ko, 'Coupling through a wall between two convective systems', *Int. J. Heat Mass Transfer*, **16**, 2087 (1973).
29. E. M. Sparrow and M. Faghri, 'Fluid-to-fluid conjugate heat transfer for a vertical pipe—internal forced convection and external natural convection', *J. Heat Transfer*, **102**, 402 (1980).
30. J. W. Rizika, 'Thermal lags in flowing systems containing heat capacitors', *Trans. ASME*, **76**, 411 (1954).
31. J. W. Rizika, 'Thermal lags in flowing incompressible systems containing heat capacitors', *Trans. ASME*, **78**, 1407 (1956).
32. J. A. Clark, V. S. Arpaci and K. M. Treadwell, 'Dynamic response of heat exchangers having internal heat sources—Part I', *Trans. ASME*, **80**, 612 (1958).
33. V. S. Arpaci and J. A. Clark, 'Dynamic response of heat exchangers having internal heat sources: Part II', *Trans. ASME*, **80**, 625 (1958).
34. V. S. Arpaci and J. A. Clark, 'Dynamic response of heat exchangers having internal heat sources: Part III', *Trans. ASME*, **81**, 253 (1959).
35. R. M. Cima and A. L. London, 'The transient response of a two-fluid counterflow heat exchanger—the gas turbine regenerator', *Trans. ASME*, **80**, 1169 (1958).
36. G. M. Dusinberre, 'Calculation of transient temperatures in pipes and heat exchangers by numerical methods', *Trans. ASME*, **76**, 421 (1954).
37. G. M. Dusinberre, *Heat Transfer Calculations by Finite Differences*, International Textbook Co, Scranton, Pennsylvania, U.S.A. (1961).
38. R. Siegel and E. G. Keshock, 'Wall temperatures in a tube with forced convection, internal radiation exchange, and axial wall heat conduction', *NASA T.N. D-2166* (1964).
39. C. S. Landram, 'Wall to fluid heat transfer in turbulent tube flow for a disturbance in inlet temperature', Sandia Labs, Livermore, U.S.A., *Report SCL-RR-69-131*. Avail. through N.T.I.S. (1969).
40. Yu. N. Kutznetzov and V. P. Belousov, 'Unsteady conjugated heat transfer in tubes', *Heat Transfer 1974*, Vol. 2, Paper FC9.3, Scripta Book Co. (1974).
41. C. R. Gane and P. L. Stephenson, 'An explicit numerical method for solving transient equations of combined conduction and convection', *Numerical Methods in Thermal Problems* (ed. R. W. Lewis, K. Morgan and O. C. Zienkiewicz), Pineridge Press, Swansea, U.K. (1979).
42. C. R. Gane and P. L. Stephenson, 'An explicit numerical method for solving transient combined heat conduction and convection problems', *Int. J. Num. Methods Eng.*, **14**, 1141 (1979).
43. E. A. Thornton, 'Application of upwind convective finite elements to practical conduction/forced convection thermal analysis', in *Numerical Methods in Thermal Problems* (ed. R. W. Lewis, K. Morgan and O. C. Zienkiewicz), Pineridge Press, Swansea, U.K. (1979).
44. E. A. Thornton and A. R. Wieting, 'Finite element methodology for thermal analysis of convectively cooled structures', *Heat Transfer and Thermal Control Systems* (ed. L. S. Fletcher), Vol. 60 of *Progress in Astronautics and Aeronautics* (1979).
45. E. A. Thornton and A. R. Wieting, 'A finite thermal analysis procedure for several temperature-dependent parameters', *J. Heat Transfer*, **100**, 551 (1978).
46. E. A. Thornton and A. R. Wieting, 'Finite element methodology for transient conduction/forced convection thermal analysis', *AIAA 14th Thermophysics Conf.*, 4–6 June, Orlando, Florida (1979).
47. E. A. Thornton and A. R. Wieting, 'Evaluation of finite element formulations for transient conduction-forced convection analysis', *1st Nat. Conf. on Numerical Methods in Heat Transfer*, Sept 24–26, Univ. Maryland, U.S.A. (1979).

48. E. A. Thornton and A. R. Wieting, 'Evaluation of finite-element formulations for transient conduction forced-convection analysis', *Num. Heat Transfer*, **3**, 281 (1980).
49. E. A. Thornton and P. Dechaumphai, 'Convective heat transport in merging flows', *3rd Int. Conf. on Finite Elements in Water Resources,* May 19–23, Univ. Mississippi (1980).
50. E. A. Thorton, P. Dechaumphai and K. K. Tamma, 'Exact finite elements for conduction and convection', in *Numerical Methods in Thermal Problems*, Vol. II, Pineridge Press, Swansea, U.K. (1981).
51. E. A. Thornton, P. Dechaumphai, A. R. Wieting and K. K. Tamma, 'Integrated transient thermal-structural finite element analysis', *AIAA/ASME/ASCE/AHS 22nd Structures, Structural Dynamics and Materials Conf.,* 6–8 April, Atlanta, Georgia (1981).
52. R. H. MacNeal, 'An asymmetrical finite difference network', *Quart. J. Appl. Math.*, **2**, 295 (1953).
53. A. M. Winslow, 'Numerical solution of the quasilinear Poisson equation in a non-uniform triangle mesh', *J. Comput. Phys.*, **2**, 149 (1967).
54. D. J. Bender, 'An explicit unlimited stability approach to the transient conduction-convection equations', *Proc. Conf. on Effective Use of Computers in the Nuclear Industry*, April 21–23, Knoxville, Tenn., U.S. Atomic Energy Comm. Conf. 690401, p. 561 (1969).
55. S. V. Patankar, 'A numerical method for conduction in composite materials, flow in irregular channels and conjugate heat transfer', *Proc. 6th Heat Transfer Conf.*, **3**, 297 (1978).
56. E. A. Thornton and P. Dechaumphai, 'Finite element analysis of plane thermal entry flows', *3rd. Int. Conf. on Finite Elements in Flow Problems*, June 10–13, Banff, Canada (1980).
57. P. J. Roache, *Computational Fluid Dynamics*, Hermoss Publishers (1972).
58. M. A. Hogge, 'A comparison of two- and three-level integration schemes for nonlinear heat conduction', in *Numerical Methods in Heat Transfer* (ed. R. W. Lewis, K. Morgan and O. C. Zienkiewicz), John Wiley (1981).
59. D. A. H. Jacobs, 'The exploitation of sparsity by iterative methods for solving systems of algebraic equations', *Conf. on Sparse Matrices and Their Uses*, Univ. of Reading (1980).
60. P. J. Roache, 'Performance of the GEM codes on nonseparable 5- and 9-point operators', *Num. Heat Transfer*, **4**, 395 (1981).
61. D. Ramkrishna and N. Amundson, 'Boundary value problems in transport with mixed or oblique derivative boundary conditions—I', *Chem. Eng. Sci.*, **34**, 301 (1979).
62. D. Ramkrishna and N. Amundson, 'Boundary value problems in transport with mixed or oblique derivative boundary conditions—II', *Chem. Eng. Sci.*, **34**, 309 (1979).
63. A. J. Oliver, 'A general method for predicting the flows and temperatures in a network of interconnecting ducts', *Numerical Methods in Thermal Problems*, Vol. II (ed.) Pineridge Press, Swansea, U.K. (1981).
64. L. Imre, Z. Jagasits and J. Barcza, 'Computer simulation of the transient warming of rotary electrical machines', in *Numerical Methods in Thermal Problems*, Vol. II (ed. R. W. Lewis, K. Morgan and B. A. Schrefler), Pineridge Press, Swansea, U.K. (1981).
65. T. J. Roberts, 'The solution of the heat flow equations in large electrical machines', *Proc. I. Mech. E.*, **3E**, 70 (1969).

66. A. F. Armor and J. J. Gibney, 'Direct conductor-cooling of large steam turbine-generator 4-pole rotors', *IEEE Trans. on PAS*, **PAS-93**, 477 (1974).
67. B. Jackson and P. E. Clark, 'An iterative technique for computing temperature distributions in rotor and stator cores', *Colloquium on Rating and ventilation of electrical machines* 22 Jan., IEE, London (1976).
68. T. K. Csillag, 'Studies on cooling of gap-pickup turbogenerator rotors with cross-flow ventilation', *IEEE paper F78-716-3* (1978).
69. K. Murata, S. Nonaka, M. Yamamoto and Y. Takada, 'Experimental study on cooling of rotor in a salient 4-pole synchronous machine', *IEEE Trans. Power Apparatus and Systems*, **PAS-98**, 310 (1979).
70. L. Imre, A. Bitai and P. Csenyi, 'Thermal transient analysis of OFAF transformers by numerical methods', in *Numerical Methods in Thermal Problems*, Vol. II (ed. R. W. Lewis, K. Morgan and B. A Schrefler), Pineridge Press, Swansea, U.K. (1981).
71. S. Nonaka, M. Yamamoto, M. Nakano and M. Kawase, 'Analysis of ventilation and cooling system for induction motors', *IEEE Trans. on Power Apparatus and Systems*, **PAS-100**, 4636 (1981).
72. J. A. Lapworth and P. L. Stephenson, 'Cooling of turbogenerator brushgear', 13th Universities Power Engineering Conf., 4–6 April, Heriott-Watt, Edinburgh, U.K. (1978).
73. C. R. Gane and D. R. Soulsby, 'Prediction of transient three-dimensional temperature distributions in externally water-cooled e.h.v. cable circuits', *2nd. Conf. on Progress in Cables and Overhead lines for 220 KV and above*, IEE Conf. Paper 176 (1979).
74. C. R. Gane and D. R. Soulsby, 'A numerical scheme for predicting transient temperature distributions in force-cooled, buried cable joint assemblies', *Numerical Methods in Thermal Problems*, Vol. II (ed. R. W. Lewis *et al.*), Pineridge Press, Swansea, U.K. (1981).
75. C. R. Gane, J. A. Hitchcock and D. R. Soulsby, 'Digital computation methods for the determination of e.h.v. cable transient ratings', *CIGRE Conf. paper 21-07* (1980).
76. A. R. Wieting, 'Application of numerical methods to heat transfer and thermal stress analysis of aerospace vehicles', in *Numerical Methods in Thermal Problems* (ed. R. W. Lewis and K. Morgan), Pineridge Press, Swansea, U.K. (1979).
77. A. R. Wieting and R. W. Guy, 'Thermal-structural design/analysis of an air-frame integrated hydrogen-cooled scramjet', *J. Aircraft*, **13**, 192 (1976).
78. E. A. Thornton, P. Dechaumphai and K. K. Tamma, 'Integrated thermal-structural analysis of aerospace structures', *Int. Conf. on Numerical Methods for Coupled Problems*, September 7–11, Swansea, U.K. (1981).
79. P. L. Stephenson and D. R. Soulsby, 'Flow and heat transfer in a buried residual fuel oil pipeline', in *Numerical Methods in Thermal Problems*, Pineridge Press, Swansea, U.K. (1979).
80. B. L. Krivoshein and V. N. Novakovskiy, 'Transient thermal processes in long gas pipelines', *Heat Transfer—Soviet Research*, **7**, 134 (1975).
81. V. L. Pokhoriler, 'Calculation of warming up of long pipes', *Thermal Engineering*, **19**, 59 (1972).
82. W. F. Phillips, 'The effects of axial conduction on collector heat removal factor', *Solar Energy*, **23**, 187 (1979).
83. P. L. Stephenson, 'ANDUIN—a general program for solving nonlinear heat conduction and convection problems—Part I: Introduction', *CEGB Program Note RD/L/P16/78* (1978).

84. E. A. Thornton, 'TAP1—a finite element program for steady state thermal analysis of convectively cooled structures', *NASA CR-145069* (1976).
85. E. A. Thornton, 'TAP2—a finite element program for thermal analysis of convectively cooled structures', *NASA Contractor Report 159038* (1980).
86. P. J. Moir and J. R. Stoker, 'THETA, a desk top computer for nonlinear time dependent heat flow problems', *Engineering Software* Vol. II, ed. R. A. Adey, CML Publications (1981).
87. T. D. A. Kennedy, '2DT, 3DT, and 2DS—two- and three-dimensional transient and two-dimensional steady-state heat-transfer finite-difference computer programs', *AERE Report 9210*, AERE, Harwell, U.K. (1978).
88. E. C. Du Fort and S. P. Frankel, 'Stability conditions in the numerical treatment of parabolic differential equations', *J. Soc. Indust. Math.* **11**(1) (1953).
89. J. L. Douglas and B. F. Jones, 'On predictor–corrector methods for nonlinear parabolic differential equations', *J. Soc. Indust. Math.*, **11**, 195 (1963).
90. G. H. Golub and R. S. Varga, 'Chebyshev semi-iterative methods, successive over-relaxation methods, and second order Richardson methods: Part 1', *Num. Math.*, **3**, 147–156 (1961).
91. R. S. Varga, *Matrix Iterative Analysis*, Prentice-Hall (1962).
92. R. M. Scarisbrick and D. H. Shroff, 'Electrical losses in an extruded cable dielectric during irradiation', *BEAMA Int. Conf. on Electrical Insulation*, Brighton, U.K., 2–5 May (1978).
93. F. J. Bayley, J. M. Owen and A. B. Turner, *Heat Transfer*, Nelson, London (1972).
94. P. L. Stephenson and D. R. Soulsby, 'Numerical solution of some one-dimensional nonlinear heat conduction problems of engineering significance', in *Numerical Methods in Thermal Problems*, Vol. II, Pineridge Press, Swansea, U.K. (1981).
95. J. H. Neher, 'The temperature rise of buried cables and pipes', *AIEE Trans*, **68**, 9 (1949).
96. H. M. Spiers, *Technical Data on Fuel*, British National Committee, World Power Conference, London (1966).
97. W. M. Rosenhow and J. P. Hartnett, *Handbook of Heat Transfer*, McGraw-Hill (1973).
98. D. R. Oliver and V. G. Jenson, 'Heat transfer to pseudoplastic fluids in laminar flow in horizontal tubes', *Chem. Eng. Sci.* **19**, 15 (1964).
99. BS 4452, *Method for Determinations of Pour Point of Petroleum Oils*, British Standards Institution (1969).
100. T. H. Kuehn and R. J. Goldstein, Correlating equations for natural convection heat transfer between horizontal circular cylinders', *Int. J. Heat Mass Transfer*, **19**, 1127 (1976).
101. J. W. Rose and J. R. Cooper, *Technical Data on Fuel*, 7th edn. British National Committee, World Energy Conference (1977).

Numerical Methods in Heat Transfer, Volume II
Edited by R. W. Lewis, K. Morgan, and B. A. Schrefler

Chapter 10

On the Accuracy of the Boundary Element Method for Three-dimensional Conduction Problems

P. L. Betts and J. S. Kang

10.1 INTRODUCTION

Problems of steady heat conduction in an isotropic medium are governed by Laplace's equation for temperature. Typically boundary conditions may vary over the surface of a body, with different regions of the surface being subjected to one of the three main types, Dirichlet (temperature specified), Neumann (normal temperature derivative or heat flux specified) or mixed (the empirical 'convection' condition which provides a linear relationship between temperature and heat flux). The boundary element method[1] is particularly well suited for the numerical solution of such potential problems either within or exterior to a bounded domain of arbitrary geometry. Similarly composite medium problems can be treated by invoking the matching condition of continuity of heat flux across common boundaries.

The boundary element method is a weighted residual formulation in which the weighting functions satisfy the governing equations, but not necessarily the boundary conditions. Green's theorem can then be used to reduce the original domain problem to a boundary integral equation form. As a consequence, discretization is required only over the boundary instead of throughout the domain. In two dimensions, the main advantage of this is found in the reduction of input data requirements and the ease with which these can be refined locally. Although the number of equations is reduced, the resulting coefficient matrix is generally full, so that one cannot take advantage of the banded structure which is typical of finite element or finite difference domain formulations. Simply in terms of storage requirements or an operation count, the domain methods are superior for long thin interior problems, whereas the boundary approach is better for regions with an aspect ratio of about unity. However, even in two dimensions, the reduction of input data for a complex shape can be a considerable advantage. In three-dimensional problems, the band width of the coefficient matrix for a finite element formulation is large, so that the advantages of a boundary method become even more attractive. For exterior problems, such as heat conduction round an insulated inclusion in two or three dimensions, a boundary method has clear advantages on all counts, since the conditions at infinity can be included automatically, without the necessity for any discretization on an exterior boundary.

In the boundary element method, both the temperature and its normal derivative are discretized over the surface. Since either one is known, or a relationship between them is given, the number of equations required is equal to the number of nodes. In recent years,[1–4] the method has been exploited widely for two-dimensional potential problems. Representation of a parameter over an element can be of any order (constant, linear, quadratic, etc.) and similarly the side of an element can be a straight or curved line. However, for linear or higher-order elements, special treatment is required when there is a discontinuity in normal temperature gradient between adjacent elements with a common node. Such discontinuities are common in conduction problems, for example when one side of a block is insulated and the adjacent side has a specified temperature. The problem can be overcome either by placing a short transitional element across the discontinuity,[1] or by using a double node at the corner.[3] When it is formulated correctly, the latter approach is the more accurate.

In this chapter, the use and accuracy of the boundary element method in three-dimensional problems is explored. The corner problem of two dimensions becomes an edge-and-corner problem in three dimensions. The additional complications which arise from this are avoided initially by considering 'constant' boundary elements first. For these, the simplest of element representations, the unknowns are considered constant over the area of an element with a node at the centroid. Since there are then no nodes on a discontinuous edge, the problem is avoided. Attention is then given the case of linear elements and the correct treatment for edges and corners is developed (cf. reference 5, where the edge problem is avoided because only values within the domain were reported). For both these representations, the element shape is a plane triangle which can be used to discretize any completely arbitrary three-dimensional surface.

Conduction in an infinite medium, exterior to a closed body, is then treated, and both zero and linear temperature gradients at infinity are considered. Finally, attention is given to ways in which storage requirements can be reduced, symmetry conditions can be treated, and integration procedures optimized or improved. Some comparisons are also given in Section 6.3 between the methodology of the boundary element method and that of the

Panel method,[6] which is an indirect boundary integral equation approach widely used for potential problems in the aeronautical field. It will be seen that the element approach provides the simpler formulation.

10.2 THE BOUNDARY ELEMENT METHOD

The analysis required to derive the boundary element method in three dimensions is included in reference 1. However for completeness, an outline is also given here.

Laplace's equation for steady heat conduction within an isotropic material of volume V (Figure 10.1) is

$$\nabla^2 T = 0. \tag{10.1}$$

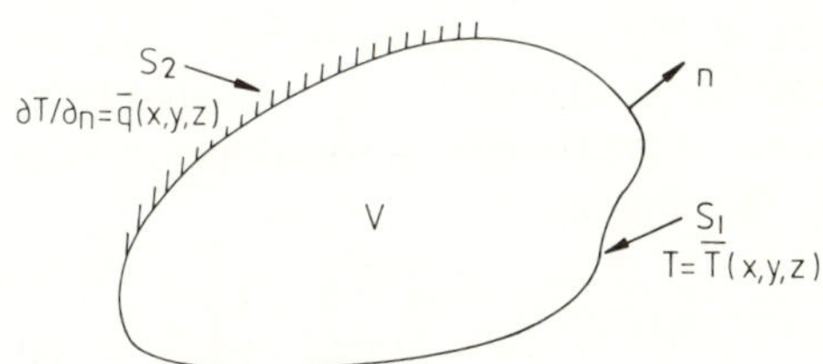

Figure 10.1 Definition of problem

Boundary conditions of two types are considered, specified temperature

$$T = \bar{T}(x, y, z) \qquad \text{on } S_1 \quad \text{(Dirichlet)}, \tag{10.2}$$

or specified normal temperature gradient

$$\partial T/\partial n = \bar{q}(x, y, z) \qquad \text{on } S_2 \quad \text{(Neumann)}, \tag{10.3}$$

where S_1 and S_2 are parts of the boundary surface, such that the total boundary $S = S_1 + S_2$ and n is the outward normal. A specified heat flux is therefore given by the product of $\bar{q}$ and the thermal conductivity of the medium. Boundary conditions of the mixed type need not be considered separately, since they may be applied later by combination. Similarly, extension of the analysis to situations of conduction in the region exterior to V will be considered in Section 10.5.

In a weighted residual formulation, Equation (10.1) may be replaced by the volume integral requirement

$$\iiint_V W \nabla^2 T \, \mathrm{d}V = 0, \tag{10.4}$$

where W is the weighting function. Equation (10.4) may then be integrated by parts twice to yield

$$\iiint_V T\nabla^2 W \, \mathrm{d}V + \iint_S \left(W\frac{\partial T}{\partial n} - T\frac{\partial W}{\partial n} \right) \mathrm{d}S = 0. \tag{10.5}$$

The weighting functions are chosen to represent singular solutions of Laplace's equation. In three dimensions, such solutions are given by

$$W = \frac{1}{4\pi r}, \tag{10.6}$$

where r is the distance between the singularity and a point in the volume. Equation (10.6) is readily recognized as the same as that for the potential produced by a three-dimensional source. Consequently, the integrand within the volume integral of Equation (10.5) is zero everywhere, except at the singularity point, and the integral may be evaluated by considering the limit within a small volume δV, surrounding the singularity, as $\delta V \to 0$. Thus if T is continuous across the singularity

$$\iiint_V T\nabla^2 W \, \mathrm{d}V = \lim_{\delta \to 0} T^i \iiint_{\delta V} \nabla^2 W \, \mathrm{d}V, \tag{10.7}$$

where T^i is the value of T at the singularity i.

In the boundary element method, the singularities are placed on the boundary surface. For a locally smooth surface, the convenient elemental volume surrounding the singularity is a hemisphere within V with its plane circle in S (Figure 10.2). The integral is then readily evaluated, after integrating by parts, as

$$\iiint_{\delta V} \nabla^2 W \, \mathrm{d}V = \iint_{\delta S} \partial W/\partial n \, \mathrm{d}S = \iint_{\delta S} -\left(\frac{1}{4\pi r^2}\right) \mathrm{d}S = -\tfrac{1}{2}, \tag{10.8}$$

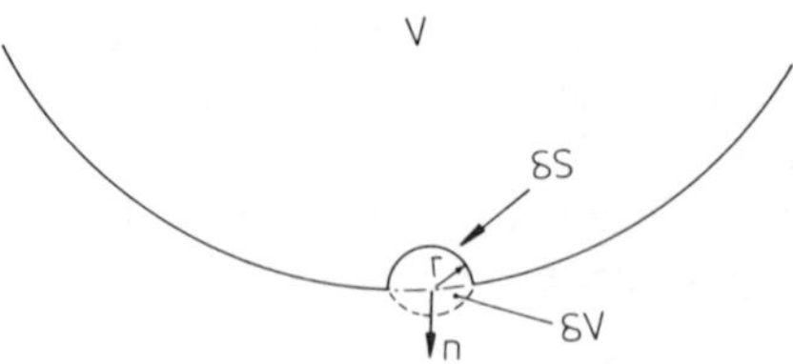

Figure 10.2 Elemental hemisphere surrounding singularity

where δS is the hemispherical surface of δV. Insertion of Equations (10.7) and (10.8) into (10.5) yields

$$\tfrac{1}{2}T^i - \iint_S \left(W\frac{\partial T}{\partial n} - T\frac{\partial W}{\partial n} \right) \mathrm{d}S = 0. \tag{10.9}$$

Equation (10.9) involves only boundary information and can be used to complete the boundary specifications. Thus in a well-posed problem, half the boundary information is given by Equations (10.2) and (10.3). $\partial T/\partial n$ on S_1 and T on S_2 can then be determined from Equation (10.9). In a discretized problem, the number of weighting functions equals the number of nodal unknowns, placed at the node or singularity positions. The integrals in Equation (10.9) are then effected by approximating in a piecewise manner over the whole surface.

Internal values of temperature can readily be calculated if required, after the boundary solution has been obtained, by the use of additional internal singularities. The only change to equation (10.9) required for these is that $\frac{1}{2}T^i$ is replaced by T^i, since the elemental volume surrounding the singularity in the domain is now a complete sphere. Similarly, if the surface of the domain is not smooth, so that the singularity is in a corner or an edge, the factor multiplying T^i is proportional to the internal solid angle.

10.3 CONSTANT ELEMENTS

10.3.1 Numerical procedures

A general three-dimensional surface is approximated by a finite number of plane triangular elements with common edges, as in a geodesic dome (Figure 10.3). Triangles are chosen because of the simplicity of plane elements, which are defined with complete generality by three points in space. For 'constant' elements, the approximations to the variables T and $\partial T/\partial n$ are constant over each element, with the values calculated at the centroid of each element.

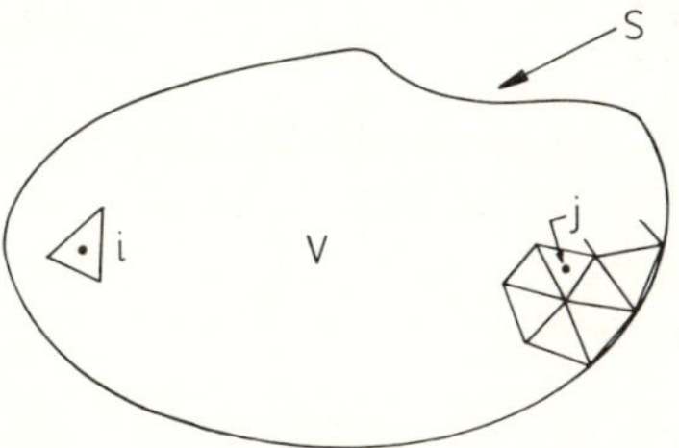

Figure 10.3 Constant elements and nodes on surface

Thus, when Equation (10.9) is applied for a singularity at the centroid of element i

$$\tfrac{1}{2}T^i = \sum_{j=1}^{N} (\partial T/\partial n)^j \iint_{S^{ij}} W\,\mathrm{d}S - \sum_{j=1}^{N} T^j \iint_{S^{ij}} (\partial W/\partial n)\,\mathrm{d}S, \tag{10.10}$$

where the superscript j indicates a value on or integral over the jth element and N is the total number of boundary elements or centroidal nodes.

For $i \neq j$, the integrals of Equation (10.10) are evaluated numerically by Gaussian quadrature, the accuracy of which is considered later. For $i = j$, the second integral in Equation (10.10) is zero, because the singular point lies in the plane surface of the element. However, the first integral is singular within element i and must be evaluated analytically.

Consider therefore the integration of W over the triangle ABC shown on Figure 10.4, with the singularity at the centroid O of the triangle. It is

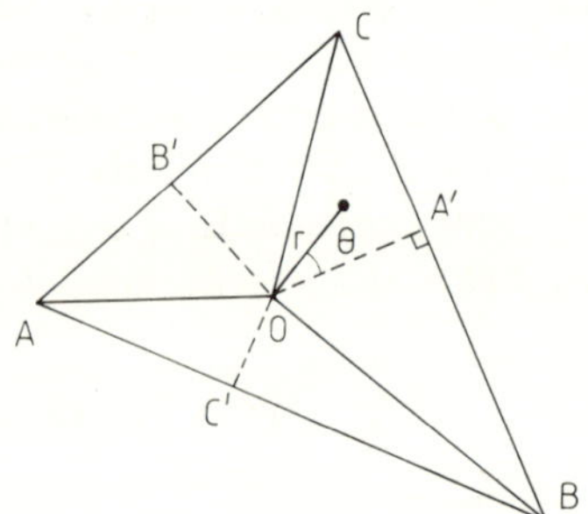

Figure 10.4 Integration over constant self-element

convenient to divide the triangle into three subtriangles by joining O to the three corners. Then

$$\iint_{\Delta\mathrm{OBC}} W\,\mathrm{d}S = \frac{1}{4\pi} \iint_{\Delta\mathrm{OBC}} \frac{\mathrm{d}S}{r} = \frac{1}{4\pi} \iint_{\Delta\mathrm{OBC}} \frac{r\,\mathrm{d}\theta\,\mathrm{d}r}{r} = \int_{\theta_B}^{\theta_C} \int_0^{h\sec\theta} \mathrm{d}\theta\,\mathrm{d}r, \tag{10.11}$$

where h is the length of the perpendicular OA′ from O to BC and θ is the angle which the vector $\mathbf{r}$ subtends with it. Hence,

$$\iint_{\Delta\mathrm{OBC}} W\,\mathrm{d}S = \frac{h}{4\pi} \int_{\theta_B}^{\theta_C} \sec\theta\,\mathrm{d}\theta$$

$$= \frac{h}{4\pi} [\ln(\sec\theta + \tan\theta)]_{\theta_B}^{\theta_C} = \frac{h}{4\pi} \ln\left(\frac{\mathrm{OC} + \sqrt{(\mathrm{OC}^2 - h^2)}}{\mathrm{OB} - \sqrt{(\mathrm{OB}^2 - h^2)}}\right) \tag{10.12}$$

since θ_B is negative. Equation (10.12) can also be written as

$$\iint_{\Delta OBC} W \, dS = \frac{h}{4\pi} \ln\left(\frac{OC + A'C}{OB - A'B}\right), \tag{10.13}$$

and the changes of sign within the natural logarithm are immediately obvious for the highly unlikely case of a long thin element with an obtuse angle. Equivalent expressions for the other two subtriangles can be written immediately, so that the integral of W over the whole triangle is given by

$$\iint_{\Delta OBC} W \, dS = \frac{h}{4\pi} \ln\left(\frac{(OC + A'C)(OA + B'A)(OB + C'B)}{(OB - A'B)(OC - B'C)(OA - C'A)}\right). \tag{10.14}$$

With the geometric integrations all performed, Equation (10.10) can be written for each singularity point in matrix form as

$$[H]\{T\} = [G]\{\partial T/\partial n\}, \tag{10.15}$$

where $[H]$ and $[G]$ are square matrices, depending only on the geometry of the boundary elements, and $\{T\}$ and $\{\partial T/\partial n\}$ are vectors of the nodal parameters. For the boundary conditions of Equations (10.2) and (10.3), in each row of (10.15) either T or $\partial T/\partial n$ is known, so that the equation can be rearranged as

$$[A]\{X\} = \{B\} \tag{10.16}$$

where $[A]$ is a coefficient matrix, $\{X\}$ is a vector of the unknown nodal parameters, and $\{B\}$ is a known vector obtained by multiplying the square matrix on the right-hand side, after rearrangement, by the vector of known nodal parameters. The procedure for dealing with a mixed (or 'convection') boundary condition can now be seen, as it simply requires eliminating one of the nodel parameters, by the specified boundary condition, and rearranging Equation (10.15) accordingly.

Solution of Equation (10.16) by standard Gaussian elimination completes the numerical procedure for calculating values on the surface of the domain. Equation (10.9), in its modified form with T^i replacing $\frac{1}{2}T^i$, can then be applied numerically for any internal points at which the temperature is required, since values of both T and $\partial T/\partial n$ are now known over the whole surface S of the domain.

10.3.2 Numerical tests

The purpose of the tests is to identify and estimate errors which arise in a computer formulation with finite-sized boundary elements. Comparisons are made with simple known analytic solutions, and note is taken of aspects which are not tested by the simple problems.

Initial tests were for a linear temperature distribution in a rectangular block with temperatures 0 and 100 over the end faces and insulated sides ($\partial T/\partial n = 0$). The ends were unit squares while the length varied from 1 to 4; the 24 elements are shown on Figure 10.5(a). Maximum percentage errors are summarized in Table 10.1, where those in T on side faces are referred to the difference between ends and in $\partial T/\partial n$ on ends to the known values there.

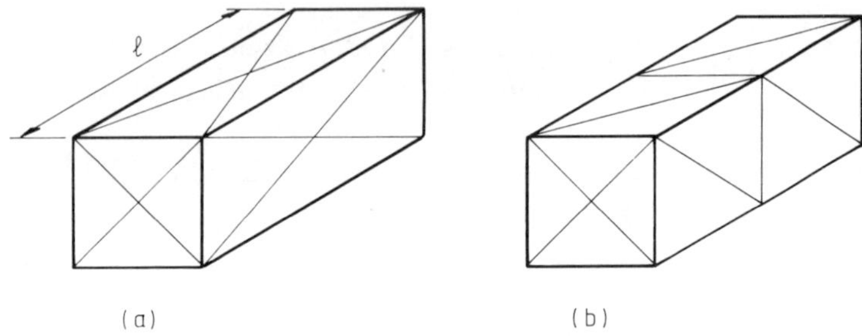

Figure 10.5 Boundary elements on rectangular block

The element representation is exact over the ends but only approximate in T along the sides. Results are given for various integration procedures; procedure A is as described in Section 10.3.1, with all numerical integrations by 7-point Gaussian quadrature. 13-point quadrature produced small erratic changes but no improvement. Errors were worst at the high-temperature end, which can be attributed to the vanishing of the second integral in Equation (10.10) when $T^j = 0$ (for end temperature ± 50, the errors were symmetric). Numerical errors in this integral for $i \neq j$ may also be more significant than in the corresponding two-dimensional case.

$\partial W/\partial n$ is a singularity of a form H/r^3, where H is the perpendicular from the node at the centroid of element i to element j. This may have a significant effect when the elements are in adjacent faces and the joining edge forms a

Table 10.1 Maximum percentage errors for Figure 10.5(a) (constant elements)

$l=1$		$l=2$		$l=4$		
T	$\frac{\partial T}{\partial n}$	T	$\frac{\partial T}{\partial n}$	T	$\frac{\partial T}{\partial n}$	Integration procedure
35	17	1.8	10	53	47	A
0.45	3	0.05	13	4.8	16	B
0.44	3	0.02	13	1.4	45	C
0.45	3	0.02	13	1.4	45	D

common side (Figure 10.6). Node i is only one third of an element height from the widest part of element j, and the effect on the numerical integration is particularly marked. The integration can be improved by subdividing (Figure 10.6) so as to take account of the singular distributions. 7-point quadrature was used for each subtriangle, and this (integration procedure B) was preferred to a vastly increased order of quadrature over the whole triangle. Later, all elements within a larger sphere of influence were subdivided (C) and finally all elements were subdivided for integration (D).

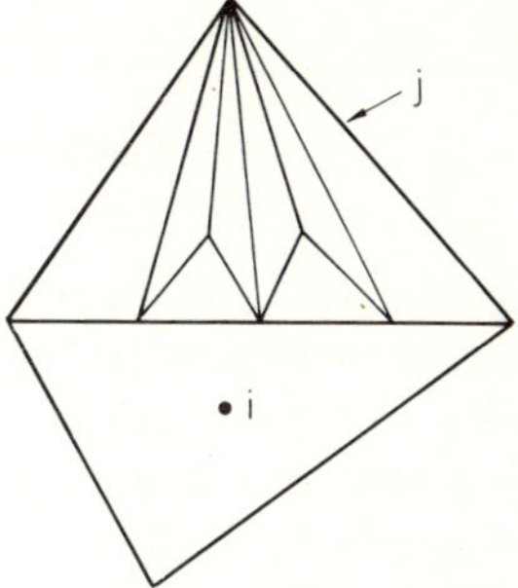

Figure 10.6 Subdivision for integration over adjacent element

For a cube ($l = 1$), procedure B reduced the errors considerably to values acceptable for such a coarse mesh (Table 10.1). Further refinements (C and D) had negligible effect. For longer blocks, errors in T reduced with integration refinements, but $\partial T/\partial n$ behaved erratically. However, results from C and D are identical, indicating a probable interaction, with A and B, between integration procedure and the crude constant-element representation of the linear distribution T along the sides. Errors in $\partial T/\partial n$ increased monotonically with l for C and D. Values of T within the block were consistently more accurate than on the surface.

The alternative arrangement of 24 elements shown on Figure 10.5(b) was also investigated. This produced an improvement in $\partial T/\partial n$ at the expense of T. For $l = 4$, errors in T and $\partial T/\partial n$ were 4.7 and 16% with procedure B and 3.3 and 24% with C. For $l = 2$ and D they were 1.5 and 7%, while for $l = 1$ and C, they were 4.9 and 2.5%.

The element representation of the linear temperature variation along the insulated sides was also investigated by refining them (Figures 10.7(a) and (b)) but leaving the ends unchanged (40 and 72 elements respectively). For $l = 1$ and procedure C, errors in T and $\partial T/\partial n$ were 1 and 3% with 40 elements and 0.1 and 0.6% with 72 elements; for $l = 2$, they were 0.07 and 2% with 72 elements. Thus, mesh refinement produces acceptable improvements in accuracy. Tests were also undertaken for a circular cylinder of length 4 and radius 2. Each end was divided into eight segments, which were joined along

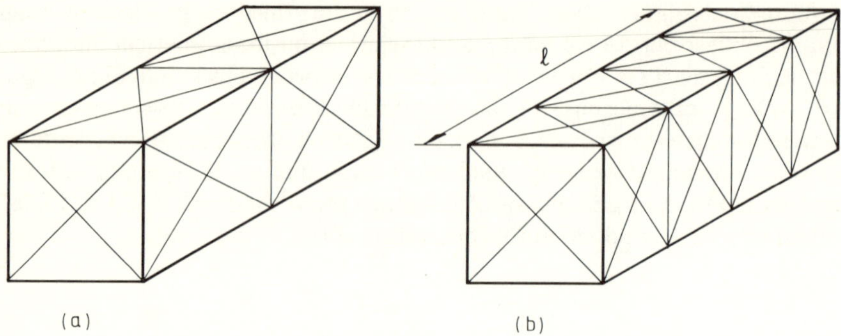

Figure 10.7 Refined elements on rectangular block

the insulated curved surface by rows of eight elements arranged in the zig-zag pattern of Figure 10.2(b). The 80 elements give maximum errors of 0.3 and 2.9% in T and $\partial T/\partial n$ with procedure C. Note that there is no error from geometric discretization here, since the analytic solution is equally valid for a polygonal cylinder.

Hollow tubes were next tested, since the effect of the numerical integrations over the close opposite faces could be significant. The 152-element arrangement on a square sectioned tube is shown on Figure 10.8. Maximum errors in T and $\partial T/\partial n$ were 14.7 and 6.7% for $l=1$, 2.0 and 10.0% for $l=2$ (both for either procedure B or C), and 0.8 and 13% for $l=4$.

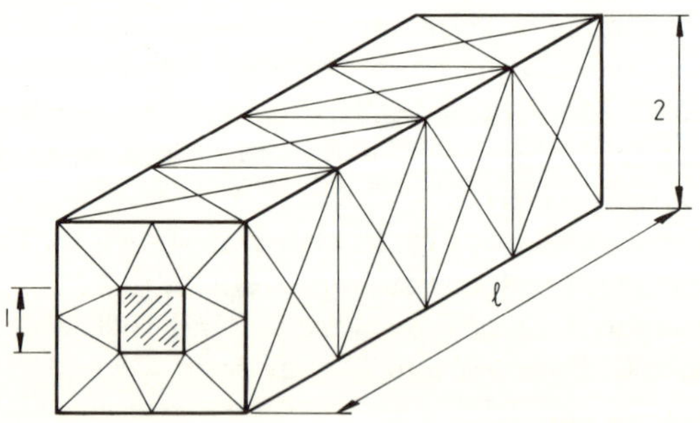

Figure 10.8 Square-sectioned tube

The increase in errors for the hollow square-sectioned tube is caused primarily by the integrals of $\partial W/\partial n$ from singularities in the inner face integrated over the close elements on the outer face, which are relatively large. For constant elements, the low-order temperature discretization combines with the rapid H/r^3 variation of $\partial W/\partial n$ to produce significant errors.

Circular tubes were also tested. These had inner and outer radii of 1 and 2, with 64 elements along each of the curved faces and 16 elements on each end (total = 160). Errors were 1.7 and 1.0% for $l = 1$, 0.4 and 4% for $l = 2$ and 1.4 and 13.5% for $l = 4$ with both procedures B and C. Note that for a comparable aspect ratio, the larger circular tube requires twice the length of that on Figure 10.8. The differences between the tubes are attributable to the difference in angles between facets of the insulated faces.

It is now possible to draw certain conclusions from the above results. When all elements had about the same area, errors in normal heat flux ($\partial T/\partial n$) were consistently higher than those in T. When the areas of elements on the insulated faces of varying temperature were reduced (by mesh refinement or reducing l), errors in $\partial T/\partial n$ were reduced, and vice versa. However, the errors in T then started to increase. These effects also occur with constant elements in two dimensions, and indeed, the magnitudes of the errors are remarkably similar. As an example, for a two-dimensional square, with one element per side, errors in T and $\partial T/\partial n$ are 0.01 and 17.5%. With two elements per insulated side (6 elements), they are 0.5 and 3.8%; and with four elements per insulated side (10 elements), they are 1.9 and 1%.

All the above tests have been for a very simple temperature distribution, with analytic values of $\partial T/\partial n$ constant over all elements and of T over some. The final test permitted nonlinear variations of both variables over the boundaries and was for an insulated inclusion, in the shape of an ovoid, in an infinite conducting medium with a uniform temperature gradient at infinity. The analytic solution for this is the same as for the analogous problem in velocity potential with axi-symmetric flow. Although the exact solution is axi-symmetric, the geometric discretization lacks this symmetry and is formulated in Cartesian coordinates, while the element integrations relate to spherical coordinates centred at the various nodes.

For an ovoid given by

$$r^2 = 0.5104\left[\frac{x+1.5}{[(x+1.5)^2+r^2]^{1/2}} - \frac{x-1.5}{[(x-1.5)^2+r^2]^{1/2}}\right] \tag{10.17}$$

and $\partial T/\partial x = 1$ at infinity, the temperature in the medium is

$$T = x + 0.2552[[(x-1.5)^2+r^2]^{-1/2} - [(x+1.5)^2+r^2]^{-1/2}]. \tag{10.18}$$

For the present purposes, an interior formulation has been retained, with integration procedure C. Elements were placed on an external cylinder and on half the ovoid, and the boundary was completed by sectional surfaces as shown on Figure 10.9 (i.e. 64 elements on the cylinder, 40 on the ovoid, 32 on the annular section and 24 on the circular one). Boundary conditions were taken as $\partial T/\partial n = 0$ on the ovoid and equal to the analytic value of $\partial T/\partial r$ on the cylinder $r = 4$, with $T = 0$ on the annular section ($x = 0$) and equal to the value from Equation (10.18) on the circular section at $x = 6$.

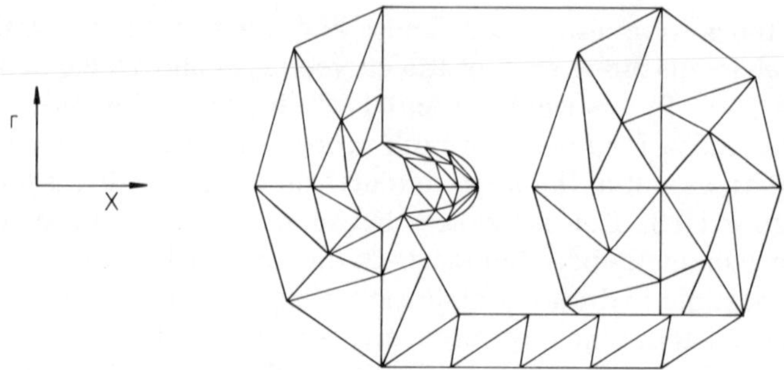

Figure 10.9 Elements for insulated ovoid

Results from this more severe test were very encouraging. Errors in T over the semi-ovoid never exceeded 1.4% of the temperature difference, ΔT, along it (2.44), while those over the external cylinder were within 0.65% of $\Delta T = 6.02$ (i.e. absolute errors of same order). Maximum errors in $\partial T/\partial n$ were 9.5% of local value over the annular section and 9.6% over the circular section; in both cases these occurred in the relatively large outer ring of elements. These errors are smaller than those from the tests with circular tubes, and the magnitude of those in $\partial T/\partial n$ must be related to the choice of element sizes in the same way and to the lower ratio of inner and outer radii. Graphs of meridional distributions of T and $\partial T/\partial n$ are shown on Figure 10.10.

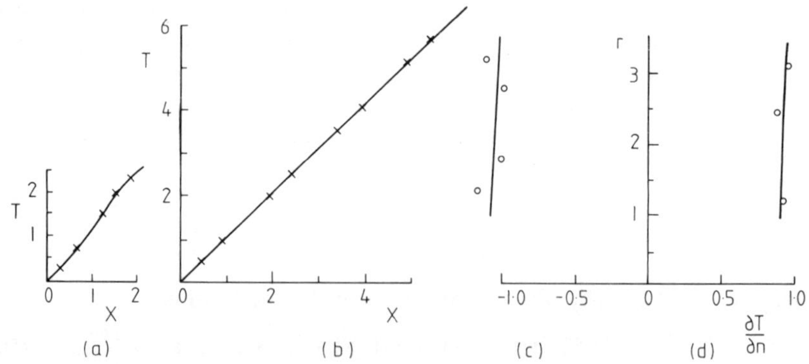

Figure 10.10 Analytic and constant element computed variations for ovoid: (a) on ovoid; (b) $r = 4$; (c) $x = 0$; (d) $x = 6$

10.4 LINEAR ELEMENTS

10.4.1 Numerical procedures

It has been shown that with sufficient mesh refinement the constant-element representation gives an acceptably accurate solution for conduction problems.

However, the required size of elements is influenced by the interactions between the accuracy of temperature and its derivative and the $1/r$ and H/r^3 variations of the weighting functions and their normal derivatives, which affect the accuracy to which the integrals of Equation (10.10) approximate the continuum integrals of Equation (10.9). Consequently, one might expect an increase in accuracy when the element representation is increased to a linear variation; moreover, this should be obtained with less computational effort, since the corners of the triangles are then the nodes and there are generally fewer nodes than elements.

For linear elements, the representations of T and $\partial T/\partial n$ over element j^e (Figure 10.11) are considered as linear functions of position within the

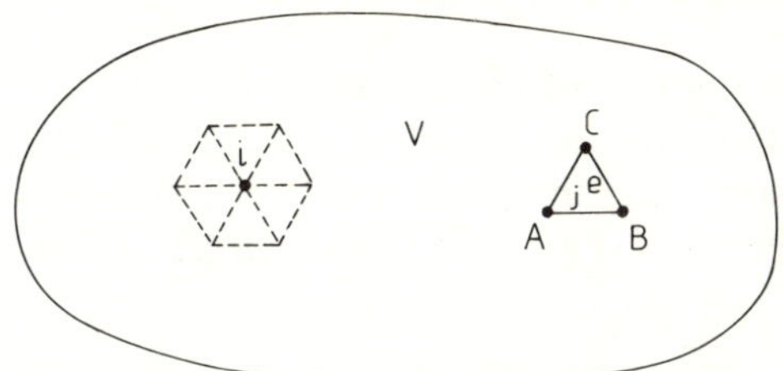

Figure 10.11 Linear elements and nodes on surface

triangular element and defined by their values at nodes A, B and C on the corners of the element. Consequently, as shown in reference 1, Equation (10.10) is replaced by the linear element representation of Equation (10.9)

$$c^i T^i = \sum_{j^e=1}^{N^e} \iint_{S^{je}} (\partial T/\partial n)^{j^e} W \, dS - \sum_{j^e=1}^{N^e} \iint_{S^{je}} (T)^{j^e} \left(\frac{\partial W}{\partial n}\right) dS, \qquad (10.19)$$

where superscript e represents an element numbering and the superscript i refers to the numbering of the nodes at the corners of the triangles. Note that in Equation (10.19) c^i has replaced the previous value of $\frac{1}{2}$, since node i is now no longer necessarily in a flat surface (Figure 10.2 and Equation (10.8)); c^i takes the value of the internal solid angle, between the elements connected to node i, divided by 4π (Figure 10.11).

$(\partial T/\partial n)^{j^e}$ and $(T)^{j^e}$ can no longer be taken outside the integrals in Equation (10.19), since they vary linearly over the element. However, their values over the element are dependent on the nodal values at A, B and C (for which j numbers are known). Consequently, numerical integration of these integrals presents no problems. However, when element j^e contains node i as one of its corners, one must again integrate analytically over the element for the first integral, while the second one is again zero. Note here that more elements are integrated analytically than for the constant element case (cf. Figures 10.3 and 10.11), with a consequent expected increase in accuracy.

With all the integrations performed for each singularity point, one again obtains a matrix equation of the same form as (10.15), where $[H]$ now contains values of c^i on the diagonal. Although these can be obtained from the solid angles, it is more convenient to follow Brebbia[1] and note that if T is constant over the boundary, $\{\partial T/\partial n\}$ is zero. Hence

$$[H]\{T\}=0 \tag{10.20}$$

for any constant T. Consequently, the sum of all elements in a row of $[H]$ must always be zero, and the diagonal coefficient

$$h_{ii}=-\sum_{j=1}^{n} h_{ij}, \tag{10.21}$$

where n equals the number of nodes and $j \neq i$.

In order to evaluate all the integrals of Equation (10.19), we require an analytic expression for the integrals over the elements surrounding node i. One of these elements is shown on Figure 10.12, where node i is at corner

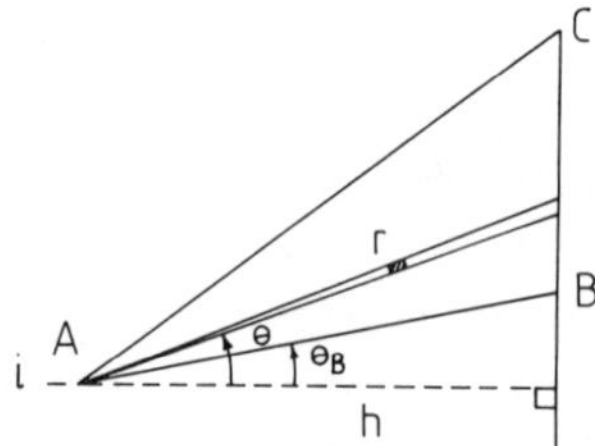

Figure 10.12 Integration over linear self-element

A of the element and θ_B would be negative if the foot of the perpendicular h were between B and C. The integral being considered

$$I_{j^e}=\iint_S (\partial T/\partial n)W\,\mathrm{d}S=\iint_S (ar\cos\theta+br\sin\theta+c)\left(\frac{1}{4\pi r}\right)\mathrm{d}S \tag{10.22}$$

where a, b and c are linearly related to the values of $\partial T/\partial n$ at the corners of the triangle. Hence,

$$\begin{aligned}
I_{j^e}&=\int_{\theta_B}^{\theta_C}\int_{r=0}^{h\sec\theta}(ar\cos\theta+br\sin\theta+c)\left(\frac{1}{4\pi r}\right)r\,\mathrm{d}r\,\mathrm{d}\theta\\
&=\frac{1}{4\pi}\int_{\theta_B}^{\theta_C}\int_{r=0}^{h\sec\theta}(ar\cos\theta+br\sin\theta+c)\mathrm{d}r\,\mathrm{d}\theta\\
&=\frac{1}{4\pi}\int_{\theta_B}^{\theta_C}\left\{\frac{h^2(a\cos\theta+b\sin\theta)}{2\cos^2\theta}+ch\sec\theta\right\}\mathrm{d}\theta\\
&=I_1+I_2+I_3.
\end{aligned} \tag{10.23}$$

Thus,

$$I_1 = \frac{h^2 a}{8\pi}[\ln(\sec\theta + \tan\theta)]_{\theta_B}^{\theta_C},$$

$$I_2 = \frac{h^2 b}{8\pi}[\sec\theta]_{\theta_B}^{\theta_C},$$

$$I_3 = \frac{hc}{4\pi}[\ln(\sec\theta + \tan\theta)]_{\theta_B}^{\theta_C} \tag{10.24}$$

as before. Values of a, b and c can be expressed in terms of the nodal values of $\partial T/\partial n$ in the usual way of finite elements as

$$\begin{Bmatrix} a \\ b \\ c \end{Bmatrix} = [C^{-1}]\begin{Bmatrix} (\partial T/\partial n)_A \\ (\partial T/\partial n)_B \\ (\partial T/\partial n)_C \end{Bmatrix},$$

where

$$[C] = \begin{bmatrix} 0 & 0 & 1 \\ (r\cos\theta)_B & (r\sin\theta)_B & 1 \\ (r\cos\theta)_C & (r\sin\theta)_C & 1 \end{bmatrix}, \tag{10.25}$$

and this completes the formulation in terms of the required nodal variables.

10.4.2 Treatment ot corners and edges

Although the above subsection completes the formal requirements for programming linear boundary elements, there remains the problem, referred to in Section 10.1, of treating edges or corners of elements where there is a discontinuity in normal temperature gradients. This treatment is easiest to understand in two dimensions, and since there still appears to be some confusion of the correct procedure even in the two-dimensional case, we will consider this first.

Consider, for simplicity, the problem of uniform conduction in a two-dimensional square block as indicated in Figure 10.13. If the elements of Figure 10.13(a) were used, one would normally take nodes 2, 3, 8 and 9 as $\partial T/\partial n = 0$, and the rest as T specified. However, $\partial T/\partial n$ does not have a unique value at the corners 1, 4, 7 and 10 but jumps from ± 1 to zero across the node. Use of any standard boundary element method computer program with these elements would then give some kind of an average value for $\partial T/\partial n$ at the corner nodes. As suggested by Brebbia,[1] the situation can be improved considerably by using the element disposition shown on Figure 10.13(b), where the length of the short additional corner elements δ is small compared with that of the others. There are, however, two disadvantages of this

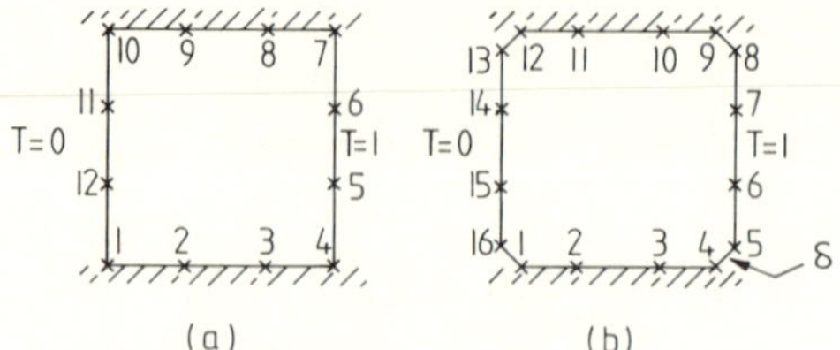

Figure 10.13 Treatment of corners for linear elements in two dimensions

approach; the optimum length δ is not specified and parameters are evaluated away from the corner, which may be of particular interest. Varying the length δ of the corner elements reveals that as $\delta \to 0$, the accuracy with which the calculated nodal parameters (at 4 and 5 say) match the correct values on the two sides (1–4 and 5–8) at first increases and then decreases. For very short lengths of δ, the limiting expression for $r \ln r$ is required in the analytic integration over δ.

The interested reader can check the above effects even on a desktop microcomputer with the aid of a standard BEM program such as in reference 1. However, in any such program only elements adjacent to the singularity node i are integrated analytically. Consequently, for a singularity at node 4 (say) elements 3–4 and 4–5 are integrated analytically whilst element 5–6 are integrated numerically, so that the error in the integration of element 5–6 increases as $\delta \to 0$. However, when $\delta = 0$, node 5 is coincident with node 4 and the integral over element 5–6 of the singularity at node 4 can be performed analytically by the same formula as for standard adjacent elements. Similarly elements such as 3–4 can be evaluated analytically. To implement this procedure requires only a minor modification to the standard program and the improvement in the results is dramatic, with errors of negligible magnitude caused only by the small errors in the remaining numerical integrations. The combined nodes 4, 5 etc. behave as nodes with two degrees of freedom for $\partial T/\partial n$ but C^0 continuous between elements in T. The fourth and fifth rows of the $[G]$ matrix (cf. Equation (10.15)) are then identical, but the internal angle term in the $[H]$ matrix alters from the fourth to the fifth column, with other columns being identical. Thus, the 'double node' should only be used at points where the values of $\partial T/\partial n$ do differ across the node. If such a node is used at an insulated corner, where both sides have $\partial T/\partial n = 0$, the resulting system matrix is singular, as would be expected on physical grounds. Indeed the only reason why Brebbia's method gave results with a small δ at the stagnation point on the front of a quadrant of a cylinder (Figure 3.14 in reference 1) is because the errors from one-sided numerical integration at the corners provided sufficient differences between rows in $[G]$ to overcome the singularity of the system matrix.

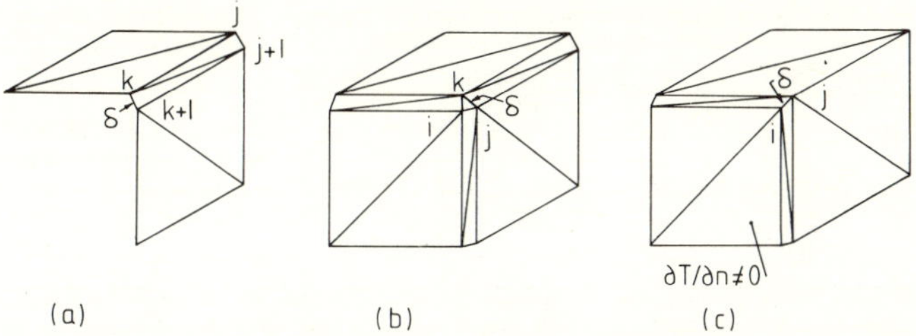

Figure 10.14 Treatment of edges and corners for linear elements in three dimensions: (a) edge between faces with different heat flux; (b) corner between three faces with different heat flux; (c) corner between two insulated faces and one with heat flux

With a clear understanding of the procedure for corners in two dimensions, it is now easy to understand those required in three dimensions. Figure 10.14(a) shows part of an edge between two faces with different values of $\partial T/\partial n$. When $\delta = 0$, the nodes k, $k+1$ and j, $j+1$ become double valued (in $\partial T/\partial n$) in the same way as in two dimensions, and all integrals over elements with nodes in the edge are performed analytically. Note that there is then naturally no need to integrate the zero area elements such as that with corners k, $k+1$, $j+1$. Similarly a corner is treated in the way indicated in Figure 10.14(b) with $\delta = 0$. The corner node ijk is now triple valued in $\partial T/\partial n$. However, if two faces are insulated and one is not, the edge joining the two insulated faces is made up of single valued nodes and the corner is represented by Figure 10.14(c) (i.e. a double-valued node ij).

10.4.3 Numerical tests

The numerical tests with linear triangular elements illustrate the superiority of this element over the constant element in a number of ways. The case of the rectangular block (Figure 10.5(a)) with a linear temperature distribution between end faces at 0 and 100 and insulated side faces was previously investigated with constant elements (Table 10.1). Linear elements require a corner treatment of the type illustrated on Figure 10.14(c), so that there are 8 double-valued nodes and 6 single ones (22 nodal unknowns) compared with the 24 nodal unknowns for constant elements. The maximum percentage errors in the computed results are given on Table 10.2, and the dramatic improvement over the constant element tests is immediately obvious.

For this test, the element and geometric representation are both exact, so that the only sources of errors are those due to numerical integration and numerical round-off errors; the integration procedures A7 and A13 refer to 7- and 13-point Gaussian quadrature without subdivision. The improvement over the constant case, where subdivision was required, is caused by the fact that with nodes at the corners of elements, instead of centroids, the exact

Table 10.2 Maximum percentage errors for Figure 10.5(a) (linear elements)

$l=1$		$l=2$		$l=4$		
T	$\frac{\partial T}{\partial n}$	T	$\frac{\partial T}{\partial n}$	T	$\frac{\partial T}{\partial n}$	Integration procedure
0	0.4	0	0.64	0	0.72	A7
0	0.4	—	—	—	—	A13

integrals extend over four elements for mid-face nodes and six elements for corner nodes. Consequently all the elements which are integrated numerically are further away from the singularity, and the effect of the H/r^3 and $1/r$ singular distribution on the numerical integrations is considerably less.

It is also worth noting that, although in this case the numbers of nodal degrees of freedom are comparable, if the number of elements had been doubled, the linear elements would have had 34 nodal degrees of freedom compared with 48 for the constant elements. Thus, the improved element representation of the linear distribution actually leads to a reduced size of numerical problem as well as greater accuracy for the same number of elements.

The second test was a repeat of the previous test for an ovoid in an infinite conducting medium (Figure 10.9) and the results for this are shown on Figure 10.15. The general disposition of the 160 elements is the same as for Figure 10.10, but they have been graded more towards the nose of the ovoid ($x=2$). For the linear elements only 106 nodes were required (counting double nodes twice), which gives an execution time of 8.8 s on the CDC7600 of UMRCC, compared with 31.8 s using constant elements, and simple 13-point Gauss quadrature was used without subdividing. Errors in T over the semi-ovoid were always less than 4.2% of $\Delta T=2.44$ and over the external cylinder less than 0.085% $\Delta T=6.02$. These are comparable with the previous errors using constant elements, especially as the worst error occurred near the nose of the ovoid, which was to some extent avoided by the constant element nodes. However, the errors in $\partial T/\partial n$ over the section $x=0$ and $x=6$ showed a marked improvement over the constant element results at maxima of 1.5% and 0.4% respectively.

Thus for the same geometric discretization, the linear elements tend to provide greater accuracy for less storage requirement, faster execution times and less need for complex integration schemes.

10.5 EXTERNAL FORMULATIONS

10.5.1 Analysis

Although the boundary element method offers advantages over domain methods for internal conduction problems, particularly in data preparation,

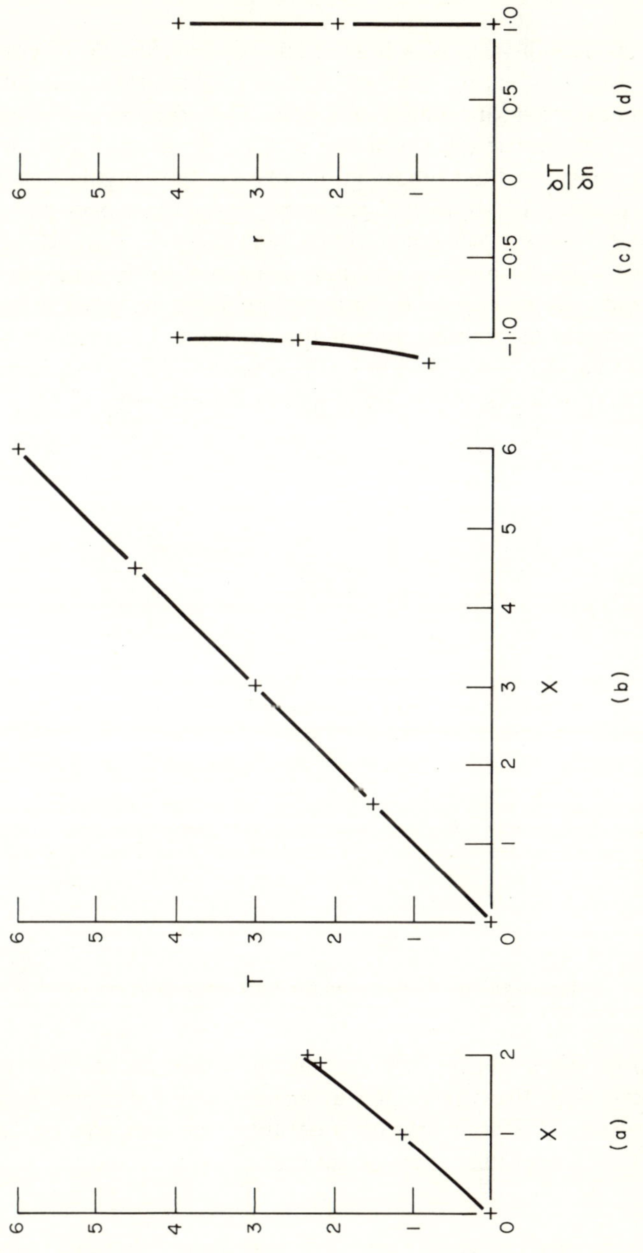

Figure 10.15 Analytic and linear element computed variation for ovoid: (a) on ovoid; (b) $r = 4$; (c) $x = 0$; (d) $x = 6$

the advantages are markedly greater for external problems when formulated directly as such.

For problems such as that of a heated enclosed surface S_i (Figure 10.16), within an infinite conducting medium and zero temperature at infinity, the extension to an external formulation is simple. The problem can be considered as an internal problem made up of the surface S_i, a small cylindrical tube extending to infinity, and the sphere at infinity S_∞. For singularities on S_i, the integrals of Equation (10.19) over S_∞ are zero, as are those over the cylindrical tube in the limit. Hence, no discretization is required over S_∞, and the problem reduces to one with the number of equations equal to the number of nodes on S_i. Thus, the only change to the internal problem program is to consider the outward normal as pointing out of the medium (i.e. towards the origin within S_i) and to add 1 ($=4\pi/4\pi$) to h_{ii} (Equation (10.21)) to account for the sphere at infinity, which is not included within the elements of H.

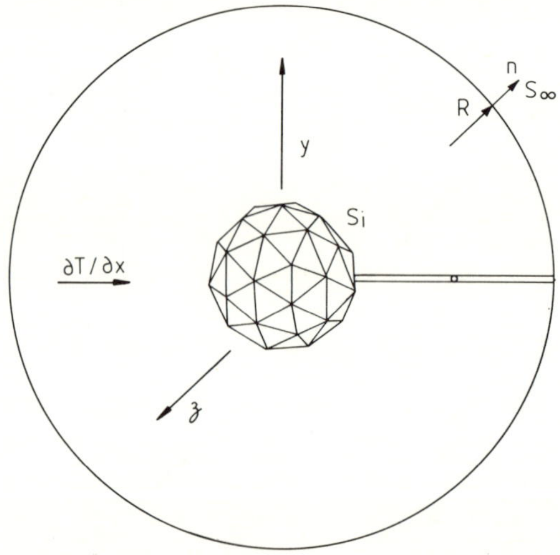

Figure 10.16 Formulation for external conduction

Consider now the situation of an inclusion within an infinite region (i.e. large compared with the size of the inclusion) and a uniform temperature gradient at infinity. Without loss of generality, the axis can be considered aligned with the temperature gradient so that

$$T = \bar{U}x \tag{10.26}$$

at infinity (Figure 10.16). We require to evaluate the integrals of Equation (10.19) over the sphere of radius $R \to \infty$ for singularity nodes on S_i. Both T and $\partial T/\partial n$ can be considered known on S_∞. Thus for each singularity node

i on S_i, we need to evaluate

$$\iint\limits_{S_\infty} \left(W\frac{\partial T}{\partial n} - T\frac{\partial W}{\partial n}\right) \mathrm{d}S = I_1 - I_2. \tag{10.27}$$

Consider a nodal point on S_i ($a \cos \theta'$, $a \sin \theta' \cos \psi'$, $a \sin \theta' \sin \psi'$) and the integral over a small element of the frustrum section of S_∞ at ($R \cos \theta$, $R \sin \theta \cos \theta$, $R \sin \theta \sin \psi$) such that the elemental area

$$\delta S = R^2 \sin \theta \, \delta\theta \, \delta\psi. \tag{10.28}$$

Then the distance r between **a** and **R** is given by the equation

$$r^2 = R^2 + a^2 - 2Ra \cos \beta, \tag{10.29}$$

where β is the angle between **a** and **R** and is given by

$$\cos \beta = \cos \theta \cos \theta' + \sin \theta \sin \theta' \cos (\psi - \psi'). \tag{10.30}$$

On the outer sphere

$$W = \frac{1}{4\pi r}$$

and

$$\frac{\partial W}{\partial n} = \frac{\partial W}{\partial \mathbf{r}} \cdot \mathbf{n} = -\frac{1}{4\pi r^2} \frac{(R - a \cos \beta)}{r} \tag{10.31}$$

by simple trigonometry.

From Equation (10.27)

$$I_1 = \iint\limits_{S_\infty} W\frac{\partial T}{\partial n} \mathrm{d}S = \int_{\psi=0}^{2\pi} \int_{\theta=0}^{\pi} \frac{1}{4\pi r} (\bar{U} \cos \theta)(R^2 \sin \theta) \, \mathrm{d}\theta \, \mathrm{d}\psi$$

$$= \frac{\bar{U}}{8\pi} \iint \frac{R^2}{r} \sin 2\theta \, \mathrm{d}\theta \, \mathrm{d}\psi$$

Moreover, from Equation (10.29)

$$\frac{1}{r} = \frac{1}{R}\left(1 - \frac{2a}{R} \cos \beta + \frac{a^2}{R^2}\right)^{-1/2} = \frac{1}{R} + \frac{a \cos \beta}{R^2} + O\left(\frac{1}{R^3}\right) \tag{10.32}$$

and

$$I_1 \simeq \frac{\bar{U}}{8\pi} \iint R\left(1 + \frac{a \cos \beta}{R}\right) \sin 2\theta \, \mathrm{d}\theta \, \mathrm{d}\psi. \tag{10.33}$$

Similarly,

$$I_2 = -\iint \frac{1}{4\pi r^2}\left(\frac{R - a\cos\beta}{r}\right)\bar{U}R\cos\theta\, R^2 \sin\theta\, d\theta\, d\psi$$

$$= -\frac{\bar{U}}{8\pi}\iint \frac{R^3}{r^3}(R - a\cos\beta)\sin 2\theta\, d\theta\, d\psi$$

$$= -\frac{\bar{U}}{8\pi}\iint \left(1 + \frac{3a\cos\beta}{R} + O\left(\frac{1}{R^2}\right)\right)(R - a\cos\beta)\sin 2\theta\, d\theta\, d\psi$$

$$\simeq -\frac{\bar{U}}{8\pi}\iint (R + 2a\cos\beta)\sin 2\theta\, d\theta\, d\psi \tag{10.34}$$

thus

$$I_1 - I_2 = \frac{\bar{U}}{8\pi}\int_{\psi=0}^{2\pi}\int_{\theta=0}^{\pi} (2R + 3a\cos\beta)\sin 2\theta\, d\theta\, d\psi$$

$$= \frac{3\bar{U}a}{8\pi}\iint \cos\beta\sin 2\theta\, d\theta\, d\psi$$

$$= \frac{3\bar{U}}{8\pi}\iint (\cos\theta\cos\theta' + \sin\theta\sin\theta'\cos(\psi - \psi'))\sin 2\theta\, d\theta\, d\psi$$

$$= \frac{\bar{U}a}{4\pi}\int_0^{2\pi} \cos\theta'[-\cos^3\theta]_0^{\pi}\, d\psi = \bar{U}a\cos\theta' = \bar{U}x_i. \tag{10.35}$$

Consequently, with the correct limits as $R \to \infty$, both integrals over the sphere at infinity can be transferred to the right-hand side in Equation (10.16) as a known constant in each row. Straightforward rotation of axes is sufficient to show that for a linear temperature gradient of arbitrary direction at infinity

$$I_1 - I_2 = \bar{U}x_i + \bar{V}y_i + \bar{W}z_i. \tag{10.36}$$

10.5.2 Numerical tests

The discretization for a spherical inclusion within an infinite medium is indicated on Figure 10.17. Only one eighth of the surface is shown; the remaining elements in the other segments are obtained by reflections in the principal planes. Tests were performed for a unit temperature gradient at infinity ($\bar{U} = 1$) and a sphere of radius 2. The analytic solution for this problem, with $\partial T/\partial n = 0$ in the sphere, is

$$T = \frac{3a\bar{U}}{2}\cos\theta. \tag{10.37}$$

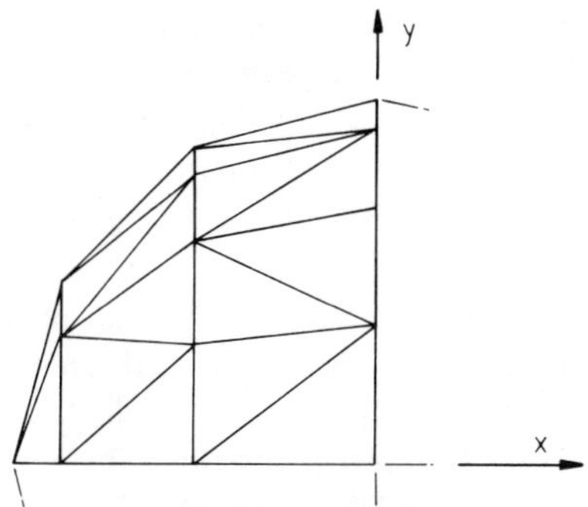

Figure 10.17 Elevation of elements over one eighth of sphere

The results for $\bar{q}=0$ on S_i are shown on Figure 10.18(a), and both constant and linear elements have been tested. Note that the overall temperature level has been specified by Equation (10.26), which enters the numerical problem via the integrals over the sphere at infinity. Hence, it is possible to specify $\partial T/\partial n$ at all the nodes on the surface S_i.

Apart from the increased accuracy of the linear elements, there is also a significant improvement in efficiency. For the same number of elements, the

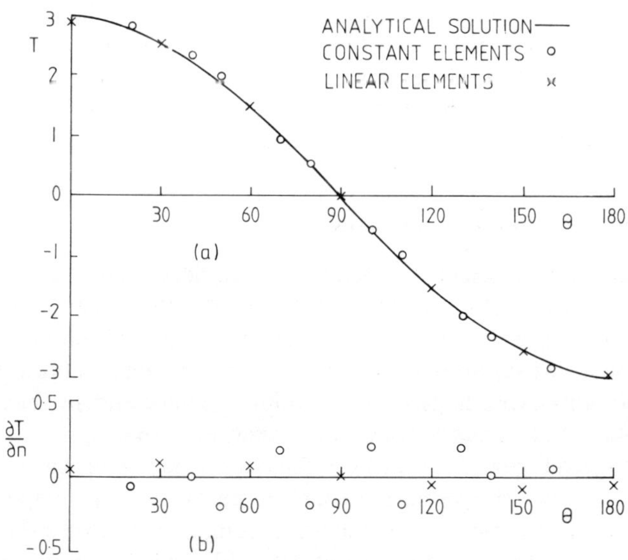

Figure 10.18 External conduction round sphere: (a) Neumann formulation; (b) Dirichlet formulation

constant representation required 128 nodes and took 40 s of CPU time on the CDC7600 of UMRCC, whereas the linear representation only needed 66 nodes and took 4.4 s. The results of an alternative test are shown on Figure 10.18(b), where T has been specified at the nodes on the surface of the sphere and the analytic solution would be $\partial T/\partial n = 0$. Again the results are highly satisfactory.

Tests were also performed for a full ovoid in an external formulation. The element pattern was the same as that indicated on Figures 10.9 and 10.15, with the elements over the other half of the ovoid ($x > 0$) being reflections in the yz plane. The resulting temperatures over the ovoid, with $\partial T/\partial n = 0$ on the surface, are shown on Figure 10.19 for both linear and constant elements. Errors between the two cases are comparable, being 4% and 3% respectively but at different locations. However, the execution time for the 80 node constant elements was 23.6 s, compared with 1.5 s for the 42 linear nodes, which again required no subdivision. In both cases the results were symmetric about $x = 0$.

10.6 SPACE AND TIME SAVING TECHNIQUES

10.6.1 Reduction of *H* array

In the above, it might appear that the G and H matrices (Equation (10.15)) have been fully assembled before the rearrangement which leads to Equation (10.16), where A is overwritten in G. Indeed, in the well-known programs in two dimensions[1] this is the case. However, since all the assembly is by rows (singularity points), it is a simple matter to reduce H to a row vector and to re-arrange after each row has been assembled. For large problems this simple procedure virtually halves the storage requirements.

10.6.2 Use of symmetry conditions

All the exact solutions against which tests have been made have been axisymmetric. However it should be stressed that the method used makes no use of this and is fully three-dimensional. Indeed, many of the element distributions did not possess this symmetry. However, there are many situations where an external problem exhibits symmetry about a plane, and in these cases it is possible to save both storage space and execution time.

Consider again Figure 10.17, where the octant may be taken as representing any arbitrary shape. If symmetries exist about all three principle planes and the linear external temperature gradient is along one of the axes (say x without loss of generality), there is no need to assemble rows for any of the nodes in the octants not shown. This is because it is known that the temperatures at nodes reflected in the xy and xz planes are identical, whereas the temperatures

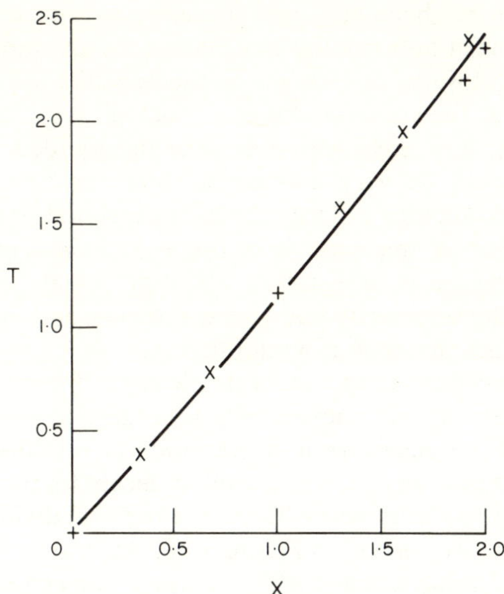

Figure 10.19 Temperature variations over ovoid with external formulations: × constant elements; + linear elements

at the nine nodes created by reflection in *yz* (plus additional reflections if needed) are the negatives of the temperatures in the octant shown. Consequently, as the elements in one of the reflected regions are assembled, they can be added (or subtracted) on top of the coefficients obtained for elements in the original octant. Thus if all three symmetries can be used, the storage requirement is reduced by a factor approaching 64, with consequent savings in execution time. In some cases even more symmetries exist and in others less. However, in all cases it is worthwhile taking advantage of all the symmetrical conditions available.

As an example of the use of symmetry, the test of Figure 10.18 was repeated with only the linear nodes shown on Figure 10.17. The results were identical, but only a 14×14 G matrix was required and the execution time was reduced to 0.78 s.

10.6.3 Integration procedures

The choice of integration procedures can have a major influence on both the accuracy and efficiency of the linear boundary element programs. The main cause of this is the H/r^3 distribution of $\partial W/\partial n$ over the element, for which Gaussian quadrature is far from ideal. In most of the results reported here

7-point quadrature has been sufficient for acceptable accuracy. However for re-entrant problems or hollow tube-like bodies, the situation arises where the singularity node is close to an element on the opposite face ($H/r \rightarrow 1$). Under these circumstances, the accuracy drops dramatically (cf. Table 10.1, A7 with constant elements). It is possible to overcome this by reducing the size of the elements. This would be very inefficient, since the time to assemble the equations, which dominates in large three-dimensional problems, is proportional to the product of the number of nodes and the number of elements. Moreover, the increase in integration accuracy is only required for those elements close to the singularity node (row number), and the additional effort for elements far from the node is wasteful.

A far better procedure is to subdivide elements for integration purposes only. Thus, the elements are chosen, without regard to integration accuracy, so as to represent the geometry and the function variation in a reasonable way. Numerical integration proceeds normally except for those elements which are too close to a singularity node. These are then subdivided for integration in ways similar to those shown on Figure 10.8. Additional subdivision can be used if a triangular subdivision is still too large. The condition for requiring subdivision has been taken as being when $A/\bar{r}^2$ is greater than a given value, where A is the area of the element and $\bar{r}$ the distance from its centroid to the singularity in use.

The effectiveness of this approach has been demonstrated by the increased accuracy with integration procedure on Table 10.1 for constant elements, where the problem is even more severe. The procedure has also been used for an aero-engine intake problem, where the velocity potential satisfies Laplace's equation. Initial computations, even with 13-point quadratures, produced erroneous results in certain regions with apparent loss of mass. Application of the local subdivisions for integration purposes immediately produced acceptable results for only a slight increase in execution time.

The above procedure can be further improved by also allowing a variable order of numerical quadrature, so that elements far from the singularity node are only integrated to a lower-order quadrature. Suggested formulae to determine the optimum order for quadrilateral elements are given for elastostatic problems by Watson.[4] An extended survey of optimization procedures for the indirect method in two-dimensions is given in reference 7 for both straight and curved line elements. Many of these could be extended to the present three-dimensional problems.

An alternative to subdivision is to use exact analytic integration over these elements. Typically such integrals require extensive use of function routines and are only preferable when either high-order quadrature or subdividing is required for numerical evaluation. Although the use of analytic integration introduces problems in extending to higher-order isoparametric elements, it has recently been introduced by Hunt[8] in the associated Panel Method.

The Panel Method as used in references 6 and 8 is an indirect boundary integral approach used extensively for external flows in the aircraft industry. Distribution of sources, doublets and vortex sheets are assumed over panels on the surface of an aerofoil, and their strengths are computed by the velocity conditions at collocation points on the surface. The required velocities and potentials (equivalent to temperature here) are then calculated afterwards by post-processing. The method requires the evaluation of more integrals than the boundary element method, since integrals equivalent to $T\,\partial W/\partial s$ in two principal in-plane directions also have to be evaluated. Moreover, the collocation point approach does not ensure continuity of normal flux (cf. $\partial T/\partial n$) between points, even though the source and doublet strengths may be continuous. Theoretically, the method requires either a source or a doublet distribution. Combination of the two, with an additional smoothness criterion, was introduced to improve the continuity of normal flux,[8] but requires increasing the matrix storage by a factor of up to four for a given number of panels. Furthermore, the method can encounter difficulties at a sharp edge such as the juncture of the intake plane and the nacelle in an aero-engine.

However, both the integrals, which appear in the more direct boundary element method, also appear in the panel method. Hunt[8] has given procedures for computing these exactly. Further work is required to see whether this approach is more efficient than subdividing elements when simple Gaussian quadrature is inaccurate. Analytic integration has been used in two-dimensional free surface flow boundary element method problems by Nakayama and Washizu.[9]

10.7 CONCLUSIONS

Programs have been developed for both constant and linear boundary elements to solve the three-dimensional heat conduction equation.

For constant elements, early tests showed that simple numerical integration was not sufficiently accurate over adjacent elements, particularly near a sharp edge. This problem was overcome by subdividing the elements for integration purposes. Comparisons with analytic results then gave acceptable errors, which were comparable with those in similar two-dimensional computations. Mesh refinement produced the expected increase in accuracy and also indicated guidelines for the choice of element arrangement, to produce a given ratio of errors in T and $\partial T/\partial n$, which are similar to those in two-dimensional problems. Note that T should be evaluated more accurately if values of $\partial T/\partial s$ are required.

The use of constant boundary elements avoids the three-dimensional edge problem of higher-order elements but requires additional computational effort from mesh refinement and subdivision for accuracy. The way in which this edge problem can be simply and accurately overcome with linear elements has been described. The results from linear elements then show an increase in accuracy with, in many cases, no need for subdivision. The linear elements also provide a lower nodal degree of freedom count for the same number of elements, with consequent savings in storage and execution time.

The boundary element method offers its maximum advantages when used for problems of conduction in an external domain surrounding one or more bodies. The formulation for this problem has been described and the tests demonstrated the efficiency of boundary element method in this context. Finally, consideration has been given to the ways in which the programs can be optimized. Apart from the efficient use of storage space and of all available symmetries in a problem, the main scope for optimization is in the integration procedures. Methods have been described for the efficient use of subdivided elements for integration. However, there is still room for further improvements in this optimization, and these have been indicated.

ACKNOWLEDGEMENTS

J. S. Kang acknowledges support through an SRC-CASE studentship in conjunction with Rolls-Royce Ltd., Aero Division.

REFERENCES

1. C. A. Brebbia, *The Boundary Element Method for Engineers*, Chapter 3, Pentech Press (1978).
2. J. C. Wu, 'Finite element solution of flow problems using integral representations', *Proc. 2nd Int. Symp. Finite Elements in Flow Problems*, ICCAD, pp. 205–216, Italy (1976).
3. G. E. Schneider and B. L. Ledain, 'The Boundary Integral Method applied to steady heat conduction with special attention given to the corner problem', *17th Aerospace Sciences Meeting*, AIAA (1979).
4. J. O. Watson, 'Advanced implementation of the boundary element method for two- and three-dimensional elastostatics', in *Developments in Boundary Element Methods*—1 (ed. P. K. Banerjee and R. Butterfield) pp. 31–63, Applied Science Publishers, London (1979).
5. M. H. Lean and A. Wexler, *A Scalar Isoparametric BEM Program for Three-dimensional Fields*, Univ. of Manitoba (1979).
6. J. L. Hess and A. M. O. Smith, *Calculation of Non-lifting Potential Flow about Arbitrary Three-dimensional Bodies*, Douglas Aircraft Co., Rep. ES. 40622 (1962).
7. O. M. S. Hamed, *The Numerical Solution of the Integral Equations of Two-dimensional Potential Theory*, Ph.D. Thesis, Univ. of Salford (1982).
8. B. Hunt, 'The mathematical basis and numerical principles of the boundary integral method for incompressible potential flow over 3-D aerodynamic configurations', in *Numerical Methods in Applied Fluid Dynamics* (ed. B. Hunt) pp. 49–135, Academic Press (1980).
9. T. Nakayama and K. Washizu, 'The Boundary Element Method applied to the analysis of two-dimensional non-linear sloshing problems', *Int. J. Num. Meth. Eng.*, **17**, 1631–1646 (1981).

Numerical Methods in Heat Transfer, Volume II
Edited by R. W. Lewis, K. Morgan, and B. A. Schrefler

Chapter 11

The Prediction of Turbulent Heat Transfer by the Finite Element Method

K. Morgan, C. E. Thomas, and C. Taylor

SUMMARY

This chapter describes how the finite element method can be used to analyse problems of turbulent heat transfer. Forced convection type flows are considered in which the momentum equations are assumed to be independent of temperature. Results are presented for the heat transfer accompanying the recirculating flow over a backward facing step and the flow within the central subchannel of a triangular rod bundle.

11.1 INTRODUCTION

The modelling of turbulent flow problems by the finite element method is an area of increasing research interest.[1–4] The computational behaviour of a variety of turbulence closure models has been investigated in the literature and a number of complex flow configurations have been successfully analysed. However, in many turbulent flows of engineering interest, a knowledge of the accompanying heat transfer is of prime importance but, to date, little use has been made of the finite element method in this area. In general, the governing momentum equations are coupled to the energy equation by a temperature-dependent buoyancy term, which takes account of the effect of variations in temperature on the fluid density. The work described in this chapter assumes that the flow is one of forced convection in which the influence of this buoyancy term on the momentum equations is negligible. Under this assumption, the momentum, continuity and turbulence closure equations are independent of temperature and may be solved as described previously,[1] and the solutions obtained can then be used to determine the accompanying temperature field from the energy equation. Here, this temperature determination will be performed using the finite element method and two turbulent flow problems which have already been analysed by the finite element method,[5–7] will be considered, viz. those of the recirculating flow over a backward facing step and the flow within the central subchannel of a triangular rod bundle.

11.2 THE ENERGY EQUATION

A time-averaged form of the equations of incompressible turbulent flow, and the various turbulence closure models which have been postulated, have been described in detail in the literature[8] and need not be quoted again here. The energy equation for the general flow of an incompressible fluid may be written as

$$\rho c\left(\frac{\partial T}{\partial t}+u_l\frac{\partial T}{\partial x_l}\right)=\frac{\partial}{\partial x_l}\left(k\frac{\partial T}{\partial x_l}\right)+\Phi, \tag{11.1}$$

where ρ, c, and k are, respectively, the density, the specific heat, and the thermal conductivity of the fluid, T is the temperature and u_l is the component of velocity in the direction x_l of a Cartesian coordinate system. The term Φ represents the rate of heat production due to viscous dissipation in the fluid and is given by

$$\Phi=\frac{\mu}{2}\left(\frac{\partial u_m}{\partial x_l}+\frac{\partial u_l}{\partial x_m}\right)\left(\frac{\partial u_m}{\partial x_l}+\frac{\partial u_l}{\partial x_m}\right) \tag{11.2}$$

in which μ is the fluid viscosity. This term can generally be ignored for viscous flows in which applied heating is the main form of heating. In the mathematical description of turbulent flows the temperature is normally written as

$$T=\bar{T}+T' \tag{11.3}$$

where $\bar{T}$ denotes a time-averaged mean value and T' is a fluctuation from the mean. Substituting this expression into Equation (11.1), ignoring the viscous dissipation and time-averaging the resulting equation produces

$$\rho\bar{u}_l\frac{\partial\bar{T}}{\partial x_l}=\frac{\partial}{\partial x_l}\left\{\left(\frac{\mu}{\sigma}+\frac{\mu_{\mathrm{T}}}{\sigma_{\mathrm{T}}}\right)\frac{\partial\bar{T}}{\partial x_l}\right\} \tag{11.4}$$

as the equation governing the distribution of the mean temperature in steady turbulent flow. Here σ $(=\rho c\mu/k)$ is the laminar Prandtl number and it has been assumed that

$$-\overline{u_l'T'}=\frac{\mu_{\mathrm{T}}}{\rho\sigma_{\mathrm{T}}}\frac{\partial\bar{T}}{\partial x_l}, \tag{11.5}$$

where μ_T is the turbulent viscosity and σ_T is a turbulent Prandtl number. It is normally assumed that σ_T is a constant in fully turbulent flows and appropriate values for σ and σ_T for various fluids and flow configurations have been tabulated elsewhere.[9] The value of σ_T varies near walls outside the fully turbulent flow region.[9,10]

11.3 SOLUTION OF THE ENERGY EQUATION

11.3.1 The finite element method

The solution of Equation (11.4) is sought by the finite element method subject to the distributions of u_l and μ_T being given by a previous numerical solution of the turbulence flow model. Here, it will be assumed that this previous solution has also been obtained using the finite element method with the same discretization of the region of interest, Ω, as is now adopted for the representation of the temperature field. If the discretization consists of E elements and M nodes, then the mean temperature is approximated by

$$\hat{\bar{T}} \cong \sum_{J=1}^{M} \bar{T}^J N^J, \tag{11.6}$$

where $\bar{T}^J$ is the approximation to the value of the mean temperature at node J and N^J is the corresponding nodal shape function. Application of the standard Galerkin weighted residual procedure[11] to Equation (11.4) then results in the matrix equation

$$\mathbf{K}(\bar{u}_l, \mu_T)\bar{\mathbf{T}} = \mathbf{f}, \tag{11.7}$$

where $\mathbf{T}$ is a vector of nodal mean temperature values, $\mathbf{f}$ is a vector arising from boundary and prescribed conditions and

$$K^{IJ} = \int_\Omega \left\{ \rho \bar{u}_l N^I \frac{\partial N^J}{\partial x_l} + \left(\frac{\mu}{\sigma} + \frac{\mu_T}{\sigma_T}\right) \frac{\partial N^I}{\partial x_l} \frac{\partial N^J}{\partial x_l} \right\} \mathrm{d}\Omega. \tag{11.8}$$

The entries in this matrix are obtained by summing, in the standard fashion, the individual contributions from the E elements used to represent the domain.

11.3.2 Wall functions for temperature

Large temperature gradients are frequently present in near wall regions of turbulent flows. An accurate determination of the temperature field can then only be obtained numerically by using a very fine mesh in this near wall region with the consequent increase in computational demands. This problem can be avoided if the finite element mesh terminates at a short distance from the wall and an analytical solution procedure[8] is used to model conditions in the near wall region. This can be achieved by assuming a Couette flow so that

Equation (11.4), in the fully turbulent near wall zone where $\mu_T \gg \mu$, reduces to

$$\frac{d}{dy}\left(\frac{\mu_T}{\sigma_T}\frac{d\bar{T}}{dy}\right)=0, \tag{11.9}$$

where y denotes the distance from the wall, and on integration

$$\frac{\mu_T}{\sigma_T}\frac{d\bar{T}}{dy}=T_c, \tag{11.10}$$

where T_c is a constant. Defining

$$T^+=\frac{(\bar{T}-T_w)}{T_c}\rho\sqrt{\tau_w} \tag{11.11}$$

where T_w is the temperature at the wall and τ_w is the wall shear stress, then standard expressions[9] for the distribution of μ_T in this region may be used and Equation (11.10) integrated to give

$$T^+=\sigma_T\left\{2.5\ln\left(\frac{y}{\mu}\sqrt{\left(\frac{\tau_w}{\rho}\right)}\right)+5.5P(\sigma/\sigma_T)\right\} \tag{11.2}$$

where $P(\sigma/\sigma_T)$ is the Spalding–Jayatallika function[12] defined by

$$P(\sigma/\sigma_T)=9.24\{(\sigma/\sigma_T)^{3/4}-1\}\{0.28\exp(-0.007\sigma/\sigma_T)+1\} \tag{11.13}$$

Within the laminar sublayer Equation (11.4) becomes

$$\frac{d}{dy}\left(\frac{\mu}{\sigma}\frac{d\bar{T}}{dy}\right)=0, \tag{11.14}$$

which on integration produces

$$\bar{T}=\left.\frac{d\bar{T}}{dy}\right|_w y+T_w, \tag{11.15}$$

where $d\bar{T}/dy|_w$ denotes the normal gradient of mean temperature at the wall.

In the intermediate layer between the laminar sublayer and the fully turbulent region a modified form of Equation (11.12) can be used.[13]

11.3.3 The numerical solution

Under standard conditions, Equation (11.7) may be solved directly when the distribution of the mean velocity components $\bar{u}_l$ and the turbulent viscosity μ_T is known and suitable boundary conditions are prescribed. However, if wall temperature functions of the type described in the previous section are employed in an attempt to avoid excessive mesh refinement near walls, then an iterative solution procedure must be adopted. The near-wall boundary condition on mean temperature is first guessed and Equation (11.7) solved for $\bar{\mathbf{T}}$. From this distribution the wall heat flux may be estimated and a new near wall boundary condition on temperature produced using the appropriate function. The method is repeated until convergence results with the modified boundary condition being under-relaxed at each stage with a relaxation factor of 0.5.

With the mean temperature $\bar{T}$ determined, it is possible to evaluate the Nusselt number, N_u, defined by

$$N_u = z \frac{\partial \bar{T}}{\partial y} \Big/ (\bar{T} - \bar{T}_0), \tag{11.16}$$

where $\bar{T}_0$ is defined to be either the bulk temperature of the fluid or its centre line value for pipe flow and z is a characteristic length.

11.4 FLOW OVER A BACKWARD FACING STEP

11.4.1 The problem

A particular characteristic of flow subject to a sudden expansion is the augmentation of the heat transfer rates within the reverse flow region, particularly in the vicinity of the re-attachment point. Evaluation of the heat-transfer coefficients within turbulent recirculating flows can therefore be of extreme importance in the practical design of many engineering devices.

To illustrate the application of the finite element method to analyse the heat transfer accompanying flows of this type, the problem of turbulent flow over a backward facing step is considered (see Figure 11.1). This problem has been the subject of much investigation by the finite difference method.[14–18]

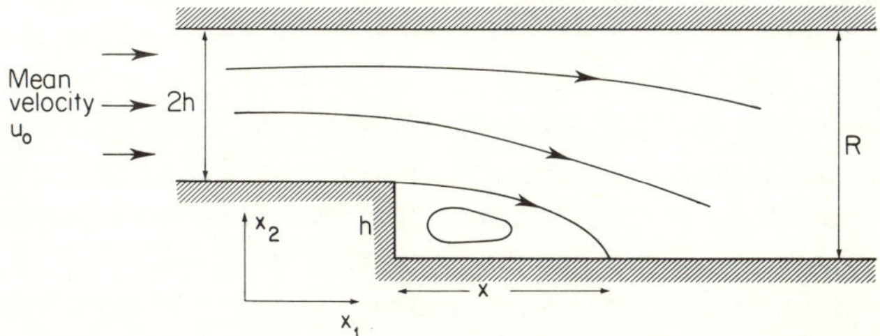

Figure 11.1 Flow domain—turbulent flow over a backward facing step

The finite element analysis is performed on the mesh shown in Figure 11.2, which was the finest mesh used in the previously reported finite element calculations of the turbulent flow field[6] using a two-equation turbulence model.

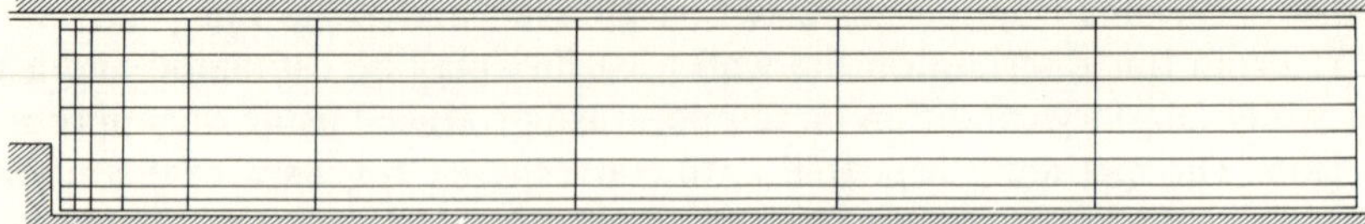

Figure 11.2 Finite element mesh for step problem

The temperature is defined using a fully developed profile across the inlet section and typical conditions along the walls prescribe either the temperature or the heat flux distribution. The iterative approach outlined in the previous section is adopted with an initial guess being made for the temperature distribution at the near wall nodes.

11.4.2 Results

The temperature calculations converge rapidly without introducing upwinding techniques. The developing value of the Nusselt number with distance from the channel at step height is evaluated in terms of the mean bulk temperature, $\bar{T}_B$, of the fluid where

$$\bar{T}_B = \frac{1}{\Omega}\int_{\Omega} \bar{T}\, d\Omega. \tag{11.17}$$

The Nusselt number is then given by

$$N_u = h\frac{\partial \bar{T}}{\partial x_2} \Big/ (\bar{T}_s - \bar{T}_B), \tag{11.18}$$

where $\bar{T}_s$ is the mean temperature of the fluid at step level. Figure 11.3 illustrates the ratio of the developing Nusselt number, N_u, to the developed Nusselt number, N_{u_D}, for a Reynolds number, R_e, of 626. The turbulent Prandtl number, σ_T, has the value 0.9 in the fully turbulent region and 1.4 at near wall nodes which lie within the laminar sublayer. In the intermediate zone, the value of σ_T is obtained by linear interpolation between these end values. In Figure 11.4 the calculated variation of N_u/N_{u_D} at $R_e = 10^5$ is compared with finite difference predictions[17] and experimental results.[18] Both numerical calculations predict that the maximum Nusselt number is attained at a distance of between four and five step heights downstream of the step whereas the experimental data indicates that this maximum is reached at a distance of about six step heights downstream. This should not be unexpected as the numerical models of the turbulent flow underpredict the recirculation zone lengths by approximately the same amount. The finite element predictions indicate maximum heat-transfer in the vicinity of the numerical calcu-

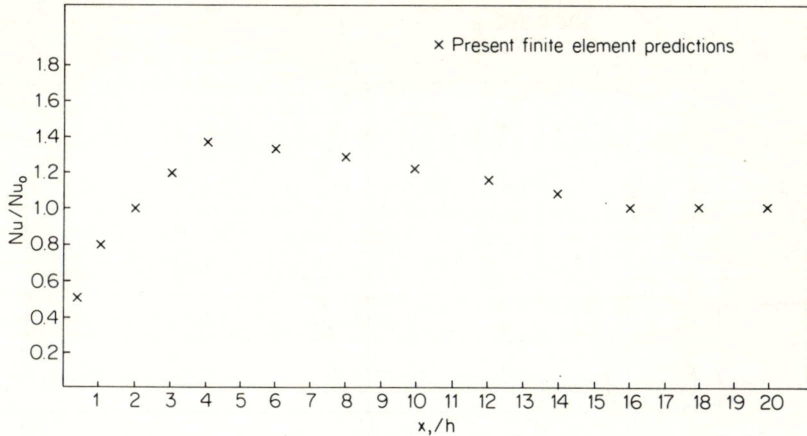

Figure 11.3 Streamwise variation of Nusselt number downstream of the step, $R_e = 626$

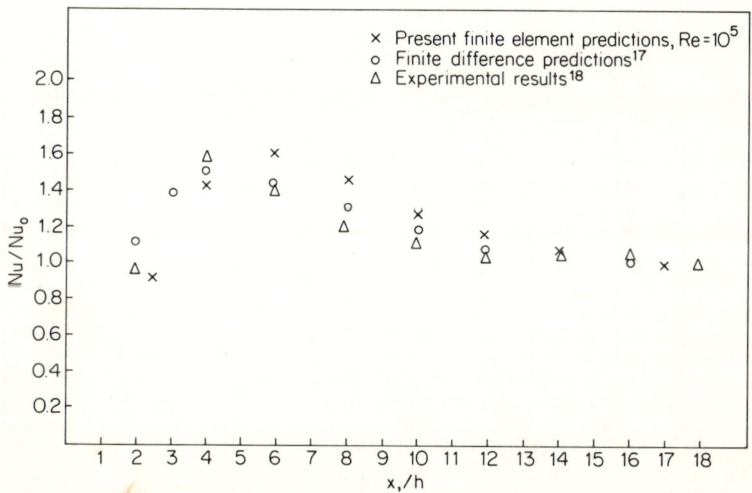

Figure 11.4 Streamwise variation of Nusselt number downstream of step, $R_e = 10^5$

lated reattachment point. Figures 11.5 and 11.6 show the calculated developing mean temperature profiles at various sections downstream of the step at $R_e = 626$ and $R_e = 10^5$ respectively. The quantity plotted is $(\bar{T} - \bar{T}_w)/(\bar{T}_\infty - \bar{T}_w)$ which denotes the ratio of the mean temperature to the mean fully developed centreline temperature, all temperatures being measured relative to the mean wall temperature. The main difference between these two figures appears in the near wall regions where, as expected, the normal temperature gradient is much greater at the higher Reynolds number. These profiles are

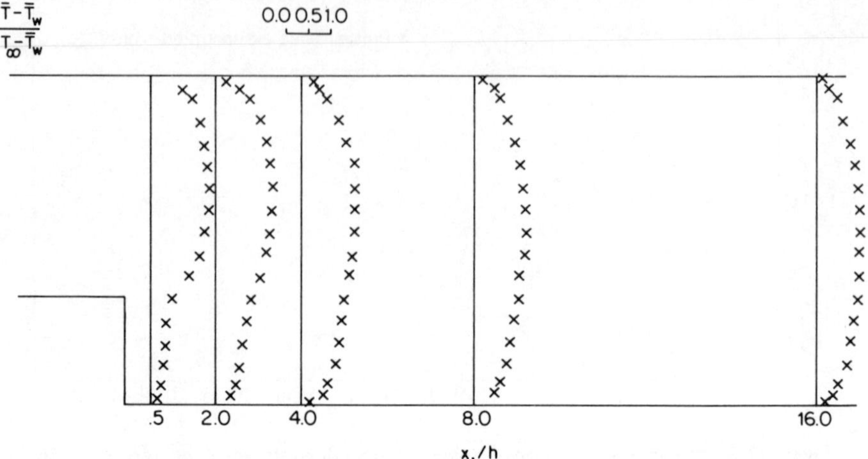

Figure 11.5 Temperature profiles downstream of step, $R_e = 626$

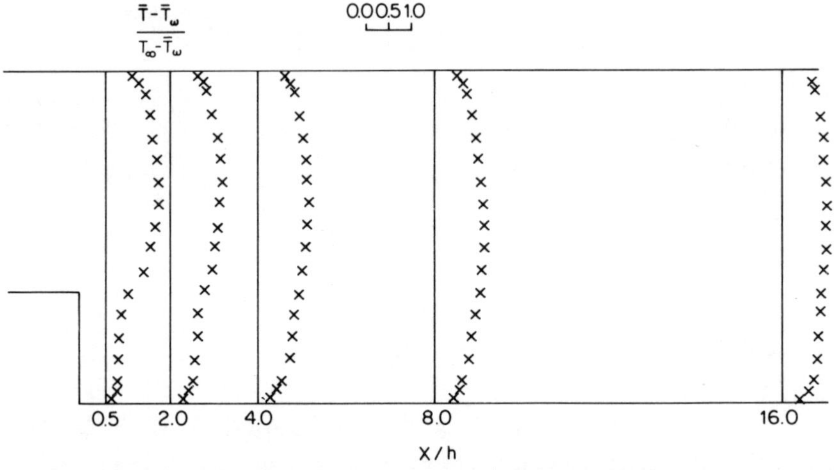

Figure 11.6 Temperature profiles downstream of step, $R_e = 10^5$

generally similar to those produced by finite difference calculations[16] at $R_e = 1330$ and with a slightly higher step to outlet diameter ratio.

11.5 HEAT TRANSFER IN ROD BUNDLES

11.5.1 The geometry

In a liquid metal cooled fast breeder reactor the core consists of an array of closely packed cylindrical fuel rods with coolant flowing axially, in the spaces between the rods, through the bundle. The optimal design of rod bundles in terms of efficient performance and operational safety requires accurate prediction of the temperature field. For present purposes, the bundles are subdivided into several subchannels, each bounded by a set of rod surfaces and imaginary lines connecting the rod centres (see Figure 11.7). With the turbulent flow calculations made for a subchannel, with suitable boundary conditions at the interfaces between subchannels, the energy equation in the fluid and the rod can then be solved for the temperature distribution in the whole flow domain, i.e. in the regions $A_r + A_c$.

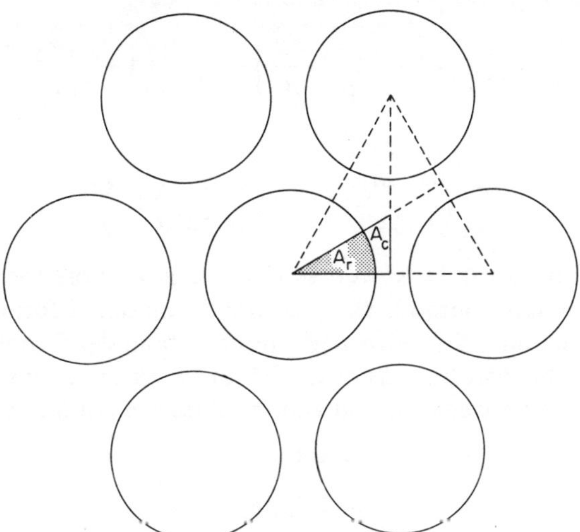

Figure 11.7 Domain for temperature calculations in rod bundle flow

11.5.2 The governing equation

The flow is three-dimensional, but we make the assumption that flows at right angles to the rod axis (the x_3 direction) are negligible relative to the axial flow.[19] The flow is also taken to be fully developed in the axial direction so that Equation (11.4) in this case becomes

$$\rho \bar{u}_3 \frac{\partial \bar{T}}{\partial x_3} = \frac{\partial}{\partial x_1}\left\{\left(\frac{\mu}{\sigma}+\frac{\mu_T}{\sigma_T}\right)\frac{\partial \bar{T}}{\partial x_1}\right\} + \frac{\partial}{\partial x_2}\left\{\left(\frac{\mu}{\sigma}+\frac{\mu_T}{\sigma_T}\right)\frac{\partial \bar{T}}{\partial x_2}\right\}. \tag{11.19}$$

Within the rod the mean steady temperature, $\bar{T}_r$, satisfies

$$\frac{\partial}{\partial x_1}\left\{k_r \frac{\partial \bar{T}_r}{\partial x_1}\right\} + \frac{\partial}{\partial x_2}\left\{k_r \frac{\partial \bar{T}_r}{\partial x_2}\right\} + Q = 0, \tag{11.20}$$

where k_r is the thermal conductivity of the rod and Q is the rate of heat generation within the rod.

Introducing dimensionless temperature coefficients, θ and θ_r, and dimensionless coordinates, X_l, defined by

$$\theta = k\bar{T}/Q^*l_0^2, \qquad \theta_r = k\bar{T}_r/Q^*l_0^2, \qquad X_l = x_l/l_0, \tag{11.21}$$

where as asterisk denotes a mean value in space and l_0 is some characteristic length. Then, on performing an energy balance in the rod and in space, it is possible to write Equations (11.19) and (11.20) as

$$\frac{\bar{u}_3}{u_3^*}\frac{A_r}{A_c} = \frac{\partial}{\partial X_1}\left\{\left(1+\frac{\mu_T\sigma}{\sigma_T\mu}\right)\frac{\partial\theta}{\partial X_1}\right\} + \frac{\partial}{\partial X_2}\left\{\left(1+\frac{\mu_T\sigma}{\sigma_T\mu}\right)\frac{\partial\theta}{\partial X_2}\right\} \tag{11.22}$$

and

$$\frac{\partial}{\partial X_1}\left(\frac{k_r}{k}\frac{\partial\theta_r}{\partial X_1}\right) + \frac{\partial}{\partial X_2}\left(\frac{k_r}{k}\frac{\partial\theta_r}{\partial X_2}\right) = \frac{Q}{Q^*}. \tag{11.23}$$

These two equations can be solved simultaneously across the whole domain by the finite element method, using a slightly modified form of the theory described in Section 11.4, provided suitable boundary conditions can be established and the distributions of Q/Q^* and k_r/k are assumed known.

At the interface between the rod and the fluid the boundary conditions are

$$\theta = \theta_r, \tag{11.24(a)}$$

$$\frac{\partial\theta}{\partial n} = \frac{k_r}{k}\frac{\partial\theta_r}{\partial n}, \tag{11.24(b)}$$

while on the remaining boundaries a zero flux condition is applied in the form

$$\frac{\partial\theta}{\partial n} = 0, \tag{11.25}$$

where n denotes the outward normal to the surface. In addition a datum for the temperature needs to be fixed and, for present purposes, it is assumed that $\theta_r = 0$ at the centre of the rod.

11.5.3 The numerical solution

A finite element analysis of the turbulent flow has been performed using an algebraic model[20] for the turbulent viscosity and also a one-equation turbulence model and the results have been presented elsewhere.[7] Previous finite difference heat-transfer analyses for this problem have taken the mesh in the fluid up to the wall and have not used wall functions[20,21] and the same

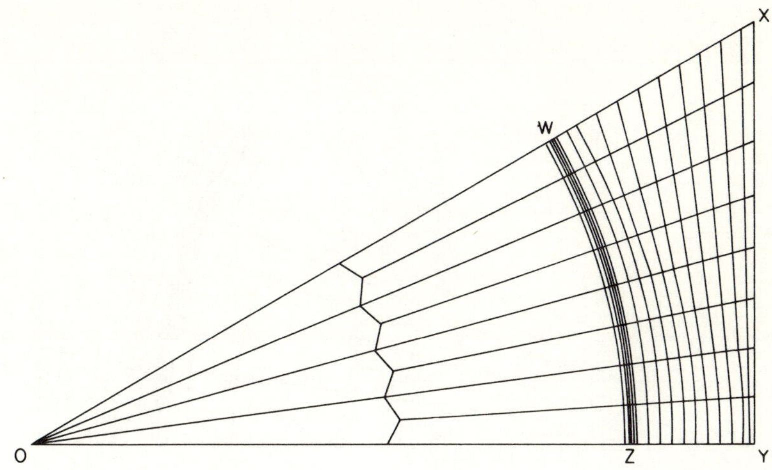

Figure 11.8 Initial finite element mesh for temperature calculation

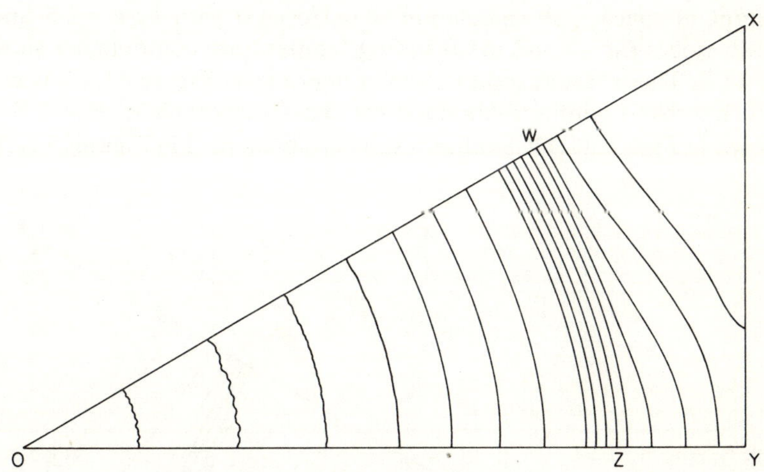

Figure 11.9 Dimensionless temperature contours with initial mesh, $k_r/k = 1.5$, algebraic eddy conductivity model

procedure is adopted here. It then follows from the previous discussion that, as the velocity and turbulent viscosity distributions are known, the governing equation is linear and may be solved without iteration.

The finite element mesh used for the initial calculation is shown in Figure 11.8 with 12 elements used to represent the section of the rod. It is assumed

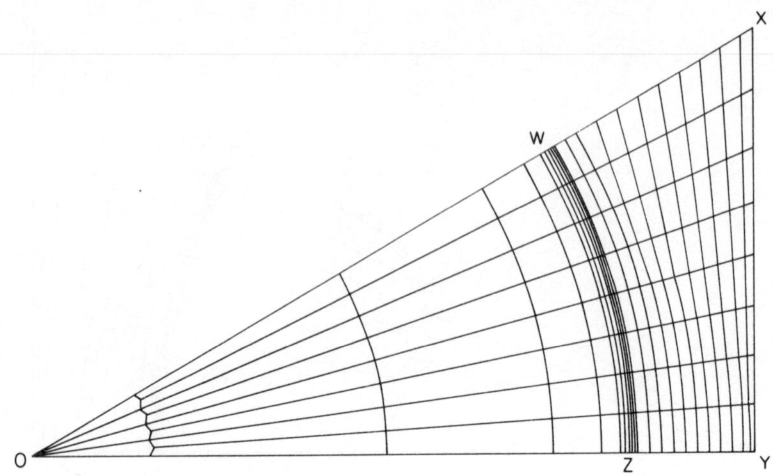

Figure 11.10 Finer finite element mesh for temperature calculations

that Q does not vary with position in A_r so that $Q/Q^* = 1$ and that k_r/k is a constant in space.[21] A calculation is performed with $k_r/k = 1.5$ and the algebraic model for μ_T and the resulting temperature contours are shown in Figure 11.9. These results indicate that a finer mesh (Figure 11.10) is needed and the improved contours obtained for the same problem, $R_e = 2.7 \times 10^5$, are shown in Figure 11.11. Similar results are presented in Figures 11.12 and

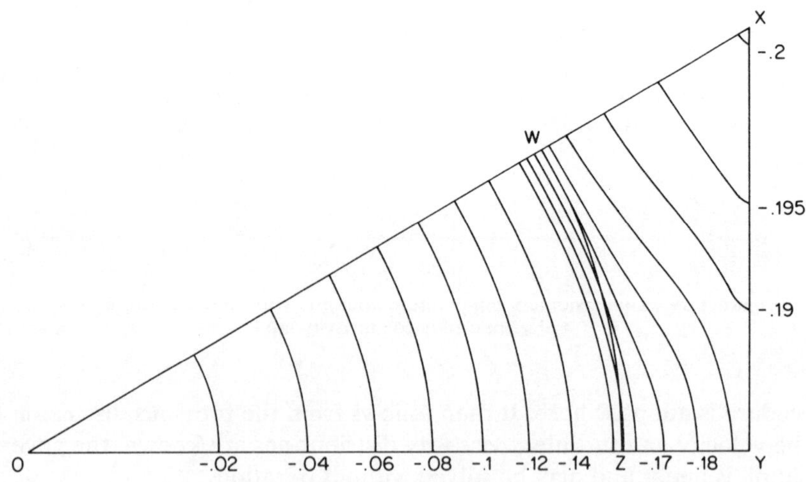

Figure 11.11 Dimensionless temperature contours with finer mesh, $k_r/k = 1.5$, algebraic eddy conductivity model

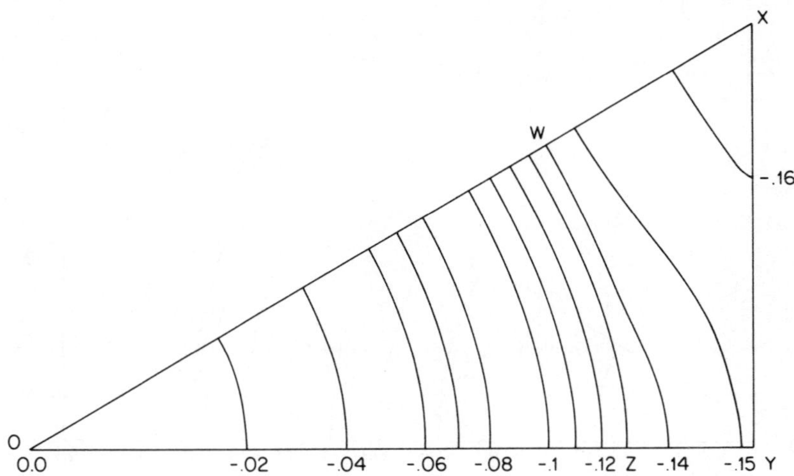

Figure 11.12 Dimensionless temperature contours with finer mesh, $k_r/k = 1.9$, algebraic eddy conductivity model

11.13 for various values of k_r/k while the one-equation turbulence model is used to produce the results shown in Figures 11.14–11.16.

The turbulent Prandtl number is 1.4 in Figure 11.14 and 2.0 in Figure 11.15. In Figure 11.16 it is evaluated from the ratio of the algebraic model for eddy viscosity (μ_T) to the algebraic model for eddy conductivity (μ_T/σ_T).[20]

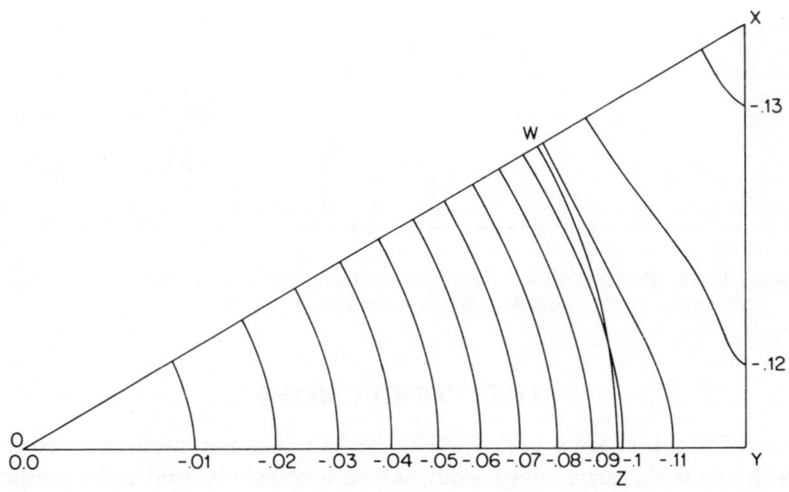

Figure 11.13 Dimensionless temperature contours with finer mesh, $k_r/k = 2.5$, algebraic eddy conductivity model

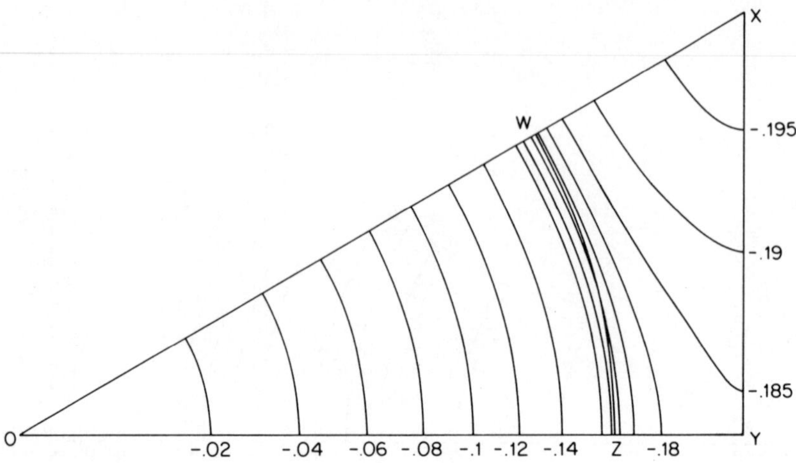

Figure 11.14 Dimensionless temperature contours with finer mesh, $k_r/k = 1.5$, one equation turbulence model, $\sigma_T = 1.4$

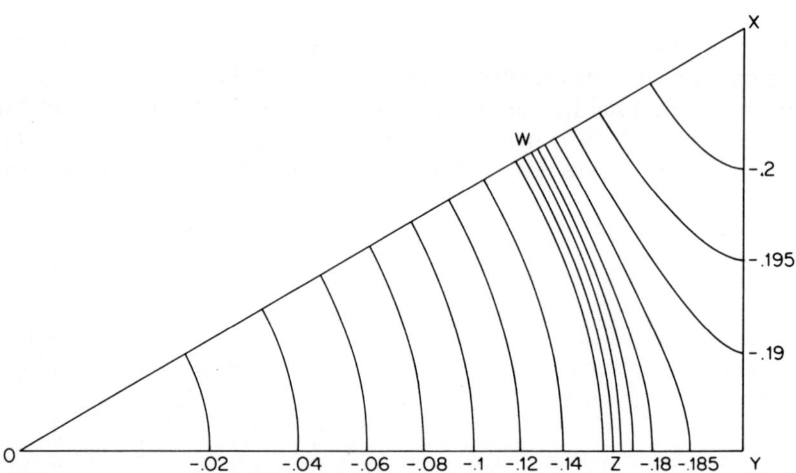

Figure 11.15 Dimensionless temperature contours with finer mesh, $k_r/k = 1.5$, one equation turbulence model, $\sigma_T = 2.0$

11.6 CONCLUSIONS

The results of this chapter indicate that the finite element method, in conjunction with a suitable turbulence model, can be successfully used in the prediction of the temperature distribution and heat-transfer coefficients for forced convective flows. The main difficulty encountered in the backward facing step

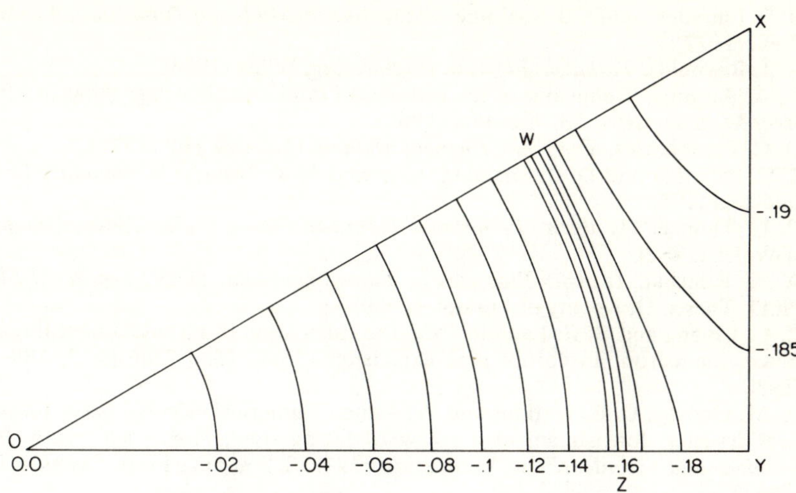

Figure 11.16 Dimensionless temperature contours with finer mesh, $k_r/k = 1.5$, one equation turbulence model for μ_T, algebraic model for σ_T

problem was the specification of suitable wall functions for temperature. The solution of the rod bundle problem indicates the ease with which a combined solid/fluid region can be modelled numerically by the finite element method.

REFERENCES

1. C. Taylor, C. E. Thomas and K. Morgan, 'Analysis of turbulent flow with separation using the finite element method', in *Computational Techniques in Transient and Turbulent Flow*', pp. 283–325, Pineridge Press, Swansea, U.K. (1981).
2. B. E. Larock and D. R. Schamber, 'Approaches to the finite element solution of two-dimensional turbulent flows', in *Computational Techniques in Transient and Turbulent Flow*, pp. 253–282, Pineridge Press, Swansea, U.K. (1981).
3. A. G. Hutton and R. M. Smith, 'On the finite element simulation of incompressible turbulent flow in general two-dimensional geometries', *Proc. Conf. Num. Meth. Laminar and Turbulent Flow*, Venice, pp. 229–242, Pineridge Press, Swansea, U.K. (1981).
4. W. Slagter and H. A. Roodbergen, 'Prediction of developing turbulent flow by the finite element method', *Proc. Conf. Num. Meth. Laminar and Turbulent Flow*, Venice, pp. 267–278, Pineridge Press, Swansea, U.K. (1981).
5. C. E. Thomas, K. Morgan and C. Taylor, 'A finite element analysis of flow over a backward facing step', *Computers and Fluids*, **9**, 265–278 (1981).
6. C. Taylor, C. E. Thomas and K. Morgan, 'Modelling flow over a backward-facing step using the finite element method and the two-equation model of turbulence', *Int. J. Num. Meth. Fluids*, **1**, 195–304 (1981).
7. C. Taylor, C. E. Thomas and K. Morgan, 'The calculation of turbulent flow in rod bundles by the finite element method', submitted to *Int. J. Num. Meth. Fluids* (1982).

8. B. E. Launder and D. B. Spalding, *Mathematical Models of Turbulence*, Academic Press (1972).
9. A. J. Reynolds, *Turbulent Flows in Engineering*, Wiley (1974).
10. R. A. Antonia, 'Behaviour of the turbulent Prandtl number near the wall', *Int. J. Heat Mass Transfer*, **23**, 906–908 (1980).
11. O. C. Zienkiewicz, *The Finite Element Method*, McGraw-Hill (1977).
12. S. V. Patankar and D. B. Spalding, *Heat and Mass Transfer in Boundary Layers*, Intertext (1970).
13. C. E. Thomas, *Analysis of Confined Turbulent Flows*, Ph.D. Thesis, University of Wales (1982).
14. A. K. Runchal, *Transfer Processes in Two-dimensional Steady Separated Flows*, Ph.D. Thesis, University of London (1969).
15. C. C. Chieng and B. E. Launder, 'On the calculation of turbulent heat transport downstream from an abrupt pipe expansion', *Num. Heat Transfer* **3**, 189–207 (1980).
16. A. M. Gooray, C. B. Watkins and W. Aung, 'Numerical calculations of turbulent heat transfer downstream of a rearward facing step', *Proc. Conf. Num. Meth. Laminar and Turbulent Flow*, Venice, pp. 639–652, Pineridge Press, Swansea, U.K. (1981).
17. A. J. Oliver, 'The prediction of turbulent flow and heat transfer over backward facing steps', in *Computer Methods in Fluids*, pp. 309–338, Pentech Press (1979).
18. R. A. Seban, 'Heat transfer to the turbulent separated flow of air downstream of a step in the surface of a plate', *J. Heat Transfer*, **86**, 259–264 (1964).
19. P. Carajilescov and N. E. Todreas, 'Experimental and analytical study of axial turbulent flows in an interior subchannel of a bare rod bundle', *Heat and Mass Transfer*, ASME, 262–268 (1976).
20. N. I. Buleev and R. Nironovich, 'Heat transfer in turbulent flow in a triangular array of rods', translated from *Teplofizika Vysokoleh Temperatur* **10**, 1031–1038 (1972).
21. M. D. Mikhailov, 'Finite element analysis of turbulent flow heat transfer in rod bundles', *Appl. Maths. Centre Sophia Report* (1979).

Numerical Methods in Heat Transfer, Volume II
Edited by R. W. Lewis, K. Morgan, and B. A. Schrefler

Chapter 12

Analysis of Heat Transfer in Complex Geometries due to Combined Conduction; Radiation and Natural Convection

R. J. Hopkirk, D. Sharma, D. J. Gilby, and P. J. Pralong

SUMMARY

A novel method of producing an approximate mathematical model of the heat transfer by conduction, convection and radiation in complex enclosures has been developed for the study of the effects of fires. The flow of heat across the boundaries separating fluid and solid domains is treated by the development of a technique previously published by the authors. The effects of the thermal boundary are accounted for in the formulation of the diffusive coefficients to fit into a standard tridiagonal matrix solution of the finite difference heat transfer equation. A particular feature, bringing large economies in computational effort into the solution of complex, three-dimensional problems, is the embedding of local, analytical subsolutions into the finite difference treatment of the whole problem domain. In the method described in this chapter the embedding technique has been used for representing the conduction in included solid regions.

NOMENCLATURE

A_i	Coefficient of the finite-difference equation
c_p	Specific heat at constant pressure
$f_{f\leftarrow\delta_i}$	View factor of the source f seen from the element of surface δ_i
h_c	Surface heat transfer coefficient
I	Intensity of turbulence
K	Kinetic energy of turbulence
k	Thermal conductivity

L	Vertical distance
l	Length scale of turbulence
N_u	Nusselt number
P_r	Prandtl number
$\dot{q}''$	Heat flux per unit area and time
R_a	Rayleigh number
s	Source
So, Sn	Linearized source terms in the finite difference equation
t	Time
T	Temperature
$\bar{U}$	Mean flow velocity u', v', w' fluctuating (turbulent velocity components
x, y, z	Dimensions in the Cartesian system of coordinates
$\alpha_{f\leftarrow\delta}$	Absorptivity of the source f for radiation from the element of surface δ_j
β	$(=h_c/k)$ auxiliary variable
Γ	Diffusion coefficient in the transport equation
ε	Surface emissivity
η	Auxiliary variable
κ	$(=k/\rho c_P)'$ thermal diffusivity
μ	Fluid viscosity
ν	Fluid kinetic viscosity
ξ	Auxiliary variable.

12.1 INTRODUCTION

Every year fires take a huge toll of human lives and property. Industrial fires in particular have a further unfortunate effect: that of interrupting supplies and services. Above all, fires are difficult to predict. The purpose of the calculational method described here is to analyse a particular type of industrial fire, namely an oil fire in the reactor building of a nuclear power plant. For reasons of both reliability and safety it was thought necessary to predict whether any important systems would be damaged by a lubricating oil fire.

One special feature of the building housing a conventional water-cooled nuclear reactor is that it contains very little which is in fact combustible. The lubricating oil for the main cooling water circulating pumps is the largest source of combustible material. The worst possible consequences of a fire are likely to be the destruction of important systems rather than of the building as a whole. The objective of the work referred to here, therefore, has been to predict the development of the temperature field within a reactor building during the course of a fire started by the ignition of leaking oil.

Figure 12.1 shows a simplified vertical section through an imaginary, but representative reactor building for a power plant using a pressurized water

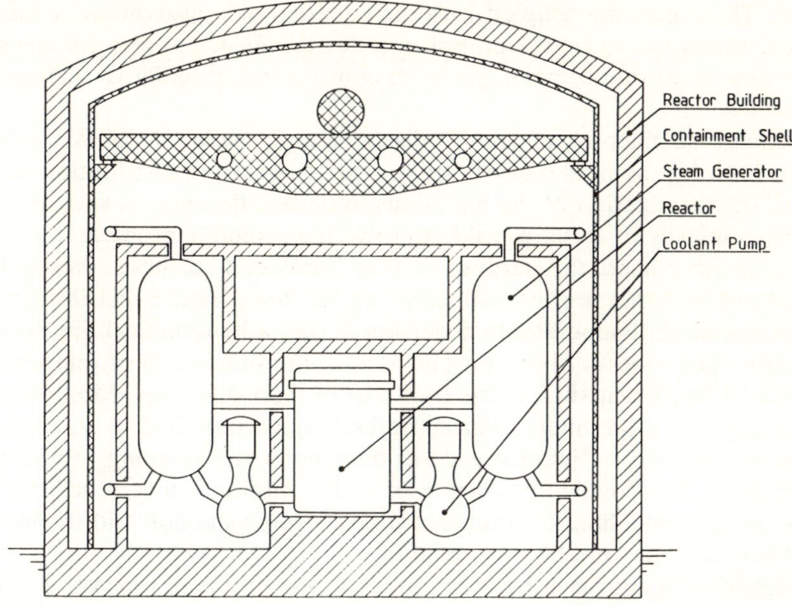

Figure 12.1 Vertical section through the reactor building showing the major components

reactor. A fire is assumed to occur in the lower part of the containment shell near one of the coolant pumps.

The problem has been solved using a development of the authors' implicit finite difference program TURF (*T*urbulent *U*nsteady *R*ecirculating *F*low Dynamics). This chapter describes the methodology and the modelling techniques used for its solution. Because of limitations on the publishing of proprietary information it is unfortunately not possible to show any calculated temperature distributions. It must be stressed that the reactor building in Figure 12.1 simply represents a composite of typical design features and is not to be attributed to any one supplier.

12.2 ANALYSIS AND FORMULATION OF THE PROBLEM

Restated in mathematical terms the task to be undertaken consists of the quantitative determination of the time-dependent temperature distribution in a large, three-dimensional domain containing concrete and steel components. The complete formal mathematical representation includes the equations of mass continuity, conservation of momentum in all three dimensions and the relationships, or equations for an appropriate turbulence model. These must be coupled with the equation of conservation of thermal energy, which has to be solved not only for the fluid, but also for the solid domains with allowance being made for convective and radiative surface heat exchanges.

Further, the fineness of spatial discretization needed to obtain reliable results both for the fluid motion and for the temperature distribution, and to permit the formulation of the equations in finite differences is such that it is not economically (if at all) possible to solve the complete problem.

On the other hand, the purpose and the framework of these investigations should not be forgotten: an estimation of the temperature distribution and of the maximum temperatures occurring during a hypothetical fire accident is sought. The motion of the air enclosed in the building is of interest only because of its participation in the transport of thermal energy. Moreover, the fluid motion in a thermally induced natural convection field in an enclosure of given geometry is implicitly dependent upon the reigning temperature distribution. The detailed prediction of such a motion is a very delicate task involving the modelling of trainsient turbulence production and dissipation, and the results may not be very reliable.

A simplifying approach has been made to the problem, rendering it more tractable by suppressing the convective terms in the equation of conservation of thermal energy and replacing convection by an augmented diffusion effect. This is achieved by use of a locally isotropic, but spatially variable effective thermal conductivity. Two possible methods are available for discovering appropriate values of effective conductivity. One of these is limited in its use to situations where heat flows across a narrow slot or cavity. It was developed by Kraussold[11] using several sets of experimental data for horizontal annuli, but starting with those of Beckmann.[5] By this method all thermal barriers are lumped together and replaced by an effective material thermal conductivity. The resulting quantity includes the effect of both the heat transfer coefficient and of the motion of the fluid in the slot. Figure 12.2 shows the correlation of Kraussold for effective conductivity against Rayleigh number based on slot width. This correlation however is only confirmed up to Rayleigh numbers of about 10^8. Further work on circulation in cavities shows that the transition to turbulence occurs at Rayleigh numbers above 10^8. In particular, Elder[8] observed transition at a cell height based Rayleigh number of 10^{10} and noted that the turbulent boundary layers behaved as those on an isolated vertical surface. He recorded further that the temperature gradients across and up his cell become extremely flat outside the boundary regions.

Raithby and Hollands[16] have proposed another approximate method for studying free convection problems, whereby attention is concentrated entirely on the solid/fluid boundaries. In view of Elder's results, which show practically the whole temperature change taking place in the boundary layers, their approach appears reasonable. Raithby and Hollands also suggest that heat-

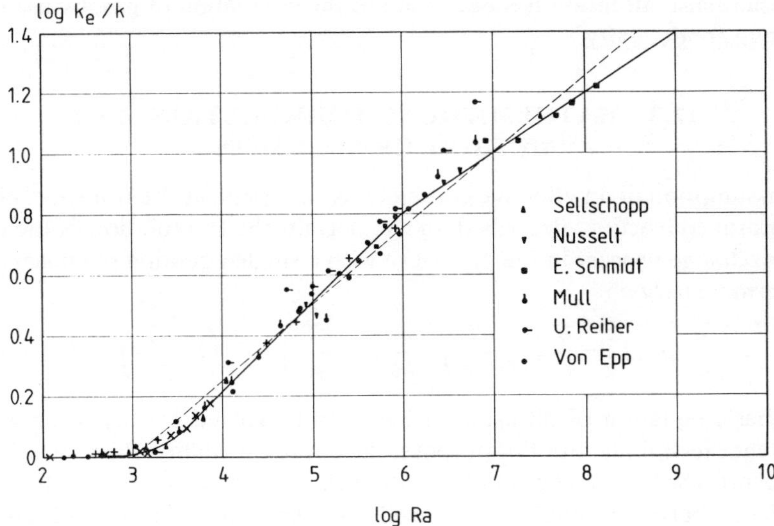

Figure 12.2 Equivalent heat conductivity for free convection through fluid layers

transfer coefficients, even in domains of complex geometry, are dependent rather upon the local rather than the global configuration. This would mean that data from the hundreds of available experimental correlations could be applied in a model at the appropriate places. In this approach the conductivity is effectively infinite in the bulk of the fluid between the boundary layers.

Now the present problem cannot be treated exactly in this way because of the presence of the fire and the approach of Kraussold is clearly inapplicable. The fire heat output will be passed to the air, which will start to flow in free convection in a loop between the lower and upper parts of the building via the two 'chimneys' formed by the steam generator spaces. Up to the time at which it burns itself out, it is largely the fire which determines the flow over surfaces exposed to convective heat exchange. This has two consequences. The first is that it becomes worthwhile to check the order and effect of the imposed flows, while still avoiding the full, coupled solution. The second is that experimental heat transfer correlations for mixed rather than only for pure natural convection must be sought.

The tests on the flow field, using a simplified two-dimensional conceptual model, are described briefly in Section 12.4. The results of these tests enabled satisfactory values of effective conductivity to be found at discrete points of the domain and at various times and employed in the one-equation, three-dimensional model.

This leaves a simplified problem, whereby the importance of the surface heat-transfer resistances, as stressed by Raithby and Hollands, is respected, but where also attention has been paid to the circulation of gas, by use of the effective conductivity.

12.3 MATHEMATICAL FORMULATION AND METHOD OF SOLUTION

The assumption of an effective conductivity to represent the transport of heat by natural convection, described above, permits the formulation of the entire heat exchange process in the enclosure with a single equation of conservation of thermal energy:

$$\frac{\partial}{\partial t}(\rho c_{\mathrm{P}} T)=\frac{\partial}{\partial x}\left(\Gamma_x \frac{\partial T}{\partial x}\right)+\frac{\partial}{\partial y}\left(\Gamma_y \frac{\partial T}{\partial y}\right)+\frac{\partial}{\partial z}\left(\Gamma_z \frac{\partial T}{\partial z}\right)+s^T. \tag{12.1}$$

This basic equation of diffusion of heat can be solved readily using a finite difference technique. For this purpose, the entire calculation domain is divided into control volumes by a set of orthogonal, unequally spaced grid lines. The interfaces between solid and gas cells or between solid cells of different properties are arranged to lie exactly at the faces of the control volume. Each control volume is thus characterized by its thermal properties (density, specific heat, thermal conductivity) and by its average temperature. Each cell has a node at its geometrical centre at which the temperature is defined.

The finite-difference form of Equation (12.1) for the control volume around point P can be written as follows, for each control volume:

$$\sum_i A_i T_{\mathrm{P}}=\sum_i A_i T_i+So_{\mathrm{P}}+Sn_{\mathrm{P}} T_{\mathrm{P}} \tag{12.2}$$

where $i=(\mathrm{E, W, N, S, B, F, O})$ ('O' = old values from previous timestep) and the corresponding control volume of P is represented in Figure 12.3.

The sets of equations represented by (12.2) are solved using an iterative TDMA (*T*hree *D*iagonal *M*atrix *A*lgorithm) in the x–y planes and a block-correction algorithm in the z-direction in order to minimize the storage requirements.

The finite difference relationships for the other heat-transfer mechanisms (convective surface heat transfer and radiation) are integrated into the diffusive coefficients (AE, AW, AN, AS, AB, AF) and into the linearized source terms (So, Sn). Details of this procedure are given in Sections 12.5 and 12.7.

12.4 NATURAL CONVECTION AND TURBULENCE

As mentioned in Section 11.2, a simple calculation of the flow around the circulation loop formed by the upper and lower parts of the building and the two 'chimneys' formed by the spaces around the steam generators has been made. Figure 12.4 shows the conceptual model of this loop. The eccentricity

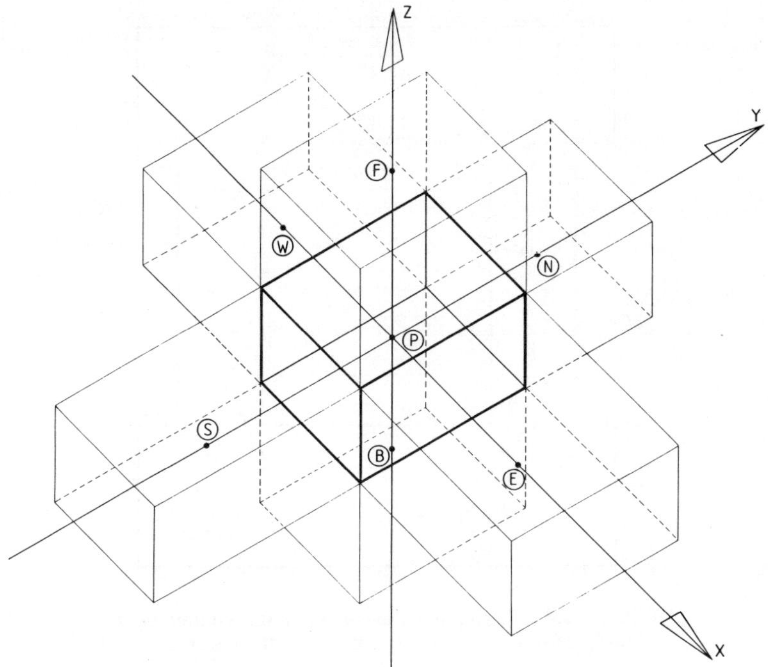

Figure 12.3 Control volume P and its neighbours

of the fire guarantees a convective flow passing up one 'chimney' and down the other. This very approximate calculation did not employ a coupled turbulence model, but respects experience with rectangular channel flows to establish the length scale of turbulence.

For the study, the analogy between thermal and momentum diffusivity has been called upon, via the Prandtl number:

$$\left.\begin{aligned} &\text{For air in laminar flow:} \quad P_r = \frac{\kappa}{\nu} = \frac{k}{\mu C_p} \sim 0.7 \\ &\text{and in turbulent flow:} \quad P_{r_t} = \frac{\kappa_t}{\nu_t} = \frac{k_t}{\mu_t C_p} \sim 1.0 \end{aligned}\right\} \tag{12.3}$$

Accepting that only small deviations of both laminar and turbulent Prandtl numbers can occur, it follows that

$$\frac{k_t}{k} \sim 0.7 \frac{\mu_t}{\mu}, \tag{12.4}$$

whereby the effective conductivity and viscosity are respectively defined as

$$k_e = k + k_t \quad \text{and} \quad \mu_e = \mu + \mu_t \tag{12.5}$$

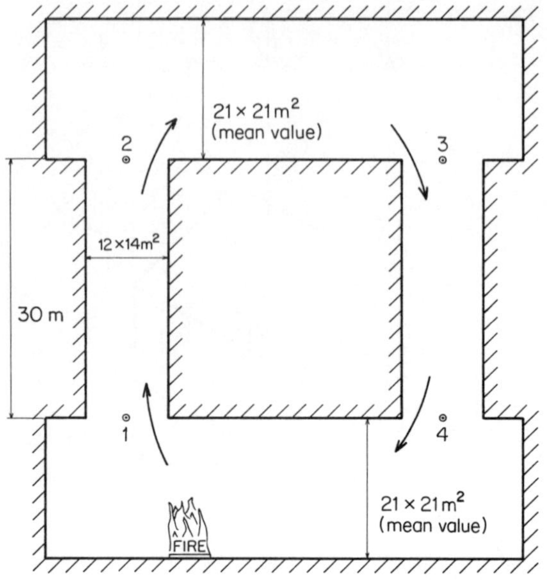

Figure 12.4 Conceptual flow loop in the containment building for a fire situated off-centre in the lower room

Elder[8] has confirmed the existence of supplementary vortices along a vertical face under natural convection, increasing in both quantity and intensity with growth of Rayleigh number. It seems reasonable to hypothesize that they will become yet more intense with the onset of turbulence, especially in view of the relationship (12.4).

In turbulent flow, the intensity of turbulence in a stream whose mean flow velocity is $\bar{U}$ is defined (see for example Schlichting,[14] as

$$I = \sqrt{[\tfrac{1}{3}(\overline{u'^2} + \overline{v'^2} + \overline{w'^2})]}/\bar{U}. \tag{12.6}$$

For natural, turbulent flows of the type to be expected over most of the domain in question, the intensity, I, can be expected to lie in a range between 5 and 20% except in the fire plume itself. Further, the intensity of turbulence is related directly to the kinetic energy of turbulence:

$$K = \tfrac{1}{2}(\overline{u'^2} + \overline{v'^2} + \overline{w'^2}), \tag{12.7}$$

so that

$$K^{1/2} = (\sqrt{\tfrac{3}{2}})\bar{U}I. \tag{12.8}$$

Using now the relationship for turbulent viscosity:

$$\mu_t = \rho l K^{1/2} \tag{12.9}$$

(see Launder and Spalding,[9] where l is the length scale of turbulence, it can be seen that, provided l can be estimated, a range of approximate values for μ_t may be derived corresponding to mean flow speed and a probable range of turbulence intensity.

Sharma[15] has shown how, using the method proposed by Buleev[13]. for geometrical determination of length scale, reasonable distributions over the cross-sections of rectangular diffusers can be obtained, at least on a coarse scale. Sharma's results permit the estimation of an appropriate mean value of length scale $\bar{l}$ (a harmonic mean), according to the aspect ratio of the cross-section. Using estimated values of $\bar{l}$, $\bar{U}$ and I one may write:

$$k_e = k\left(1 + \frac{0.7}{\mu}\rho\bar{l}(\sqrt{\tfrac{3}{2}})\bar{U}I\right). \tag{12.10}$$

The flow calculation was performed as a quasi-steady-state problem using temperatures obtained from initial assumptions of k_e in the thermal transport model. By the above procedure new estimates of k_e were obtained and the whole process iterated. This was repeated at successive time steps. The converged values of k_e/k ranged between 5×10^3 to 5×10^4 and average section Reynold's numbers between 120×10^3 to 235×10^3 whilst turbulence intensity was varied between 5 and 20%.

The effective conductivity, as can be seen from the relationship (12.10), varies linearly with I. It was found that variations of the order tried (4:1) produced negligible variations in the temperature distributions in the solid.

The lack of sensitivity to the value of k_e, even in this time-dependent problems, reinforces the validity of the approximate approach of Raithby and Hollands[16] for turbulent flows.

12.5 MATHEMATICAL MODELLING OF HEAT-TRANSFER MECHANISMS

Although the basic heat-transfer mechanism considered is diffusion in both solids and gas (the convective heat transfer in the gas being represented by an effective thermal conductivity) other mechanisms, such as convective surface heat transfer and radiation, play an important role in the overall energy distribution. Furthermore, the loss of heat to the outside of the enclosure and the strongly nonlinear temperature distributions within the solid walls, due principally to the transient nature of the fire, both have to be carefully analysed in order to obtain a representative picture of the overall energy distribution and of the peak temperatures occurring in the different parts of the enclosure.

12.5.1 Convective surface heat transfer

The heat-transfer mechanisms from a naturally induced convective flow to vertical, horizontal, or inclined surfaces have been thoroughly investigated by many authors. These are usually expressed as relationships between Nusselt and Rayleigh numbers, as for instance in references 4, 14, 17 and 20. In mixed convection situations the practice is usually to involve Nusselt, Rayleigh and Reynold's numbers, as demonstrated for instance by Lloyd and Sparrow[13] or Nakajima *et al.*[15] The cited relationships are in fact those which have been used for the present problem.

Having determined which relationship is locally appropriate, the corresponding surface heat-transfer coefficient is defined as

$$h_c = N_u \frac{k}{L}, \tag{12.11}$$

where L is the streamwise distance from the point considered to the point of initiation of the natural convection. According to the local value of the Rayleigh number and the inclination of the surface considered, the local surface heat-transfer coefficient is estimated and introduced into the general formulation for heat conservation (12.2) as follows:

Consider S to be a solid cell at T_S and G to be a gas cell at T_G. The corresponding temperature distribution takes the form given in Figure 12.5.

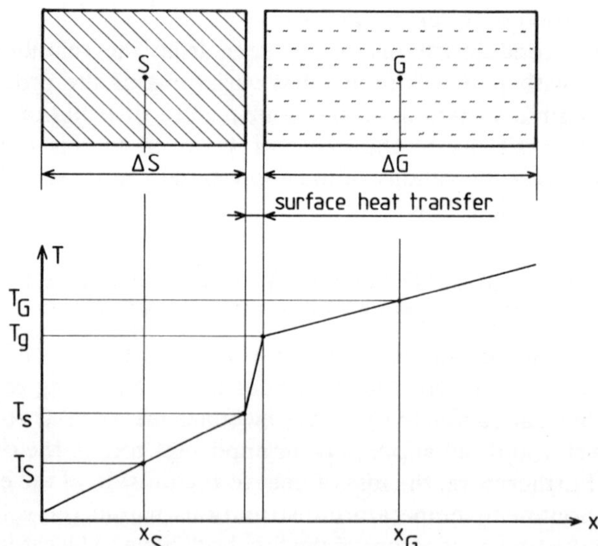

Figure 12.5 Idealized temperature distribution at a solid/gas interface

The condition for heat flux continuity across the interface is given by

$$\dot{q}'' = \frac{k_S}{\frac{1}{2}\Delta S}(T_s - T_S) = h_c(T_g - T_s) = \frac{k_G}{\frac{1}{2}\Delta G}(T_G - T_g) \tag{12.12}$$

or, upon eliminating the temperature at the interface, namely T_s, the solid surface temperature and T_g the gas temperature at the edge of the surface film layer, which is assumed thin with respect to ΔG:

$$\dot{q}'' = \Gamma_g(T_G - T_S) \tag{12.13}$$

where Γ_g is defined by

$$\frac{1}{\Gamma_g} = \frac{\frac{1}{2}\Delta S}{k_S} + \frac{\frac{1}{2}\Delta G}{k_G} + \frac{1}{h_c}. \tag{12.14}$$

The effective diffusion coefficient is thus replaced by the harmonic mean of the three participating coefficients.

12.5.2 Conduction in solids

The distribution of heat in the solid walls is controlled by conduction alone, and therefore its calculation should not present any special difficulty. As an example of what to expect, the penetration of the temperature perturbation into a semi-infinite slab of concrete at whose surface heat is exchanged with a gas subjected to a sudden temperature change, is represented in Figure 12.6.

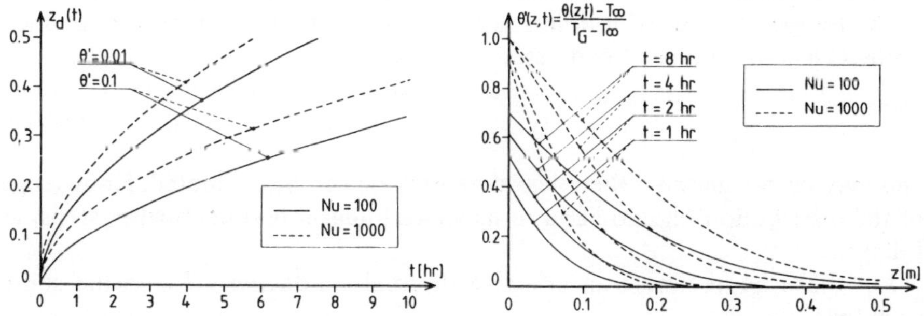

Figure 12.6 Penetration of the temperature perturbation in a semi-infinite slab of concrete

It can be seen from this figure how, for a typical range of Nusselt number, the penetration develops with time. Use of a first-order finite difference method demands a fine grid spacing in the disturbed regions close to the outside face in order to match the steep temperature distributions at the start of the transient. This would imply the use of at least 5 grid nodes across the distance likely to be disturbed during the fire. In a nuclear reactor building, for instance, with many interior walls the total number of grid nodes would need to be between 50,000 and 100,000, which would be impracticable.

This difficulty may be overcome by the use of a technique of 'embedding' analytical solutions into numerical solutions. In the application under discussion in this chapter the procedure commences with the assumption of a one-dimensional temperature distribution within the solid walls. The temperature distribution results from surface heat transfer between gas and solid. For walls, therefore, whose surface dimensions are much larger than their thickness, this is a reasonable assumption. It becomes especially so in the present case, which concerns an intense fire of relatively short duration (less than 1 hour), since the penetration of heat into the solid (see Figure 12.6) can only be of the order of centimetres.

Having decided to accept the validity of local one-dimensional temperature distributions, the solid conduction problem is vastly simplified. One is then able to make use of one or other of those analytical solutions to the heat conduction problem in slabs or semi-infinite bodies which treat the desired boundary conditions (see reference 6). Suppose for the moment that in the direction x, normal to a solid/fluid interface, the instantaneous temperature distribution in the solid at time t is

$$\theta = f(x, t)$$

and has been caused by a history of heat exchange with a neighbouring gas, whose instantaneous temperature is T_g. The same terminology and model as in the previous section and as in Figure 12.5 are employed here.

If the thickness ΔS of the solid cell is so large that a linear approximation to the temperature distribution is not valid, the average temperature defined at the cell centre must be expressed rigorously as

$$T_S = \bar{\theta} = \frac{1}{\Delta S}\int_0^{\Delta S} \theta \, dx \tag{12.15}$$

and may be calculated if the integral expression can be evaluated. Knowledge of the distribution function and of its space integral permits further steps, as follows:

Let there be at time t an effective thermal conductivity k_{eff} in the solid, such that

$$\frac{k_{eff}}{\frac{1}{2}\Delta S}(T_s - T_S) = k_s \left.\frac{\partial \theta}{\partial x}\right|_{x=0}, \tag{12.16}$$

enabling the discretized left-hand side of the expression to represent correctly the surface heat flux. It is apparent that k_{eff} is a time-dependent property of the solid. Further, looking at the surface heat flux:

$$\dot{q}'' = \frac{k_{eff}}{\frac{1}{2}\Delta S}(T_s - T_S) = h_{eff}(T_G - T_s), \tag{12.17}$$

where h_{eff} is the effective surface heat-transfer coefficient and in general is defined as

$$h_{eff} = \frac{h_c k_G}{k_G + \frac{1}{2} h_c \Delta G}$$

(eliminating T_g from Equation (12.12)). The effective, time-dependent conductivity is therefore defined as

$$\left.\begin{aligned} k_{\text{eff}} &= \tfrac{1}{2}\Delta S\, h_{\text{eff}} \frac{(T_G - T_s)}{(T_s - T_S)} \\ &\tfrac{1}{2}\Delta S\, h_{\text{eff}} \frac{(T_G - \theta(0,t))}{(\theta(0,t) - \theta)}. \end{aligned}\right\} \tag{12.18}$$

The ability to compute k_{eff} and T_S from an analytical solution enables both fluid and solid cells in the domain to be treated alike, thus avoiding complicating the numerical solution algorithm. It is interesting to note here that whilst it is convenient in the numerical solution (in fact, as seen in Section 12.3, an alternating direction, tridiagonal matrix algorithm is used) to eliminate the difficult-to-measure surface temperatures from the calculation, they do in fact appear in the embedded analytical solution.

In the light of the approach described here it becomes possible to discretize the solid into finite difference cells of the same order of size as those employed for the gas domain. In the containment of a nuclear power plant, the walls are frequently of the order of 1 m thick or more. These may be modelled by just one layer of cells, giving a very worthwhile saving in computational effort.

The Appendix to this chapter describes some analytical formulations and their behaviour in more detail.

12.5.3 Radiative heat transfer

Radiation between the flames and the neighbouring (i.e. visible) walls represents a very high contribution to the overall heat exchange process in the vicinity of the fire. The wall-to-wall radiative heat transfer on the other hand, is of secondary importance.

The radiative heat flux from a source (f) on to an element (area δ_j) of the surface of a solid body, see References 10 and 14, is defined as

$$\dot{q}''_{f \rightleftarrows \delta_j} = \sigma f_{f \leftarrow \delta_j} (\varepsilon_f T_f^4 - \alpha_{f \leftarrow \delta_j} T_j^4). \tag{12.19}$$

Over the modest temperature range, $\alpha_{f \leftarrow \delta_j}$ can be written as

$$\alpha_{f \leftarrow \delta_j} = C_1 T_f^{C_2} T_j^{C_3} \qquad \text{and} \qquad \varepsilon_{f \leftarrow f}. \tag{12.20}$$

The net flux can thus be written as

$$\dot{q}''_{f \rightleftarrows \delta_j} = h_r (T_f - T_j) \tag{12.21}$$

with:

$$h_r = \sigma f_{f \leftarrow \delta_j} \left(1 + \frac{C_3}{4}\right) \varepsilon_{\text{avg}} \{(T_f + T_j)(T_f^2 + T_j^2)\} \tag{12.22}$$

and, from (12.20),

$$\varepsilon_{avg} = C_1 T_{avg}^{C_2+C_3}.$$

View factors are first estimated from standard relationships for the surfaces in question and then corrected using the law of reciprocity and the fact that the sum of the reciprocal view factors (in an enclosure) is unity.

12.6 PHYSICAL LIMITS AND BOUNDARY CONDITIONS

12.6.1 The source of heat

Specific to any fire study is the type of fire involved. In the problem under discussion here the fire is localized to the combustion of a broad, shallow pool of oil on a horizontal concrete floor. A model of this fire has been constructed using data from an experiment conducted specifically for the purpose. The experiment indicated the rate of burning of the lubricating oil in question, and the expected height of the flames. Since the approximate flame temperature is also known, the total rate of heat release as a function of time and the proportion radiated to neighbouring absorbing bodies may be determined.

In each of the cases studied it was possible to assume that the average shape of the radiating flame front could be represented by a vertical, hollow cylinder with a flat upper face. It is for this shape that standard relationships for the view factors affecting heat exchange with elements of the surrounding walls have been used (previous section).

That portion of the total heat released, which does not participate in radiative heat exchange, is conducted to the fluid (according to the simplified convection model). This has been achieved by employing a very high effective thermal conductivity across the flame front, and also using an augmented value, fitted to the experimental studies of Alpert[1,2] for the immediate fire plume region—the space between the top of the flame and the ceiling of the room in which it is situated.

The fire has thus been modelled phenomenologically, without any attempt to simulate the combustion kinetics. Such a model is possible only when the data requirements are precisely known and experiments can be formulated or reanalysed to provide these data in the desired form.

12.6.2 Heat loss to the environment

The reactor building of a nuclear power plant nowadays generally consists of a thick cylindrical concrete shell within which a further steel shell acting as safety containment is situated. These two shells, but mainly the concrete one, offer a high resistance to heat exchange with the outside environment.

In the case of a fire which is violent but of short duration, the thermal inertia of the overall building is such that a zero heat flux boundary condition may safely be assumed. However, this would be too conservative an assumption in the case of a fire lasting several hours. In such a situation, heat flow from the boundary cells to the outside environment must be taken into account, for instance introducing a one-dimensional thermal resistance between each boundary cell and the outside. The estimation of this thermal resistance has to be made carefully, taking into account the true surface area available for heat transfer. A semi-analytical approach as described in Section 12.5.2 and in the Appendix has been found to be suitable.

12.7 IMPLEMENTATION

The formulations of the different heat transfer mechanisms presented above have now to be integrated into the general finite difference form of the equation of conservation of thermal energy (12.2).

Consider the solid cell S, whose bulk temperature is T_S and whose surface temperature is T_s, subjected to thermal radiation from the fire cell F at T_F, with a flame temperature T_f and the neighbouring gas cell G at T_G, with the heat-transfer coefficient h_c at the interface (see Figure 12.7).

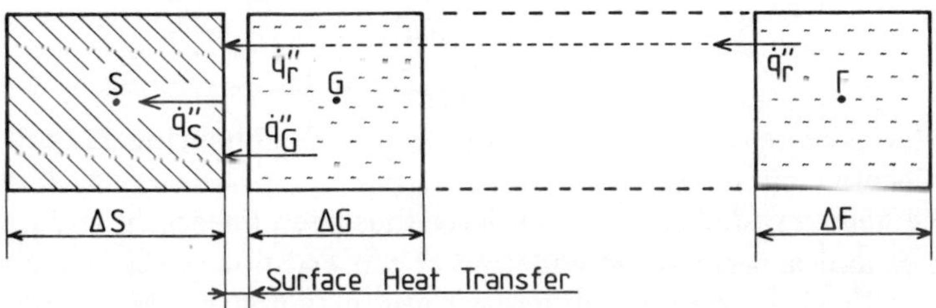

Figure 12.7 Heat Flux in a solid cell

The following heat fluxes can be defined:

$\dot{q}''_S = \Gamma_S(T_s - T_S)$: total heat flux entering the solid cell,

where

$$\Gamma_S = \frac{k_{\text{eff}}}{\frac{1}{2}\Delta S}$$ (as in Section 12.5.2, Equation (12.17);

$\dot{q}''_G = \Gamma_g(T_G - T_s)$: heat flux from the gas cell into the solid cell, including the influence of the surface heat transfer coefficient,

where

$$\frac{1}{\Gamma_g}=\frac{\frac{1}{2}\Delta G}{k_G}+\frac{1}{h_c};$$

$\dot{q}''_r = h_r(T_f - T_s)$: linearized radiative heat flux (see Section 12.5.3);

and

$$\dot{q}''_S = \dot{q}''_G + \dot{q}''_r$$

or

$$\dot{q}''_S = \frac{\Gamma_g \Gamma_s}{\Gamma_S + \Gamma_g + h_r}(T_G - T_S) + \frac{h_r \Gamma_S}{\Gamma_S + \Gamma_g + h_r}(T_f - T_S). \tag{12.23}$$

Thus the coefficient linking the cells S and G becomes

$$A_E = \frac{\Gamma_g \Gamma_S}{\Gamma_S + \Gamma_g + h_r} \tag{12.24}$$

and the following terms have to be introduced into the source coefficients:

$$\left.\begin{aligned} So &= \frac{h_r \Gamma_S}{\Gamma_S + \Gamma_g + h_r} T_f, \\ Sn &= -\frac{h_r \Gamma_S}{\Gamma_S + \Gamma_g + h_r}. \end{aligned}\right\} \tag{12.25}$$

Similar terms appear in the coefficients for the neighbouring gas cell and for the cell located within the fire.

All the heat-transfer mechanisms have thus been linearized and integrated into the canonical form of the equation of conservation of thermal energy. In addition, all cells in the finite difference grid may now be treated identically, whether they contain gas or solid. Hence, the solution of the single equation (12.1) over the whole domain proceeds as described in Section 12.3.

In reality, the solution is further complicated by the temperature dependency of both the thermal conductivity and of the thermal capacity of the gas. The implicit formulation of the conservation equation and its iterative solution permit successive corrections to be made for these variable properties.

Again, no change in the standard solution procedure is required. However, the variation with temperature of these properties does increase the number of iterations necessary for convergence of the solution.

12.8 CONCLUDING REMARKS

The method described here has been used for the prediction of fire effects in a nuclear power plant reactor building. It represents the use of a proven numerical technique to treat a problem of such complexity and of such a nature that experimental validation of the whole model is not reasonable.

The very fact that an experimental simulation of the accident is out of the question however, means that prediction is necessary. At the same time the complexity of the geometrical situation and the spatial resolution demanded mean that the physical problem must be simplified. This has been achieved principally by combining a maximum of empirical data into the model, for instance:

— Experimental determination of oil burning rate and flame temperature,
— Use of augmented, effective thermal diffusion, shown experimentally to be suitable for representing convective transport in a natural convection field,
— Use of experimental data on convective surface heat-transfer coefficients in mixed and natural convection fields,
— Use of empirically derived surface absorptivities to radiant heat.

The general nature of the latter three groups of relationships made it necessary to check the sensitivity of solutions to variations in the resulting parameters.

One relevant physical process in particular has been omitted in the preceding sections, namely the participation of the gas in radiative heat transfer. The reason for this was a lack of data concerning radiative heat absorption and emission per unit volume for the type of combustion products expected from the postulated fire. The omission brings to light one domain in which usable data are lacking.

A further call for data stems from the authors' attempt, as related here, to simplify the modelling of natural and mixed convection heat transfer in enclosures. The approach taken addresses the problem directly. The solution of the complete, formal complex of coupled equations needed to represent a turbulent flow field with heat transfer is not impossible, given satisfactory models for the turbulence. However, when details of the flow field itself are perhaps only of secondary importance in themselves, such a complete solution may be interpreted as being heavy-handed, especially when it can offer little or no increase in reliability or prediction, but demands considerably more computing effort.

In effect, this is the case at present, especially when geometrically complex domains are to be analysed. It would be, therefore, both useful and interesting if experimental investigations of both turbulence and effective thermal conductivity in natural convection flows could be combined.

A finite difference solution was adopted because of the existence of the basic code and its extreme flexibility in accepting the auxiliary relationships for radiation and coupled gas/solid heat transfer. However, a problem of this nature does present some difficulties when using a finite difference grid made up of rectangular cells.

The need to represent bodies and barriers, which only traverse part of the calculational domain means that a large number of strictly redundant small cells are created in the gas-filled regions. In the event the total domain of 55 m × 55 m × 60 m was modelled with approximately 14,000 cells of which approximately 20% were redundant.

The use of embedded analytical relationships for the determination of temperature distributions in solid components avoided the use of several tens of thousands of further cells, many of which would have been redundant in the same way.

This study then has demonstrated the need and some possibilities for reducing and simplifying complex, real problems to make predictions possible and has shown the practicability of the implicit, finite difference treatment, but at the same time one of its chief limitations.

REFERENCES

1. R. L. Alpert, 'Fire induced turbulent ceiling jet', Factory Mutual Research Corp., *Technical Report No. 19722-2,* (1971).
2. R. L. Alpert, Response time of ceiling mounted fire detectors', Factory Mutual Research Corp., *Technical Report No. 19722-3,* (1972).
3. N. I. Buleev, 'Theoretical Model of the Mechanism of Turbulent Exchange in Fluid Flows', *Teploperedacha* (Heat Transfer), pp. 64–98, USSR Academy of Sciences, Moscow (1962). Translation by J. J. Cornish, AERE, Harwell (HL 63/3797(C.7)), AERE Trans. 957.
4. S. W. Churchill and H. H. S. Chu, 'Correlating equations for laminar and turbulent free convection from a vertical plate', *Int. J. Heat Mass Transfer,* **18**, 1323–1329 (1979).
5. W. Beckmann, *Forschung,* Vol. 2, p. 165 (1931).
6. H. S. Carslaw and J. C. Jaeger, *Conduction of Heat in Solids,* Clarendon Press, Oxford (1959).
7. E. R. G. Eckert and R. M. Drake, *Analysis of Heat and Mass Transfer,* McGraw-Hill, Kogakusha (1972).
8. J. W. Elder, 'Turbulent free convection in a vertical slot', *J. Fluid Mech.,* **23**(1), 99–111 (1965).
9. R. J. Hopkirk, D. Sharma and P. J. Pralong, 'Coupled convective and conductive heat transfer in the analysis of hot dry rock geothermal sources', in *Numerical Methods in Heat Transfer,* (ed. R. W. Lewis, K. Morgan and O. C. Zienkiewicz) Wiley Chichester (1981).
10. H. C. Hottel and A. F. Sarofim, *Radiative Transfer,* McGraw-Hill, (1967).
11. H. Kraussold, Equivalent heat conductivity for free convection through fluid layers. *Forsch. Gebiete Ingenieurw.,* **5**, 83 (1934).
12. B. E. Launder, and D. B. Spalding *Mathematical Models of Turbulence,* Academic Press (1972).
13. J. R. Lloyd, and E. M. Sparrow, 'Combined forced and free convection flow on vertical surfaces', *Int. J. Heat Mass Transfer,* **13**, 434–438 (1970).
14. W. H. McAdams, *Heat Transmission,* McGraw-Hill, Kogakusha (1954).
15. M. Nakajima, K. Fukui, H. Ueda and T. Mizushina, 'Buoyancy effects on turbulent transport in combined free and forced convection between vertical parallel plates', *Int. J. Heat Mass Transfer,* **23**, 1325–1336 (1980).

16. G. D. Raithby, and K. G. T. Hollands, 'A general method of obtaining approximate solutions to laminar and turbulent free convection problems', *Advances in Heat Transfer*, Vol 11, Academic Press, London (1979).
17. B. Roux *et al.* 'Natural convection in inclined rectangular cavities, *Int. Conf. on Numerical Methods in Thermal Problems*, Swansea, (1979).
18. H. Schlichting, *Boundary Layer Theory*, Translation by J. Kestin, 6th Edn., McGraw-Hill (1968).
19. D. Sharma, 'Turbulent Convective Phenomena in straight, rectangular-sectioned diffuser's, *PhD. Thesis*, Mech. Eng. Dept., Imperial College, London, September 1974.
20. R. A. Wirtz and W. F. Tseng, 'A finite difference simulation of free convection on tilted enclosures of low aspect ratio', *Int. Conf. on Numerical Methods in Thermal Problems*, Swansea, (1979).

APPENDIX

This appendix discusses aspects of the embedded analytical solution technique, as described in Section 12.5.2, relevant to the application under discussion.

Analytical solutions to the one-dimensional heat conduction equation are available for a variety of boundary conditions. However, it is generally only the relatively straightforward problems that yield a tractable analytical expression. It is in consequence usually necessary to introduce certain simplifying assumptions into the mathematical model.

The nature of these assumptions is dictated principally by the physical processes being modelled, so that the analytical expressions used tend to be problem specific. In the following paragraphs some general comments on the application of the technique to problems, significantly different from the current one, are introduced where appropriate.

Carslaw and Jaeger[5] present a comprehensive summary of analytical solutions and solution methods, including one for the following problem, which most closely approximates the current application.

The slab $-l<x<l$, initially at zero temperature, is in contact with a fluid at temperature rt (ramp function). Heat transfer between the fluid and the slab is defined by the boundary condition equations:

$$\frac{\partial v}{\partial x}-h(v-rt)=0, \qquad \text{at } x=-l$$

$$\frac{\partial v}{\partial x}+h(v-rt)=0, \qquad \text{at } x=+l$$

Also:

$$\frac{\partial v}{\partial x}=0 \qquad \text{at } x=0$$

and, as stated,

$$v=0 \qquad \text{at } t=0.$$

For h = constant, this problem has the following solution:

$$v = rt + \frac{r(hx^2 + l^2h - 2l)}{2\kappa h}$$

$$+ \frac{2hr}{\kappa} \sum_{n=1}^{\infty} e^{-\kappa\phi_n^2 t} \frac{\cos(\phi_n x)}{\alpha_n^2[(h^2+\phi_n^2)l + h]\cos(\phi_n l)} \quad (12.26)$$

where κ is the thermal diffusivity and ϕ_n are the positive roots of $\phi \tan(\phi l) = h$.

This leads to expressions for $\bar{v}$ and $v(\pm l, t)$ as follows:

$$\bar{v} = rt - \frac{rl}{\kappa h}\left(\frac{hl}{3} + 1\right) + \frac{2h^2 r}{\kappa} \sum_{n=1}^{\infty} \frac{e^{-\kappa\phi_n^2 t}}{\phi_n^4[(h^2+\phi_n^2)l + h]} \quad (12.27)$$

$$v(\pm l, t) = rt - \frac{rl}{\kappa h} + \frac{2hr}{\kappa} \sum_{n=1}^{\infty} \frac{e^{-\kappa\phi_n^2 t}}{\phi_n^2[(h^2+\phi_n^2)l + h]} \quad (12.28)$$

These correspond to $\bar{\theta} = T_S$ and $\theta(0, t) = T_s$ in Equations (5.15)–(5.18) with the difference that v is measured from the initial uniform temperature at time $t = 0$.

It can be seen that the use of this solution implies two principal assumptions. The first concerns the variation with time of the air temperature and the second the constant nature of h.

In the case of the first assumption it should be noted that in the 'embedded' solution the rate of air temperature increase (r in Equation (12.26)) is recalculated at each time step, or iteration within a time step, so that the product rt is always identically equal to the current level of air temperature above the initial value. This of course means that the value of r used in each call to the analytical solution is different, as the air temperature variation is in fact not perfectly linear with time. The validity of this procedure has been tested by calculating the surface temperature of a 1 m thick slab of concrete exposed to radiation, using a fine grid and small time-step finite difference model as a control and increasing the fluid temperature according to a third-order polynomial law. Surface heat exchange is by radiation. The results are summarized in Table 12.1, from which it can be seen that there is good agreement between the analytical approximation and the control calculation.

It appears that this approximation is acceptable in the present case. So long as the air temperature is rising monotonically, it would seem reasonable to use the 'ramp' solution in the way described. However, over a longer time scale the air temperature can be expected to fall eventually and at this point the above solution method would cease to give reliable results. For such problems, an alternative to Equation (12.26) must be used. Other useful solutions available include that for a sinusoidally varying fluid temperature from which, using representation by Fourier series, a wide range of boundary

Table 12.1 Comparison of ramp approximation to fluid temperature variation with precise numerical representation

Time (s)	Fluid temperature (°C)	Numerical solution surface temperature (°C)	Analytical solution surface temperature (°C)	Ramp rate (degrees/s)
200	93.8	3.60	3.25	0.469
400	144.0	7.94	6.96	0.360
600	197.7	12.80	11.60	0.330
800	269.6	19.02	18.11	0.337
1000	342.1	26.45	25.52	0.342

temperature patterns can be modelled. Also for curved surfaces it may be considered appropriate to use solutions derived in cylindrical or spherical coordinates.

From the boundary condition equation it can be seen that the value of h is the overall surface heat-transfer coefficient divided by the thermal conductivity of the solid at the surface. In deriving the expression (12.26) it is assumed that h remains constant over the calculation period (the second major assumption). This assumption is reasonable over a limited temperature range, but ceases to be accurate when, as in the present case, heat exchange is dominated by radiation and there is a substantial change in the air temperature.

Radiation heat exchange over limited temperature ranges can be 'Linearized' as mentioned in Section 12.5.3, to yield an effective heat-transfer coefficient. Recapitulating, the method is as follows.

$$\begin{aligned} \dot{q}'' &= \sigma f \varepsilon (T_1^4 - T_2^4) \\ &= \{\sigma f \varepsilon (T_1^2 + T_2^2)(T_1 + T_2)\}(T_1 - T_2) \\ &= h_R (T_1 - T_2), \end{aligned} \tag{12.29}$$

where σ is the Stefan–Boltzmann constant
f is the appropriate view factor
ε is the surface emissivity/adsorptivity
T_1, T_2 are fluid and surface temperatures respectively (K)
$\dot{q}$ is the heat flux per unit area.

Over small ranges of T_1 and T_2 the value of h_R can safely be assumed constant. When either or both temperatures vary significantly, it is necessary to use either a varying value of h_R or some representative mean value. In order to maintain a tractable analytical solution the latter course has been chosen, with a representative mean based on the total quantity of heat flowing into the solid.

At any instant the surface heat flux is

$$\dot{q}'' = \sigma\varepsilon[T_1(t)^4 - T_2(t)^4],$$

where $T_1(t)$ and $T_2(t)$ are functions of time.

Over the period $t=0$ to $t=\tau$ the total quantity of heat flowing into the solid is given by

$$\int_{t=0}^{\tau} \dot{q}''\,\mathrm{d}t = \sigma\varepsilon \int_{t=0}^{\tau} [T_1(t)^4 - T_2(t)^4]\,\mathrm{d}t.$$

Dividing by τ to yield time-averaged values leads to

$$\overline{\dot{q}''} = \overline{h_R}\,\overline{\Delta T},$$

where

$$\overline{h_R} = \frac{\sigma\varepsilon}{\tau}\int_{t=0}^{\tau} [T_1(t)^4 - T_2(t)^4]\,\mathrm{d}t/\overline{\Delta T}$$

and:

$$\overline{\Delta T} = \tfrac{1}{2}\{[T_1(0) - T_2(0)] + [T_1(\tau) - T_2(\tau)]\}.$$

Since the instantaneous value of h_R is not sensitive to small changes in T_1 or T_2 a simple linear relationship between T and t is assumed. This permits the derivation of an expression for $\overline{h_R}$, which can be recalculated at each time step for use in the analytical solution.

Figure 12.8 depicts the results of some experiments with the averaging of h_R, carried out to determine the optimum approach: The first set of calculations was made using a constant value for h_R based only on the initial temperatures. With the fluid temperature increasing at a steady 1 degree/s, the temperature distribution close to the surface of the 1 m thick concrete slab after 300 s was calculated to be as shown in Figure 12.7 (the curves labelled 'constant heat transfer coefficient'). It can be seen that the analytical solution and a finite difference control calculation based on the same conditions yield results generally differing by less than 0.35 degrees. The computed surface temperature was 13.30 °C for the numerical method. The numerical method was then used to calculate the temperature distribution with the value of h_R being re-evaluated at each 5 second time step using the latest values of air and surface temperatures. This resulted in a much higher surface temperature of 40.9 °C. Knowing in advance the nature of the variation of the fluid temperature, it was possible to calculate a time averaged $\overline{h_R}$ using the expressions (12.27) and neglecting the effect of the much smaller change in the surface temperature. This yielded an analytically derived surface temperature of 32.4 °C and a temperature profile labelled in the diagram 'averaged heat transfer coefficient'. Finally, taking account of the change in the surface

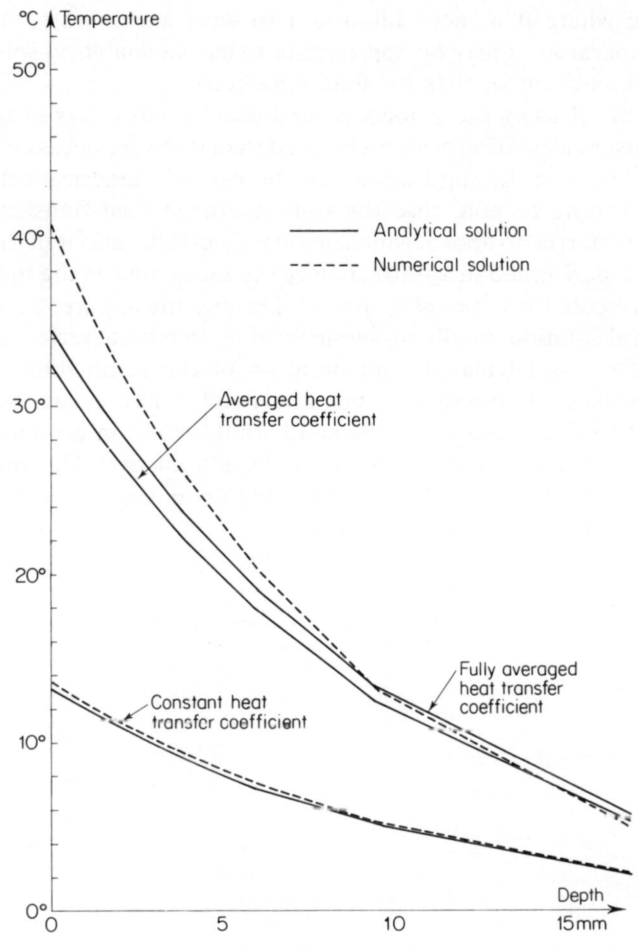

Figure 12.8

temperature in the estimation of h_R (which involves an iterative calculation with the analytical solution) to obtain a 'fully averaged' heat-transfer coefficient results in a slightly higher surface temperature of 34.5 °C.

Although both the analytical solutions appear to significantly underestimate the surface temperature, in each case the mean temperature across the thickness of the slab is the same as that estimated from the numerical solution. This is to be expected bearing in mind the criterion for averaging the heat transfer coefficient and does mean that, in using the embedded analytical solution in this way the quantity of heat remaining in the air is correctly calculated.

In a case where it is more important to have an accurate value for the surface temperature it may be appropriate to use an analytical solution based on this parameter rather than the fluid temperature.

The benefit of using the embedded analytical solution lies in the fact that a much coarser calculation grid can be used than if it were necessary to include the heat transfer in the solid as part of the overall numerical calculation. It is also interesting to note that the fully averaged heat-transfer coefficient calculations referred to above, which involve several iterations of the analytical solution, are performed in approximately the same time as the fine grid finite difference calculations used as a control. Despite the apparent complexity of the analytical solution, involving summation of an infinite series (some 10–20 terms need to be calculated) and iteration of the result, there is no time penalty involved. Furthermore, the analytical solution can be localized, whereas a fine discretization in the solid would result in a multiplication in the number of grid nodes across the whole calculation field. One may conclude therefore that computational efficiency is greatly enhanced by the use of the embedded analytical solution technique.

Numerical Methods in Heat Transfer, Volume II
Edited by R. W. Lewis, K. Morgan, and B. A. Schrefler

Chapter 13

Coupled Heat and Groundwater Flow in Porous Rock

J. Rae, P. C. Robinson, and L. M. Wickens

13.1 INTRODUCTION

There are a number of technical areas where coupled heat and flow problems occur for water in porous rock. The area of most interest to the authors has been the possible disposal underground of high-level radioactive waste. High-level waste can emit enough heat to drive significant flows by buoyancy effects[1] and groundwater flow is expected to be the chief transport process for solute leached from such a repository. The possible disposal of radioactive waste under the seabed[2] raises many similar questions and needs similar techniques to find answers. Other areas where related questions arise are the storage and retrieval of hot water in underground reservoirs,[3] the attempts to extract useful geothermal energy by pumping water into fracture systems in hot rock[4] and in certain thermal techniques for persuading oil to flow in tight reservoirs.[5] We will address the questions in a rather general way and give examples which lie more in the area of waste disposal.

Numerical simulations of flows were dominated till recently by finite difference methods and this still holds for multiphase flow in porous media as encountered, for example, in oil reservoir problems.[6] However, since the original work with finite elements for flow in porous media[7,8,9] applications to single-phase flow have flourished and there is now a large body of research.[9,10] Some advantages of finite elements over finite differences are quite evident. The former's ability to model in a natural way complex geometrical shapes,[12–14] the ease with which its meshes can be refined or coarsened locally, and the possibility of including quite complicated patterns of rock faults are all features of use in modelling groundwater flow. In addition, the method allows in a very simple way discretizations which are equivalent to very high-order complicated finite difference schemes. The authors have written such a finite element program, NAMMU, for coupled heat and groundwater flow problems[14–16] and this was used for the examples in this article. (NAMMU stands for Numerical Assessment Methods for Migration Underground; also the Sumerian goddess of the abyssal waters held immobile by the spell of Ea, Jacobson[17]).

13.2 THE FLOW EQUATIONS

In modelling groundwater flow on a large scale, with thermal effects, the basic equations arise from four sources:

(a) the continuity equation, which expresses mass conservation;
(b) the Darcy flow laws for porous media, which effectively express momentum conservation;
(c) the temperature transport equation, or conservation of energy;
(d) thermodynamic relationships giving material properties as functions of pressure and temperature.

13.2.1 Mass conservation

The continuity equation for large-scale motion in a porous media can be written

$$\frac{\partial}{\partial t}(\phi\rho_f)+\nabla\cdot(\rho_f\mathbf{q})=Q, \tag{13.1}$$

where ρ_f is the fluid (water) density (kg m^{-3}),
ϕ is the rock porosity (dimensionless),
$\mathbf{q}$ is the Darcy flow velocity, (m s^{-1}),
t is time(s),
Q the source or sink strength of fluid (kg m^{-3} s^{-1}).

In general both ϕ and ρ_f are functions of temperature and pressure, although the effects of this dependence are often negligible. An example where the temperature dependence of ρ_f in Equation (13.1) is significant is that of a heat source in clay (see Section 13.5).

13.2.2 Darcy's equation

The equation of motion commonly adopted for flow in a porous medium is Darcy's equation[18]

$$\mathbf{q}=-\frac{K}{\mu(T)}(\nabla P-\rho_f\mathbf{g}+\rho_0\mathbf{g}), \tag{13.2}$$

where $\mathbf{q}$ is the volume averaged fluid velocity,
P is the total pressure minus the hydrostatic part (N m^{-2}, Pa),
K is the rock permeability tensor (m^2),

μ the fluid viscosity, which may vary with temperature T (N s m $^{-2}$, Pl)
ρ_0 an average fluid density (kg m^{-3})
and $\mathbf{g}$ gravitational acceleration (m s^{-2}).

For most porous media this equation is found to hold well provided the Reynolds numbers of the flows are small; since these are typically in the range 10^{-2} to 1 this is well satisfied. There is more doubt however about the correctness of treating some rocks as porous media at all. To fit the requirements the 'pores' of medium have to be small on the scale of interest in the flow so that many of them are averaged in the variables appearing in (13.2). If significant geological features appear in the rock on the scale of interest they ought to be taken into account explicitly. There are formidable problems raised by such a lack of clear separation of scales, and in common with most workers in this field we have chosen to adopt (13.2) for the time being, keeping in mind its limitations.

13.2.3 Energy conservation

The equations which describe heat transport in the system are,[18] in the fluid

$$\rho_f c_f \frac{\partial T_f}{\partial t} = \Gamma_f \nabla^2 T_f + H - \rho_f c_f \mathbf{V}_f \cdot \nabla T_f + h_f(T_s - T_f), \tag{13.3}$$

and in the rock

$$\rho_s c_s \frac{\partial T_s}{\partial t} = \Gamma_s \nabla T_s + H + h_s(T_f - T_s). \tag{13.4}$$

In (13.3) the terms on the right-hand side can be interpreted as combined conduction and dispersion in the fluid, heat production in the fluid, flow convection of heat and, lastly, heat transfer between the rock and fluid. The interpretation of (13.4) is similar. The quantities appearing which have not been defined till now are

c specific heat (J kg^{-1} k^{-1})
Γ conduction/dispersion constant (W m^{-1} K^{-1})
H heat source strength (W m^{-3})
$\mathbf{V}_f$ mean flow velocity in pores (m s^{-1})
h a 'volume' heat transfer coefficient from fluid to rock (W m^{-3} K^{-1}).

The subscripts f and s indicate fluid and solid.

Since the transport equations (13.3) and (13.4) are meaningful only when averaged over a large pore volume (Bear[18]) there are the further relations between the transport coefficients

$$h_{\mathrm{f}}=\frac{h}{\phi} \qquad \text{and} \qquad h_{\mathrm{s}}=\frac{h}{1-\phi}, \tag{13.5}$$

where ϕ is the porosity.

It is common in groundwater flow to find circumstances such that the heat transfer h is large and $T_{\mathrm{f}}-T_{\mathrm{s}}$ small in such a way that $h(T_{\mathrm{f}}-T_{\mathrm{s}})$ is not negligible. We assume these conditions and eliminate the term from the equation by adding ϕ times Equation (13.3) to $(1-\phi)$ times (13.4) to get

$$(\rho c)_{\mathrm{a}}\frac{\partial T}{\partial t}=\Gamma_{\mathrm{a}}\nabla^2 T+H-\rho_{\mathrm{f}}c_{\mathrm{f}}\mathbf{q}\cdot\nabla T. \tag{13.6}$$

This is a single equation for the common rock/fluid temperature T in terms of the averaged transport coefficients

$$(\rho c)_{\mathrm{a}}=\phi\rho_{\mathrm{f}}c_{\mathrm{f}}+(1-\phi)\rho_{\mathrm{s}}c_{\mathrm{s}}, \tag{13.7}$$

$$\Gamma_{\mathrm{a}}=\phi\Gamma_{\mathrm{f}}+(1-\phi)\Gamma_{\mathrm{s}}, \tag{13.8}$$

and a volume averaged flow velocity

$$\mathbf{q}=\phi\mathbf{V}_{\mathrm{f}}. \tag{13.9}$$

This last velocity is generally taken identical to the Darcy velocity,[18] which is a good approximation if ϕ is the porosity in which flow takes place and diffusive heat fluxes are smaller than convective ones.

13.2.4 Thermodynamics

As water is only slightly compressible its equation of state can be taken to a good approximation to be

$$\rho_{\mathrm{f}}(P,T)=\rho_0(1+\alpha P-\beta T), \tag{13.10}$$

where α, β are the compressibility and volume coefficient of expansion, respectively, at the reference density ρ_0. Although the intact rock may be taken as incompressible its porosity will vary with pressure changes and of course with the thermal expansion of the rock. The pressure and temperature dependence in Equation (13.1) is therefore quite complicated.

13.2.5 Boundary conditions

The set of governing equations need suitable boundary conditions to specify a problem. For the fluid flow equations these are mostly of three types:

(a) pressure specified at a fixed boundary as a function of time (e.g. flow into a large lake);

(b) a specified flux of water normal to a fixed boundary as a function of time (e.g. zero flux at impermeable barriers);
(c) pressure prescribed on a boundary which may move in time, with its position to be determined as part of the solution (e.g. at a phreatic surface);

For the temperature equation two types of boundary condition are common:

(d) a prescribed temperature on a surface;
(e) a prescribed heat flux normal to a surface. Note however that depending on the circumstances the proper heat flux may be diffusive ($-\Gamma_a \nabla T$) or convective ($\rho_f c_f \mathbf{q} T - \Gamma_a \nabla T$.)

13.3 THE FINITE ELEMENT FORMULATION

Most authors in this field have chosen to take the nonhydrostatic pressure P and temperature T as the dependent variables to be found. The equations being solved are then

$$\frac{\partial}{\partial t}(\phi \rho_f) - \nabla \cdot \left(\frac{K \rho_f \nabla P}{\mu}\right) + \nabla \cdot \left\{\frac{K \rho_f}{\mu}(\rho_f - \rho_0)\mathbf{g}\right\} = 0 \qquad (13.11)$$

and

$$(\rho c)_a \frac{\partial T}{\partial t} + \rho_f c_f \mathbf{q} \cdot \nabla T - \Gamma_a \nabla^2 T = H, \qquad (13.12)$$

where the Darcy velocity is given by

$$\mathbf{q} = -\frac{K}{\mu}\nabla P + \frac{K}{\mu}(\rho_f - \rho_0)\mathbf{g}. \qquad (13.13)$$

There are many ways to form finite element versions of these equations and they are well documented.[13,19] For NAMMU we chose what is probably the most common method, the Galerkin finite element method. In this, as is well known, the fields P, T are expanded in a series of finite element basis functions and the equations multiplied by each basis function in turn and integrated over the flow region. The basis functions ϕ_i are piecewise polynominals defined on elements and usually associated with nodes i of these elements. This gives a set of nonlinear ordinary differential equations in time for the nodal variables, there being as many equations as unknowns. On any nodal boundary or reference point where a value of P or T is specified the corresponding equation is deleted from the set and the known value used wherever it occurs. The equations are discretized in time in a way described in the next section. The resulting algebraic equations have to be linearized, usually by the Newton–Raphson method, and solved as a system of linear equations, in our case by the well-known frontal solution technique.[20]

The above approach throws up a number of problems some of which should be discussed further. The first problem concerns the Darcy velocity $\mathbf{q}$. If P is discretized as a piecewise polynomial of low order and equation (3.13) is used pointwise to define the Darcy flow whenever it is needed, the resulting $\mathbf{q}$ may have a discontinuous normal component at element boundaries. With care and a respectably refined mesh this can be tolerated (see the example in Section 13.5) but it can cause problems, for example, when trying to plot flow paths. Several authors recommend treating $\mathbf{q}$ as a third dependent field with its own finite element representation calculated from the weak form of Equation (13.13). This gets around some difficulties but may lead to others; for example in cases where rock permeability varies discontinuously the velocity component tangential to the discontinuity ought to be discontinuous too.

In many groundwater flow problems it is common to have to calculate fluxes through parts of the boundary where prescribed pressure or temperature values are given. This also raises a problem. The normal derivative of the finite element P or T field is usually a very poor approximation to the true flux, even to the extent of not satisfying a conservation law. The difficulty occurs because equations are 'thrown away' at nodes where the values are prescribed. One way around this is to keep *all* the equations and supplement the list of unknowns by consistent fluxes $\hat{C}$ defined at the troublesome nodes by, for example,

$$\int_{\text{boundary}} \hat{C}\phi_j = \Gamma_a \int_{\text{boundary}} \frac{\partial \hat{T}}{\partial n} \phi_j, \tag{13.14}$$

where $\hat{T}$ is the finite element approximation to T and j runs over the prescribed nodes. A conservation law can now be formed using $\hat{C}$ showing that it is a useful way of defining flux on the boundary. It is not equal to a pointwise $\Gamma_a\, \partial\hat{T}/\partial n$, this usually being a very poor expression to use for flux. If a very thin (order ε) elements are used on the boundary the conventional method of rejecting equations can be used with success. To first order in ε, $\Gamma_a\, \partial\hat{T}/\partial n$, can play the role of a useful flux. There can be problems, however with element Jacobians if the boundary is curved.

13.4 THE TIME-STEPPING METHOD

Although it is possible to use finite elements in time it is seldom done and time stepping is usually by differencing methods. Many methods are available and we must face the usual questions of ensuring stability, deciding between implicit and explicit methods etc.[21,22] For the present class of problems there are some very clear pointers which suggest the best method. First, many coupled heat and flow problems tend eventually to an equilibrium, often over very long time scales. This makes explicit methods, with their time step sizes restricted by stability requirements, look very unattractive. Implicit methods are then favoured. Second, there is often a wide range of time scales, representing different components in the sources or different physical processes coming into play. The problem thus exhibits 'stiffness' and a stiffly stable time stepping scheme is called for. As automatic time step adjustment is desirable we have been attracted to Gear's method (Byrne and Hindmarsh,[23] Gear[24]). This is a general predictor–corrector method which uses differencing schemes of up to fifth order. The schemes used are based on polynomial extrapolation from previous values for the predictor and the qth order backward difference schemes for the corrector ($q = 1, \ldots, 5$). At each stage the maximum allowed time step at various orders of scheme is calculated, keeping the time truncation error within user-controlled limits, and the scheme used which gives the largest step. All the orders of the scheme it uses are stable for all time step sizes, so once the high frequency components have decayed from the true solution it can take large steps without worry. The error on each step is $O(h^{q+1})$ for step size h, so the greatest benefits are found when asking for high accuracy. If an accuracy of 0.01 is attained with a first-order method, then the fifth-order method would give 10^{-6} without extra cost. However the fifth-order method would give only a factor 2 gain in speed at an accuracy of 0.01.

The heat sources used in this kind of work are usually of the form $f(t)\,h(x)$ with f and h known. It is possible to improve the accuracy of the time stepping by doing explicit integrals of $f(t)$ over time. For illustration consider the equation

$$\frac{\partial T}{\partial t} = f(t)h(x) + D\nabla^2 T. \tag{13.15}$$

The solution of the simpler equation

$$\frac{\partial T_1}{\partial t} = f(t)h(x) \tag{13.16}$$

is known, namely

$$T_1 = h(x)\int_0^t f(t)\,\mathrm{d}t \tag{13.17}$$

In cases where the source is important it is better to solve the equation

$$\frac{\partial T_2}{\partial t} = D\nabla^2 T_2 + D\nabla^2 T_1 \tag{13.18}$$

for $T_2 = T - T_1$ than to solve the original Equation (13.15) with the source present.

13.5 THE RESPONSE OF CLAY PORE WATER TO A HEAT SOURCE

Given the large cost of conducting on-site tests it has been proposed that laboratory scale experiments be used to establish the response of clays to heating. As part of this approach we have modelled the pore water response to a 100 W heater embedded in the centre of a cylindrical tank of clay 5.5 m high by 3 m radius. Since the problem is cylindrically symmetric only a radial section is modelled, the centreline corresponding to zero heat and fluid flow from symmetry considerations. Note that the clay is assumed to be a rigid matrix, so that yield of the clay is not considered. The properties assumed for the clay are listed in Table 13.1.

Table 13.1 Physical properties of clay and water (at 40 °C)

Physical quantity	Symbol	Value
Average thermal conductivity	Γ_a	$1.0\ \mathrm{W\,m^{-1}\,°C^{-1}}$
Average heat capacity per unit volume	$(\rho c)_a$	$3.0\times10^{6}\ \mathrm{J\,m^{-3}\,°C^{-1}}$
Permeability	K	$6.5\times10^{-17}\ \mathrm{m^2}$
Porosity	ϕ	0.5
Density of water	ρ_{f0}	$9.92\times10^{2}\ \mathrm{k\,g\,m^{-3}}$
Specific heat of water	C_f	$4.18\times10^{3}\ \mathrm{J\,kg^{-1}\,°C^{-1}}$
Thermal conductivity of water	Γ_f	$6.23\times10^{-1}\ \mathrm{W\,m^{-1}\,°C^{-1}}$
Viscosity of water	μ	$6.53\times10^{-4}\ \mathrm{kg\,m^{-1}\,s^{-1}}$
Volumetric thermal expansion coefficient of water	β	$3.85\times10^{-4}\ \mathrm{°C^{-1}}$

Figure 13.1(a) shows a schematic cross-section of this notional experiment. Note that the bottom and outer surface of the cylindrical tank are both thermally insulated and impermeable to fluid flow, corresponding to no heat or fluid flow boundary conditions across these surfaces in the numerical model. The top surface is bathed in a constant temperature water reservoir maintained at 10 °C, so that the relevant boundary conditions on this surface are constant temperature (10 °C) and pressure (here taken as zero). Note that throughout this section (and Section 13.6) pressure is taken to mean nonhydrostatic pressure; that is, total pressure minus the hydrostatic component. The volume (0.5 m height by 0.06 m radius) containing the 100 W heater is small in comparison with the volume of clay so that the error introduced by using the same physical properties for the heater volume as for the clay volume, as is done here, is slight.

The grid used is shown in Figure 13.1(b). It consists of a rectangular grid of nine-noded quadrilateral elements. There are 132 elements in all. The elements are smaller around the heater to model the sharp temperature gradients expected there. Both the pressure and temperature fields are modelled by the full quadratic functions. As was explained in Section 13.2, we use temperature and pressure as our basic dependent variables.

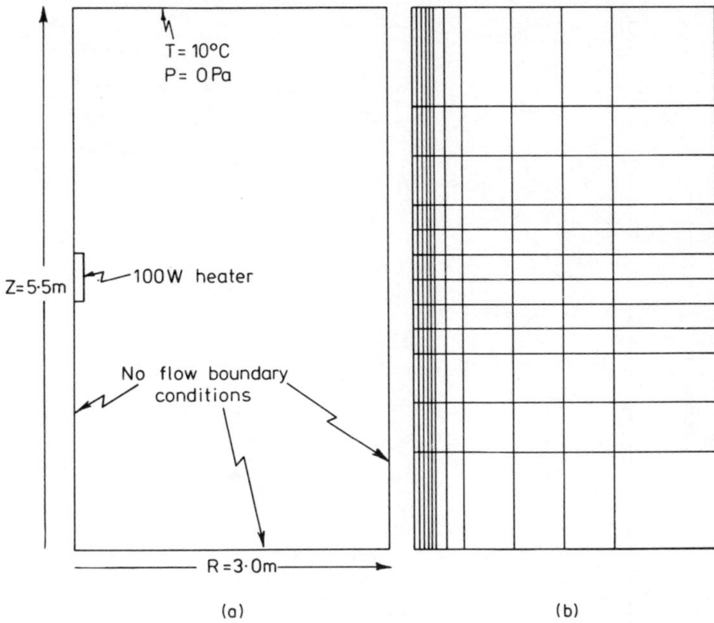

Figure 13.1 (a) Schematic radial section of the tank containing the clay; (b) finite element discretization of the tank

The problem described above is time-dependent, so we use a Gear's method solution routine. We chose an initial time step of 10^4 s and allow Gear's method to select the most suitable subsequent time steps (for our error bounds) as the solution proceeds.

The program was run for 45 time-steps, at which point 37.1 years of simulated experiment had elapsed and the steady-state solution had been reached. By this time the Gear's method had increased the time step to 5.76×10^8 s.

Figure 13.2 shows how the temperature at the centre of the heat source increases with time. The temperature continues to rise until heat loss from the surface of the tank is balanced by the 100 W heat production rate, when a steady-state temperature distribution is attained. From Figure 13.2 it can be seen that a steady-state temperature distribution develops at times greater than approximately 10^8 s (or 3.2 years), when the centre of the heater reaches a temperature of 96 °C. The points marked correspond to the temperature attained at time-steps 2 to 45. The exponential increase in timestep size as the solution progresses can clearly be seen.

The pore-water response to the heat source can be divided into two convenient regimes. These are:

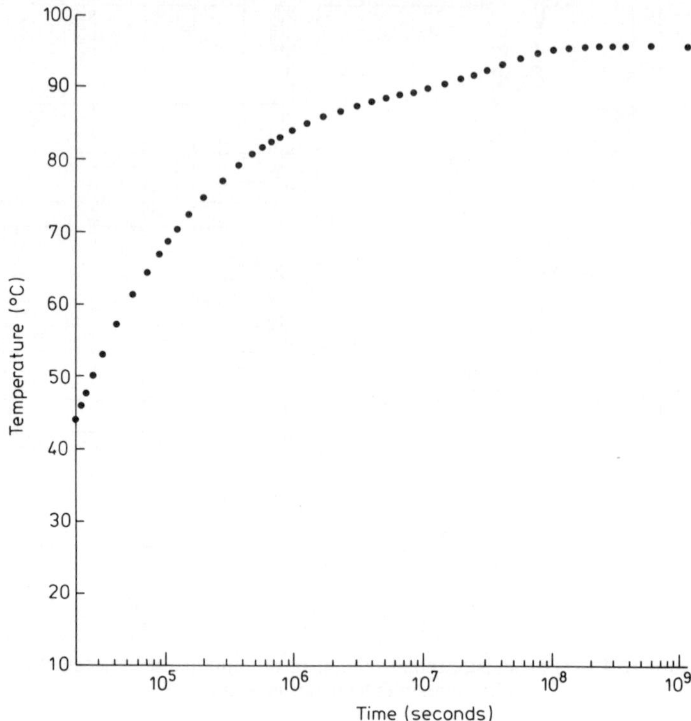

Figure 13.2 Temperature evolution at the centre of the heat source at the successive time steps 2–45

(i) At early times, prior to heat loss from the top surface becoming significant (i.e. $t<10^7$ s), pore water velocity and the concomitant pressure field are generated by the net expansion of the water contained within the clay pores. This is proportional to the net heat input into the clay. The effect is significant because of the large porosities and low permeabilities of most clays.

(ii) At large times (i.e. $t>10^8$ s) a steady-state temperature profile is attained and the net heat flow into the clay is zero. Since effect (i) disappears under these circumstances, buoyancy force (cf. the Boussinesq approximation) is now the dominant pore water driving term. This is the more conventional driving force that always dominates in low-porosity media.

Figure 13.3 shows a plot of vertical pore-water velocity through the heat source as a function of time that clearly illustrates these two regimes. Regime (i) corresponds to times $t<10^7$ s. At early times in this regime ($t<5.10^5$ s) the passage of the heat front causes the vertical velocity to at first increase, and then to decrease to a constant value once the heat front has passed

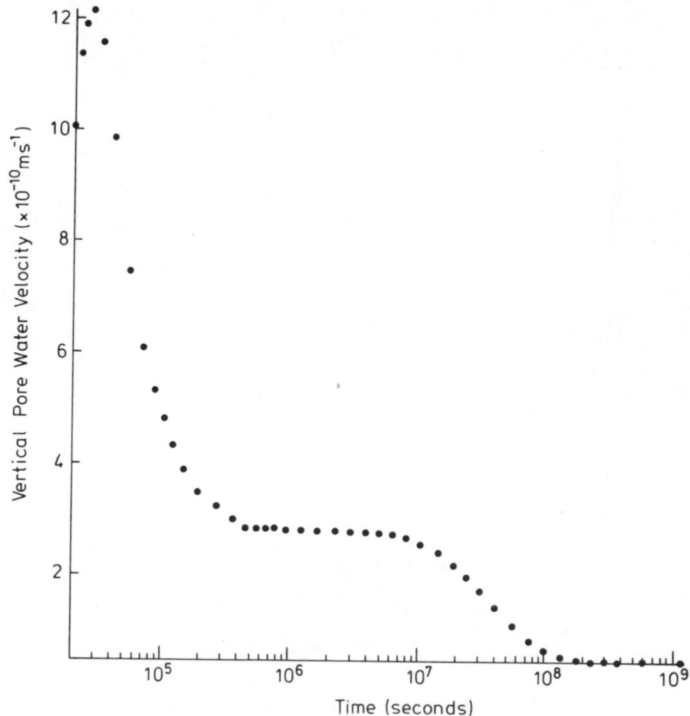

Figure 13.3 Vertical pore-water velocity through the centre of the heat source at the successive time steps 2–45

($t > 5.10^5$ s) but not yet reached the tank surface ($t < 10^7$ s). Regime (ii) corresponds to times $t > 10^8$ s when a steady-state temperature profile has been established and the buoyancy force drives the fluid at substantially lower velocities than was the case in regime (i). $10^7 < t < 10^8$ s corresponds to a transition between regimes (i) and (ii) when the heat front has reached the surface but has not yet resulted in sufficient heat loss for a steady-state temperature profile to be established. Figure 13.4 shows radial plots of temperature against distance from the centre of the heat source, at successive times, and clearly shows that the heat front does not reach the walls of the tank until times $t > 10^7$ s.

Figure 13.5 shows radial pore water pressure profiles at successive times, corresponding to two of the radial temperature profiles of Figure 13.4. It is especially interesting to note that although at early times the temperature profiles drop to the ambient value a short distance from the heat source, this is not the case for the pressure profiles. This is due to the fact that the effect of localized pore-water expansion is communicated throughout the tank instantaneously (the compressibility of water being neglected) and is not

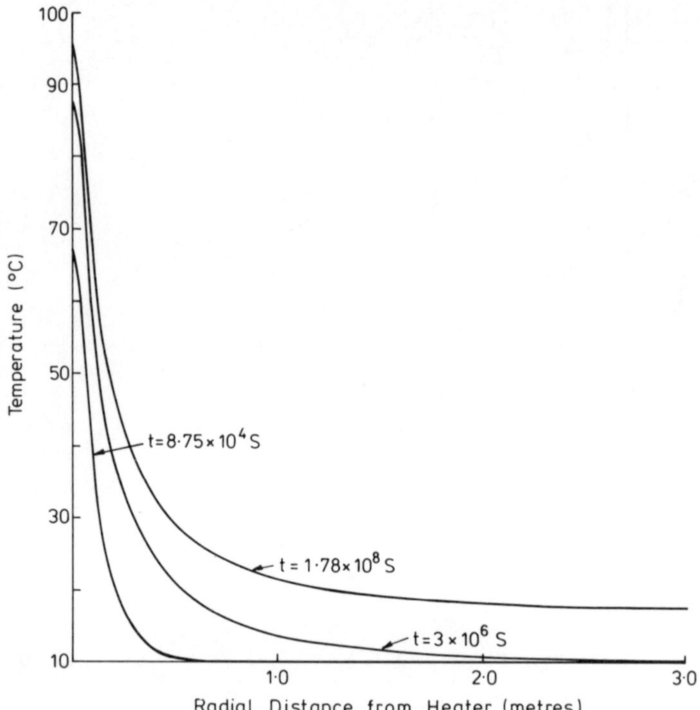

Figure 13.4 Radial temperature profiles from the centre of the heat source at $t = 8.75 \times 10^4$, 3×10^6 and 1.78×10^8 s, corresponding to time steps 10, 25 and 40 respectively

dependent on the precise location of the heat front. This contrasts with buoyancy forces, which are confined to act within the heat front. Notice how the centre pressure falls with time more quickly than pressures at points away from the heater. This causes a broadening of the peak and eventually there is almost no pressure change along this radial line.

The pore-water response is most easily understood by recourse to a simple analysis of the equations. By estimating the magnitude of the terms in Equations (13.12) and (13.13) we can see that heat transport by convection will not be important so the equations decouple. The temperature equation is then just the usual conduction equation with a source

$$\frac{\partial T}{\partial t} = \frac{\Gamma_a}{(\rho c)_a} \nabla^2 T + \frac{H}{(\rho c)_a} \tag{13.19}$$

and by combining Equations (13.10) and (13.11) we get for the pressure

$$\nabla^2 P = -\frac{\mu \beta \phi}{K} \frac{\partial T}{\partial t} + \rho_0 g \beta \frac{\partial T}{\partial z}. \tag{13.20}$$

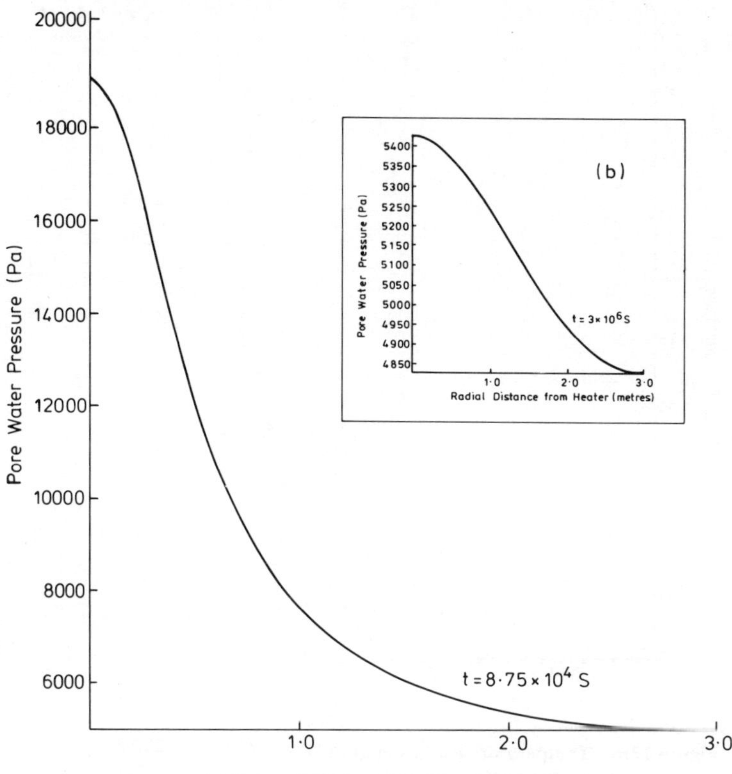

Figure 13.5 Radial pore-water pressure profile at $t = 8.75 \times 10^4$ s (time step 10). The inset shows the corresponding curve at the later time of $t = 3 \times 10^6$ s (time step 25)

During regime (i) the $\partial T/\partial t$ term in this equation dominates. Thus, for the constant power heat source used here, the pore water response outside the heat front and during regime (i) is independent of both the temperature field and time. This point is well illustrated by Figures 13.6 and 13.7, which show the time evolution of the temperature and pressure, respectively, at a point below the heat source and 0.25 m from the tank bottom. The arrival of the heat front at 2×10^6 s clearly corresponds to the onset of a decrease in pressure.

Figures 13.8–13.13 show a sequence of pore-water pressure and velocity maps starting at early times (Figures 13.8(a) and 13.8(b)) and proceeding through intermediate states to the latter stages of regime (i) (Figures 13.13(a) and 13.13(b)).

In order to visualize the velocity field we draw arrows at a regular array of points. The direction corresponds to the direction of the local velocity and

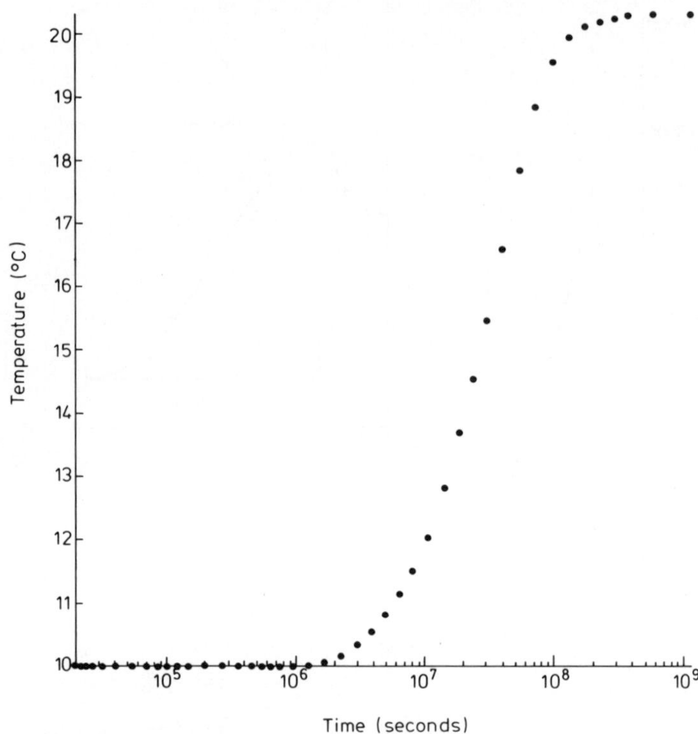

Figure 13.6 Temperature evolution at the point $r = 0.0$, $z = 0.25$ vertically below the heater and at the successive time steps 2–45

the length is equal to the magnitude of the velocity at the point multiplied by a time which is given on each figure.

The pore water response $8.75 \times 10^{+4}$ s (one day after the heater is switched on) is shown in Figures 13.8(a) and 13.8(b). The characteristic feature at this early time is a rapid rise in pore water pressure (Figure 13.8(a)) that results in essentially radial flow (Figure 13.8(b)) in the vicinity of the heater. Away from the heater the flow turns to accommodate the boundary conditions.

Figures 13.9(a) and 13.9(b) show the development of the flow after 7.64×10^5 s (8.8 days). The most important features to note here are that the maximum in the pressure field has decreased in magnitude by a factor of two and shifted to a point below the heater. This downward shift in the pressure maximum continues till the time of 3×10^6 s (34.7 days) shown in Figure 13.10(a), when it is close to the bottom of the tank, At this stage the velocity field (Figure 13.10(b)) appears to emanate from the centre of the tank bottom, corresponding to the position of the pressure maximum. The pressure maximum essentially follows the path of the heat front below the heater,

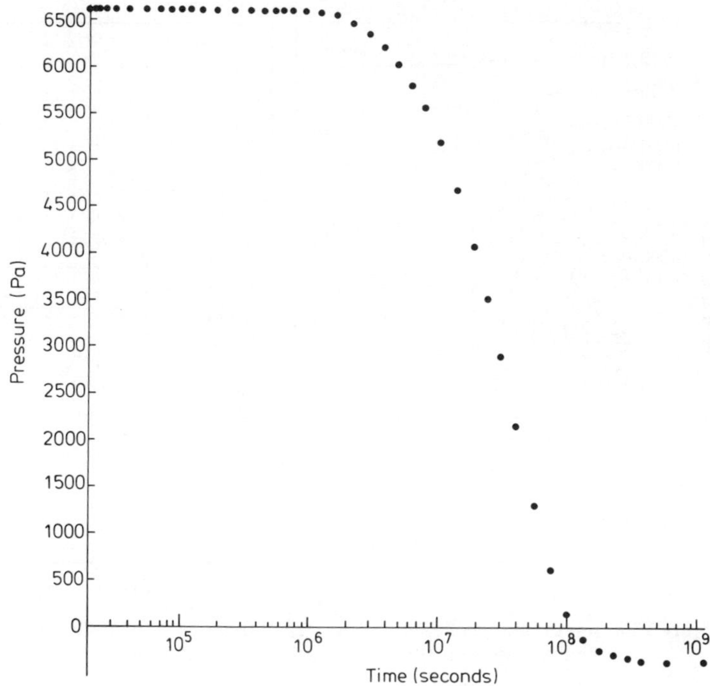

Figure 13.7 Pressure evolution at the point $r=0.0$, $z=0.25$ vertically below the heater and at the successive time steps 2–45

representing a balance between the downward flow component induced by pore-water expansion within the heat front and the general upflow caused by the no-flow boundaries on the bottom and side of the tank. Thus the arrival of the heat front at the tank bottom at approximately 2×10^6 s, as shown in Figure 13.6, corresponds to the arrival of the pressure maximum at the tank bottom, as shown in Figure 13.10(a).

Figures 13.11(a) and 13.11(b) show the pore-water pressure and velocity fields at $t=10^7$ s (115.7 days), by which time the boundaries are influencing the temperature field and in particular heat is beginning to be lost from the top surface of the tank. Throughout the pressure maps shown in Figures 13.12(a), 13.12(b) and 13.13(a) at $t=4\times10^7$ s, 10^8 s and 1.34×10^8 s respectively, this heat loss process continues. The pressure field steadily decreases in magnitude, as does the pore-water velocity, throughout this time period. Note that the pore-water velocity is directed towards the top surface of the tank. The influence of the equilibrating temperature field on the pore-water pressure can be seen to generate a pressure maximum that passes up the tank as time progresses. The progress of this pressure peak is clearly shown in the successive-time vertical pressure profiles of Figure 13.14. The pressure at the

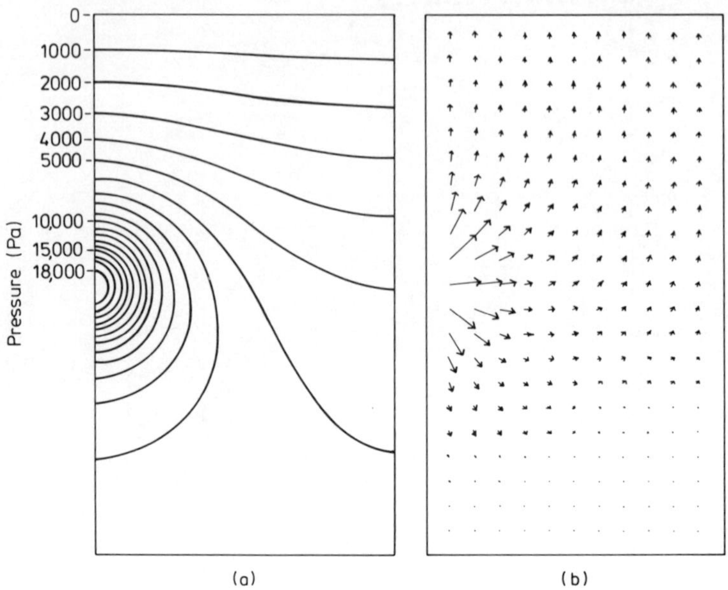

Figure 13.8 (a) Pressure contours at time step 10($t = 8.75 \times 10^4$ s); (b) velocity field at time step 10($t = 8.75 \times 10^4$ s); $\Delta t = 1.0 \times 10^8$ s

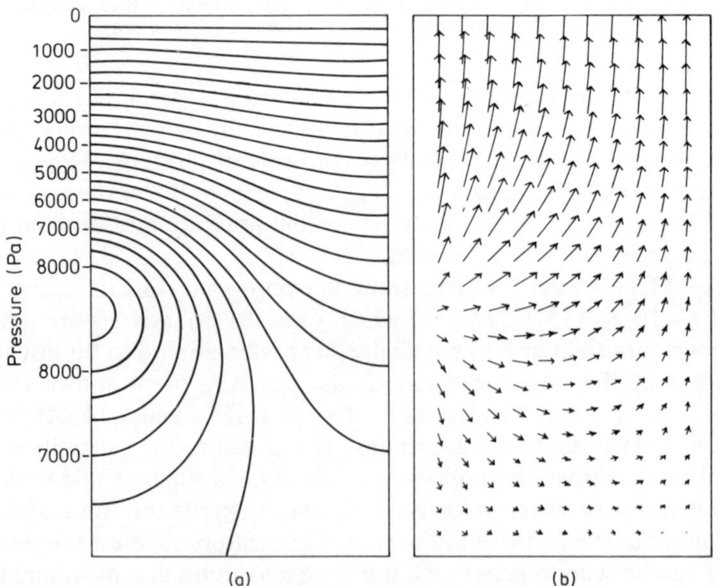

Figure 13.9 (a) Pressure contours at time step 20 ($t = 7.64 \times 10^5$ s); (b) velocity field at time step 20 ($t = 7.64 \times 10^5$ s); $\Delta t = 6.0 \times 10^8$ s

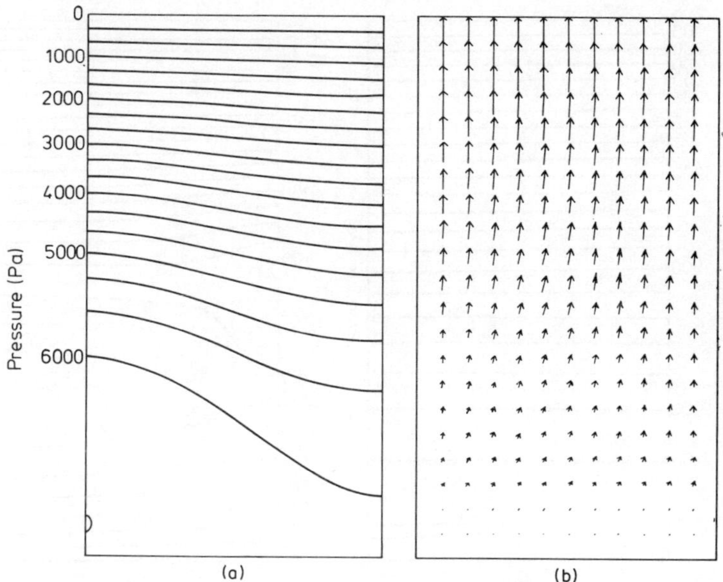

Figure 13.10 (a) Pressure contours at time step 25 ($t = 3.0 \times 10^6$ s); (b) velocity field at time step 25 ($t = 3.0 \times 10^6$ s); $\Delta t = 6.0 \times 10^8$ s

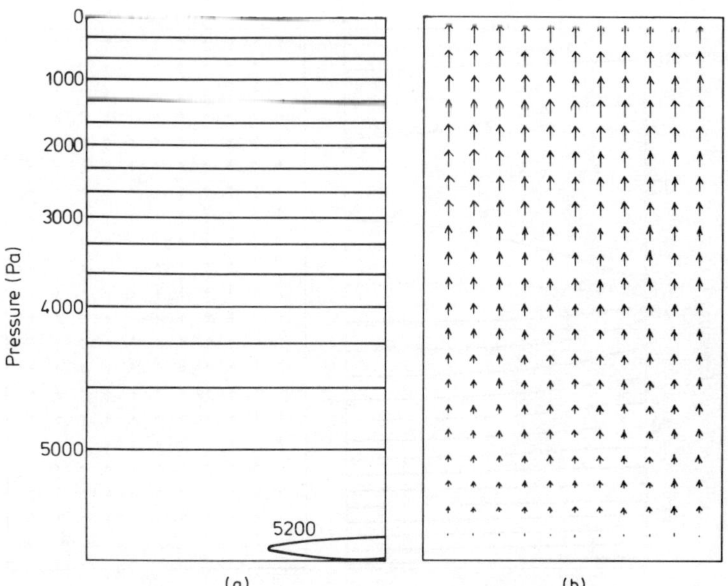

Figure 13.11 (a) Pressure contours at time step 30 ($t = 1.0 \times 10^7$ s); (b) velocity field at time step 30 ($t = 1.0 \times 10^7$ s); $\Delta t = 6.0 \times 10^8$ s

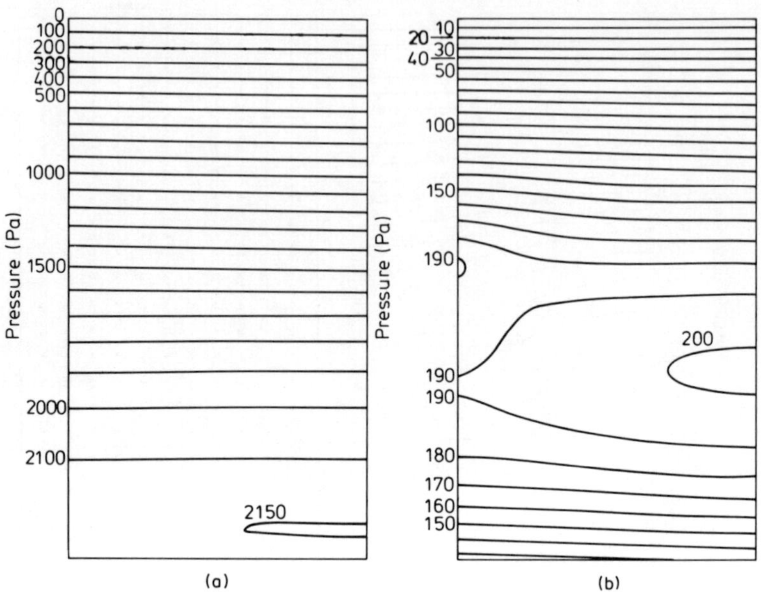

Figure 13.12 Pressure contours (a) at time step 35 ($t = 4.0 \times 10^7$ s); (b) at time step 38 ($t = 1.0 \times 10^8$ s)

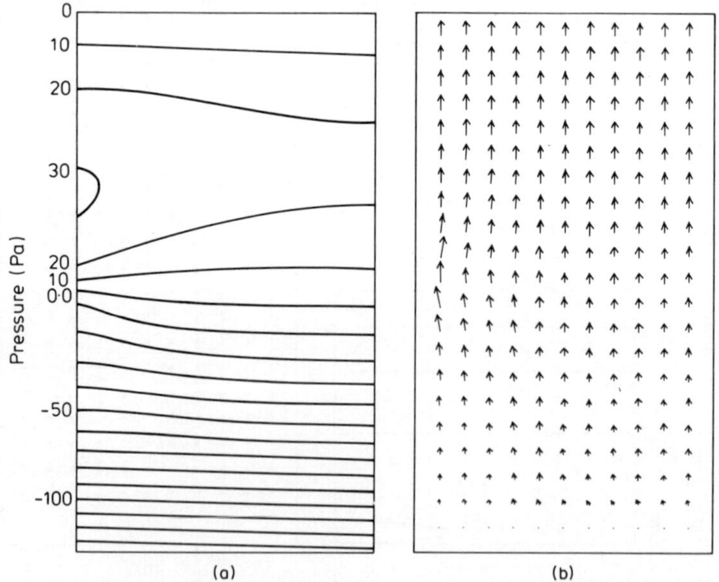

Figure 13.13 (a) Pressure contours at time step 39 ($t = 1.34 \times 10^8$ s); (b) velocity field at time step 39 ($t = 1.34 \times 10^8$ s); $\Delta t = 120.0 \times 10^8$ s

bottom of the tank can be seen to decrease with time till eventually the direction of the pressure gradient has completely reversed throughout the section.

At the time of $t = 1.34 \times 10^8$ s shown in Figures 13.13(a) and 13.13(b), the pore-water velocities have decreased sufficiently for the effects of buoyancy forces on the pressure and velocity fields to be clearly visible in the immediate vicinity of the heat source (see the $\partial T/\partial z$ term in Equation (13. 20)).

Figures 13.15(a) and 13.15(b) show the pressure and velocity fields that result at large times (here $t = 11.7 \times 10^8$ s) when a steady-state temperature profile has been attained (corresponding to the regime (ii) previously mentioned) and only buoyancy forces act to drive the pore-water flow. The convection cell that results is clearly visible in Figure 13.15(b). Note, however, that the pore-water velocities are at least a factor of two down on those present at $t = 1.34 \times 10^8$ s (see Figure 13.13(b)) and an order of magnitude less than the velocities at earlier times (see Figures 13.8–10 and 13.11(b)). Figures 13.16(a) and 13.16(b) show the steady-state temperature distribution and streamlines that result at large times. The centre of the heater reaches 96 °C with the tank base at 20 °C. The top is fixed at 10 °C. Note that it is

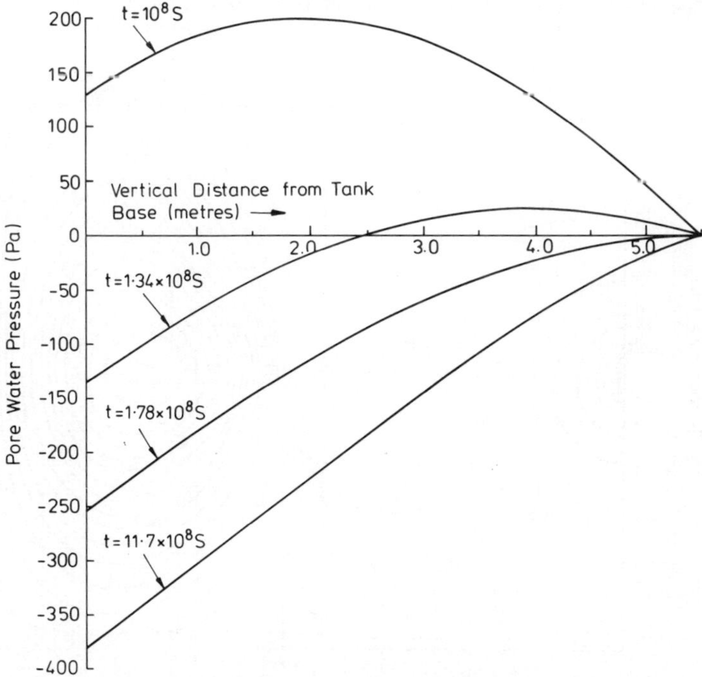

Figure 13.14 Vertical pressure profiles along the tank axis ($r = 0.0$) at the successive time steps 38, 39, 40 and 45

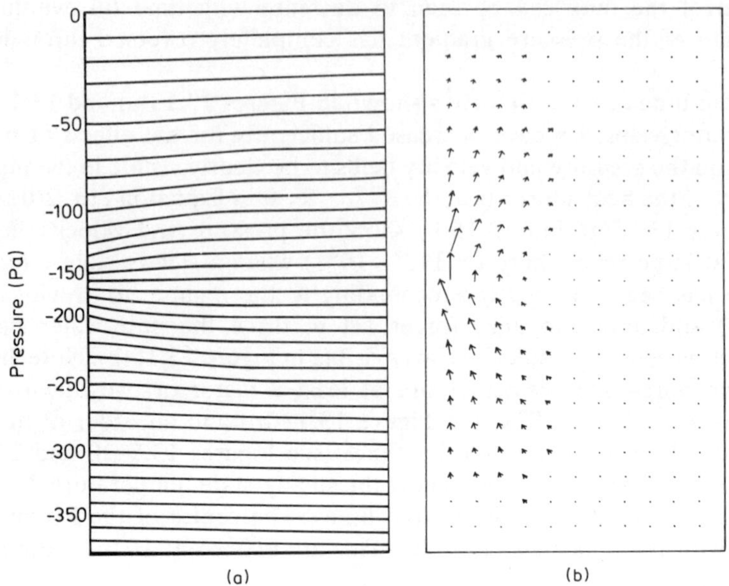

Figure 13.15 (a) Pressure contours at time step 45 ($t = 11.7 \times 10^8$ s); (b) velocity field at time step 45 ($t = 11.7 \times 10^8$ s); $\Delta t = 300 \times 10^8$ s

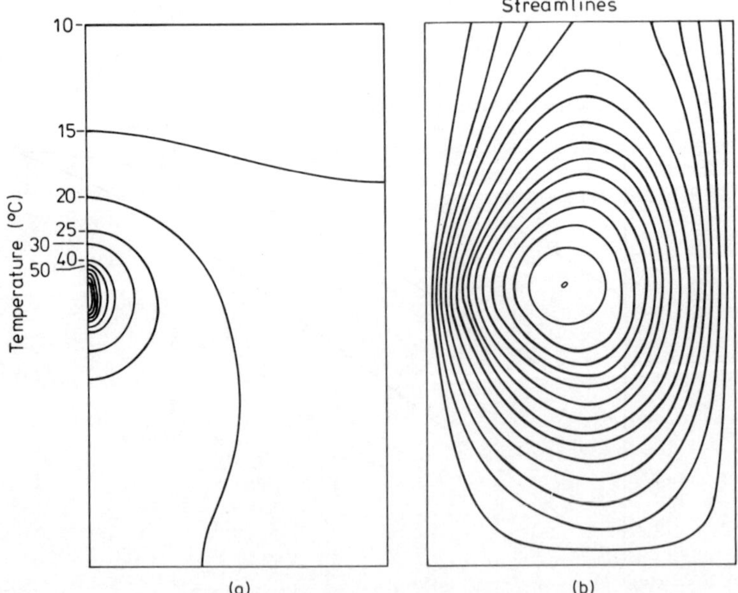

Figure 13.16 Temperature contours at time step 45 ($t = 11.7 \times 10^8$ s); (b) streamlines at time step 45 ($t = 11.7 \times 10^8$ s)

only in the steady state that streamlines can be drawn since only then is the divergence of the velocity field zero.

Figure 13.17 shows the velocity normal to the annulus running vertically up the cylinder at $r = 1.5$ m. This shows the convection cell very clearly, illustrating the reversal of the flow direction. The inset to Figure 13.17 shows the variation of fluid velocity tangential to the same annulus. This is discontinuous because the vertical velocity field is made up of two components, one quadratic and one linear. The density variation term is proportional to temperature and is therefore modelled quadratically, but the pressure gradient term is only linear, being the derivative of a quadratic function. These two terms almost cancel and so leave a somewhat discontinuous result (see Section 13.3).

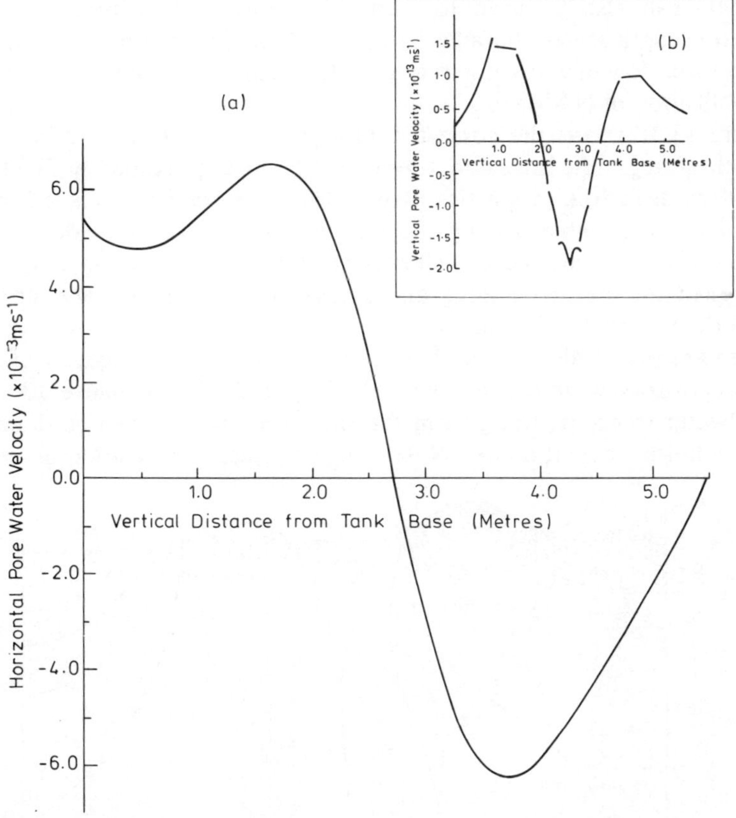

Figure 13.17 The pore-water velocity normal to the annulus at $r = 1.5$ m and at time step 45 ($t = 11.7 \times 10^8$ s). The inset shows the pore water velocity tangential to the same annulus and at the same time; note its discontinuous nature

This simple example has served to bring out many of the aspects of finite element modelling of heat transfer and fluid flow in porous media. We have seen that with a fairly simple grid the time evolution of a system can be successfully modelled and that by using Gear's method for time stepping both the initial short time-scale phenomena and the longer time-scale events can be followed without using a large number of time steps. It is also interesting that, using totally plausible parameters, the dominant flow initially is physical pore-water expansion.

13.6 THE RESPONSE OF GRANITE PORE WATER TO A HEAT SOURCE

The application of the program NAMMU to coupled heat and fluid flow in a high-porosity, small-scale experiment was discussed in Section 13.5. NAMMU can also be used to study groundwater flow problems on the kilometre length scales characteristic of geologic formations. An example of groundwater flow in a low-porosity granite intrusion is given here to illustrate this application of NAMMU.

Figure 13.18 shows the nonhydrostatic pressure field in a section through a hypothetical granite intrusion, the section being approximately 7100 m long by 3000 m deep (the properties assumed for the granite are listed in Table 13.2). The water table driving the flow is taken to be coincident with the granite surface and minimal crossflow is assumed. No flow boundary conditions are applied on the lower edge and sides. The section is made up of 160 nine-node quadrilateral elements.

There are proposals in a number of countries to place a high-level nuclear waste repository within a section such as that shown in Figure 13.18. The groundwater transport times from the site to the surface can be determined from the finite element model. Note that radionuclide migration times would

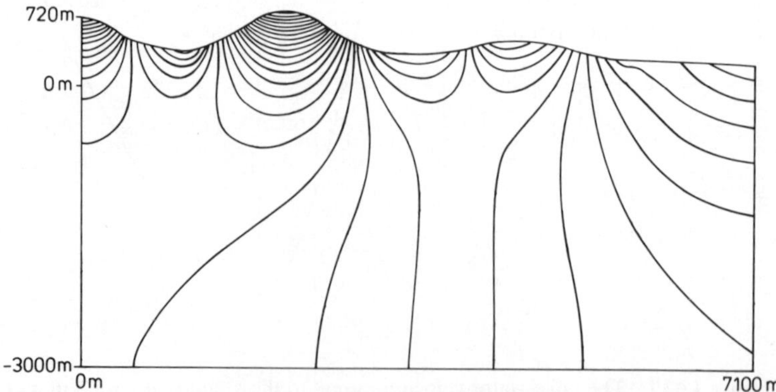

Figure 13.18 Pressure contours in a section through a hypothetical granite intrusion at pressure intervals of $\Delta p = 1.84 \times 10^5$ Pa

Table 13.2 Physical properties of granite

Physical quantity	Symbol	Value
Thermal conductivity of granite	Γ_s	$2.51\ \mathrm{W\,m^{-1}\,°C^{-1}}$
Specific heat of granite	C_s	$8.79\times10^2\ \mathrm{J\,kg^{-1}\,°C^{-1}}$
Density of granite	ρ_s	$2.60\times10^3\ \mathrm{kg\,m^{-3}}$

exceed the times obtained for water because of the neglect of other effects such as radionuclide adsorption by minerals in the granite and diffusion and dispersion effects.[25] Figure 13.19(a) shows an enlargement of the pore water velocity field around one possible site in the section of Figure 13.18. The local finite element grid is superposed on the velocity field, the arrow lengths being the pore-water velocity at the foot of an arrow multiplied by $\Delta t = 4\times10^8$ s. The repository is assumed to be situated in the section covered by the central element (250 m × 200 m) with a depth of 500 m into the plane of the figure and an initial heat output of 3.4 MW. This heat output decays with the characteristics of 70 year-old reprocessed high-level waste.[26] Figure 13.19(b) shows the effect of such a repository heat output on the pore-water velocity. Comparing Figures 13.19(a) and 13.19(b) it can be clearly seen that buoyancy forces increase the upflow of water through the repository. For the lower regional groundwater flows generated by a less mountainous section (as might be chosen for a real repository site) the effect of buoyancy-induced flow will be proportionately larger. Note that for low-porosity granite the effect of direct expansion of pore water on the flow field is negligible.

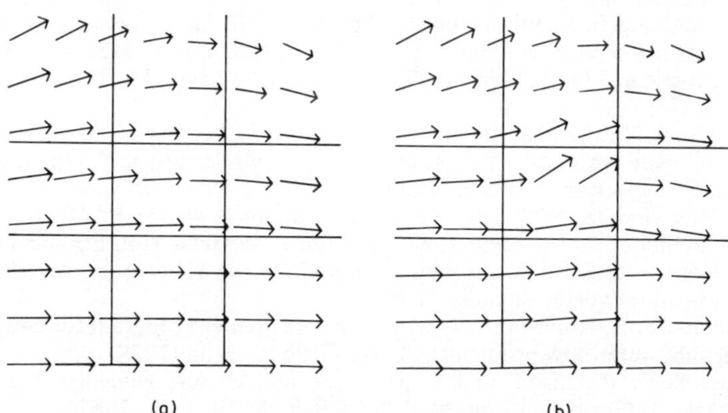

Figure 13.19 Pore-water velocity fields around a possible repository site centred on $x = 4125.0$ m, $y = -1000.0$ m in the section of Figure 13.18. The local finite element grid is shown, the central 250 × 200 m element corresponding to a repository with an initial heat output of 3.4 MW. (a) is at $t = 0.0$ and (b) at $t = 2.8\times10^9$ s. $\Delta t = 4\times10^8$ s. The effect of the repository heat output on the flow can clearly be seen in (b)

ACKNOWLEDGEMENTS

This work has been commissioned by the Department of the Environment, as part of its radioactive waste management research programme. The results will be used in the formulation of Government policy, but at this stage they do not necessarily represent Government policy.

The work was performed under contract with the European Atomic Energy Community in the framework of its R & D programme on Management and Storage of Radioactive Waste.

REFERENCES

1. P. J. Bourke and P. C. Robinson, 'Comparison of thermally induced and naturally occurring water-borne leakages from hard rock depositories for radioactive waste', *Radioactive Waste Management*, **1**, 365–380 (1981).
2. K. R. Hinga, G. R. Heath, D. R. Anderson and C. D. Hillister, 'Disposal of high-level radioactive wastes by burial in the sea floor', *Environ. Sci. Technol.*, **16**, 28A–37A (1982).
3. C. F. Tsang and J. Claeson, 'Energy storage in aquifers', in *Subsurface Space, Proc. of the Int. Symp. Rockstore '80*, **2**, 771–778 (1980).
4. R. G. Cummings, G. E. Morris, J. W. Tester and R. L. Bevins, 'Mining earth's heat: hot dry rock geothermal energy', *Technology Review*, February, 58–78, Pergamon Press (1979).
5. Society of Petroleum Engineers 'Thermal recovery techniques', *SPE Reprint Series*, **10** (1972).
6. H. B. Crichlow, *Modern Reservoir Engineering, a Simulation Approach*, Prentice-Hall, Englewood Cliffs, N.J. (1977).
7. O. C. Zienkiewicz and Y. K. Cheung, '*Finite elements in the solution of field problems, The Engineer*', **220**, 507. (1965).
8. I. Javandal and P. A. Witherspoon, 'Application of the finite element method to transient flow in porous media, *Soc. Pet. Eng. AIME Trans.*, **243**, 241–252 (1968).
9. D. H. Norrie and G. de Vries, *A Finite Element Bibliography*, Plenum Press, New York (1976).
10. C. S. Desai, 'Finite element methods for flow in porous media', Chapter 8 of *Finite Elements in Fluids*, Vol. 1, (ed. by R. H. Gallagher, J. T. Oden, C. Taylor and O. C. Zienkiewicz, Wiley, London (1975).
11. O. C. Zienkiewicz, 'Why finite elements?' Chapter 1 of Desai[10](1975).
12. O. C. Zienkiewicz, *The Finite Element Method*, McGraw-Hill, London (1977).
13. J. J. Connor and C. A. Brebbia, *Finite Element Techniques for Fluid Flow*, Newnes–Butterworth, London (1976).
14. J. Rae and P. C. Robinson, 'NAMMU: Finite element program for coupled heat and groundwater flow problems', *AERE-R*9610, 1–38 (1979).
15. J. Rae, P. C. Robinson, and L. M. Wickens 'A users guide for the program NAMMU. 1. General information', *AERE-R*10120, 1–89 (1981).
16. L. M. Wickens 'A user's guide for the program NAMMU. 2. An example problem', *AERE-R*10274, 1–26 (1981).
17. T. Jacobson, *Before Philosophy*, Penguin Books (1949).
18. J. Bear, *Dynamics of Fluids in Porous Media*, Elsevier, New York (1972).
19. R. H. Gallagher, J. T. Oden, C. Taylor and O. C. Zienkiewicz, (eds.) *Finite Elements in Fluids*, Vols. 1–3, Wiley, London (1975).

20. B. M. Irons 'A frontal solution program for finite element analysis', *Int. J. Num. Math. Eng.* **2**, 5(1970).
21. G. Hall and J. M. Watt, *Modern Numerical Methods for Ordinary Differential Equations*, Clarendon Press (1976).
22. R. A. Willoughby, *Stiff Differential Systems*, Plenum Press (1934).
23. G. D. Byrne and A. C. Hindmarsh, 'A polyalgorithm for the numerical solution of ordinary differential equations', *A. C. M. Trans. Math. Software*, **1**, 71(1975).
24. C. W. Gear, 'Ordinary differential equation techniques for partial differential equations', in *Formulations and Computational Algorithms in Finite Element Analysis*, (ed. K. J. Bathe, J. T. Oden and W. Wunderlich) MIT, Cambridge (1977).
25. M. D. Hill and P. D. Grimwood, 'Preliminary assessment of the radiological protection aspects of disposal of high-level waste in geological formations', *NRPB-R*69, 1–75 (1978).
26. D. P. Hodgkinson, 'A mathematical model for hydrothermal convection around a radioactive waste repository in hard rock', *Annals of Nuclear Energy*, **7**, 313–334 (1980).

Numerical Methods in Heat Transfer, Volume II
Edited by R. W. Lewis, K. Morgan, and B. A. Schrefler

Chapter 14

Environmental Effects of Fluid Injection into Geothermal Reservoirs

M. Borsetto, G. Carradori, and A. Peano

SUMMARY

Subsidence and microseismicity effects due to geothermal energy exploitation are investigated by numerical simulations. Attention is focused on aspects determined by injection of cool fluid into a hot formation. The above phenomena, which are common also in oil and gas reservoir engineering and in man-made lakes, are usually interpreted on the basis of pressure variations of the reservoir fluid. On the other hand, this chapter indicates that in geothermics temperature effects are at least as important as pressure effects.

Ground subsidence is investigated, taking into account various cultivation schemes of an idealized reservoir.

Moreover, the probability of microseismic activity triggered by the *in situ* stress perturbation in the vicinity of an injection well is discussed for elastic and elastoplastic behaviour of a naturally fractured layer.

14.1 INTRODUCTION

Geothermal energy is considered 'cleaner' and 'safer' than fossil and nuclear fuel energies, respectively; nevertheless some environmental problems may arise in connection with intensive industrial exploitation. The environmental impact of geothermal energy exploitation includes various effects, which may be subdivided into three groups: thermal effects, chemical effects, rock mechanics effects.

Thermal effects are connected with the large discharge of vapour in the atmosphere. The energy dispersed by cooling towers is estimated eight times the electric energy produced and, for the same electric output, is about five times the heat discharge of usual power stations.[1] This effect may modify the local climate and favour the formation of fog. Moreover, when hot fluids are discharged in rivers, all the biological life may be heavily affected even for a few degrees of temperature increase.

Chemical effects are represented by the release of many components from the produced geothermal fluid; for instance, noncondensable gases, such as carbon dioxide and hydrogen sulphide, and water-soluble components, such as chlorides, silica, sulphates, calcite or traces of toxic ingredients. Many consequences may be brought about by these substances; for instance, deposits and corrosion of power-plant machineries, and, above all, air and water pollution, with negative effects on the ecological system.[2]

Possible rock mechanics effects are ground subsidence and induced seismicity. Interest on these consequences has increased in recent years in connection with the practice of reinjecting spent fluids into the reservoir formation.

Reinjection is motivated by environmental and economical considerations. In fact, both thermal and chemical pollution are considerably contained by this practice. Moreover, reinjection is expected to prolong or improve energy recovery through additional fluid supply and reservoir pressure sustainment. On the other hand, rock mechanics effects of fluid injection are still under investigation, but there is reason to believe that this practice may contribute both to ground subsidence and to induced seismicity.

Horizontal strains and tilts, due to relative or differential displacements determined by subsidence, may cause serious problems to surface structures and to power plants, or even to wells. On a larger scale, subsidence may reverse natural drainage directions and activate fault movements.[3,4]

Early theoretical work on ground subsidence computations is based on the soil consolidation theory of Terzaghi,[5] which provides the fundamentals of the poroelasticity theory of rocks.[6] Practical computations are often based on the nucleus of strain method developed by Geertsma.[7,8] Recently, more emphasis was given to refined finite element simulations, which model the coupling between fluid transfer and mechanical behaviour.[9,10]

Seismic activity induced by exploitation and injection covers a broad range of energy releases. Nanoseismic activity is usually recorded in connection with intact rock failure produced by hydraulic fracturing operations. An increase in natural microseismic activity is often recorded in many geothermal fields. Also some seismic events with magnitudes ranging from 3 to 5 of the Richter scale were recorded as a consequence of massive injection programmes.[11,12]

Early work on induced seismicity is based on Terzaghi's theory of effective stress,[5] that was applied to the fault behaviour by Hubbert and Rubey.[13] Experiments performed in the oil field of Rangely, Colorado,[14] confirmed the assumptions of the above theory and clearly indicated that seismicity could be induced by fluid pressure variations. The same explanation was suggested in case of seismicity induced by the filling of a man-made lake.[15]

This chapter is focused on the study of ground displacements and fault slips caused by industrial cultivation of geothermal reservoirs. Particular attention is paid to the importance of temperature effects in comparison with pressure effects. In fact, as subsidence and induced seismicity studies first started in engineering fields different from geothermics, thermal effects were disregarded and the observed phenomena were explained on the basis of pressure influence only. In geothermics, however, the presence of injected fluids with a temperature of 30–60 °C in hot formations at 200–300 °C causes important additional effects that may sometimes be comparable and often more important than pressure variation effects.

Ground subsidence of some idealized reservoir structures is examined, and the role of reinjection practice is pointed out. A novel numerical approach is developed by extending the methods proposed by Geertsma in oil engineering. Then, microseismic effects are analysed in connection with induced stress variations and rock-mass strength properties. At first, an elastic rock behaviour is assumed and the size of the out-of-safety region surrounding the injection well is evaluated in order to forecast possible dislocations of faults. Finally, an elastoplastic behaviour is considered in order to investigate the progressive failure phenomena of a naturally fractured rockmass.

14.2 APPROXIMATE TEMPERATURE AND PRESSURE DISTRIBUTIONS DUE TO AN INJECTION WELL

Temperature and pressure fields determined by the injection of cool water into a geothermal reservoir are here discussed. Approximate analytical solutions are presented for some simplified geometries; these solutions are used throughout the following sections. As the results depend considerably on the reservoir structure, some hypotheses are introduced for simplification.

The reservoir is considered to be formed by a porous medium, with local thermal equilibrium between rock and fluid. In case of highly fractured formations, the above hypothesis is questionable: the rapid flow in the fractures prevents uniform cooling of the rock blocks.[16] This fact, added to the anisotropy of permeability determined by fractures, may lead to early and unexpected delivery of cool water from production wells.

As a further simplification, gravity effects will be disregarded. This implies that the following results are valid only for a horizontal single-layer reservoir in which fluid is injected uniformly over the formation thickness. In case of a vapour-dominated reservoir, gravity segregation leads to saturation and temperature distributions as the ones represented in Figure 14.1: the effect is especially significant for a formation partially penetrated by injection wells.

Some further effects, such as the dependence of permeability on pressure and temperature, or the hydrodynamic dispersion, are disregarded. Also injection into dry steam reservoirs is not considered. Readers interested in this problem are referred to other works.[16–18]

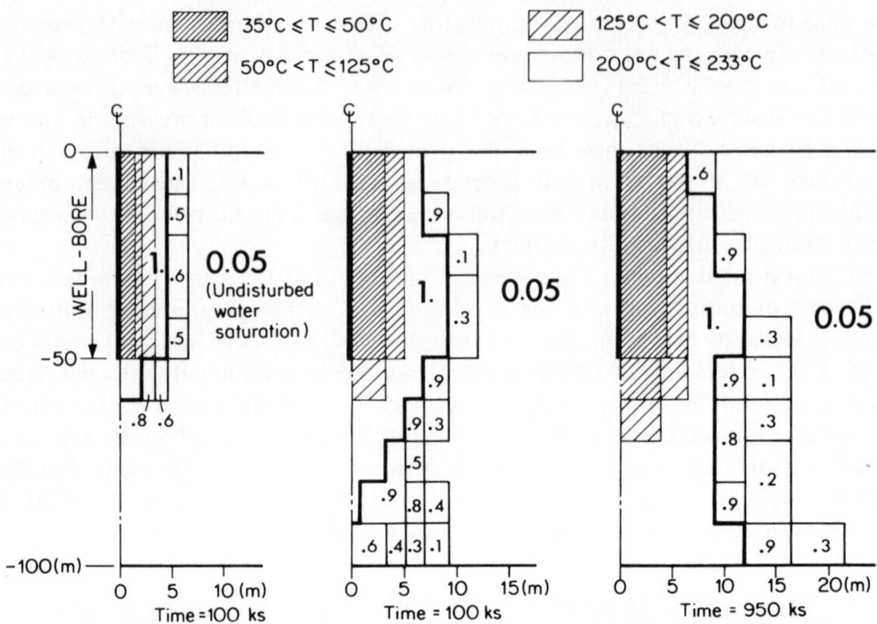

Figure 14.1 Water saturations and temperatures produced by a well half penetrating the formation F_c of Figure 14.5 filled by a two-phase water–steam mixture. Undisturbed values: pressure $p = 3$ MPa; temperature $T_2 = 233\,°C$; water saturation $S_w = 5\%$; residual water saturation in Corey's law $S_{wr} = 30\%$; intrinsic permeability $k_p = 2.5 \times 10^{-13}\,m^2$; porosity $n = 15\%$; rock thermal capacity $\rho^* c^* = 2\,MJ\,°C^{-1}\,m^{-3}$; rock thermal conductivity $\lambda^* = 1\,W\,m^{-1}\,°C^{-1}$; rock-mass bulk compressibility $C = 0$; injected water flow $q = 8.35\,kg\,s^{-1}$; temperature of the injected water $T_1 = 35\,°C$. Results of Figures 14.1–14.3 have been evaluated with the computer code GEOTHER[20]

14.2.1 Injection into a hot-water reservoir

Cooling of the rock mass is governed by the heat-transport phenomenon. Since thermal conductivity is negligible, the propagation of a step cool front must be modelled.[16] A simple representation of the radial temperature field $T(r)$ is given by the following equation:

$$\begin{aligned} T(r) &= T_1, \qquad \text{for } a \leq r \leq R_c, \\ T(r) &= T_2, \qquad \text{for } R_c < r, \end{aligned} \tag{14.1}$$

where T_1 and T_2 are the temperatures of the injected fluid and of the undisturbed formation, respectively; a is the well-bore radius and R_c is the radial coordinate of the cooled region. R_c is determined by the following energy balance:

$$(qt - \pi R_c^2 F_c n \rho_w) c_w\, \Delta T = \pi R_c^2 F_c (1-n) \rho^* c^*\, \Delta T, \tag{14.2}$$

where $\Delta T = T_2 - T_1$, q is the injected mass-flow rate, t is time, F_c is reservoir thickness, n is porosity, ρ is density, c is specific heat. Subscript w refers to water, and the asterisk to rock matrix.

The pressure profile may be determined by assuming quasi-static seepage (steady-state conditions for every instant and for every radius R_c of the thermal front). Taking into account a large viscosity decrease due to the temperature step,[19] the integration of Darcy's law for constant volumetric injection rate Q and constant pressure at the boundary $r = r_e$, yields

$$\left.\begin{aligned} &\Delta p = \Delta p_a\left(1 + M \ln \frac{r}{a}\right) && \text{for } a \leqslant r \leqslant R_c, \\ &\Delta p = \Delta p_a M \frac{\mu_2}{\mu_1} \ln \frac{r}{r_e} && \text{for } R_c < r \leqslant r_e \\ &\Delta p_a = \frac{Q\mu_1}{2\pi F_c k_p M}; && M = 1 \bigg/ \left(\frac{\mu_2}{\mu_1} \ln \frac{R_c}{r_e} - \ln \frac{R_c}{a}\right) \end{aligned}\right\} \qquad (14.3)$$

where Δp is pressure increment, μ_1, μ_2 are viscosities in the cooled and in the undisturbed regions, respectively, and k_p is the intrinsic permeability of the formation.

The results obtained from the above simplified analytical equations are compared in Figure 14.2 with those obtained by means of a finite difference simulator.[20]

14.2.2 Injection into a two-phase reservoir

The location of the temperature front is again determined by Equation (14.2). The cooled region is occupied by water with density ρ_w corresponding to temperature T_1. A circular annulus with water at the reservoir temperature T_2 and density ρ'_w surrounds the cooled region and determines the hydrodynamic front. The latter is, in many cases, characterized by a rapid decrease in fluid density to the low undisturbed values. An estimation of the radius R_h of the moving hydrodynamic front is given by[16]

$$R_h = \sqrt{\left[\frac{(qt - n(\rho_w - \rho'_w)\pi R_c^2 F_c}{n S_v \rho'_w \pi F_c}\right]}, \qquad (14.4)$$

where S_v is the reservoir vapour saturation.

An approximation of pressure behaviour for constant injection rate and constant pressure at $r = r_e$ is again based on Equations (14.3). The boundary distance r_e is now replaced by the hydrodynamic radius R_h, however. This simplified model with an undisturbed pressure field for $R_h \leqslant r \leqslant r_e$ is justified by the very high compressibility of vapour, compared with that of liquid water. Two typical results for injection in a two-phase reservoir are given in Figure 14.3.

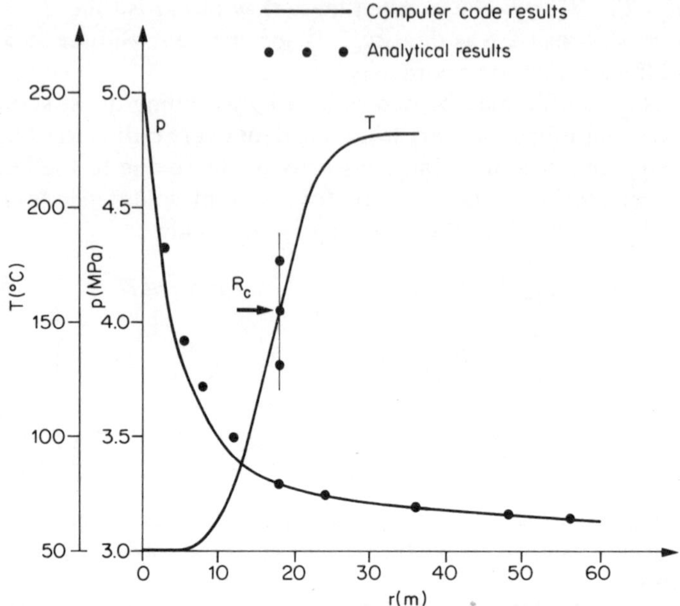

Figure 14.2 Pressures and temperatures produced after 21 days by a well totally penetrating the formation F_c of Figure 14.5 filled with hot water. Data of the problem: $p = 3$ MPa, $T_2 = 233$ °C, $k_p = 5 \times 10^{-14}$ m^2, $n = 15\%$, $\rho^* c^* = 2.1$ MJ °C^{-1} m^{-3}, $\lambda^* = 3$ W m^{-1} °C^{-1}, $C = 2.4 \times 10^{-4}$ MPa^{-1}, $q = 29.6$ kg s^{-1}, $T_1 = 50$ °C

The above examples indicate that the cooled region is the same for both hot-water and two-phase reservoirs. However, the hot-water situation appears more severe, owing to pressure behaviour; therefore, only this case is considered later on.

14.3 GROUND SUBSIDENCE MODELLING

A complete numerical model of phenomena producing subsidence has to take into account the coupling of pressure and thermal field with the state of stress.[21] However, for practical purposes and for complex exploitation schemes, it is reasonable to compute subsidence on the basis of a known distribution of pressure and temperature variations in the reservoir.[22] This distribution may be provided by the interpretation of field measurements or by numerical simulations based on more or less sophisticated mathematical models.

Due to the large computational effort required and to the lack of a detailed description of the system, a completely general deformation analysis (e.g. by means of a three-dimensional program) has never been attempted, and a

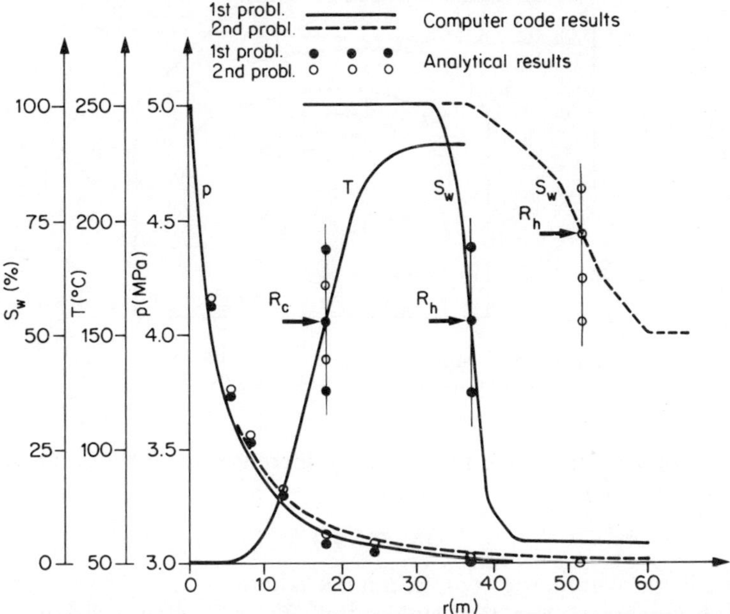

Figure 14.3 Pressures, temperatures and water saturations produced after 21 days by a well totally penetrating the formation F_c of Figure 14.5 filled by a two-phase water-steam mixture. Two families of results are presented, in connection with two different initial undisturbed water saturation values. Data of the problems: $p = 3$ MPa, $T_2 = 233$ °C, S_w (first problem) = 5%, S_w (second problem) = 50%, $S_{wr} = 30\%$, $k_p = 5 \times 10^{-14}$ m^2, $n = 15\%$, $\rho^* c^* = 2.1$ MJ °C^{-1} m^{-3}, $C' = 2.4 \times 10^{-4}$ MPa^{-1}, $q = 29.6$ kg s^{-1}, $T_1 = 50$ °C

number of very simplified analysis procedures are currently used. One of the most efficient is the nucleus of strain method developed by Geerstma and Van Opstal for homogeneous elastic subsoils.[7,8] An extension to horizontally layered reservoirs was presented by Williamson.[23] His analytical method, however, evaluates only the total volume of subsidence and does not provide the actual surface deformations.

In this chapter, results are evaluated using a new variant of the nucleus of strain method, that makes possible the study of systems with any number of horizontal layers, each one formed by a different homogeneous material.

14.3.1 The nucleus of strain method

Subsidence is a consequence of deformations in the subsoil (Figure 14.4). The volumetric strain due to a temperature increase ΔT of a unit volume can be written through the volumetric thermal expansion coefficient α_v, as

$$\Delta V_T = \alpha_v \Delta T. \tag{14.5}$$

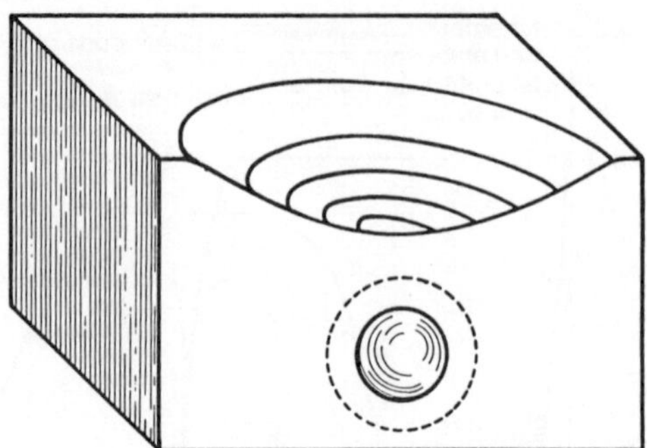

Figure 14.4 Surface subsidence due to the shrinkage of a spherical volume in the subsoil

The volumetric strain due to a pressure increase Δp of the saturating fluid may be expressed as

$$\Delta V_{\mathrm{p}} = C_{\mathrm{v}}\, \Delta P, \tag{14.6}$$

where C_{v} is the volumetric compaction coefficient.

There is an obvious analogy between the effect of pressure and temperature variations. Therefore, for subsidence evaluation, pressure variations can be replaced by equivalent temperature variations (assuming $\Delta T = C_{\mathrm{v}} \Delta p/\alpha_{\mathrm{v}}$): for this reason pressures are disregarded in the subsequent notation.

Assuming a linear elastic behaviour of the soil, a displacement component D at a surface point I may be computed, using the superposition principle, as

$$D(I) = \int_V \delta(I, U)\alpha_{\mathrm{v}}(U)\, \Delta T(U)\, \mathrm{d}V, \tag{14.7}$$

where $\delta(I, U)$ is the displacement of I caused by a unit volumetric strain at the reservoir point U. Computational advantages accrue if δ does not depend on the two particular points, but only on some relationship between their coordinates, as

$$\delta(I, U) = \delta(u, h) = \delta_{\mathrm{h}}(u), \tag{14.8}$$

where h is the depth of U and u is the distance of U from the vertical axis passing through I. Such a situation occurs in horizontally layered soils. Analytical expressions of $\delta_{\mathrm{h}}(u)$ are available for homogeneous soils.[8]

The fundamental solution $\delta_{\mathrm{h}}(u)$ for more complex geological stratigraphies can be computed by a numerical procedure, such as the finite element method. In fact, $\delta_{\mathrm{h}}(u)$ can closely be approximated by solving an axisymmetric problem (as shown in the next paragraph) with an initial unit volumetric strain in a finite volume V_n (nucleus of strain). Of course, the maximum diameter of the volume V_n must be small with respect to the average depth H. The same displacement solution $\delta_{\mathrm{H}}(u)$ is approximately valid for any volume V_j, of similar magnitude and located at the same depth, provided that the fundamental solution is multiplied by the ratio V_j/V_n. The evaluation of the integral in Equation (14.7) is therefore reduced to the summation

$$D(I)=\sum_{j=1}^{J}\delta_{\mathrm{H}}(u_j)\alpha_{\mathrm{v}j}\,\Delta T_j(V_j/V_n), \tag{14.9}$$

where J is the number of subdomains into which the reservoir is divided.

A convenient way of operating is to use the interpolation properties of the finite element method in order to compute the contributions to the right-hand side of Equation (14.9), especially when pressure and temperature fields are previously calculated by the same method. In this case, V_j is the volume corresponding to the integration weight of a quadrature formula. Moreover, displacements $D(I)$ may be evaluated at the nodes of a finite element mesh, in order to utilize the differentiation properties of the interpolating polynomials in the calculation of strains, tilts and curvatures.

14.3.2 Computation of nucleus fundamental solutions

In this section fundamental solutions are computed for the geological formation of Figure 14.5. The fundamental solution, used in the following, is that due to the uniform shrinkage of a discoidal volume located on the axis of symmetry. The formation is modelled with axisymmetric finite elements up to a radial distance of 11 km. Solutions are computed for different model problems in order to point out the influence of the stiffness of the various layers. A summary of the analysed problems is given in Table 14.1.

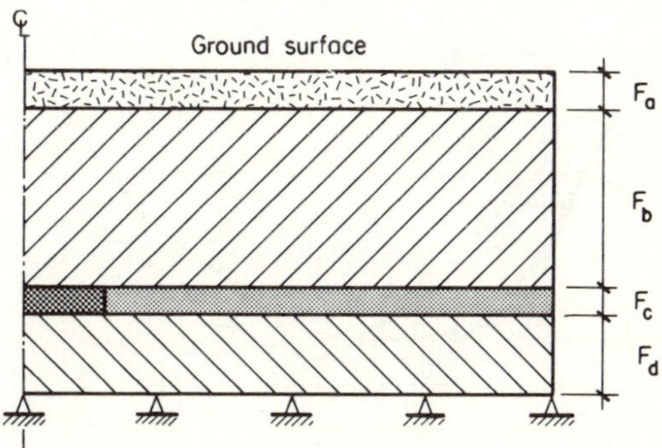

Figure 14.5 Layered subsoil of infinite extent. Formations F_a, F_b, F_d are impervious. In the pervious layer F_c a discoidal nucleus of strain is shaded; $F_c=100$ m. $F_a=200$ m, $F_b=1800$ m, $F_d=1000$ m for results in Figures 14.6–14.11

Table 14.1 Elastic moduli E of layers F_a, F_b, F_c, F_d for five different model problems

Layer	Problem 1	2	3	4	5
F_a	0.01	0.01	0.01	1.0	0.01
F_b	0.1	0.25	0.5	1.0	0.25
F_c	1.0	1.0	1.0	1.0	1.0
F_d	1.0	1.0	1.0	1.0	10.0

Surface displacements for the first three cases are shown in Figure 14.6(a). The problems correspond to three different stiffness values of the cap formation F_b. The maximum displacements decrease with increasing stiffness of the formation. The curves for Problem 2 are again shown in Figure 14.6(b) to be compared with the results obtained for uniform subsoil and for a stiff formation underlying the reservoir. In this case, the results appear to be more sensitive to the varying stiffness values.

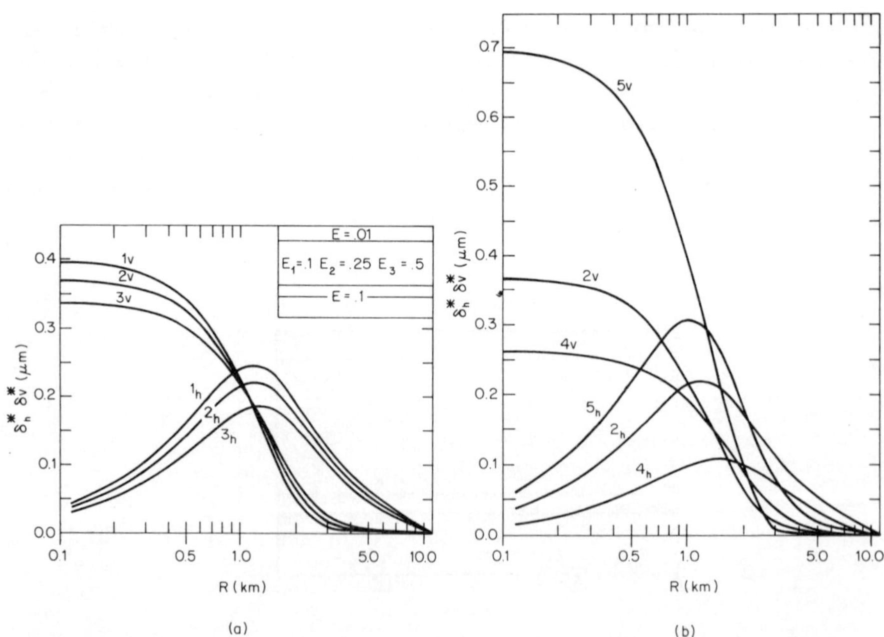

Figure 14.6 Surface displacements for a nucleus of strain with radius = 34 m and uniform volumetric strain = -2.4×10^{-5}. Poisson's ratio is 0.2, δ_h^* = centripetal horizontal displacement, δ_v^* = downward vertical displacement. Numbering of curves refers to the different cases in Table 14.1

14.3.3 Subsidence in simple geothermal reservoirs

In subsidence studies published in technical literature, reservoirs are idealized by very simple shapes, such as discs or parallelepipeds.[8] This simplification is here adopted for studying the effect on subsidence of different geological stratigraphies and reservoir dimensions.

Two disc-shaped geothermal reservoirs are considered for the same geology of Figure 14.5. The first reservoir has a radius $R_1 = 1.2$ km, the second $R_2 = 3.0$ km. Both are subjected to a temperature drop of $\Delta T = -1$ °C. Using the superposition method of equation (14.9), the results of Figure 14.7 are computed. They suggest a different influence of geological conditions for 'small' and 'large' reservoirs (so defined according to the ratio of the radius in relation to the depth). This distinction is proposed by other authors as well.[14] In the first case (small, Figure 14.7(a)) the sequence of maximum vertical displacements is the same as in Figure 14.6(a), while in the second case (large, Figure 14.7(b)) the sequence correspond to the displacements plotted in Figure 14.6(a) at a distance greater than 1 km. This indicates that the contribution of the interior disc of radius equal to 1.2 km (about the distance of the intersection of the curves in Figure 14.6(a)) is overcome by the contribution of the circular annulus with radii $R_1 = 1.2$ km and $R_2 = 3.0$ km. On the contrary, the sequence of maximum horizontal displacements is the same for small and large reservoirs, as no intersection occurs in their corresponding curves of Figure 14.3(a). Similar considerations are valid for results of Figures 14.6(c) and 14.6(d).

14.3.4 Subsidence for some production schemes in geothermal reservoirs

Three different production schemes in hot-water reservoirs are studied for the geological stratigraphy of Figure 14.5 and for the second set of elastic moduli in Table 14.1. The selected schemes are shown in Figure 14.8, where only a quadrant is considered because of symmetry. A total production rate $Q = -63 \times 10^{-2}$ m^3/s, equally supplied by production wells, is assumed. For the schemes of Figures 14.8(b) and 14.8(c), where reinjection wells—as in the secondary oil recovery schemes—are present, the total amount of water produced is injected at the same rate for each well. A constant pressure boundary is located at a distance r_e of about 9 km from the centre.

Pressure variations are evaluated by superimposing the stationary solutions given by Equations (14.3), where, for simplicity, the viscosity of the fluid is assumed constant ($\mu_1 = \mu_2$). Actually, pressure transients are very short, and may be disregarded.

Pressure for the scheme of Figure 14.8(a) is plotted in Figures 14.9(a), 14.9(b) and main subsidence effects at the surface are shown in Figures 14.9(c) and 14.9(d). Surface effects are nearly axisymmetric, owing to the regular

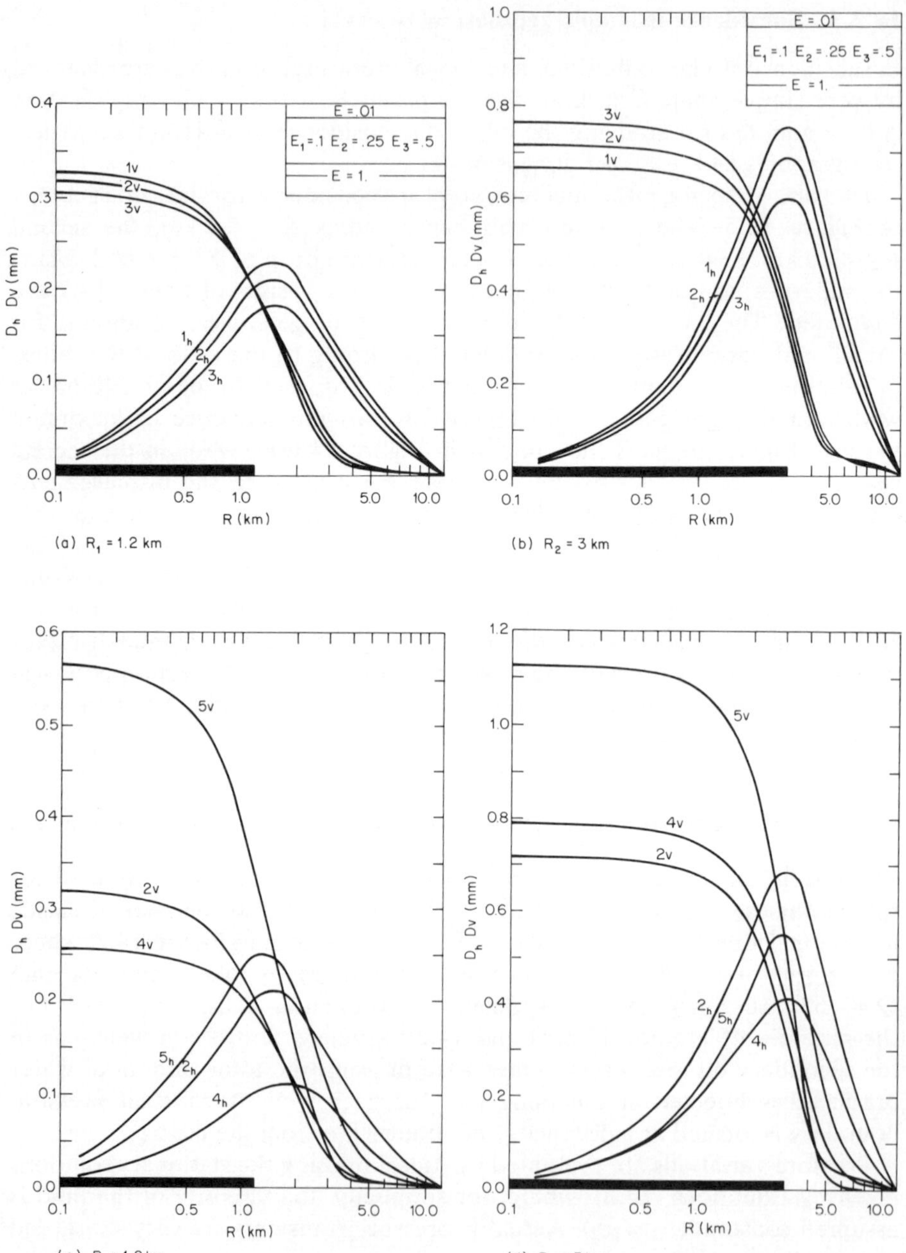

Figure 14.7 Disc-shaped reservoirs: surface displacements due to an uniform volumetric strain = -2.4×10^{-5}. D_h = centripetal horizontal displacement, D_v = downward vertical displacement. Numbering of curves refers to the different cases in Table 14.1

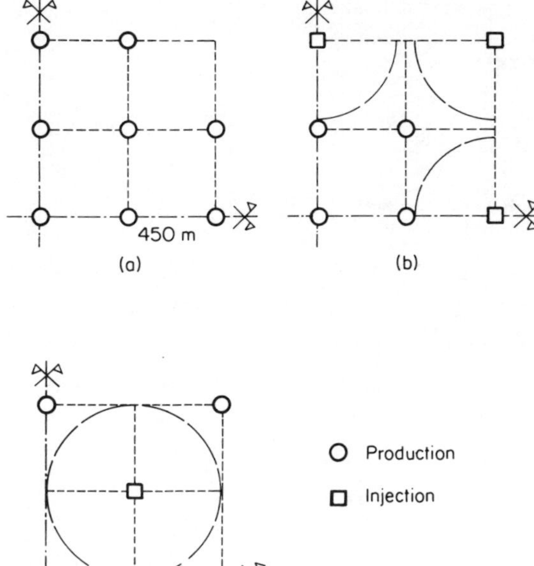

Figure 14.8 Aerial view of three different exploitation schemes. Only a quadrant is represented. Circles around injection wells show the cooled region after 11 years (b) and 7 years (c). Data of the problem: $T_2 = 200\,°C$, $k_p = 5 \times 10^{-14}\,m^2$, $C_v = 1.75 \times 10^{-4}\,MPa^{-1}$, $\alpha_v = 2.4 \times 10^{-5}\,°C^{-1}$, $a = 15$ cm

disposition and the small distance between wells, in comparison with the depth. Stationary pressure distributions for the second and third scheme (Figures 14.8(b) and 14.8(c)) are reported in Figures 14.10(a), 14.10(b), 14.11(a), 14.11(b) and the corresponding displacements are presented in Figures 14.10(c), 14.10(d), 14.11(c), 14.11(d). Pressure variations and consequent subsidence are lessened with respect to the first scheme (Figure 14.8(a)); moreover, some upward displacement arises (Figure 14.11(c)). As expected, reinjection pressures have a positive effect on subsidence, due to poroelastic phenomena.

In the last two schemes (Figures 14.8(b) and 14.8(c)), however, reinjection determines the growth of two cooled regions, which may be considered circular with a reasonable approximation. The following cooling radii, $R_{c,b} = 400$ m and $R_{c,c} = 450$ m, are considered (Figure 14.8), which correspond approximately to 11 and 7 years of injection. In Figures 14.10(e), 14.10(f), 14.11(e), and 14.11(f), thermal subsidence is shown for a reference temperature drop $\Delta T = -100\,°C$ that yields values comparable to those of Figures 14.9(c),

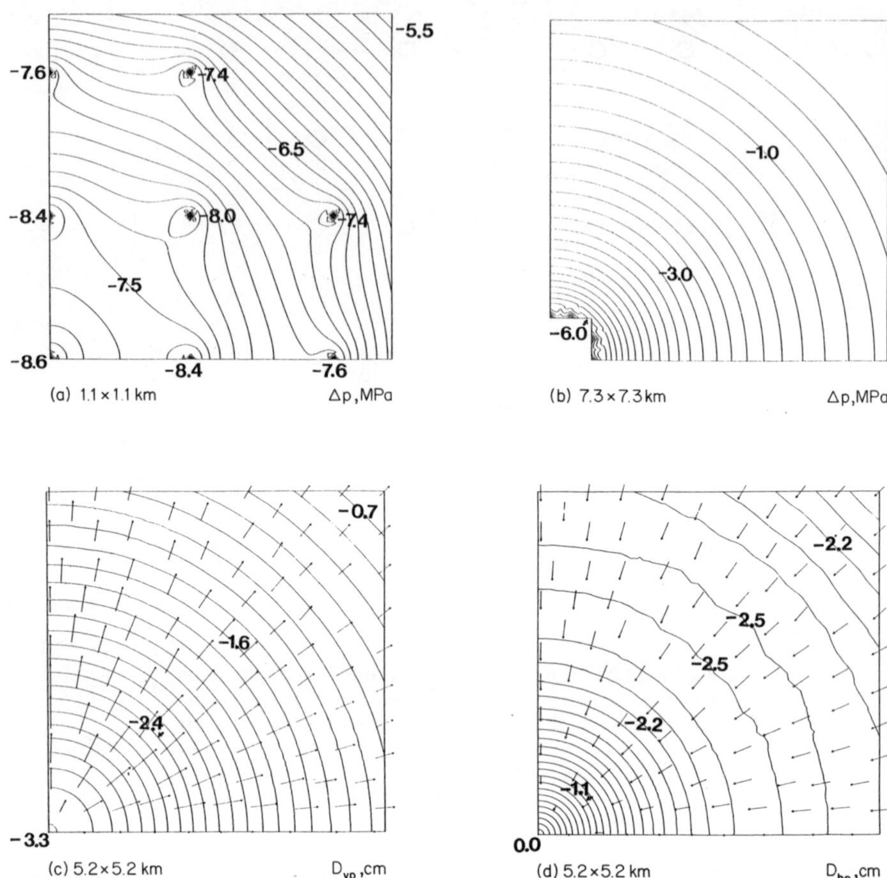

Figure 14.9 Pressure variations and displacements for production scheme in Figure 8(a). (a, b) Contours of pressure Δp; (c) contours of vertical displacements D_{vp} and vectors of tilt due to pressure (maximum tilt = 5.5 μrad); (d) contours of horizontal displacements D_{hp} due to pressure. Overpressures, upward and centrifugal displacements are positive

14.9(d). If the injection is interrupted, pressure becomes uniform quite rapidly, while the aforementioned thermal subsidence remains unchanged for a long time.

A more correct evaluation of pressure field takes into account temperature effects on viscosity, according to Equation (14.3). This computation—not reported here—has been made for the case of Figure 14.8(c). A noticeable increase in pressure with respect to the previous distribution in Figure 14.11(a) occurs only in the cooled region, and does not reduce the thermal subsidence effect by more than 20%, however.

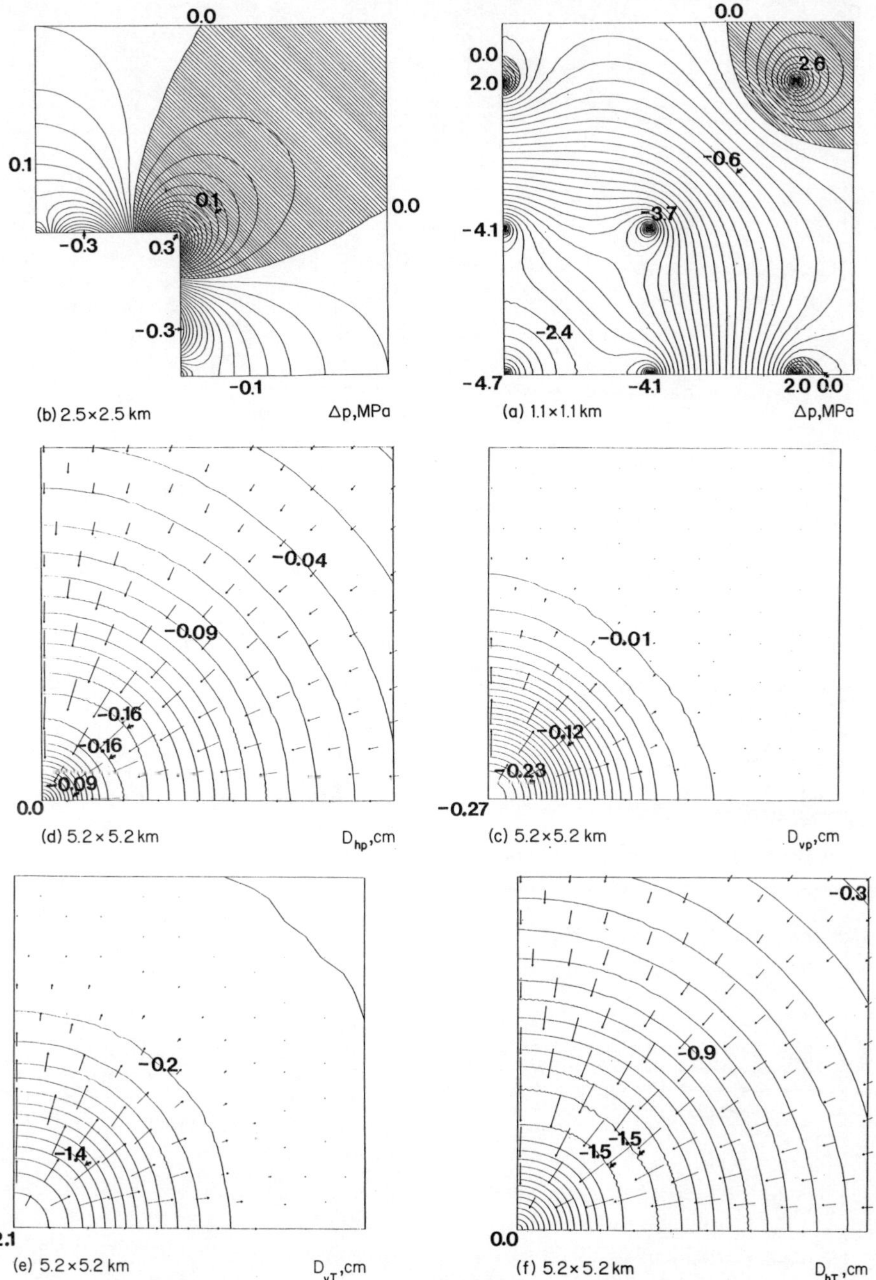

Figure 14.10 Pressure variations and displacements for production scheme in Figure 14.8(b). (a, b) Contours of pressures Δp; (c, e) contours of vertical displacements D_{vp}, D_{vT}, and vectors of tilt due to pressure (c) and to temperature (e). Maximum tilt due to pressure = 1.46 μrad. Maximum tilt due to temperature = 9.7 μrad. Contours of horizontal displacements D_{0p}, D_{0T} due to (d) pressure and (f) temperature

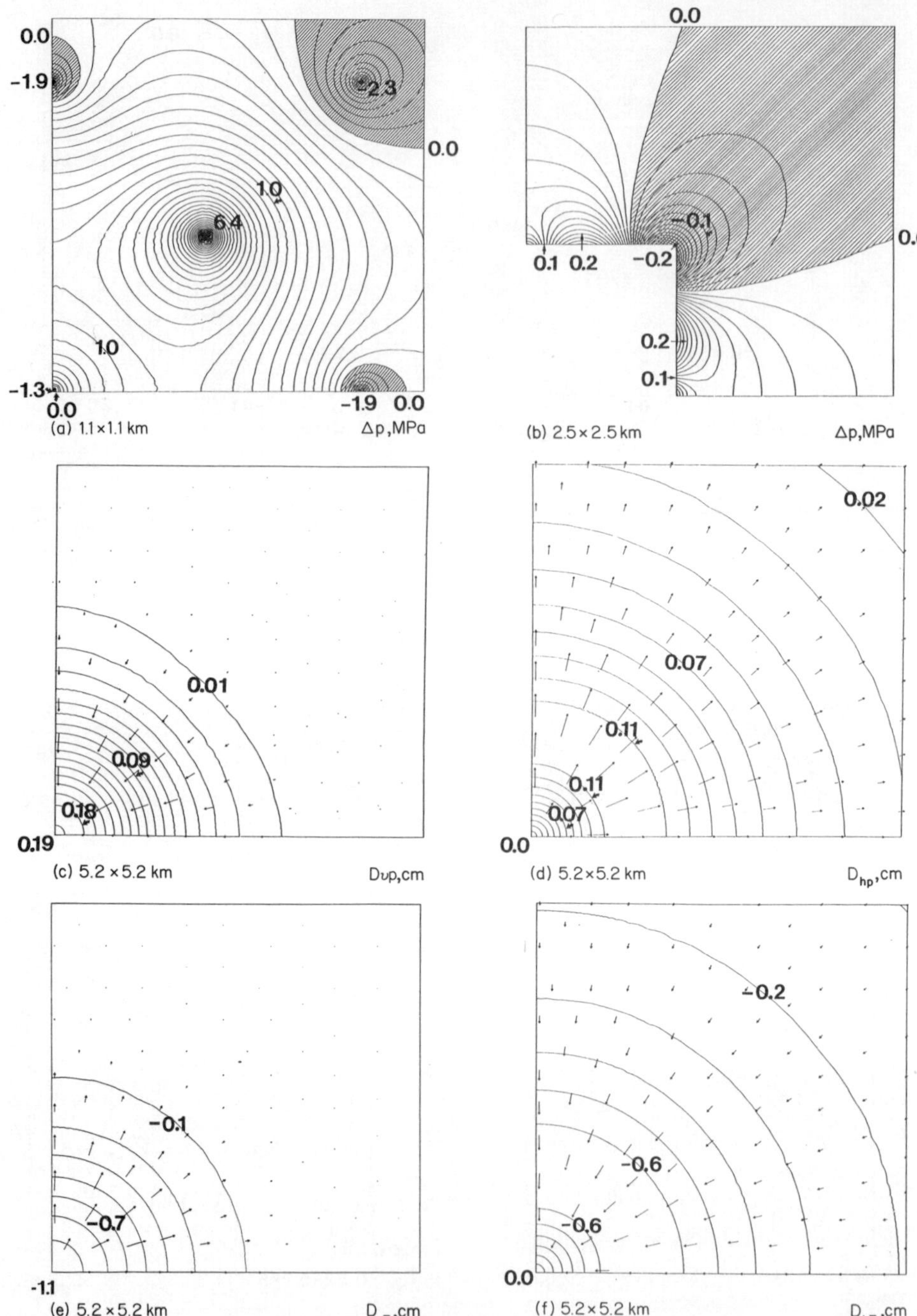

Figure 14.11 Pressure variations and displacements for production scheme in Figure 14.8(c). (a, b) Contours of pressure Δp; (c, e) contours of vertical displacements D_{vp}, D_{vT} and vectors of tilt due to pressure (c) and to temperature (e). Maximum tilt due to pressure = 0.97 μrad. Maximum tilt due to temperature = 5.6 μrad. (d, f) Contours of horizontal displacements D_{0p}, D_{0T} due to pressure (d) and to temperature (f)

14.4 DYNAMIC EFFECTS CAUSED BY FLUID INJECTION

The lithosphere is subjected to deformations that produce stresses. In some cases, they may be absorbed by creep strains; otherwise, they may accumulate and lead to failure of intact rock. As a consequence, discontinuities arise along planes determined by the original stresses at failure.[24]
Faulting planes are present in the earth crust at different scales, ranging from systems of hundreds of kilometres that divide continental and oceanic plates, to regional and local systems with dimensions of kilometres or metres. Many geological structural associations are characterized by faulting. Also the formation of joint families is governed by similar causes.[24] When intact rock fails, the associated state of stress is partially released; however, the geological causes may produce new stress increments and a related accumulation of elastic energy. Generally, the weakest areas are represented by the cohesionless fault planes, where shearing resistance is due to friction only. When resistance is overcome, the elastic energy is transformed into frictional heat and into seismic waves.[25]

Water pressure plays an important role in the phenomenon, as the shearing resistance of fault planes depends on the normal effective stress (defined as the total stress minus pore pressure), as first suggested by Hubbert and Rubey.[13] This simple conceptual model supports the possibility of controlling fault movements by gradual injection of water. Experiments performed in the oil field of Rangely, Colorado, indicated a quantitative correlation between pressure increase caused by fluid injection and the start of seismic events. Earthquake focuses clustered along an active fault in the overpressurized region. Similar experiences were repeated in Japan, where water was injected into an aquifer near the city of Matsushiro.[26] Recent seismical surveys in The Geysers geothermal field, California, relate the growth of seismic activity to the increase in steam production, and relate some earthquakes up to degree 3 of the Richter scale to seasonal peaks in the amounts of injected water.[11]

It must be noted that induced seismicity appears more likely in regions with a relevant natural activity, and this is often the case in geothermal areas.

While pressure-induced fault movements are usually interpreted according to the above theory, less attention has been paid to thermal effects as responsible for triggering earthquakes. Rangely and Matsushiro experiences obviously disregarded this problem and the Geysers data are not currently interpreted by any specific model. Only a few other records are available from geothermal fields.[27]

In the following, both temperature and pressure effects produced by an injection well are investigated and mutually compared, in order to predict, for some simple cases, the time evolution of a region of possible induced seismic events.

14.4.1 Stresses induced by injection

Temperature and pressure variations induced by an injected fluid are described in Equations (14.1) and (14.3). The horizontal total thermal stresses are given by the following analytical formulae:

$$\sigma_{r,T}=\sigma_{\theta,T}=-\frac{\alpha_T E\,\Delta T}{2(1-\nu)}\left(1-\frac{a^2}{r^2}\right),\qquad \text{for } a\leqslant r\leqslant R_c,$$
$$\sigma_{r,T}=-\sigma_{\theta,T}=-\frac{\alpha_T E\,\Delta T}{2(1-\nu)}\left(\frac{R_c^2}{r^2}-\frac{a^2}{r^2}\right),\qquad \text{for } R_c<r, \tag{14.10}$$

where the symbols σ_r and σ_θ represent the principal radial and circumferential stresses, positive in compression; α_T, E, ν are the linear thermal expansion coefficient, the Young's modulus and the Poisson's ratio, respectively. The above equations hold for the boundary condition $\sigma_r=0$ at the wellbore and at infinity; they can be easily obtained on the basis of the thermoelasticity theory.[28]

Disregarding the local pressure effect on the wellbore, the total stress variation due to pressure is obtained in a similar way and is given by

$$\left.\begin{aligned}\sigma_{r,p}&=\frac{\alpha_p E\,\Delta p_a}{2(1-\nu)}\left[\left(1-\frac{a^2}{r^2}\right)\left(1-\frac{M}{2}\right)+M\ln\frac{r}{a}\right]\\ \sigma_{\theta,p}&=-\sigma_{r,p}+\frac{\alpha_p E\,\Delta p_a}{1-\nu}\left(1+M\ln\frac{r}{a}\right)\end{aligned}\right\}\quad \text{for } a\leqslant r\leqslant R_c \tag{14.11(a)}$$

$$\left.\begin{aligned}\sigma_{r,p}&=\frac{\alpha_p E\,\Delta p_a}{2(1-\nu)r^2}\left\{(R_c^2-a^2)\left(1-\frac{M}{2}\right)+MR_c^2\ln\frac{R_c}{a}\right.\\ &\quad\left.+\frac{\mu_2}{\mu_1}M\left[r^2\left(\ln\frac{r}{r_e}-\frac{1}{2}\right)-R_c^2\left(\ln\frac{R_c}{r_e}-\frac{1}{2}\right)\right]\right\}\\ \sigma_{\theta,p}&=-\sigma_{r,p}+\frac{\alpha_p E\,\Delta p_a}{1-\nu}\frac{\mu_2}{\mu_1}M\ln\frac{r}{r_e}\end{aligned}\right\}\quad \text{for } R_c<r \tag{14.11(b)}$$

where

$$\alpha_p=\left(1-\frac{C^*}{C}\right)\frac{1-2\nu}{E},\qquad M=1\Big/\left(\frac{\mu_2}{\mu_1}\ln\frac{R_c}{r_e}-\ln\frac{R_c}{a}\right).$$

Moreover, C^* and C are the matrix and rock-mass bulk compressibilities, respectively.

The induced effective stresses are obtained by subtracting the pressure from the summations of σ_r and σ_θ:

$$\begin{aligned}\sigma_r'&=\sigma_{r,T}+\sigma_{r,p}-\Delta p,\\ \sigma_\theta'&=\sigma_{\theta,T}+\sigma_{\theta,p}-\Delta p.\end{aligned} \tag{14.12}$$

14.4.2 Elastic analysis of seismic potential

A simplified criterion for predicting the seismic activity at every point of a faulted or jointed medium is used in the following. It is based on the elastic computation of the local state of effective stress (due to original and induced stresses) and of a related index, the safety factor; this is defined as the ratio of the maximum admissible shear stress to the actual one, as represented in Figure 14.12. The same figure indicates that factors greater than unity characterize the stresses compatible with the available strength; on the other hand, values less than unity are indices of possible slipping. This event depends on the probability that a discontinuity with appropriate orientation and strength exists at the considered location. From this point of view, a lower safety factor indicates a higher probability of seismic activity.

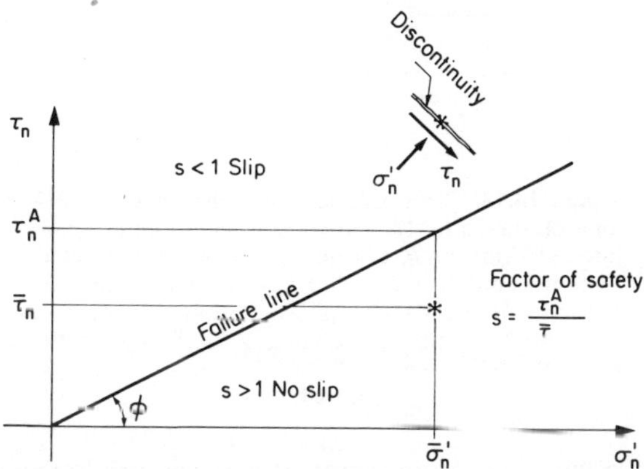

Figure 14.12 Strength criterion for fault or joint slip. The state of effective stress acting on the discontinuity is represented by the point $(\bar{\sigma}'_n, \bar{\tau}_n)$. σ'_n = normal effective stress; τ_n = shear stress; τ_n^A = maximum admissible shear stress; ϕ = frictional angle of the discontinuity

In the following, the problem of an injection well is considered. The stress perturbation after 135 days of constant injection is given in Figure 14.13. In addition, the time history of the induced effective stresses at a point located 45 metres from the well is plotted in Figure 14.14.

The results indicate that pressure and stress behaviour are governed by the thermal front position. In the hot region, the stationary pressure field is rapidly reached, owing to the high pressure diffusivity, while in the cooled area pressure increases logarithmically with time and exhibits a radial logarithmic distribution.[19] Thermal stresses are strictly deviatoric and decrease as (R_c^2/r^2) in the hot region, while an isotropic traction field, constant with time, is generated in the cooled area.

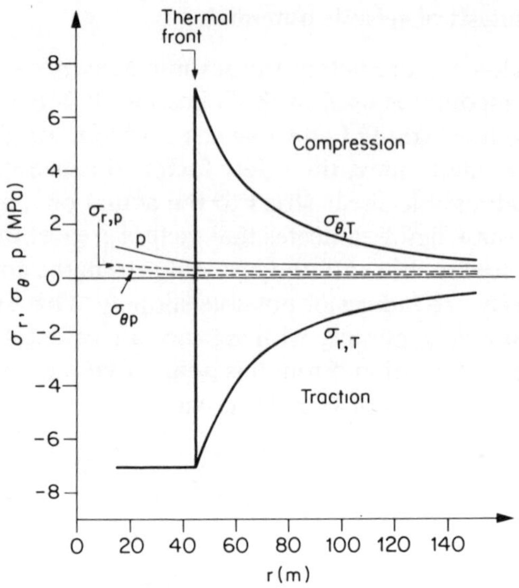

Figure 14.13 Total stresses induced by temperature and pressure, after 135 days of constant water injection into the formation F_c of Figure 14.5 filled by hot water. Data of the problem: $T_2 = 200\,°C$, $k_p = 5 \times 10^{-14}\,m^2$, $n = 15\%$, $\rho^* c^* = 2.1\,MJ\,°C^{-1}\,m^{-3}$, $C^* = 2.4 \times 10^{-5}\,MPa^{-1}$, $C = 2.4 \times 10^{-4}\,MPa^{-1}$, $\nu = 0.2$, $E = 7.5 \times 10^3\,MPa$, $\alpha_T = 10^{-5}\,°C^{-1}$, $q = 3 \times 10^{-2}\,m^3\,s^{-1}$, $a = 15\,cm$, $r_e = 13\,km$

In order to evaluate the safety factor, it is necessary to specify the normal and shear original stresses acting on the discontinuity, and the angle of friction, in the following assumed equal to 16.5 MPa, 7.8 MPa and 30°, respectively. The above data imply an original value of the safety factor equal to 1.2 all over the formation.

After 135 days of constant injection rate the unsafe area is given in Figure 14.15; it presents two axes of symmetry (x and y); axis x in inclined at $(\pi/4 + \varphi/2)$ with respect to the direction of discontinuities. The unsafe regions at different injection times are similar and their area is proportional to the total injected fluid. This fact is easily explained by noting that temperature effects prevail, and therefore the unsafe area grows with cooling.

The stress paths of the most significant regions are exhibited in Figure 14.16.

14.4.2.1 *Induced seismicity in case of a single fault*

In the following, the behaviour of a single fault is discussed. Initially, fault activity can be triggered only by pressure effects; however, the assumed value

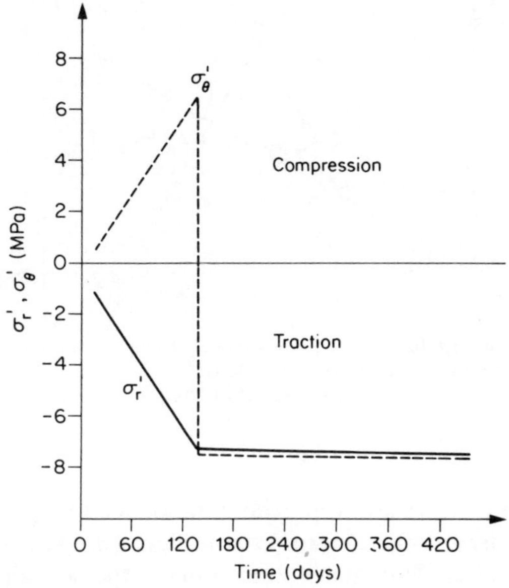

Figure 14.14 Time evolution of the induced effective stresses at a distance of 45 m from injection well

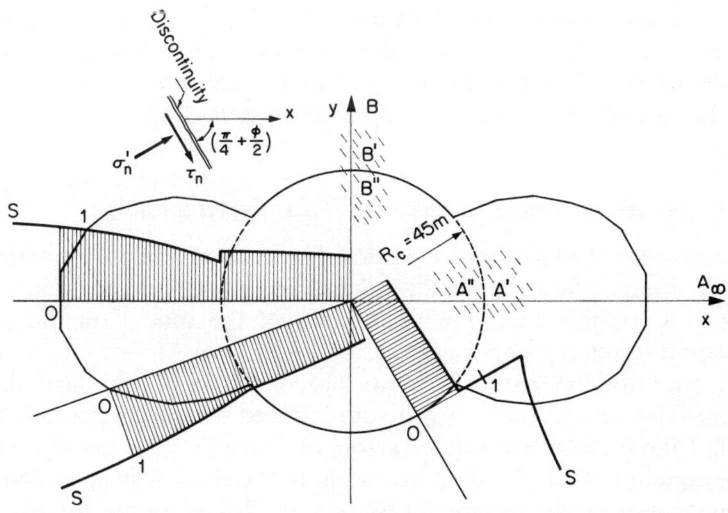

Figure 14.15 Unsafe region after 135 days of injection. Safety factors are reported along 3 directions. Letters A and B represent different typical regions for microseismic activity

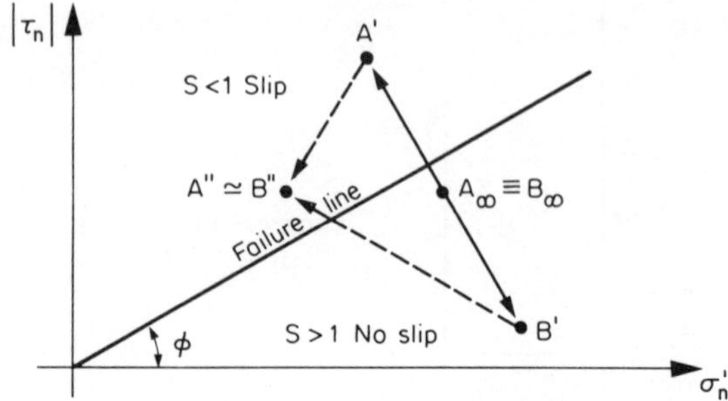

Figure 14.16 Stress paths determined by the expansion of the unsafe area, for the regions A′, A″, A_∞, B′, B″, B_∞ of Figure 14.15

of the safety factor 1.2 is not appreciably influenced by pressure variations. Moreover, the hydraulic characteristics considered are reasonable for a long-term injection project. This means that initially the seismic activity may be triggered by pressure, only if the fault is in a very critical state. However, due to the high permeability of the fault, failure may occur at considerable distances from the injection well in some weaker region of the fault itself.

The behaviour under cooling effects is very different. Thermal disturbance takes a long time to develop, as shown by the slow propagation of the unsafe region in the example problem. However, the effects are very strong and may easily reduce the safety factor to values less than unity. Thermal effect in a fault region larger than that forecast by the present model appears possible only in the case of a high fluid flow in the fault itself.

14.4.2.2 Induced seismicity in the case of distributed faulting

In some geological situations, a natural microseismic activity, related to a diffused faulting in the rock mass, produces a number of microevents all over the reservoir. In this case, the verification of the model on the basis of experimental results appears easier.

The records of the natural activity are likely to be modified during a long-term injection. Initially, a pressure-induced activity is possible around the well, if the undisturbed safety factors of discontinuities are close to one. In a subsequent period, the time evolution of the records is governed by the slow propagation of the unsafe region, mainly driven by the thermal effect.

The present model indicates that the safety factor increases ahead of the cooled front, along the direction of the symmetry axis y of Figure 14.15 (region B′), so that no seismic activity should be recorded in this area. On the other hand, the activity must increase in the region A′, still ahead of the front, but along the symmetry axis x (see also Figure 14.16).

When a new circular annulus is added to the cooled region, the activity should be greater than normal in the y-direction (B″ region), while in the x-direction the phenomena largely depend on the pre-front behaviour. Actually, the maximum slipping probability of a region along the axis x occurs in the A′ area. If slip occurs, stresses cannot be evaluated any longer on the basis of the elastic solution and the computed safety factor in the A″ region does not have a clear meaning.

14.4.3 An elastoplastic model for joint families slipping

Joint slipping determines an overall stress redistribution, which affects the slip probability at later times. Therefore, a time incremental nonlinear process has to be simulated. The complete simulation of the slip mechanism involves much more knowledge than the simple peak strength criterion previously stated. Dynamic effects, kinematic friction, dilatancy stabilization, crack propagation properties and other phenomena determine the amplitude of the dislocation and the stress redistribution. Again, a simplified model is formulated.

Since the spacing of the discontinuties is assumed small in comparison with the scale of the problem, an equivalent continuum model is used. The rock mass is considered perfectly elastic except along the planes of weakness, where the previous strength criterion holds. The failure is simulated as a quasi-static process according to an ideally plastic and nondilatant behaviour (some brittleness criterion may also be included).[29] The region where rock fails during the elastoplastic computation provides an indication of the area where microseismic events may occur.

The injection period is subdivided into a discrete number of time intervals; nonlinear phenomena are allowed only at the end of every time step. Consequently, the stress variations induced by injection are easily evaluated analytically by means of Equations (14.10)–(14.12). On the other hand, nonlinear phenomena are simulated by a finite element program based on the viscoplastic algorithm.[30]

The total failed region, after 180 days and five time steps, is represented in figure 14.17(a); as before, the region grows linearly with time, but now with polar symmetry. For comparison, the unsafe area evaluated by the elastic model is shown in the same figure. The shape of the area failed in a time step is also given in Figure 14.17(b), together with the approximate intensity of the dissipated energy. The development of a failed area does not prevent a subsequent slipping of the surrounding region. In addition, the subsequent failure produced by the advancing cold front does not induce a renewed activity in the previously broken region.

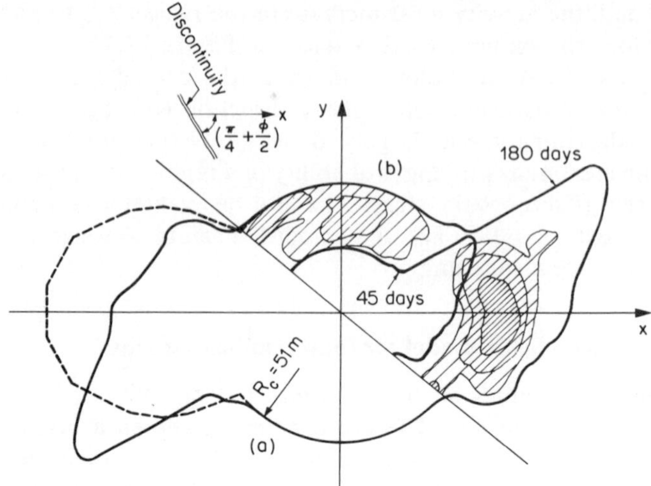

Figure 14.17 (a) Comparison between the total failed area according to the elastoplastic model (solid line) and the unsafe area evaluated elastically (dashed line), after 180 days of constant injection; (b) time evolution of the failed area. The intensity of the dissipated energy is represented by shading

14.5 CONCLUDING REMARKS

The relatively recent subject of geothermal reservoir engineering takes advantage of some numerical methods first developed in other engineering fields. Also, experience gained in the problem of controlling other environmental negative consequences can be made use of. The presence of important thermal effects, however, has to be considered in addition to pressure effects, as demonstrated in this chapter.

A numerical method for subsidence evaluation, which represents a generalization of earlier analytical procedures, is presented for layered formations.

In the case of induced seismicity, the problem appears very complex and is affected by many undetermined factors. At present, two objectives may be reasonably tackled. The first is related to forecasting in which area induced seismicity may occur. The second is the interpretation of the available seismic surveys by numerical models, in order to gain a more detailed understanding of the recorded phenomena. The work presented here should be viewed as a first step in the development of the required numerical tools.

ACKNOWLEDGEMENTS

This work has been supported by the Italian National Research Council (C.N.R.). The authors thank Dr. R. Di Bacco and Mr. L. Giambarini for their contribution to the implementation of the computer programs and to the computation of the results.

REFERENCES

1. D. J. Moore, 'Atmosphere and water pollution from power plants', in *Geophysical Aspects of the Energy Problem*, Elsevier, Amsterdam (1980).
2. H. C. H. Armstead, *Geothermal energy*, Spon, London (1978).
3. N. D. G. Salomon, 'Surface subsidence', in *Advances in Rock Mechanics*, Natl. Academy of Sciences, Washington (1974).
4. I. Miller, *et al.* (1980), 'Simulation of geothermal subsidence', *Lawrence Berkeley Laboratory Report LBL-10794, GSRMP-6* (1980)
5. K. Terzaghi and R. B. Pech, Soil mechanics in Engineering Practice, 2nd edn, Wiley, New York (1967).
6. J. Geertsma, 'The effects of fluid pressure decline on volumetric changes of porous rocks'. *Trans. AIME*, **210**, 331–340 (1957).
7. J. Geertsma, 'Numerical treatment of some geomechanical problem areas in oil and natural gas production', in *Numerical Methods in Geomechanics*, (ed. C. S. Desai) American Society of Civil Engineers, New York (1976).
8. G. M. C. Van Opstal, 'The effect of base-rock rigidity on subsidence due to reservoir compaction', in *Advances in Rock Mechanics*, Natl. Academy of Sciences, Washington (1974).
9. R. W. Lewis and B. A. Schrefler, 'A fully coupled consolidation model of the subsidence of Venice', *Water Resources Research*, **14**, 223–230 (1978).
10. B. A. Schrefler, R. W. Lewis and V. A. Norris, 'A case study of the surface subsidence of the Polesine area', *Int. J. Num. and Analyt. Meth. in Geomech.*, **1B**, 377–386 (1977).
11. R. S. Ludwin and C. G. Bufe, 'Continued Seismic monitoring of the Geysers Geothermal area, California'. *Open-File Report Number 80*, U.S. Geological Survey, Menlo Park, California (1980).
12. P. A. Hsien and J. D. Bredehoeft, 'A reservoir analysis of the Denver earthquakes: a case of induced seismicity, *J. Geophys. Res.*, **86**, 903–920 (1981).
13. M. K. Hubbert and W. W. Rubey, 'Role of fluid pressure in mechanics of overthrust faulting', *Geol. Soc. Amer. Bull.*, **70**, 115–166 (1959).
14. J. H. Dietrich, C. B. Raleigh and J. D. Bredehoeft, 'An experiment in earthquake control at Rangely, Colorado', *Science*, **191**, 1230–1237 (1976).
15. H. K. Gupta and G. K. Bastogi, *Dams and Earthquakes*, Elsevier, Amsterdam (1976).
16. R. C. Schroeder, M. J. O'Sullivan, K. Pruess, R. Celati, and C. Ruffilli, 'Reinjection studies of Vapor-Dominated Systems', *Proc. 2nd DOE-ENEL Workshop for Cooperative Research in Geothermal Energy*, Berkeley (January 1981).
17. G. F. Pinder, H. J. Ramey Jr., A. Shapiro, and L. Abriola, *Block Response to Reinjection in a Fracturated Geothermal Reservoir.* Civil Engineering Department, Princeton University (1979).
18. G. Cappetti, A. Giovannoni, C. Ruffilli, C. Calore and R. Celati, 'Reinjection in the Larderello geothermal field', *Proc. Intl. Conf. on Geoth. Energy*, Florence, (May 1982).
19. M. Borsetto, G. Carradori and R. Ribacchi 'Thermal effects in well-testing for geothermal purpose', paper presented at the *1st Int. Conf. on Num. Meth. in Thermal Problems*, Swansea, July 1979. ISMES Paper No. 134, Bergamo (1979).

20. C. R. Faust and J. W. Mercer, 'Geothermal reservoir simulation: 1. Mathematical models for liquid- and vapor-dominated hydrothermal systems', *Water Resources Research*, **15**, 23–30 (1979).
21. M. Borsetto, G. Carradori and R. Ribacchi, 'Coupled seepage, heat transfer, and stress analysis with application to geothermal problems', in *Numerical Methods in Heat Transfer*, (ed. R. W. Lewis, K. Morgan and O. C. Zienkiewicz) Wiley, Chichester (1981).
22. G. F. Pinder, 'State of the art review of geothermal reservoir modelling, *Lawrence Berkeley Laboratory Report LBL-9093, GSRMP-5* (1979).
23. A. S. Williamson, 'Volumetric subsidence-compaction ratios for a horizontal stratified elastic half-space' in *Advances in Rock Mechanics*, Natl. Academy of Sciences, Washington (1974).
24. B. E. Hobbs, W. D. Means and P. F. Williams, *An Outline of Structural Geology*, Wiley, New York (1976).
25. J. N. Brune, 'Tectonic stress and the spectra of seismic shear waves from earthquakes', *J. Geophys. Res.*, **75**, 4997–5002 (1970).
26. M. Ohtake, 'Seismic activity induced by water injection at Matsushiro, Japan', *J. Phys. Earth.*, **22**, 163–176 (1974).
27. *Proc. 2nd DOE-ENEL Workshop for Cooperative Research* in *Geothermal Energy*, Berkeley (January 1981).
28. S. Timoshenko and J. N. Goodier, *Theory of Elasticity*, McGraw-Hill, New York (1951).
29. O. C. Zienkiewicz and I. C. Cormeau, 'Viscoplasticity-plasticity and creep in elastic solids. A unified numerical solution approach', *Int. J. Num. Meth. Eng.*, **8**, 821–841 (1974).
30. NOLIVP Computer Code, *ISMES Internal Report*, Bergamo (1979).

Numerical Methods in Heat Transfer, Volume II
Edited by R. H. Lewis, K. Morgan, and B. A. Schrefler

Chapter 15

Analysis of the Transient Thermal Performances of Composite Devices by Network Methods

L. Imre and L. I. Kiss

SUMMARY

Transient thermal processes are of decisive importance in the course of solving a number of engineering problems.

In the design and utilization of certain types of machines problems of heating, cooling, and thermal protection have to be solved. These problems often involve the determination of the loadability of the machines, at the operation time permissible with given loading conditions. The solution of problems of this type requires the transient temperature field to be determined.

In the cases of technological equipment and heating systems the transient thermal analysis is indispensable for the optimization of the energy consumption or of the technological result.

When designing or using thermal energy converters (e.g. solar systems), the prediction of efficiency, utilizable energy and operational properties to be expected requires the transient thermal analysis to be carried out.

The machines and equipment in question are composite devices built up from a great number of parts and elementary units, in which one or several working media are flowing for heating, cooling, or some other technological purpose. Their transient thermal condition is produced by temporary variations of the operational conditions. The thermal and flow processes take place in them simultaneously and in interaction with each other; therefore the transient thermal analysis requires a simultaneous thermohydrodynamical model.

The complicated geometry of composite devices, the utilization of inhomogenous materials, and the nonlinearity of the thermal and flow phenomena greatly restrict or even exclude the possibility of applying continuum models, and numerical solution methods based on discretization, which meet the requirements of economy and time saving, come into prominence.

A simple variant of the methods based on discretization and meeting the above-mentioned requirements of modelling composite devices is the network method, the principles of which are given by transport processes,[1,2] which are substantially analogous.

The network method consists of a peculiar application of discretization with respect to space, which results in a clear-cut network equation system suitable for computer methods.

The simultaneous thermohydrodynamical network model contains mass flow network (MFN) and heat flow network (HFN) part models.

The MFN part model serves to determine the mass flow distribution, i.e. the time functions of local mass flow rates. There are as many mass flow network part models required for the description, as there are working media used in the machine or equipment. The MFN part model of simultaneous thermohydrodynamical models is not isothermic, therefore some of the network elements (e.g. the so-called dual-function elements) are temperature dependent and can be identified if the local temperature of the media is known.

The temperature variation of the flowing media can be modelled by the use of the so-called modifying elements of the simultaneous model. (Such are, e.g. the heat exchangers and the thermal mixing spaces.) The temperature variations of the flowing media, occurring along the heat transferring surfaces of the device, can be described by the help of the HFN part model. This possibility is offered by the application of the mass-flow-dependent (dual function) elements of the HFN. In this respect the HFN part model can be regarded as a modifying element of the simultaneous model. The final result of the analysis (e.g. the transient temperature distribution) will be given as the solution of the HFN equation system.

The solution of the nonlinear equation system of the simultaneous network model can be produced by means of various computing strategies based on discretization with respect to time, with the application of suitably chosen schemes of finite differences or finite time elements.

The application of simultaneous network models is demonstrated by means of examples of electric rotary machines, oil transformers and solar systems.

15.1 TRANSIENT THERMAL ANALYSIS OF ELECTRIC ROTARY MACHINES WITH INTERNAL VENTILATION

15.1.1 Aim of the transient thermal analysis

The failure of electric rotary machines working under varying load and overloaded temporarily is often caused by overheating. As the service life of insulations is dependent also on the duration of overload, depending on the momentary temperature conditions of the motor, overloads of short duration may be permissible without danger of damage.

A requirement of economy is that the motor destined for varying load be not overdimensioned. Both for the designer selecting the motor and its thermal protection and for the user it is desirable to apply a calculation method that, in addition to satisfying the technical precision, permits the determination of the transient thermal conditions to be expected for arbitrary load combinations.

15.1.2 Simultaneous network model of motors

In closed electric machines with internal ventilation, the role of the thermal environment of the structural parts is taken over by the flowing medium (air in most cases). By virtue of the heat input, the temperature of the air varies as a function of place. The temperature variation is determined by the local mass flow rate and the local heat flux getting into the air. The local temperature of the air and the local flow rate exert an effect on the local heat transfer conditions.[7] At the same time, the temperature variations of the air affect also the value of mass flow rates. A complete calculation of the warming process requires also the determination of the local mass flow rate of the cooling air.

To take the thermal and flow interactions into account, a simultaneous network model can be applied that is built up from HFN and MFN part models.

Linear heat flow network modelling[8–14] has proved to be an economical and satisfactorily accurate method of modelling the transient warming of electrical rotary machines. Considering the temperature dependence of heat radiation and convective couplings, the HFN part model becomes nonlinear.[3] As the flow resistances are dependent on the mass flow rates, the MFN part model is not linear either. Therefore, numerical methods appear suitable for the solutions.[1,2] The application of the simultaneous heat and mass flow network model and the construction of the part models are demonstrated on a 100 kW asynchronous motor according to Figure 15.1.

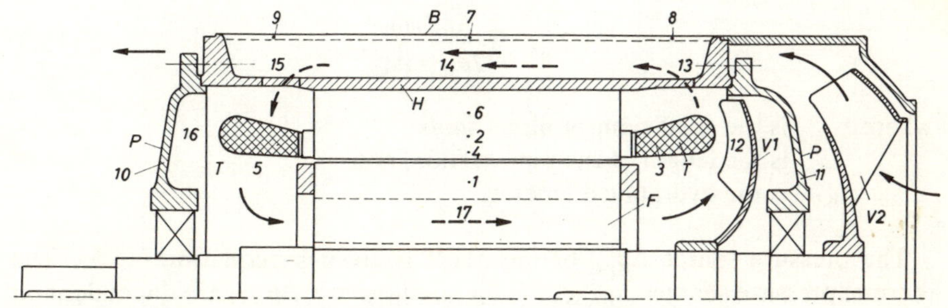

Figure 15.1 Sketch of a 100 kW asynchronous motor

15.1.3 Mass flow network part model of the motor

The motor according to Figure 15.1 has both external and internal cooling. The external cooling system is open, ventilator V2 has practically unrestricted blowing, and the mass flow can be considered given, therefore no mass flow network modelling is needed. The air has a large mass flow rate and it has a moderate warming effect on the cooling fins.

In the internal cooling system the flow is maintained by ventilator V1. The motor has $b_v = 6$ cooling ducts in the wall of the house and $b_F = 12$ cooling ducts in the rotor.

The main model conditions are the following:

MM1. The ducts in the rotor are considered to be equivalent thermally and hydraulically, and so are the ducts in the house.

MM2. The ventilator V1 having an open impeller can be characterized by a shunt flow path with a shape resistance $R_{m1}(\dot{m}_1)$ determined by measurement.

MM3. The characteristic curve of the ventilator for a given r.p.m. is known from measurement data:

$$\Delta p_{g\phi} = A + B\dot{m}_g + C\dot{m}_g^a. \tag{15.1}$$

The MFN model of the motor is built on the duct system of the cooling medium (Figure 15.2).

$R_{m1}, R_{m2}, R_{m4,1}, \ldots, R_{m4,6}, R_{m5}, R_{m6}, R_{m8,1}, \ldots, R_{m8,12}$ and R_{m9} are shape resistances that can be calculated by the following relationship:

$$R_{m\zeta,i} = \frac{\zeta_i \dot{m}_i}{2\rho_i A_{ci}^2}, \tag{15.2}$$

where ζ, the shape resistance factor, is an experimental constant.

$R_{m3,1}, \ldots, R_{m3,6}, R_{m7,1}, \ldots, R_{m7,12}$ are resistances of pipe friction:[1,4]

$$R_{m\lambda,i} = \frac{\lambda_{c,i} L_i \dot{m}_i}{2 d_{h,i} \rho_i A_{c,i}^2}, \tag{15.3}$$

where $\lambda_{c,i}$ is the coefficient of pipe friction,
L_i is the length of the pipe section, and
$d_{h,i}$ is the hydraulic diameter.

The pressure source $\Delta p_{g\phi}$ of the MFN is given by equation (15.1). The thermal pressure sources $\Delta p_{u,1}, \ldots, \Delta p_{u,6}, \Delta p_{u,8}, \ldots, \Delta p_{u,19}$ are dependent on heat flux $\phi_{h,i}$ getting into the medium of mass flow rate $\dot{m}_i$, i.e. they are values dependent on the temperature variation of the cooling medium, and are the

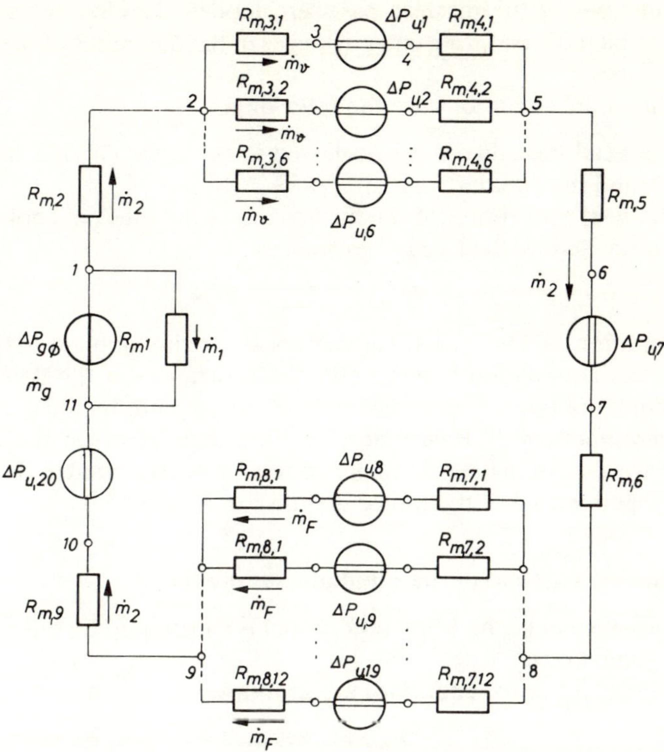

Figure 15.2 MFN part model of the machine shown in Figure 15.1

dual-function elements of the mass flow network:[1,4]

$$\Delta p_{u,i} = \frac{R}{c_{pa}}\left[\rho_{a,i}\frac{\phi_{h,i}}{\dot{m}_i} - \frac{\dot{m}_i^2}{2A_{c,i}^2}\left(\frac{1}{\rho_{out}} - \frac{1}{\rho_{in}}\right)\right], \tag{15.4}$$

where R is the gas constant,
c_{pa} is the average specific heat on constant pressure,
$\rho_{a,i}$ is the average density,
$\rho_{in,\,out}$ are the densities before and after heat absorption, respectively.

15.1.4 Heat flow network part model of the motor

For construction of the HFN model, the motor is viewed as divided into discrete parts, the thermal state of which can approximately be characterized by a single temperature. The number of discrete parts is chosen in compliance with the requirements of accuracy and the economy of the calculation. The thermal functions of the discrete parts are modelled by concentrated HFN elements coupled by heat transfer resistances in the directions of the supposed main heat fluxes.[1,2]

The main model conditions are the following:

MH1. The axial heat fluxes are neglected in the rotor (1), the stator tooth (4), and the crown (6).

MH2. The heat absorption of the air flowing in the internal cooling system is modelled by the formal resistances

$$R_{c,i,12}, \ldots, R_{c,i,17} = R_{c,i,j} = \pm(\dot{m}_{i,j}c_p)^{-1}.$$

A precondition of the formal resistances $R_{c,i}$ being applied is that, when writing the nodal equation system, only the resistances in the flow direction of the medium are taken into consideration. As a result, the heat conduction matrix is asymmetrical.[10] Resistances $R_{c,i}$ are dependent on the mass flow rate; they are dual-function elements of the HFN part model. The scheme of the HFN part model of the motor is shown in Figure 15.3.

15.1.5 Equation system of the simultaneous model

The equation system of the MFN part model is constituted by the nodal, loop and branch equations.[1–3]

The independent nodal equations are as follows:

$$\dot{m}_g = \dot{m}_1 + \dot{m}_2, \tag{15.5}$$

$$\dot{m}_2 = b_v \dot{m}_v, \tag{15.6}$$

$$\dot{m}_2 = b_F \dot{m}_F. \tag{15.7}$$

The independent loop equations, substituted on the basis of the branch equation $\dot{m}_i R_{mi} = \Delta p_i$ are

$$\Delta p_{g\phi} = \dot{m}_1 R_{m,1}, \tag{15.8}$$

$$\Delta p_{g\phi} - \Delta p_{u,1} + \Delta p_{u,7} + \Delta p_{u,8} + \Delta p_{u,20} = \dot{m}_2(R_{m,2} + R_{m,5} + R_{m,6} + R_{m,9}) + \dot{m}_v(R_{m,3,1} + R_{m,4,1}) + \dot{m}_F(R_{m,7,1} + R_{m,8,1}). \tag{15.9}$$

The nodal equation written for the ith node of the HFN part model, with substitution of the branch equation

$$\phi_{h,i,j} = K_{i,j}(T_i - T_j)$$

$$K_{i,j} = R_{i,j}^{-1}, \qquad i = 1, \ldots, n, \quad j = 1, \ldots, n, \quad i \neq j \text{ is}$$

$$C_i \frac{dT_i}{dt} + \sum_{j=1}^{n} [K_{i,j}(T_i - T_j)] + K_{i,w}(T_i - T_w) = \phi_i^0 [1 + \beta(T_i - T_0)]. \tag{15.10}$$

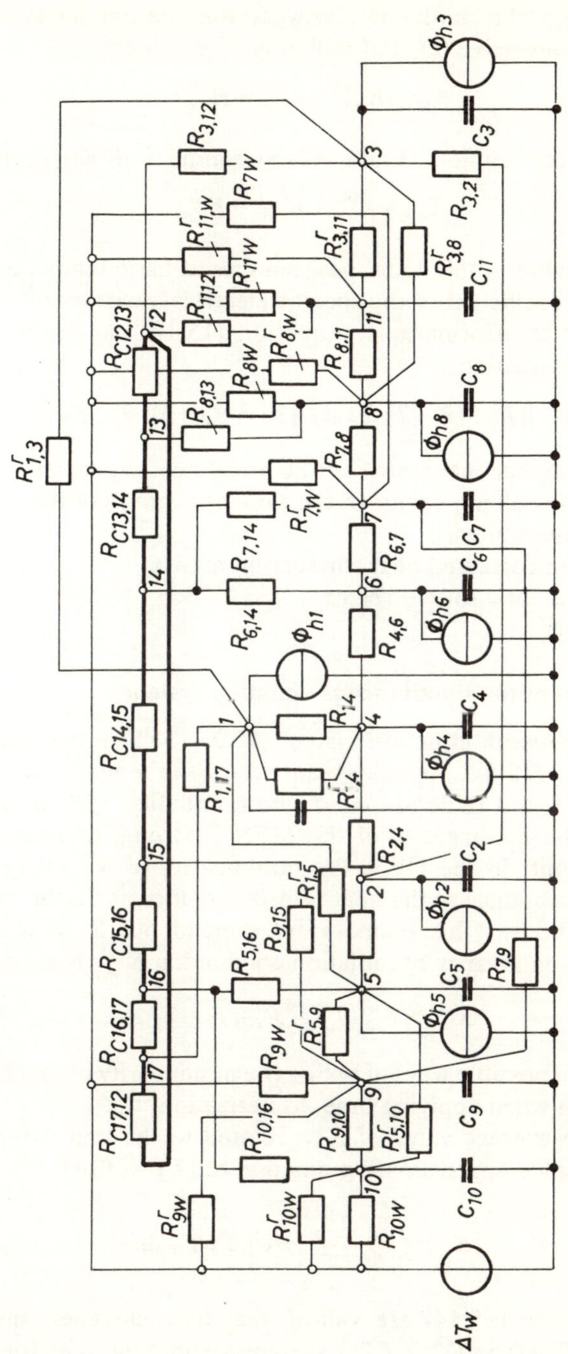

Figure 15.3 HFN part model of the machine shown in Figure 15.1

In this relationship, the conduction between the internal nodes is the sum of conductive (k), convective (c), and radiating (r) conductions:[3]

$$K_{i,j} = K_{i,j}^{(k)} + K_{i,j}^{(c)} + K_{i,j}^{(r)}; \tag{15.11}$$

the conduction between node 1 and the environment of temperature T_w

$$K_{i,w} = k_{i,w}^{(k)} + K_{i,w}^{(c)} + K_{i,w}^{(r)} \tag{15.12}$$

ϕ_i^0 is the basic value of the nodal heat sources at basic temperature T_0; β is the temperature coefficient of the specific electrical resistance.[13]

After suitable transformations, Equation (15.10) can be written as the following matrix equation:[3]

$$[C]\{\dot{T}\} + [K](T_i, T_j, \phi_i^0 \{T\} = \{Q\}(\phi_i^0, T_w, T_0), \tag{15.13}$$

where $[C]$ is the diagonal matrix of the nodal heat capacities,
$\{\dot{T}\}$ is the column vector of the time derivatives of the nodal temperatures,
$[K]$ is the corrected heat conduction matrix,
$\{Q\}$ is the total source vector.

15.1.6 Solution of the simultaneous equation system

The mass flow network equation system (15.5)–(15.9) is not linear (see also Equations (15.1)–(15.4)). As a consequence of the temperature dependence of the convective and radiation conduction, equation system (15.13) is not linear either. The nonlinearity of the MFN is strong, therefore it must be taken into account. In the HFN, the nonlinearity of $K_{i,j}^{(c)}$ is caused by the temperature dependence of the heat transfer coefficient. In the case of forced convection of the air, this temperature dependence is weak and can be neglected. The nonlinearity of radiation conduction $K_{i,j}^{(r)}$ is strong:[3]

$$K_{i,j}^{(r)} = b_{i,j}^{(r)}(T_i^2 + T_j^2)(T_i + T_j). \tag{15.14}$$

There are several possible ways of taking the nonlinearity of radiation conduction into account when applying time discretization.[3,16,17]

Producing the average value of $K_{i,j}^{(r)}$, related to the time interval,[3] on the basis of the assumed approximating function of $\{T\} = \{T\}(t)$:

$$[\hat{K}] = \frac{1}{2\,\Delta t}\int_{t-\Delta t}^{t+\Delta t} [K](T_i, T_j)\,dt. \tag{15.15}$$

If $T_i^{(k)}$ and $T_j^{(k)}$ in (15.14) are values near to each other, then, with the substitution of $T_a = 0.5\,(T_i^{(k)} + T_j^{(k)})$, the approximating value for the (k+1)th time step,

$$K_{i,j}^{(r)} \cong b_{i,j}^{(r)} 4T_a^3, \tag{15.16}$$

can be applied with small error.

In the exceptional case when during the transient the value of T_a does not change essentially and, in addition, the share of the radiation in the heat flux balance is small, $K_{i,j}^{(r)}$ can be considered constant with its mean value (linear case).

Considering the dual-function elements $\Delta p_{u,i}$ and $R_{c,i,j}$ of the part networks, the equation systems of the MFN and of the HFN have to be solved simultaneously. According to a possible strategy for the solution, the main steps of the procedure are as follows:

1. Input of data, preliminary calculations.
2. In the case of a start from the cold state,

   $$T_i(0) = T_w(0), \qquad i = 1, \ldots, n$$

   generation of the MFN for the first time step.
3. Solution of the MFN equation system (e.g. by the Newton–Raphson method), determination of $\{\dot{m}\}^{(1)}$.
4. Generation of the HFN based on $\{\dot{m}\}^{(1)}$, $\{T\}(0)$ and $\{\phi\}(0)$.
5. Solution of the HFN equation system, determination of $\{T\}^{(1)}$, with using one of the time difference schemes to be detailed later.

In the further time steps, the computation cycles are repeated from the second step on.

If the speed of the motor being examined is not constant in the transient phase, the time function of the speed as well as the parameters A, B, C and a of Equation (15.1) belong to the input data system and have to be identified by time steps. Similarly to the input data system belong the time functions of the load, of the source heat fluxes and of the environmental temperature. If the r.p.m. is constant and the influence of $\Delta p_{u,i}$ is small, the branch mass flow rates of the MFN vary to only a small extent during the transient. In such a case it is not necessary to generate the MFN by time steps and to produce the solution of the MFN equation system in each time step. The necessity of repeated solution can be prescribed according to the norm stated in the program.

Various numerical methods have been applied to solve the HFN equation system.

The finite difference time scheme according to Crank and Nicolson,[18] with the notation

$$\begin{aligned} \{T\}_2 &= \{T\}(t+\Delta t), \qquad & \{Q\}_2 &= \{Q\}(t+\Delta t), \\ \{T\}_1 &= \{T\}(t), \qquad & \{Q\}_1 &= \{Q\}(t) \end{aligned} \tag{15.17}$$

is

$$[C]\frac{\{T\}_2-\{T\}_1}{\Delta t}+[K]\frac{\{T\}_2+\{T\}_1}{2}=\frac{\{Q\}_2+\{Q\}_1}{2}. \tag{15.18}$$

The scheme gives an unconditionally stable approximation of second order in time.

The time marching scheme using the least-squares method is:[19]

$$\left[[K]^{\mathrm{T}}[K]\frac{\Delta t}{3}+\frac{[K]^{\mathrm{T}}[C]+[C]^{\mathrm{T}}[K]}{2}+[C]^{\mathrm{T}}[C]\frac{1}{\Delta t}\right]\{T\}_2$$
$$=\left[-[K]^{\mathrm{T}}[K]\frac{\Delta t}{6}+\frac{[K]^{\mathrm{T}}[C]-[C]^{\mathrm{T}}[K]}{2}+[C]^{\mathrm{T}}[C]\frac{1}{\Delta t}\right]\{T\}_1$$
$$+\frac{[K]^{\mathrm{T}}}{\Delta t}\int_0^{\Delta t}\{Q\}t\,\mathrm{d}t+\frac{[C]^{\mathrm{T}}}{\Delta t}\int_0^{\Delta t}[Q]\,\mathrm{d}t. \tag{15.19}$$

The scheme is unconditionally stable and its coefficient matrix is always symmetrical, even for an asymmetrical $[K]$.

The linear Galerkin finite time-element scheme[2,23] is

$$\left[\tfrac{1}{3}[K]+\frac{1}{2\,\Delta t}[C]\right]\{T\}_2=-\left[\tfrac{1}{6}[K]+\frac{1}{2\,\Delta t}[C]\right]\{T\}_1+\frac{1}{\Delta t^2}\int_0^{\Delta t}\{Q\}t\,\mathrm{d}t. \tag{15.20}$$

The scheme is unconditionally stable and gives an approximation of first order in time.

Based on the known initial value of $\{T\}_0$, the quadratic Galerkin finite time-element scheme for determining $\{T\}_1$ and $\{T\}_2$ can be given by the relationship

$$\left\{\frac{1}{15}\begin{bmatrix}8[K] & [K]\\ [K] & 2[K]\end{bmatrix}+\frac{1}{12\,\Delta t}\begin{bmatrix}[Q] & 4[C]\\ -4[C] & 3[C]\end{bmatrix}\right\}\begin{bmatrix}\{T\}_1\\ \{T\}_2\end{bmatrix}$$
$$=\left\{-\frac{1}{15}\begin{bmatrix}[K]\\ -0.5[K]\end{bmatrix}+\frac{1}{12\,\Delta t}\begin{bmatrix}4[C]\\ -[C]\end{bmatrix}\right\}\{T\}_0+\frac{1}{15}\begin{bmatrix}10\\ 2.5\end{bmatrix}\{Q\}. \tag{15.21}$$

The scheme is self-starting, unconditionally stable, and its approximation is of second order for $\{T\}_1$ and of first order for $\{T\}_2$. Its coefficient matrix is of the size $2n\times 2n$ and always asymmetrical.[24]

The quadratic Galerkin finite time-element three-level scheme, $\{T\}_0$ and $\{T\}_1$ are known values:

$$\left[\frac{2}{15}[K]+\frac{3}{12\,\Delta t}[C]\right]\{T\}_2=\left[\frac{1}{30}[K]-\frac{1}{12\,\Delta t}[C]\right]\{T\}_0$$
$$+\left[\frac{4}{12\,\Delta t}[C]-\frac{1}{15}[K]\right]\{T\}_1+\frac{5}{45}\{Q\}. \tag{15.22}$$

The scheme is not self-starting, unconditionally stable and its approximation is of first order in time.

15.1.7 Applications, results

The application of the simultaneous network model is demonstrated on the asynchronous motor sketched in Figure 15.1, for operational conditions under various load combinations.

The values of the coefficients of flow and heat transfer, as well as the pressure function of the ventilator according to Equation (15.1) have been verified by measurements. In the motor in question the r.p.m. of the ventilator is constant, except for the few seconds of starting from stand still. For this reason, and because the variation and influence of the pressure source Δp_u according to Equation (15.4) is small, it is superfluous to solve the MFN equation system by time steps. Values of the branch mass flow rates in the air temperature range between 65 °C and 80 °C are:

$$\dot{m}_g = 0.36 \text{ kg s}^{-1}, \dot{m}^2 = 0.146 \text{ kg s}^{-1},$$

$$\dot{m}_v = 2.434 \times 10^{-2} \text{ kg s}^{-1}, \dot{m}_F = 1.267 \times 10^{-2} \text{ kg s}^{-1}.$$

I Transient of starting, with full load

Starting takes place from a standstill, the motor being at the temperature of the environment ($T_i = T_w$). A current surge of $0 < t < 5$ s and $I_0 = 5I_r$ is taken into account.

To compare the different numerical schemes, the analysis has been performed for the 'linear case', with different time steps.

The results obtained for the temperature of the winding head on the V1 side (3rd node) are summarized in Table 15.1. In the comparison of the results, the values obtained by the method of least-squares, with time steps of $\Delta t = 100$ s, are considered bases of comparison, in respect of accuracy.

As can be seen from Table 15.1, the quadratic Galerkin scheme $\{T\}_0$, $\{T\}_1 \rightarrow \{T\}_2$ satisfies the accuracy requirements for the solution of the given problem only in the case of $t \leq 100$ s. As regards accuracy, the other schemes examined are practically equivalent in the case of $t \leq 1200$ *s*. At the beginning of the transient, the linear Galerkin scheme according to (15.20) has the highest sensitivity to the time step (it appears with $\Delta t = 1200$ s), while it begins at $\Delta t = 3600$ s with the other schemes.

To compare the time consumption of the schemes, the HFN equation system was solved with identical time steps by placing the program part of each scheme as an independent subroutine in the main program. The subroutine used for solving the linear equation system performs the elimination of the coefficient matrix by partial selection of the main elements. As a starting

Table 15.1 Results obtained using different schemes and differing time steps, for the start period with full load. (Temperatures in °C; initial temperature: 18 °C.)

Time t (s)	Time step t (s)	Schemes: Crank Nicolson	Least Squares	Linear Galerkin	Quadratic Galerkin 2→1	Quadratic Galerkin 1→2
600	100	20.0	19.97	19.87	18.81	19.99
	600	20.11	19.79	19.37	—	19.77
1200	100	31.12	31.11	30.95	30.29	31.11
	600	31.36	31.11	30.32	27.32	31.02
	1200	31.99	31.17	29.86	—	30.34
3600	100	53.15	53.22	53.03	52.87	53.14
	600	53.24	53.82	53.54	51.16	53.09
	1200	53.51	54.83	52.12	49.02	52.88
	3600	60.61	58.58	53.09	—	51.72
7200	100	64.34	64.38	64.28	64.25	64.34
	600	64.37	64.70	64.02	63.76	64.32
	1200	64.45	65.13	63.76	62.92	64.27
	3600	64.32	68.50	63.30	60.91	63.09
10800	100	68.26	68.28	68.23	68.23	68.26
	600	68.28	68.44	68.09	68.05	68.25
	1200	68.34	68.63	67.94	67.78	68.22
	3600	69.01	69.42	67.61	66.06	67.93
14400	100	69.75	69.76	69.74	69.74	69.75
	600	69.76	69.84	69.66	69.68	69.75
	1200	69.77	69.92	69.59	69.58	69.74
	3600	70.04	70.09	69.42	68.71	69.62
18000	100	70.33	70.33	70.32	70.32	70.33
	600	70.33	70.37	70.28	70.30	70.33
	1200	70.35	70.51	70.25	70.27	70.32
	3600	70.36	70.63	70.15	69.90	70.26

scheme to the scheme (15.22), the scheme based on the method of least squares according to (15.19) was applied. The CPU times of the schemes were determined relative to 100 time steps. The CPU time of scheme (15.22) relates to 50 'double' steps and includes also the period of the starting scheme. The CPU times (in seconds) for each scheme are as follows:

Crank–Nicolson:	13.58
Least squares:	23.74
Linear Galerkin:	14.52
Quadratic Galerkin, ($\{T\}_0 \rightarrow \{T\}_1, \{T\}_2$)	14.54
Quadratic Galerkin, ($\{T\}_0, \{T\}_1 \rightarrow \{T\}_2$)	15.40

As can be seen from the comparison, the method of least squares is the most time-consuming, as might be expected. The Crank–Nicolson scheme is remarkable for its accuracy and low time consumption.

The results obtained for the 'linear case' and those which take into consideration the nonlinearities of the heat radiation by (15.15) are compared in Figure 15.4.

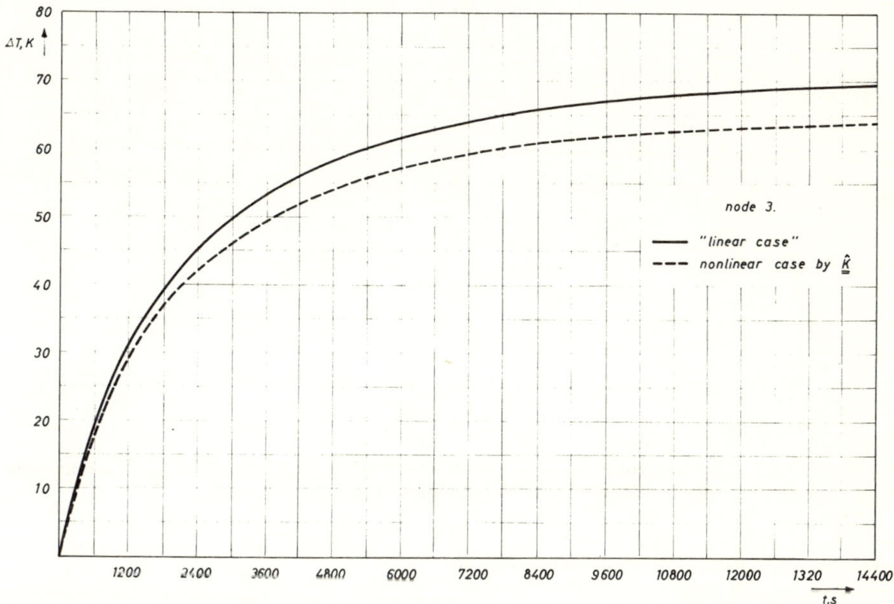

Figure 15.4 Comparison of the $T_3(t)$ functions calculated as a 'linear case' and by the use of (15.15) respectively

II Intermittent operation with varying load

The load combination examined is as follows:

$$
\begin{array}{rll}
0^{+} < t < 5\ \text{s}, & I = 5I_r & \text{(start)}, \\
5\ \text{s} < t < 300\ \text{s}, & I = 1.5I_r, & \\
300\ \text{s} < t < 480\ \text{s}, & I = I_r, & \\
480\ \text{s} < t < 600\ \text{s}, & I = 0.14I_r & \text{(idle run)} \\
600\ \text{s} < t < 660\ \text{s}, & & \text{(standstill)}.
\end{array}
$$

After this, the load combination is repeated.

The examination was aimed at establishing after how many loading cycles the motor will be switched off by the thermal protection adjusted to an overtemperature of $(\Delta T_3)_{\text{perm}} = 80\ °\text{C}$. The curve of T_3(t), obtained as the result of the calculation using Equation (15.18), for the 'linear case' and using (15.15) is shown in Figure 15.5. Switching off takes place in the 6th and 8th loading cycle, respectively.

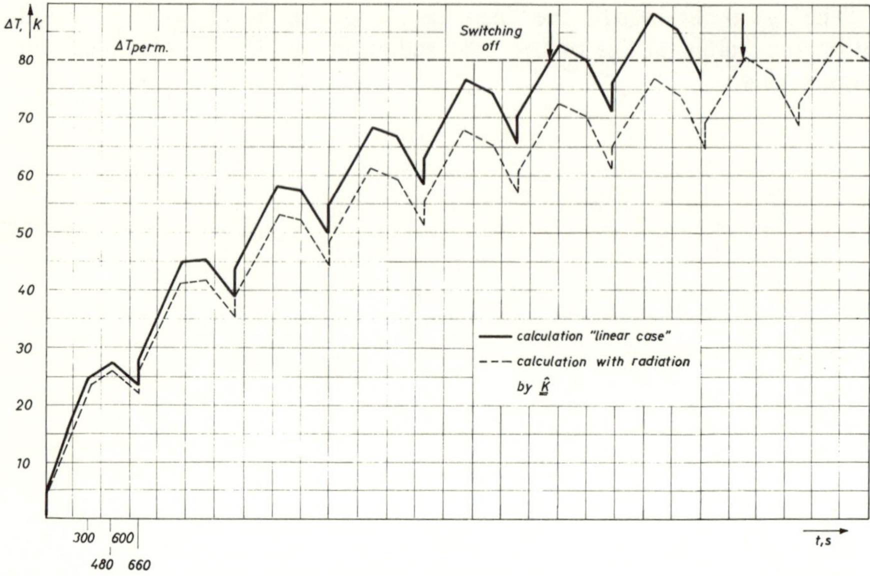

Figure 15.5 Temperature–time function of the third node for the intermittent operation with varying load calculated by the use of (15.15)

III Operation with periodical loading

Data of the load variation:

$$\begin{aligned} 0^{+} < t < 5\ \text{s}, &\quad I = 5I_r(\text{start}) \\ 5\ \text{s} < t < 900\ \text{s}, &\quad I = 1.5I_r \\ 900\ \text{s} < t < 3600\ \text{s}, &\quad I = 0.14I_r \quad (\text{idle run}). \end{aligned}$$

After this, the loading cycle is repeated. The results obtained from computation were verified by measurements in the course of testing the motor protection. For the measurements, temperature sensors had been built into the winding heads. The weighted mean values of the measured temperatures were considered decisive.

The calculated and measured temperatures are shown in Figures 15.6 and 15.7, respectively. (Figure 15.7 is a larger-scale variant of the third cycle of Figure 15.6.)

The curves of $T_3(t)$ obtained by calculation using Equations (15.18) and (15.15), and by measurements, respectively, intersect each other. In spite of the strongly simplifying assumptions, the calculated and measured values exhibit fairly good agreement in the warming phase. In the cooling phase, the largest instantaneous deviation is 8 °C. The deviations do not accumulate in the subsequent loading cycles. The results have been used for the solution of motor protection problems.[25]

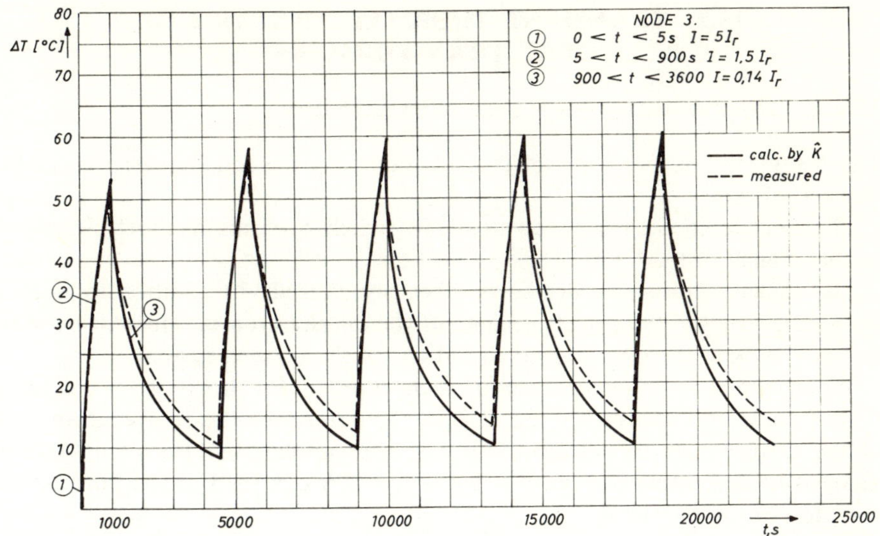

Figure 15.6 Calculated and measured temperatures of the third node for operation with periodical loading

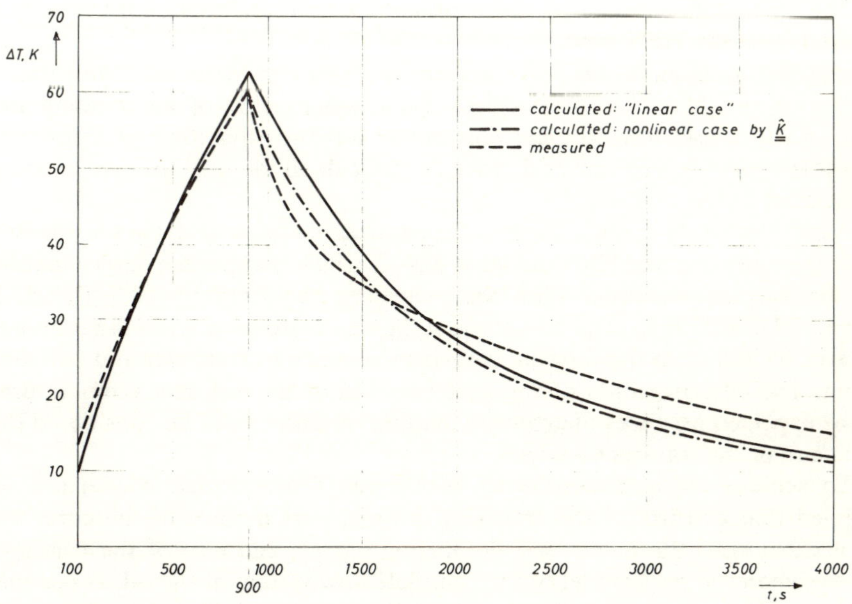

Figure 15.7 Larger-scale variant of the third cycle of Figure 15.6

15.2 TRANSIENT THERMAL ANALYSIS OF OFAF TRANSFORMERS

15.2.1 The transient thermal state of transformers

The transient thermal state of transformers is produced by electric and/or thermal effects.

The electric transients caused mostly by switch-over, mains short, switch-off or the change of load, are of short duration.

The thermal transients are slow. They are caused—in addition to electric transients—by the temporary change of the ambient temperature or by variations arising in the operational conditions of the cooling system.

When dimensioning transformers or checking up for short-circuit warming, it is of great importance for the designer to determine the temperature field of the transformer to be expected. The user applies the transient thermal analysis to determine the permissible temporary overload or the permissible duration of the overload.

15.2.2 The transient thermal model of OFAF transformers

There have been several initiatives in the past two decades for the computer analysis of the steady-state warming of oil transformers.[25–33]

The recommendations for the transient thermal analysis of transformers started from the HFN model with two heat capacitances.[34] The international standard recommendation IEC *Loading Guide for Oil-immersed Transformers* mentions also two concentrated heat capacitances, that of the winding and that of the whole transformer. This model has the advantage of simplicity, but this causes inaccuracy and makes it difficult to identify the temperature calculated by the method.

A more accurate consideration of the physical phenomena are made possible with the multi-storage HFN model of Kiss,[35] which permits the determination of the average warming. With the increasing requirements of accuracy it appeared desirable to take into account the effects deriving from the transient flow of oil and from the thermal couplings between transformer and environment. The determination of the time function of the hot-spot temperatures to be expected for cases of arbitrary loading conditions can be considered the final aim of the transient analysis.

To achieve the aim mentioned, a thermohydrodynamical model can be applied that consists of the transient thermal part models of the core, the heat exchanger, the house and the mixing spaces, and also of the transient hydrodynamical part model of the complete flow system of the oil. A possible variant of the simultaneous thermohydrodynamical model consists of network part models.[36]

15.2.3 The mass flow network part model of the transformer

The simplified block scheme of the MFN part model of the transformer with disc winding sketched in Figure 15.8 is shown in Figure 15.9. The MFN part model serves to determine the instantaneous oil mass flow rates of each oil duct.

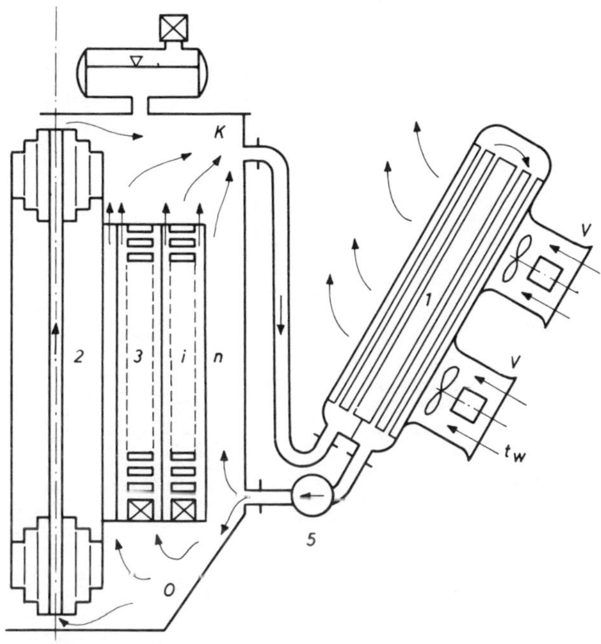

Figure 15.8 Simplified scheme of an OFAF cooled oil transformer—the arrows indicate the flow direction of oil and air respectively

The iron core (2), the winding columns $(3, \ldots, i)$ and the shunt channel (n) between the winding and the housing constitute a parallel connected flow path system. From the upper space K the oil flows into the heat exchanger (1), where it is recooled by the air of temperature T_w kept flowing by ventilators V. The oil pump (5) delivers the oil from the heat exchanger into the lower dividing space (O), from where it flows into the channels $2, \ldots, i, \ldots, n$.

The oil is kept flowing in the flow resistances R_i of the channels by the pressure source Δp_0 of the pump and the hydrostatic pressure differences Δp_i arising on the flow loops. The accelerating pressure sources in the branches are notated as Δp_{ai}. All the temperature-dependent elements of the MFN are dual-function network elements.

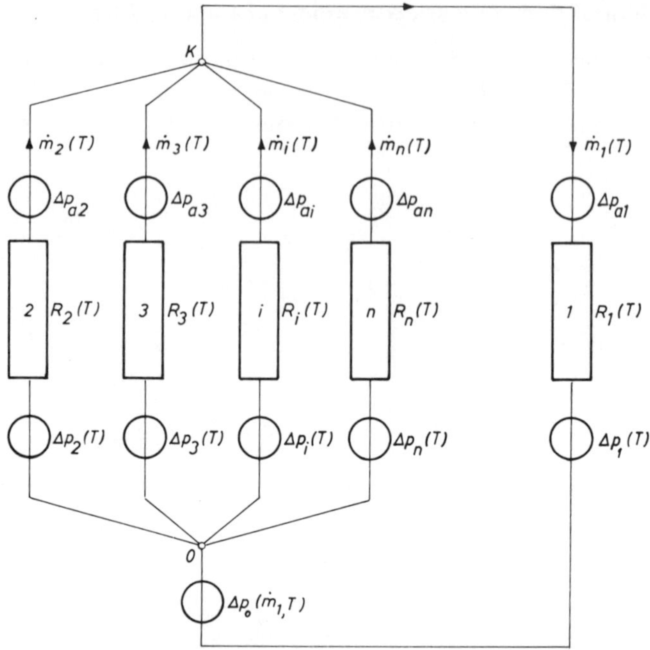

Figure 15.9 Schematic diagram of the transient MFN part model

The main model conditions are the following:[36]

TM1. In the description of the flows, spaces K and O are considered isothermal and to have zero flow resistance.

TM2. The three phase windings, and also the heat exchangers and their couplings, are equivalent from thermal and flow viewpoints.

TM3. In the pipes of the heat exchangers and pumps, coupled to the housing, the variation of oil temperature is neglected.

TM4. In the shunt flow path the local natural convective effects are disregarded.

TM5. The characteristic curve of ventilators V and pump[5], and the temperature function of the latter are known from measurement data.

The state functions of the elements of the mass flow network are as follows: Full flow resistance on the side of the heat exchanger:

$$R_1 = R_C + R_H + R. \tag{15.23}$$

In the equation, the flow resistance R_C of the connecting pipe of the heat exchanger, taking TM3 into consideration,[1,4] is

$$R_C = \frac{8C_{DC}l_C\dot{m}_1}{d_C^5\pi^2 s^2 \rho_C} \tag{15.24}$$

where C_{DC} is the flow resistance coefficient of the connecting pipe,
d_C is the diameter of the pipe,
l_C is the length of the pipe,
s is the number of the parallel connected heat exchangers,
$\dot{m}_1$ is the full mass flow rate on the heat-exchanger side.

R_H in Equation (15.23) is the flow resistance of the heat exchanger:

$$R_H = \frac{C_{DH}\dot{m}_1}{2d_H A_H^2 s^2 b^2}\int_0^{l_H} \rho^{-1}[T(l)]\,dl, \tag{15.25}$$

where C_{DH} is the coefficient of flow resistance in the oil ducts of the heat exchanger,
d_H is the hydraulical diameter,
A_H is the flow cross section of the duct,
b is the number of the parallel connected ducts in the heat exchanger,
l_H is the length of the oil ducts.

In Equation (15.23) R_3 is the full shape resistance of the couplings. The kth shape resistance is

$$R_{\zeta,k} = \zeta_k \frac{\dot{m}_1}{2\rho_k A_{Ck} s^2} \tag{15.26}$$

where ζ_k is the coefficient of shape resistance,
A_C is the decisive cross section of flow.

In the network according to Figure 15.9, R_i is the flow resistance of the winding columns, the iron core and the shunt oil ducts ($i = 2, \ldots, n$):

$$R_i = \frac{C_{Dw,i}}{d_{w,i}A_{w,i}}\int_0^{l_i} \frac{\mu}{\rho}[T(l)]\,dl, \tag{15.27}$$

where $C_{Dw,i}$ is the flow resistance of the oil ducts, determined experimentally,
$d_{w,i}$ is the hydraulic diameter,
$A_{w,i}$ is the flow cross section of the duct,
l_i is the length of the duct,
μ is the dynamic viscosity of the oil.

The pressure source of the pump:

$$\Delta p_0(\dot{m}_1, T) = A + B\dot{m}_1 + C\dot{m}_1^a, \tag{15.28}$$

where A, B, C and a are values dependent on the r.p.m. and the temperature, for the given pump.

The hydrostatic pressure source Δp_i is

$$\Delta p_i = g\int_0^{z_i} \rho[T(z)]\,dz, \tag{15.29}$$

where z_i is the vertical dimension of the ith branch,
g is the gravitational acceleration.

The acceleration pressure source is

$$\Delta p_{a,i} = a_i\int_0^{l_i} \rho[T(l)]\,dl, \tag{15.30}$$

where a_i is the acceleration of the oil in the ith branch,
l_i is the length of the oil column.

Most elements of the MFN are dependent on the mass flow rate, thus the network is not linear. To identify the temperature-dependent MFN elements, it is necessary to know the temperature variation of the oil according to place and time.

15.2.4 Thermal part model of the transformer

The transient thermal part model of the transformer serves to determine the time function of the temperature of the oil, the winding, the iron core, the housing and the heat exchanger. The model is built up from elementary blocks discretized according to place. The elementary blocks are coupled into block columns in the direction of the oil flow by suitable coupling conditions[36,37] (Figure 15.10).

The main model conditions are the following:

TH1. The heat transfer between the iron core and the winding, as well as between the winding columns can be neglected.

TH2. The temperature of the housing cover is assumed to be equal to the temperature of the warmest oil in the space K.

TH3. The loss by eddy current in the wall of the housing passes into the oil flowing in the shunt flow path and into the outside air.

TH4. The heat-transfer conditions are identical on the lower and upper surfaces of the winding discs.

TH5. In the oil ducts adjacent to the discs, the temperature of the oil is considered to be equal to the mixed-mean temperature of the oil arriving at the disc.

TH6. In the iron core and in the heat exchanger, the conduction heat fluxes in the direction of oil flow are disregarded.

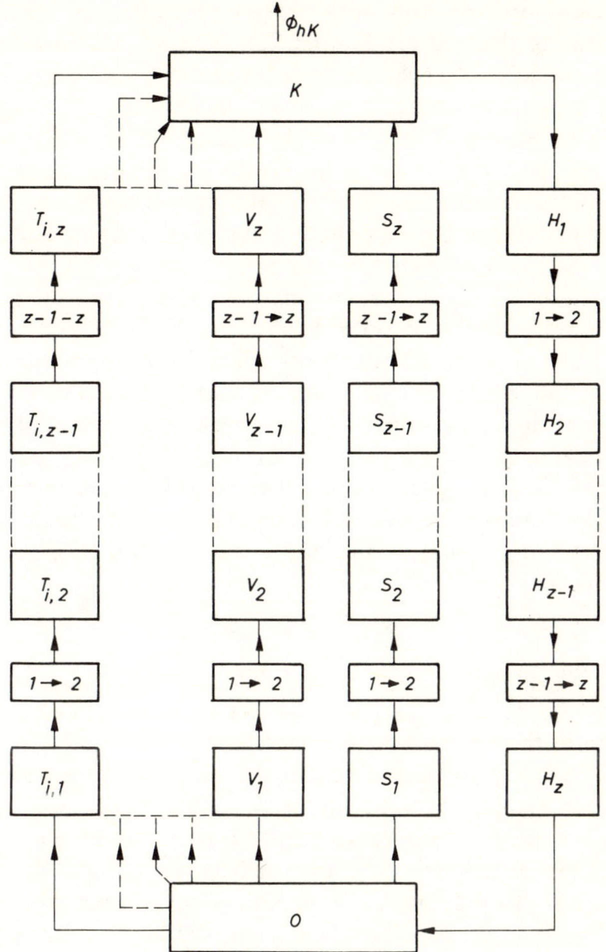

Figure 15.10 Scheme of the transient thermal part model of the transformer built up from elementary blocks

TH7. In the model blocks of the heat exchanger, the temperature of the oil and of the air is considered to be equal to the mixed-mean temperature.

TH8. The heat transfer resistance between the heat exchanger and the external air is considered to be constant during the transient.

The model blocks shown in Figure 15.10 may be elementary HFN models of different details in compliance with the requirements of calculation. For a thermal transient process, sufficient details are generally contained in the information stating the average temperature of each winding disc. In such a case, the heat capacity of each winding disc is taken into account by a single concentrated heat capacitance in the HFN model block.[36] If more detailed information is necessary, the HFN of the model blocks of the winding discs can be prepared with a finer lumping in which a single turn or pair of turns constitutes a concentrated element.[3,37] In the case of moderate requirements, a single block may serve for modelling a group of discs consisting of two or more discs. It is also possible to use a coarser lumping for the HFN models of the lower model blocks of the winding discs and a finer lumping for those of the upper model blocks, in order to determine the hot-spot temperature.

Figure 15.11(a) shows a sketch of the HFN model of simple lumping for the elementary block of a disc (or group of discs).[36] The temperature-dependent source heat flux ϕ_h is equal to the sum of the ohmic and eddy current losses. The heat capacity of the disc (disc group) is C_1, that of the oil surrounding the disc (disc group) is C_2. The resulting heat-transfer resistance between the disc and the oil is R_{12}. The formal thermal resistance R_2 models the heat transporting action of the oil; it is a dual-function element of the heat flow network:

$$R_{2,i} = -(\dot{m}_i c_{p,i})^{-1} \tag{15.31}$$

where $c_{p,i}$ is the specific heat of the oil. The temperature sources ΔT_1 and ΔT_2 serve to state the initial temperatures of the nodes, while the temperature source ΔT_0 states the entering temperature of the oil.

The model blocks of the iron core (blocks marked with V in Figure 15.10) have a construction agreeing with that of the one sketched in Figure 15.11(a). In this case C_1 is the heat capacity of the discrete part of the iron core, and R_{12} is the resultant heat-transfer resistance between the iron core and the oil.

Figure 15.11(b) shows the HFN model of an elementary block of the transformer housing and the shunt flow path (the block column marked S in Figure 15.10). In the figure, node 1 represents the discrete part of the housing; ϕ_h is the heat flux source of the loss deriving from stray fluxes; R_w is the heat-transfer resistance between the housing and the external environment; ΔT_w is the temperature source serving to state the ambient temperature. Node 2 corresponds to the oil flowing in the shunt flow path.

The HFN block of a discrete part of the heat exchanger is sketched in Figure 15.11(c). In the figure, C_1 is the heat capacity of the discrete part of the heat exchanger, C_2 is the heat capacity of the oil in the discrete part, R_w is the heat transfer resistance on the air side, R_{20} is the formal thermal resistance modelling the heat-transporting action of the oil (with a positive sign in the case of cooling oil in Equation (15.31)). Temperature sources ΔT_1 and ΔT_2 serve to state the initial temperatures of nodes 1 and 2, while ΔT_w and ΔT_0 mean the entering temperature of the air and of the oil, respectively.

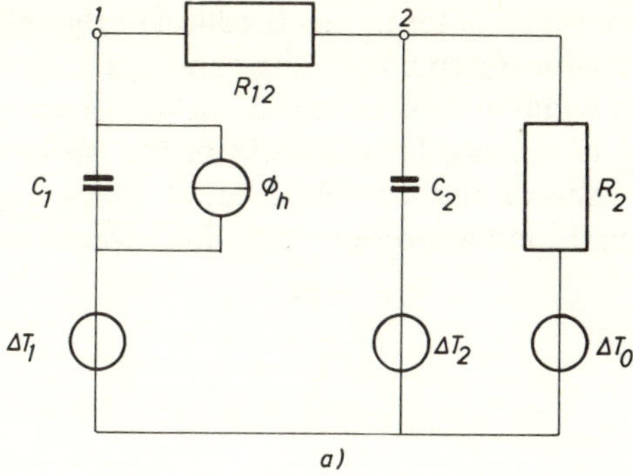

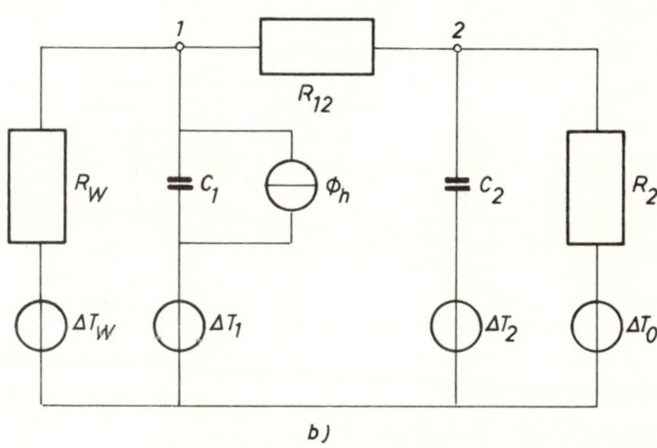

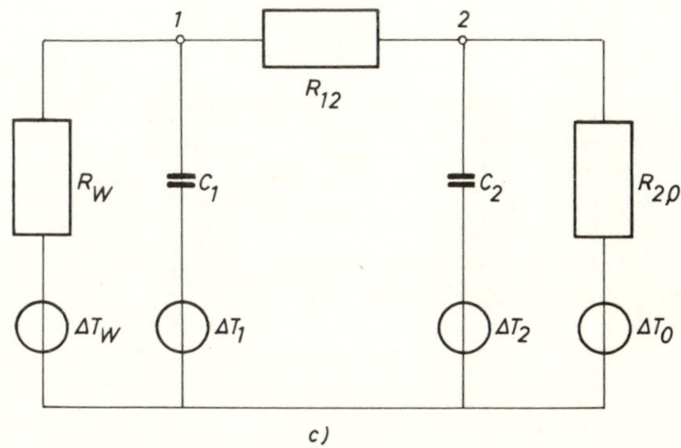

Figure 15.11 Simplified scheme of the HFN model block of (a) the winding column; (b) the shunt; and (c) the heat exchanger

If blocks lying over each other in the direction of the oil flow are connected into block columns with suitable coupling conditions, the HFN model of the complete iron core, winding column and shunt will be obtained. Figure 15.12 shows the coupling of the two lower blocks of the *i*th winding block column and Figure 15.13 shows the complete HFN block column of the heat exchanger. Coupling is, in the present case, downwards.

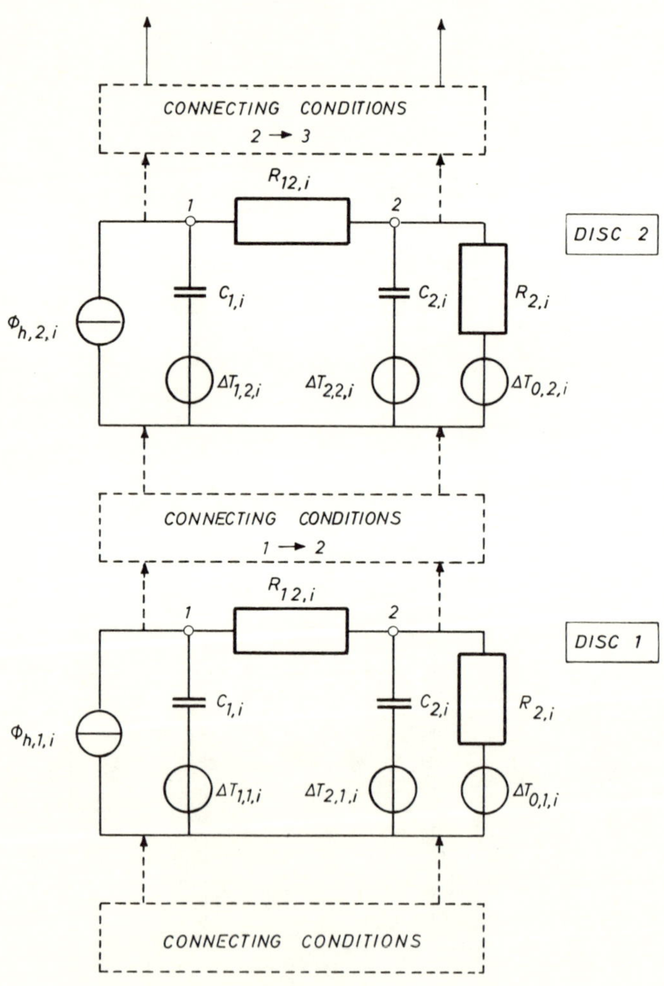

Figure 15.12 Block model of a winding column or of the iron core built up from elementary HFN blocks

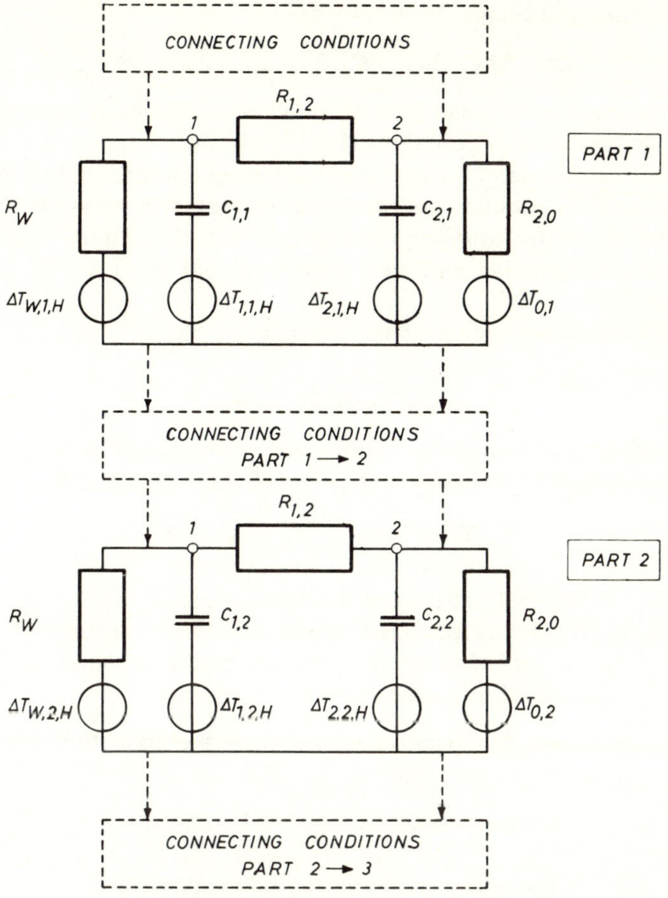

Figure 15.13 HFN block model of the heat exchanger

15.2.5 Equation system of the simultaneous model

For the MFN part model according to Figure 15.9, one independent nodal equation and $n-1$ independent loop equations may be written.[36] The nodal equation is

$$\dot{m}_1 - \sum_{i=2}^{n} \dot{m}_i = 0. \tag{15.32}$$

The independent loop equations can be written for the flow loops marked with 1 and $i\,(i=2,\ldots,n)$, by substituting the branch equation ($\Delta p_i = R_i \dot{m}_i$).

With the notation of Figure 15.9,

$$\Delta p_0 + \Delta p_1 - \Delta p_{a1} - R_1 \dot{m}_1 = \Delta p_i + \Delta p_{ai} + R_i \dot{m}_i. \tag{15.33}$$

The pressure sources and flow resistances in the loop equation can be identified on the basis of Equations (15.23)–(15.30).

The equation system of the thermal part model according to Figure 15.10 consists of the HFN equation system of the elementary blocks of block columns V, T, S and H, of the coupling equations of the elementary blocks, further of the transient heat flux balance equation of blocks O and K (as thermal mixing spaces).

The HFN equation system of the elementary blocks[36] (see also Equation (15.13)) is

$$[C]\{\dot{T}\} + [K]\{T\} = \{Q\}^{\mathrm{U}}. \tag{15.34}$$

The coupling equation of the blocks will be discussed in the next section.

The transient heat flux balance equation of dividing space O:

$$C_O \dot{T}_O + \sum_{i=2}^{n} \dot{m}_i c_{pO} T_O = \dot{m}_1 c_{p1} T_1, \tag{15.35}$$

where C_0 is the heat capacity of the oil in space O,
T_0 is the mixed mean temperature of the oil in space O,
T_1 is the mixed mean temperature of the oil entering from block column H into space O.

The transient heat flux balance equation of the collecting space marked with K is

$$C_K \dot{T}_K + \dot{m}_1 c_{pK} T_K + \phi_{hK} = \sum_{i=2}^{n} (\dot{m}_i c_{pi} T_i)_{in}, \tag{15.36}$$

in which C_K is the heat capacity of the oil in space K
T_K is the mixed mean temperature of the oil in space K,
ϕ_{hK} is the heat flux passing between space K and the external environment (see Figure 15.10).

The subscript 'in' notates the enthalpy flux of the oil passing from the oil ducts into space K.

15.2.6 Strategy for the solution of the simultaneous equation system

The simultaneous equation system describing the transient thermal processes of the transformer is nonlinear. Most elements of the MFN are dependent on the mass flow rate, and strongly temperature dependent. The corrected heat conduction matrix $[K]$ of the HFN equation system (15.34) is mildly temperature dependent, and its $R_{2,i}$ elements according to (15.31) are dependent on the mass flow rate.

The interaction of the thermal and flow phenomena taking place in the transformer is very strong. The construction discretized according to place of the model sketched in Figure 15.10 is aimed at determining the local temperatures of the oil and identifying the temperature dependent elements of the MFN. For this reason, the branches of the MFN according to Figure 15.9 are similarly built up as block columns (Figure 15.14).

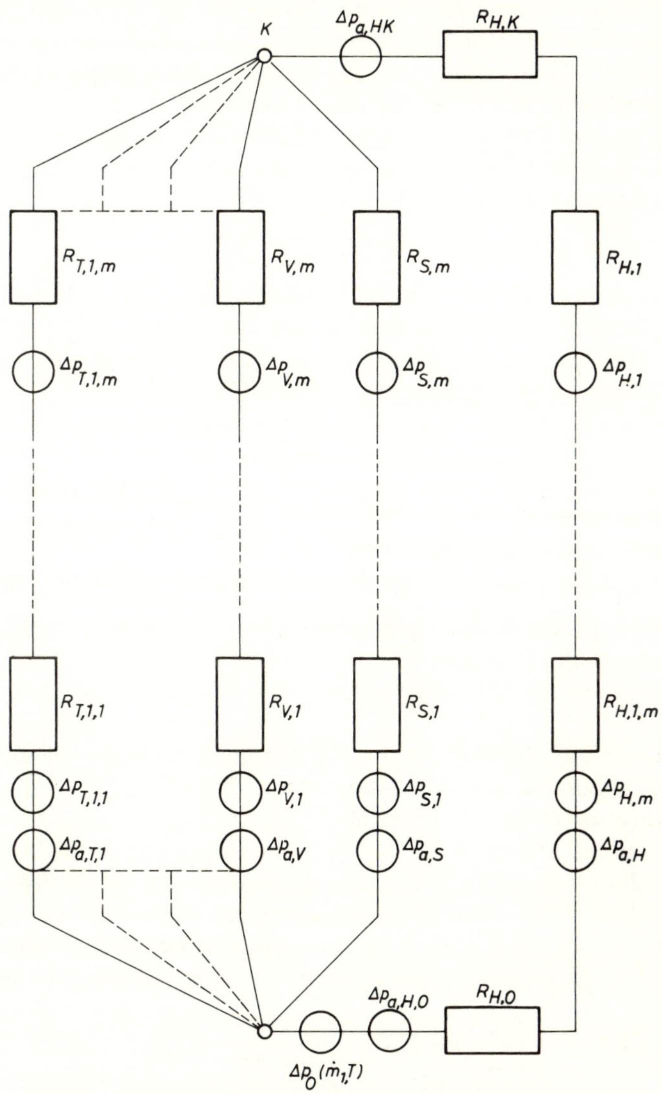

Figure 15.14 MFN part model built up from elementary blocks

The elementary blocks of the block columns in Figures 15.10 and 15.14 are coupled with suitable coupling conditions. It can easily be proved that the equation system describing the complete block columns can be separated at the coupling points into equations for the elementary blocks and coupling equations. This separation is justified by the endeavour to find a simple strategy for the solution.

Several calculation strategies can be applied to solve the equation system of the simultaneous model.[36,37] Considering the nonlinearities, numerical methods based on time discretization appear to be advantageous.

Two possible variants of the strategy based on the solution of the network equation system will briefly be discussed below.

The first variant uses a mean-time substitution[36] according to which the mass flow rate of the oil in the time interval $k\,\Delta t < t < (k+1)\,\Delta t$ is taken into account with the mean value

$$\dot{m}_i^{(k+0.5)} = 0.5(\dot{m}_i^{(k)} + \dot{m}_i^{(k+1)}). \tag{15.37}$$

The dual function elements of the HFN models $R_{2,i}$ of the elementary blocks, according to (15.31), are identified on the basis of $\dot{m}_i^{(k+0.5)}$. In the overall mean-time substitution, the temperatures of the oil entering the elementary blocks are identified in a similar way ($\Delta T_0 = T_0^{(k+0.5)}$).

The production of $\dot{m}_i^{(k+0.5)}$ requires to be checked step by step and must be made more accurate by iteration if necessary.[36]

A simpler strategy based on the solution of the network equation system uses the initial values of the mass flow rates and applies the mean-time coupling condition only to the oil temperatures.

In the case of the latter strategy the main steps of the solution will be:

(a) preliminary calculations, generation of the elements of constant value,
(b) selection of the time step,
(c) identification of the mass flow network elements, based on the initial state $\{T\}^{(k)}$,
(d) solution of the equation systems (15.32) and (15.33), e.g. by the Newton–Raphson method, and calculation of the vector $\{\dot{m}\}^{(k)}$.
(e) identification of the R_{ci} elements of the HFN,
(f) in the time interval $k\,\Delta t < t < (k+1)\,\Delta t$—starting from block O and proceeding from block to block in the direction of the oil flow—solution of the heat flow network equation system (15.34) of the elementary blocks, e.g. by the Crank–Nicolson time scheme, and calculation of the desired temperature column vector $\{T\}^{(k+1)}$ (see also Equation (15.18)):

$$\{T\}^{(k+1)} = \left[\frac{[C]}{\Delta t} + 0.5[K]^{(k)}\right]^{-1} \left\{\left[\frac{[C]}{\Delta t} - 0.5[K]^{(k)}\right]\{T\}^{(k)} + 0.5[\{Q\}^{(k+1)} + \{Q\}^{(k)}]\right\}. \tag{15.38}$$

The mean-time coupling condition of the elementary blocks coupled to space O is stated by extrapolation (see also Figures 15.10 and 15.12):

$$\Delta T_{O,1,i} = T_O^{(k+0.5)} \cong T_O^{(k)} + 0.5(T_O^{(k)} - T_O^{(k-1)}). \qquad (15.39)$$

$T_O^{(k)}$ can be produced from Equation (15.35) discretized with respect to time:

$$C_O \frac{T_O^{(k)} - T_O^{(k-1)}}{\Delta t} + \sum_{i=2}^{n} \dot{m}_i c_{p,0} \frac{T_O^{(k)} + T_O^{(k+1)}}{2} = \dot{m}_1 c_{p,1} \frac{T_{2,z,1}^{(k)} + T_{2,z,1}^{(k-1)}}{2}. \qquad (15.40)$$

The predictive values $\Delta T_{O,1,i}$ must be checked at the end of each cycle, and in case of deviations greater than permissible the calculation has to be repeated with the more accurate values obtained.

Since the calculation is performed in each cycle from block to block, the result of the previous block is available for the next cycle when the coupling conditions of the $2, \ldots, z$th blocks are stated. The initial and coupling conditions of the elementary block networks of the block columns marked with $i = 2, \ldots, n$ are (see also Figures 15.10 and 15.12):

$$\begin{aligned}
0 \to 1:&\quad \Delta T_{1,1,i} = T_{1,1,i}^{(k)}; \quad \Delta T_{2,1,i} = T_{2,1,i}^{(k)}; \quad \Delta T_{0,1,i} = T_0^{(k+0.5)};\\
1 \to 2:&\quad \Delta T_{1,2,i} = T_{1,2,i}^{(k)}; \quad \Delta T_{2,2,i}^{(k)} = T_{2,2,i}^{(k)}; \quad \Delta T_{0,2,i} = 0.5(T_{2,1,i}^{(k+1)} + T_{2,1,i}^{(k)});\\
&\quad \vdots \qquad\qquad \vdots \qquad\qquad \vdots\\
(z-1) \to z:&\quad \Delta T_{1,z,i} = T_{1,z,i}^{(k)}; \quad \Delta T_{2,z,i} = T_{2,z,i}^{(k)}; \quad \Delta T_{0,z,i} = 0.5(T_{2,z-1,i}^{(k+1)} + T_{2,z-1,i}^{(k)}).
\end{aligned} \qquad (15.41)$$

The first block of heat exchanger H is coupled to space K (branch marked with $i = 1$ in the MFN). Mean-time statement of $\Delta T_{0,1,1}$ (see also Figure 15.13):

$$\Delta T_{0,1,1} = T_K^{(k+0.5)} = 0.5(T_K^{(k)} + T_K^{(k+1)}), \qquad (15.42)$$

where $T_K^{(k+1)}$ can be calculated from the transient heat flux balance equation (15.36), discretized according to time:

$$C_K \frac{T_K^{(k+1)} - T_K^{(k)}}{\Delta t} + \dot{m}_1 c_{p,K} \frac{T_K^{(k+1)} + T_K^{(k)}}{2} + \frac{\phi_{hK}^{(k+1)} + \phi_h^{(k)}}{2} = \sum_{i=2}^{n} \dot{m}_i c_{p,i,z} \frac{T_{2,i,z}^{(k+1)} + T_{2,i,z}^{(k)}}{2}. \qquad (15.43)$$

Initial and coupling conditions of the elementary blocks of the heat exchanger block column are (see also Figure 15.13):

$$\begin{aligned}
K \to 1:&\quad \Delta T_{1,1,1} = T_{1,1,1}^{(k)}; \quad \Delta T_{2,1,1} = T_{2,1,1}^{(k)}; \quad \Delta T_{0,1,1} = T_K^{(k+0.5)}\\
1 \to 2:&\quad \Delta T_{1,2,1} = T_{1,2,1}^{(k)}; \quad \Delta T_{2,2,1} = T_{2,2,1}^{(k)}; \quad \Delta T_{0,2,1} = 0.5(T_{2,1,1}^{(k+1)} + T_{2,1,1}^{(k)});\\
&\quad \vdots \qquad\qquad \vdots \qquad\qquad \vdots\\
(z-1) \to z:&\quad \Delta T_{1,z,1} = T_{1,z,1}^{(k)}; \quad \Delta T_{2,z,1} = T_{2,z,1}^{(k)}; \quad \Delta T_{0,z,1} = 0.5(T_{2,z-1,1}^{(k+1)} + T_{2,z-1,1}^{(k)})
\end{aligned} \qquad (15.44)$$

15.2.7 Applications and results

In the course of the thermal transient processes of transformers, the hydrostatic pressures Δp_i in the flow loops of the MFN (see Equation (15.29)) may considerably differ from each other temporarily. In the oil ducts having small heat sources and small flow resistance coefficient (e.g. in the core and shunt channels), the oil warms up more slowly and therefore, in certain stages of the transient, loop equation system (15.33) is only satisfied for these branches in the case of negative (downward) oil mass flows. After switching on or a sudden increase of the load, a negative oil mass flow necessarily develops in the shunt flow path. This circumstance has a considerable effect on the transient warming of the transformer.

Owing to the negative oil mass flows, a warmer oil flows back from space K into the ducts in question until, because of this, the ratio of the hydrostatic pressures changes to an extent sufficient to satisfy the loop equation system for positive oil mass flows too.

The simultaneous model and the calculation strategy discussed above are suitable also for a global modelling of this phenomenon. If in any of the time steps the calculation results in a negative value for the mass flow rate of a branch, the progress on the elementary blocks of the block columns in question has the reverse direction in the calculation. This can be achieved by a corresponding instruction in the computer program.

In reality, the local natural convective effects may also maintain an upward motion of the oil along the heated duct walls, while in the core of the channel the bulk of the oil moves downward. According to model condition TM4, the model makes it possible to determine the resultant mass flow rate of the oil.

The model conception and the application of the computation strategy will be presented for the operation of a 200 MVA OFAF transformer with three differing load combinations.

I. Switching from cold state to rated load and at $t = 17{,}350$ s to an overload of 50%, until the permissible winding temperature 105 °C is reached. It is desired to determine the possible duration of the overload operation.

The results obtained are presented in Figure 15.15.

The notation used is as follows:

1. mean temperature of the upper winding disc of the high voltage winding column,
2. temperature of the oil in the vicinity of the winding disc mentioned,
3. temperature of the oil in space K,
4. temperature of the oil in space O,
5. mass flow rate ($\dot{m}_1$) of the oil in the heat exchanger,
6. oil mass flow rate in the channel of the high voltage winding column,
7. oil mass flow rate in the shunt channel.

As can be seen in Figure 15.15, curve 1 reaches the permissible value at $t = 20{,}620$ s. (The possible duration of overload operation is 3270 s.)

After switching on ($0 < t < 2950$ s), the value of the temperature-dependent flow resistances decreases as a consequence of the increase of oil temperature, the pressure source of the pump and the values of the thermosiphon pressure sources increase, and also the oil mass flow rate $\dot{m}_1$ increases (curve 5). As the laminar–turbulent transition takes place in the heat exchanger, $\dot{m}_1$ decreases temporarily, then increases again slowly for the reason mentioned.

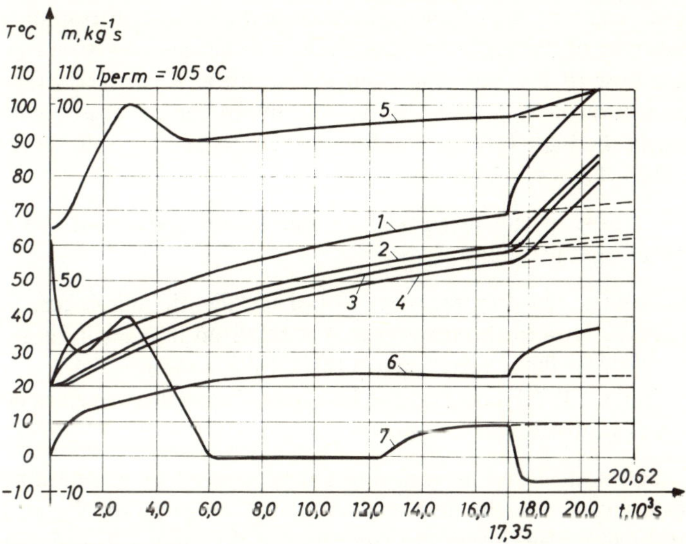

Figure 15.15 Calculated temperature and oil mass flow rate transients of a 200 MVA OFAF transformer (load combination I)

The oil mass flow rate of the shunt flow path ($\dot{m}_n$, curve 7) exhibits a peculiar variation feature. Following switch-on, the whole mass flow of the oil passes through the shunt channel of low flow resistance. As a consequence of the fast warming of the oil mass flows of the core and of the windings (curve 2), the thermosiphon effects get stronger and the oil mass flow rate in the shunt channel starts decreasing strongly. From the 1200th second on, the decrease is stopped by the viscosity decrease accompanying the fast warming of the oil in the windings. After this, owing to the effects mentioned, the oil mass flow rate of the shunt channel decreases monotonically and reaches a small negative value at $t = 6000$ s. The slight back-flow from space K adds to modifying the temperature and the ratio of the hydrostatic pressures, and the negative oil mass flow stops at $t = 12{,}200$ s, then the oil mass flow rate of the shunt channel increases slowly until the beginning of the overload. When the overload starts, the previous effect is produced in the oil ducts of the iron core and the winding columns by the quick warming of the oil, and oil flows provisorily into the oil channels of the iron core and also the windings from the shunt channel.

II. After switching the transformer from cold state to rated load, at $t = 17{,}350$ s the oil pumps stop working by break-down. The permissible duration of the operation has to be established.

The results of calculation are presented in Figure 15.16. After the failure of the pumps a practically negligible oil mass flow rate $\dot{m}_1$ passes through the heat exchanger of high flow resistance (curve 5). At the same time, the oil mass flow rate of the high-voltage winding increases slowly (curve 6). Thus the oil mass flow of the iron core and the winding is covered practically by the negative mass flow of the shunt. The mean temperature of the upper winding disc of the high-voltage winding column (curve 1) reaches the permissible value (105 °C) at $t = 24{,}600$ s.

III. Simulation of the test-room measurement state with the following operational program.

(a) $0 < t < 9000$ s: the oil pumps are out of service, the whole loss pertaining to the rated load of the transformer is released in the windings.
(b) $9000 < t < 16{,}200$ s: switching on the OFAF cooling.
(c) $t > 16{,}200$ s: OFAF cooling; the loss heat flux in the windings is identical with the rated value.

The calculation is aimed at verifying the procedure by measurements. The parameters measured in the test room are: oil temperature T_K and T_O as a function of the time; loss powers as functions of the time; average temperature of the windings in the quasi-steady state.

The temperatures obtained by calculation and measurements, respectively, are indicated in Figure 15.17. As shown by the figure, the calculated and measured values exhibit a satisfactory agreement.

15.3 TRANSIENT THERMAL ANALYSIS OF SYSTEMS UTILIZING SOLAR ENERGY

15.3.1 Object of the analysis

Owing to the time dependence of the intensity of solar radiation and of the meteorological characteristics, the operation of solar systems is inherently transient. Consumption similarly varies with time. The operation of the controlling equipment of the system involves further effects varying with time.

Solar energy as an energy source is of low density (dilute). Because of the small heat flux density large-surface collector systems are needed and to compensate for temporary fluctuations it is necessary to store the energy.

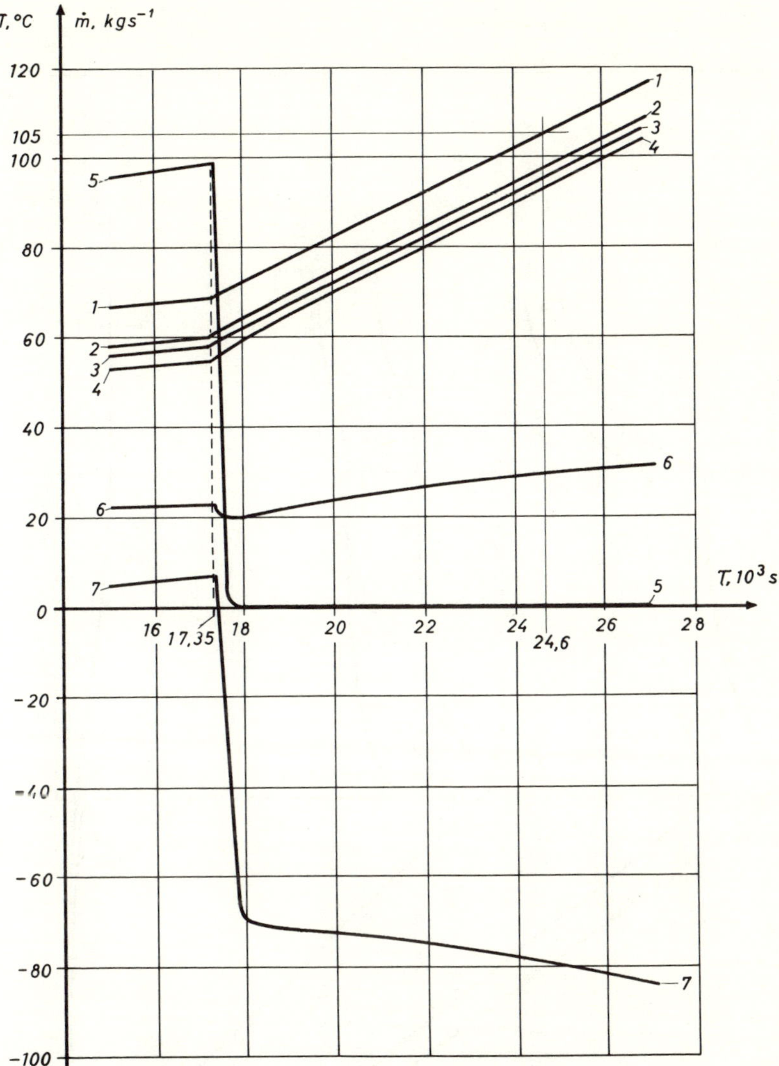

Figure 15.16 Calculated results for load combination II

Systems utilizing solar energy are of varying construction. Comparative evaluation of the differing types is an important part of the work of design and development. In the course of designing the system components, constructing the system and working out the optimum controlling methods, examination through simulation by numerical methods plays an important part.

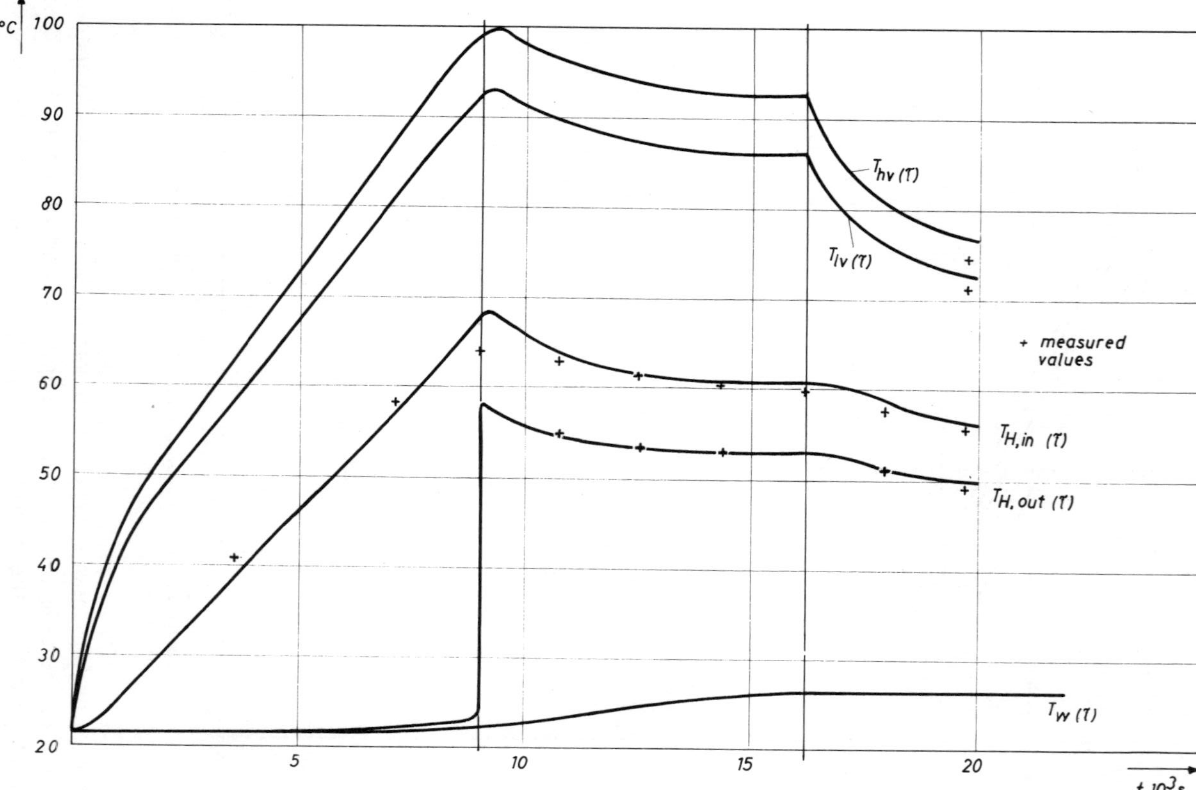

Figure 15.17 Comparison of the calculated and the measured values for test-room measurement state (load combination III)

The most important parameter of solar systems is the energy utilized, characterized qualitatively by the temperature level of the heat-carrying medium and quantitatively by the efficiency. The instantaneous and long-term efficiency are interpreted for the whole system and for the collectors.

The service life of the system utilizing the solar energy is generally longer than 15 years.

Their economic evaluation requires an examination of the performance over a long period of time, for which a simulation technique can similarly be applied advantageously.

A number of procedures have been elaborated for the numerical analysis of systems utilizing solar energy.[38–45] These techniques differ from each other considerably as regards the field of applicability, the accuracy of approximation, as well as the consumption of work, time and the needs of memory.

15.3.2 The construction and the main parameters of solar systems

A considerable number of solar systems serving for the direct utilization of heat are equipped with collectors of nonconcentrating type. Examples are domestic hot-water supplying systems, solar heating equipment, and solar driers. Their main elements are the collector system, the pump, the storage unit, and the technological unit serving as consumer. Except for the simplest cases, the system is completed by a supplementary heat source (heating equipment), a heat exchanger, and a control device.

Figure 15.18 shows the scheme of two characteristic solar systems. In the hot-water supply system shown in Figure 5.18(a) the elements serving for collecting and storing the solar energy are clearly separated.

Figure 15.18(b) shows the block scheme of a solar convective drier. The system contains a collector combined with a heat storage unit.

15.3.3 Network model of the solar systems

As can be seen from Figure 15.18, the elements of solar systems are coupled into an integrated thermohydrodynamical system by the mass flow flowing through them.

The simultaneous network model concept described in Sections 15.1 and 15.2 can be applied to their modelling. The simultaneous model is built up from thermal and hydrodynamical part models connected by dual-function elements.

The hydrodynamical part model is a mass flow network model, modelling also the executional functions of the control system and serving to determine the branch mass flows. The temperature variations of the flowing medium are determined by means of the thermal part model built up from the heat flow network model of the collectors and from the models of the storage, heat exchanger and consumption.

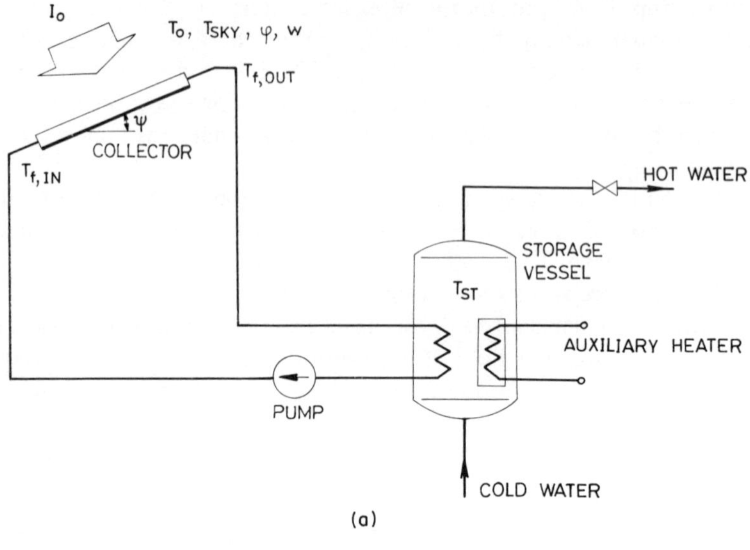

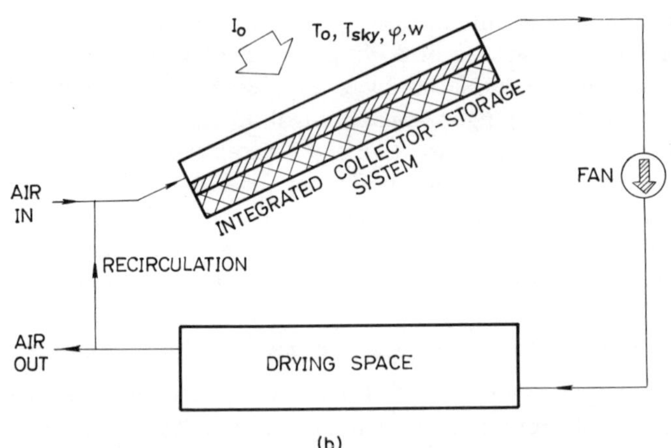

11057 **Figure 15.18** Scheme of (a) solar hot-water supply system, and (b)
11058 a solar drier

15.3.4 Part model of the collector

The construction of typical flat-plate collectors is shown in Figure 15.19. The physical processes taking place in flat-plate collectors are clarified qualitatively,[38] but their numerical analysis requires certain approximating assumptions to be applied. These assumptions are necessary, owing in the first place to the spectral and directional distribution of the radiation, and also to the

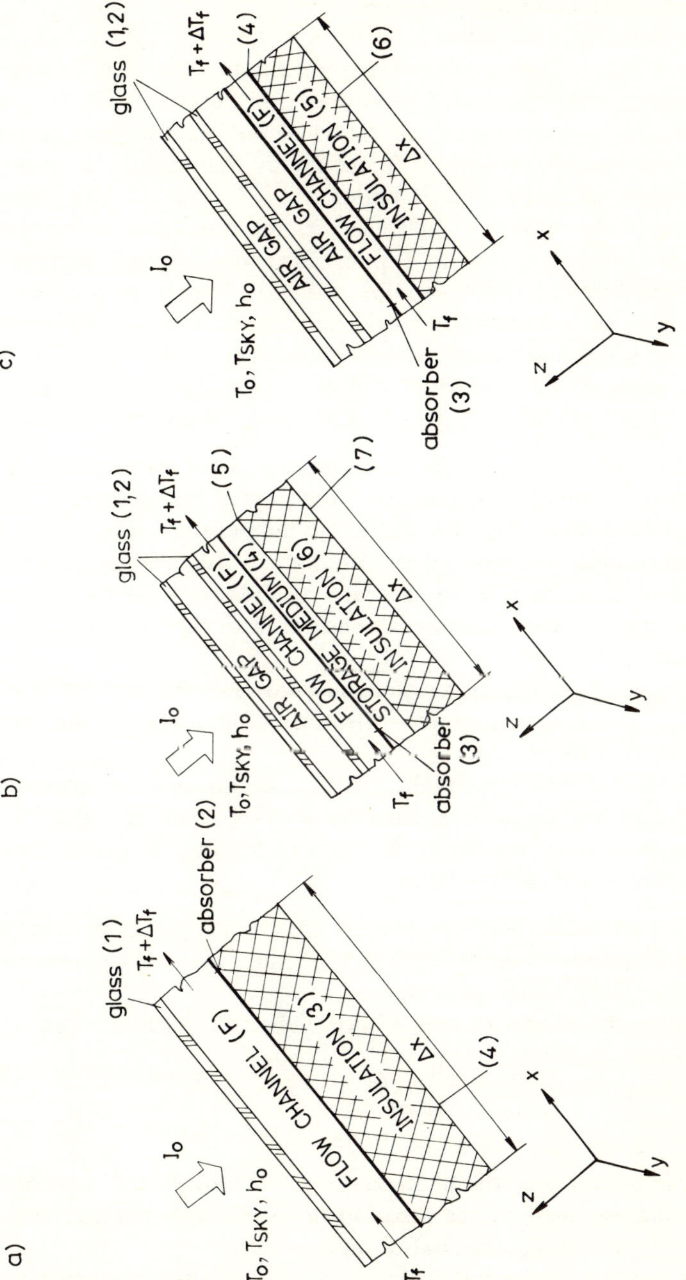

Figure 15.19 Elements of different solar collectors

temperature dependence of the radiation, and convective heat-transfer resistances as well as the material properties.

The part model of the collector serves for determining the instantaneous outlet temperature of the heat-carrying medium in cases of arbitrary temporary changes in the outside air temperature, the equivalent sky temperature, the wind velocity and the inlet temperature of the heat-carrying medium.

Many of the procedures known for collector simulation are based on the classical method of Hotter and Whillier.[38,40,41,43] The method neglects the heat capacity of the collector and describes the temperature distribution of the absorber in the form of an explicit function. These functions can be produced if the heat-transfer resistances along the length and the cross section are supposed to have constant values. The procedure is of iterative type and can further be simplified by semi-empirical relationships.[38]

The simulation of the collectors will be accomplished with the use of a heat-flow network model. The main model assumptions are as follows:

KT1. In the collector plane, the temperature distribution is homogeneous in the direction perpendicular to the flow of the medium.

KT2. For collectors having pipe ducts, the temperature inhomogeneity perpendicular to the flow of the medium is taken into account on the basis of an average temperature, with the use of the fin efficiency.

KT3. The frame of the collector is considered to be completely heat-insulated.

KT4. Along the flow direction of the medium, only the convective enthalpy flux is taken into account—the heat conduction in the cover and absorber is neglected.

KT5. The spectral variation of absorption and transmission relative to heat radiation are taken into consideration by average values weighted with the solar spectrum and with the distribution of the low-temperature Planck radiator.

Keeping in view model assumption KT4, the collector is lumped into discrete parts in the direction of the flow of the medium, and the discrete parts are replaced by their HFN model blocks.

The HFN model blocks of two adjacent discrete parts are connected by the temperature source

$$\Delta T = \frac{\phi_{hi}}{\dot{m} c_p} \tag{15.45}$$

modelling the warming of the heat-carrying medium. In the relationship ϕ_{hi} is the heat flux assumed by the medium in the discrete part, $\dot{m}$ is the mass flow rate and c_p is the specific heat of the medium.

The heat-flow network model blocks of the collectors sketched in Figure 15.19 are shown in Figure 15.20.

The describing equation system of the HFN model blocks, taking (15.10) and (15.12) into account, is

$$[C]\{\dot{T}\}+[K](T_0, T_s, T_i, \tau)\{T\}=\{Q\}(I_0, \tau). \qquad (15.46)$$

Matrices $[C]$, $[K]$ and $\{Q\}$ can be generated on the basis of the heat-flow networks of Figure 15.20.

In the heat flow networks, $R_{i,j}^{(k)}$ denote the heat-transfer resistance of heat conduction, $R_{i,j}^{(c)}$ that of convection, and the source fluxes ϕ_s denote the energy fluxes actually absorbed from radiation.

The computational relationships of specific heat conductions are summarized as follows ($K_{i,j}=R_{i,j}^{-1}$):

$$K_{i,j}^{(k)}=\frac{k}{\delta}, \qquad (15.47)$$

where k is the heat conduction factor and
δ is the layer thickness in direction z;

$$K_{i,0}^{(c)}=h, \qquad (15.48)$$

where $h=5.7+3.8\,w$ if $w<5$ m/s
and $h=7.6\,w$ if $w\geqslant 5$ m/s
where w is the velocity of wind;

$$K_{i,j}^{(c)}=k_{eq}, \qquad (15.49)$$

where k_{eq} is the so-called equivalent heat conduction of the air gap;

$$K_{s,i}^{(r)}=\varepsilon_g\sigma(T_i+T_s)(T_i^2+T_s^2), \qquad (15.50)$$

where ε_g is the emissivity of glass at low temperature level,
σ is the Stefan–Boltzmann constant,
T_s is the equivalent sky temperature;

$$K_{i,j}^{(r)}=\varepsilon_{i,j}\sigma(T_i+T_j)(T_i^2+T_j^2), \qquad (15.51)$$

where $\varepsilon_{i,j}^{-1}=\varepsilon_i^{-1}+\varepsilon_j^{-1}-1$,
ε_i, ε_j are the emission factors of the glass panes and of the absorber, respectively.

The heat flux sources can be identified on the basis of the following relationships:

in the jth glass or in the absorber

$$\phi_j^{(\tau)}=a_jI_0(\tau)\prod_{n=1}^{j-1}d_n, \qquad (15.52)$$

where $I_0(\tau)$ is the intensity of solar radiation,
a_j is the absorptivity of glass,
d_n is the transmissivity of glass or absorber.

The heat transfer coefficients and the equivalent heat conduction factors are calculated from the known equations $Nu = Nu(Re, Gr, Pr)$.

The statement of the external parameters $T_0(\tau)$, $T_s(\tau)$, $\phi_s(\tau)$ is one of the key issues of the simulation of solar systems.

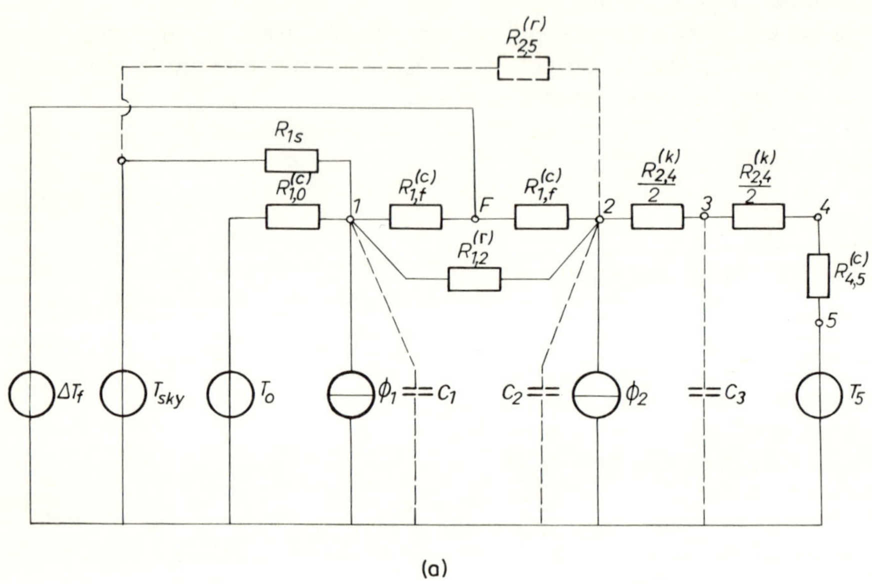

(a)

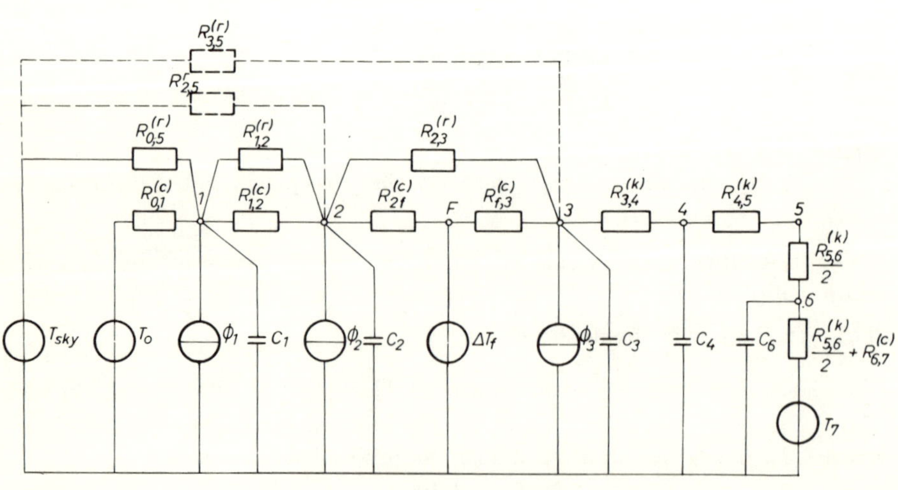

(b)

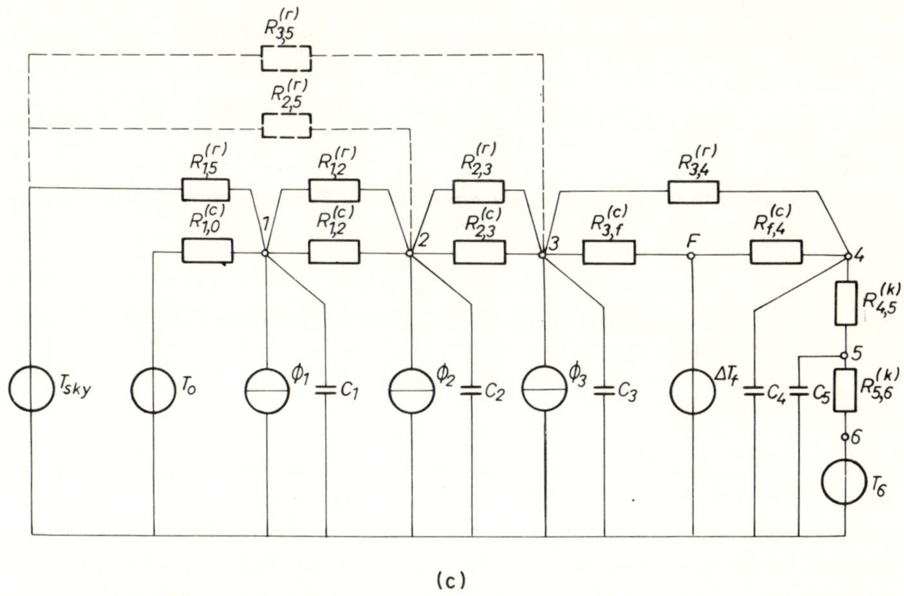

Figure 15.20 HFN block models of collector elements shown in Figure 15.19

The I_0 intensity of solar radiation and its directional distribution varies with time. Owing to the effect of clouds and to the contamination of the atmosphere, a stochastic variation is also superposed upon the regularly periodical changes by day and year. The humidity of external air, the degree of cloudiness and the velocity of the wind exhibit variations similarly irregular in time. When the meteorological data at disposal are not satisfactory in every respect, one must try to determine the required detailed data by measurements.

The direct and diffuse components of solar radiation is characterized for the simulation by a single function of intensity/time, while the time dependence of absorption and transmission due to directional characteristics are taken into account in the form of averages.[39] The radiation into the environment—in the case of an unknown site—is calculated for a hemispherical sky characterized by equivalent temperature. The relationships used in the calculation will be detailed below.

If the diffuse radiation can be considered completely isotropic, the intensity of the solar radiation incident on the collector plane is stated as follows:

$$I_0(\tau) = I_{\mathrm{HORTOT}}(\tau)[\beta(\tau)\chi(\tau) - \chi(\tau) + 1], \tag{15.53}$$

where $I_{\mathrm{HORTOT}}(\tau)$ is the total radiation arriving on the horizontal surface at the given moment,

$$\chi = \frac{I_{DIR}}{I_{TOT}}$$ is the ratio of direct and total radiations,

$$\beta = \frac{\cos(\varphi - \psi)\cos\delta\cos\omega + \sin(\varphi - \psi)\sin\delta}{\cos\varphi\cos\delta\cos\omega + \sin\varphi\sin\delta} \tag{15.54}$$

is a factor characterizing the relative position of the collector and of the Sun,

where φ = latitude,
ψ = inclination angle of the collector,
$\delta = 23.45 \sin(360(284+n)/365)$: the angle of declination,
n = serial number of the day with the year,
$\omega = 180-15\tau$ the so-called 'clock angle',
τ = time in hours.

The factor χ itself, too, is dependent on the time and on the weather and its value is affected also by the nature of the installation site. The daily course of the temperature T_O of the environmental air is fairly well known; the equivalent temperature of the sky is significantly influenced in cloudless weather by the air temperature and the humidity. It was found satisfactory to use the so-called Swinbank formula:[38]

$$T_s = 0.0552(T_O)^{1.5}. \tag{15.55}$$

In cloudy weather the equivalent temperature of the sky is assumed to be identical with the temperature of the air.

For giving the wind velocity in the calculations, a few time sequences based on own measurements have been used.

In the course of solving the heat-flow network equation system, the values of heat capacities are of interest from the viewpoint of conditioning the equation system. In the case of collectors without storage the heat capacity of the cover and of the absorber + heat insulation have to be taken into account. These are values practically of the same order. In the collector type according to Figure 15.19(c) the own 'time constant' of the outlet temperature variation of the heat carrier is affected in the first place by the heat capacity of the absorber + insulation, since the heat-transfer resistance between the absorber and the glass is usually large. This statement cannot be maintained with certainty in the case of the construction shown in Figure 15.19(a), since the medium is flowing between the glass and the absorber.

If a collector with storage (solid-wall, fluid bag, phase-changing package) is applied, the heat capacity of the collector elements can be neglected, except that of the heat storage unit. In the simulation of the phase-changing package, the effect of latent heat can be simulated by building it into the volumetric heat capacity.[22]

15.3.5 Thermal part model of the storage

The storage unit may be a fluid tank, a solid body (wall or rock-bed) or a phase change material. In fluid storage units the fluid usually gets mixed if the heat is introduced from below. In such cases the storage can be modelled as a concentrated heat capacity. The transient heat flux balance equation of the thermal mixer (see also e.g. Equation (15.36)) is

$$(Mc)_{st}\frac{dT_{st}}{d\tau}=\phi_{h,in}(\tau)+\phi_{h,aux}(\tau)-\phi_{h,out}(\tau)-\phi_{h,loss}(\tau). \tag{15.56}$$

$\phi_{h,in}$ is the heat flux passing from the heat carrier medium through the heat exchanger into the storage unit; its value is furnished by a heat exchanger subroutine after the determination of the heat loss in the pipe duct between the collector and the storage. $\phi_{h,aux}$ is the intensity of the heat flux source produced by the auxiliary electric heater. In order to keep the temperature of the supplied energy at the desired level, the electric heating is switched on by control of the system whenever the solar energy is not sufficient. In practice, the control of heating is almost exclusively a two-point system—the source intensity varies according to a step function.

The time function $\phi_{h,out}(\tau)$ of the consumption is considered to have a known history. The loss heat flux of the storage is

$$\phi_{h,loss}=K_{st}A_{st}(T_{st}-T_0), \tag{15.57}$$

where K_{st} is the resultant conductance of the insulated wall of the storage,

A_{st} is the heat transmitting surface of the storage.

15.3.6 Application to the simulation of a solar hot-water supply system

The scheme of the solar hot-water supply system in question is shown in Figure 15.21. The thermal part models of the simultaneous model were discussed in Sections 15.3.4 and 15.3.5. A mass flow network model (Figure 15.22) is applied as hydrodynamical part model. The describing equation system of the mass flow network part model consists of the nodal equation written for node 1 and two independent loop equations. With the notation of Figure 15.22 (disregarding the hydraulic transients, $\Delta p_a=0$):

$$\dot{m}=\dot{m}_1+\dot{m}_2 \tag{15.58}$$

$$\Delta p_{h1}+\Delta p_p=\dot{m}(R^{(1)}_{m,L}+R_{m,c}+R^{(2)}_{m,L})+\dot{m}_1(R^{(1)}_{m,V}+R^{(3)}_{m,L})+\Delta p_{h,3} \tag{15.59}$$

$$\Delta p_{h,3}+\dot{m}_1(R^{(1)}_{m,V}+R^{(3)}_{m,L})=\dot{m}_2(R^{(2)}_{m,V}+R^{(4)}_{m,L}+R_{m,\zeta}/+\Delta p_{h2}, \tag{15.60}$$

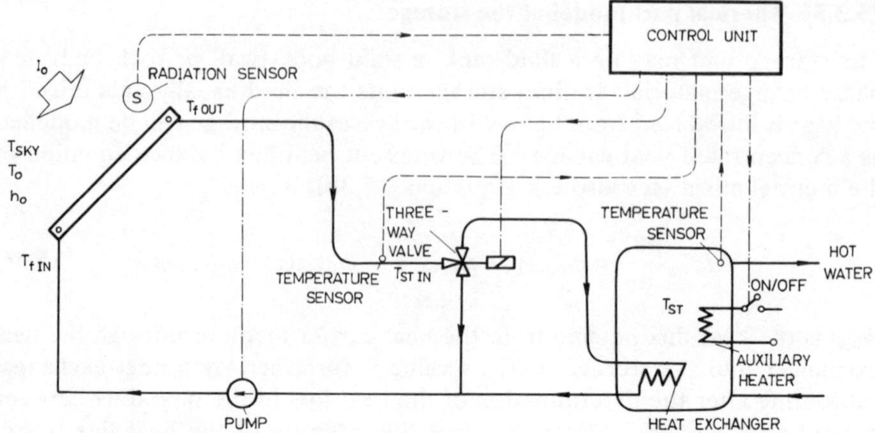

Figure 15.21 Scheme of a solar hot-water supply system with control

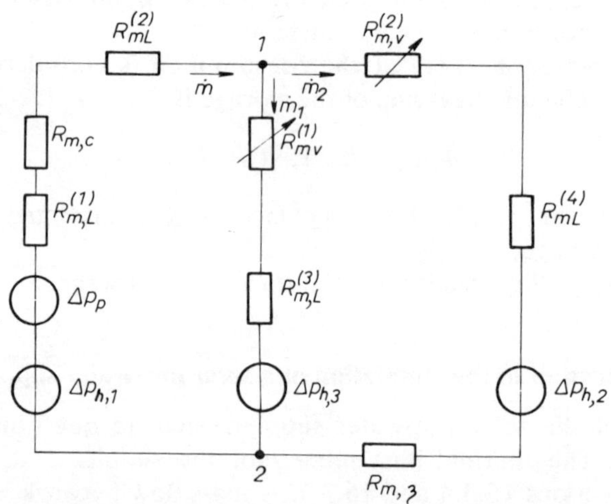

Figure 15.22 MFN part model of solar system shown in Figure 15.21

where the flow resistances are

$$R_{m,\text{c}} = R_{m,\text{L}} = \frac{\lambda L \dot{m}}{2 d_{\text{h}} \rho A_{\text{c}}^2}, \tag{15.61}$$

$$R^{i}_{m\text{V}} = \frac{\zeta^i \dot{m}}{2 \delta A_{\text{c}}^2}, \tag{15.62}$$

for the pump:

$$\Delta p_{\mathrm{p}} = A + B\dot{m} + C\dot{m}^{a}. \tag{15.63}$$

A, B, C and a are functions of the temperature and the speed.
The hydrostatic pressure source

$$\Delta p_{\mathrm{h}} = g\int_{0}^{H} \rho[T(h)]\,\mathrm{d}h. \tag{15.64}$$

For the solution, the HFN equation system (15.46) of the collector and (15.57) of the storage unit is discretized with respect to time, time averages $\hat{K}$ and $\hat{\phi}$ are produced according to (15.15), and the solution is produced by means of a properly chosen difference scheme (for details see Section 15.1.6). If the heat capacities of the collector without heat storage are neglected, even a simple explicit scheme of the solution gives an acceptable approximation.
The main steps of the simulation procedure are as follows:

(a) Determination of the values of the meteorological parameters, valid at the given moment (intensity of the total radiation incident on the collector plane, angles of directions, ratio of direct and diffuse radiation, air temperature, equivalent temperature of the sky, wind velocity).
(b) Determination of the mass flow passing through the collector. In the scheme according to Figure 15.21, the control unit starts the pump above a threshold value of the intensity of solar radiation, and the heat-carrying medium will circulate in the bypass branch until its temperature in the branch returning from the collector surpasses the collector temperature by a prescribed value; then the three-way valve switches over. A prerequisite for solving the MFN equation system is thus to determine the operational condition of the control unit.
(c) Determination of the outlet temperature of the collector from the inlet temperature of the working medium and the meteorological data. The inlet temperature acts as thermal pressure source in each block. After solving the HFN equation system of the first block, the inlet temperature of the next element is will be calculated and, proceeding in this way along the collector, the outlet temperature is found. As an intermediate characteristic, the instantaneous value of the collector efficiency is also determined.
(d) Determination of the state of the storage unit. For a storage unit of mixed temperature, its state is characterized by a single function of temperature vs. time. In the solution of the HFN equation of the storage unit, the heat flux entering from the heat exchanger, the environmental temperature, and the function of heat-source intensity vs. time of the additional heating serve as input data.

(e) Determination of the energy delivered during the elapsed time, the solar energy arriving, the energy covered from the additional heating, time stepping.

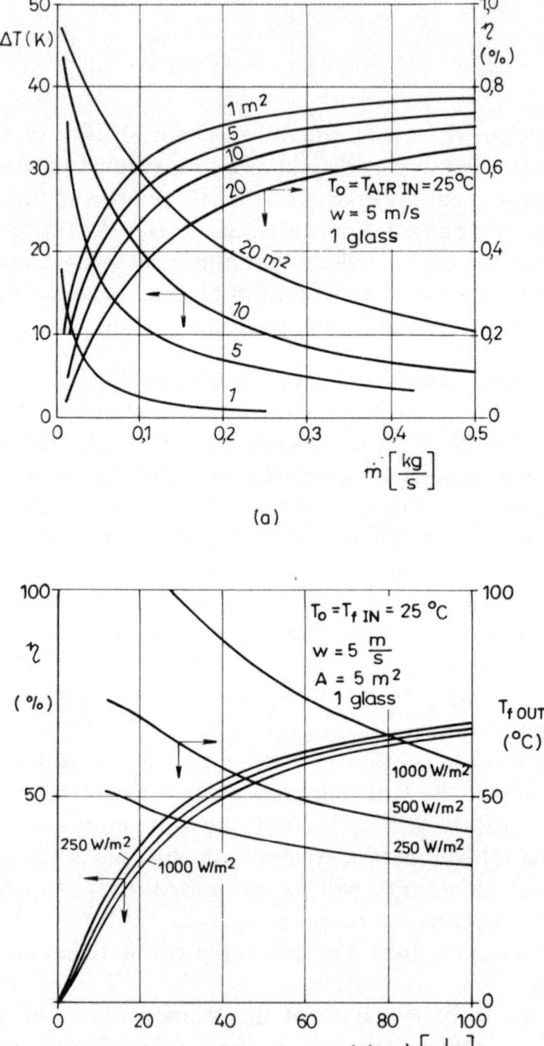

Figure 15.23 Instantaneous efficiencies of (a) air and (b) fluid type collectors

15.3.7 Some results

The examinations were aimed at selecting a suitable collector type, adjusting the optimum mass flow rate during operation, determining the collector length, the outlet temperature (series connection) and the total required surface (parallel connection). Figure 15.23 shows the curves of the instantaneous efficiency and the outlet temperature for a simple air collector and a fluid

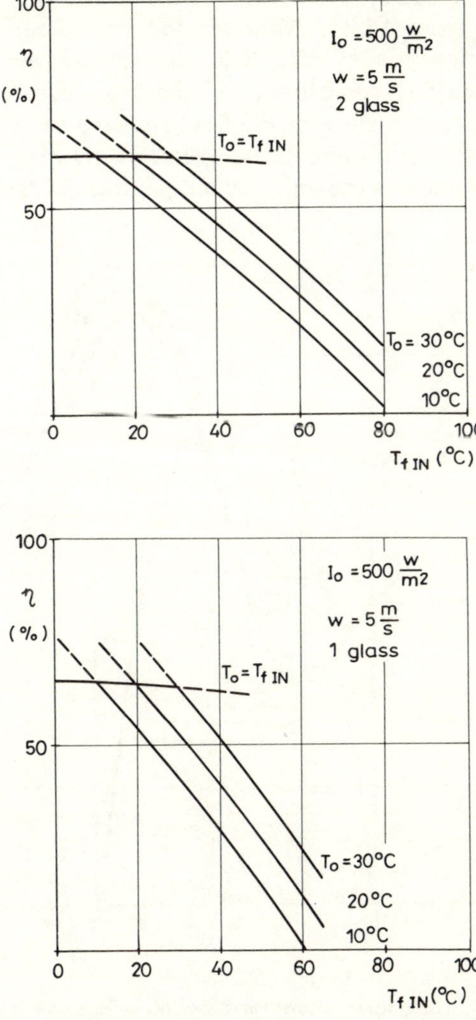

Figure 15.24 Effect of inlet temperature on collector efficiency

collector. There is a contradiction between the quantitative and qualitative characteristics of collectors: a higher efficiency is associated with a lower temperature level. As can be seen from Figure 15.23(a), increasing the surface by series connection results in a tendency similar to that produced by the decrease of the mass flow rate.

Changing the intensity of solar radiation is accompanied by a change of the nonlinear thermal resistances in the collector; this is the reason for the separation of the efficiency curves in Figure 15.23(b). The heat losses increase with increase of the outlet temperature. This is illustrated more clearly by the diagrams in Figure 15.24. Although the curves serve to indicate the instantaneous efficiency, they also help to evaluate the efficiency of the collectors under varying conditions, i.e. the long-term efficiency. On the horizontal axis the temperature of the fluid entering the collector is plotted, the parameter is the environmental temperature. During the day, with the gradual charging of the storage unit, the temperature of the medium entering

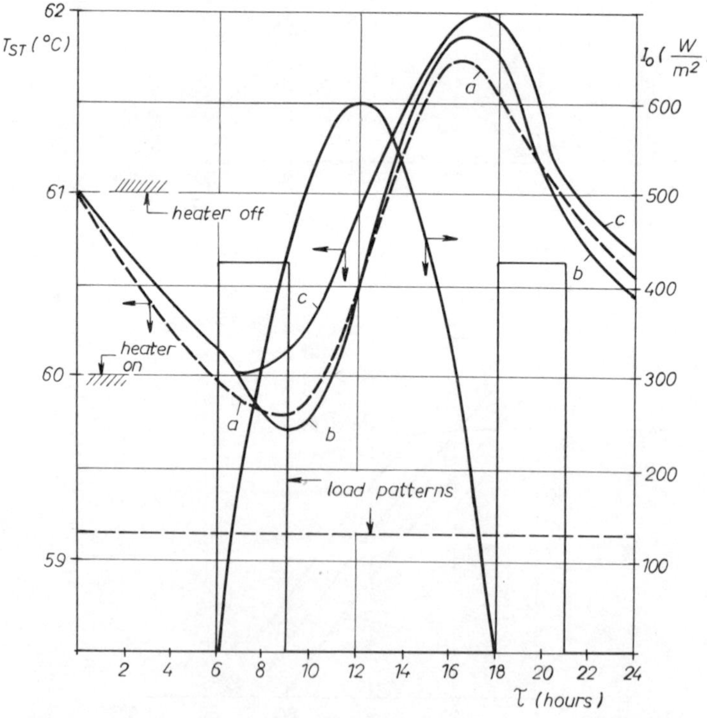

Figure 15.25 Daily temperature variations of storage tank on 16 March: (a) constant consumption, without auxiliary heater; (b) intermittent consumption, without auxiliary heater; (c) intermittent consumption with auxiliary heater

the collector rises, and the instantaneous efficiencies pertaining to it decrease steeply, thus producing a situation that a collector capable of an efficiency of 50–60% has an average daily efficiency of only 25–30%. This is the point where the optimum size of the storage unit and the desired temperature level of consumption must be included in the analysis of economical operation.

Two characteristic examples from the operation of the system according to Figure 15.21 will be presented, namely the cases of permanent and intermittent consumption.

Figure 15.25 pertains to 16 March and Figure 15.26 to 16 September. The peak values of the radiation intensities are nearly the same, but the spring curve is 'thinner'. Thus, the temperature of the storing unit is about 1 °C

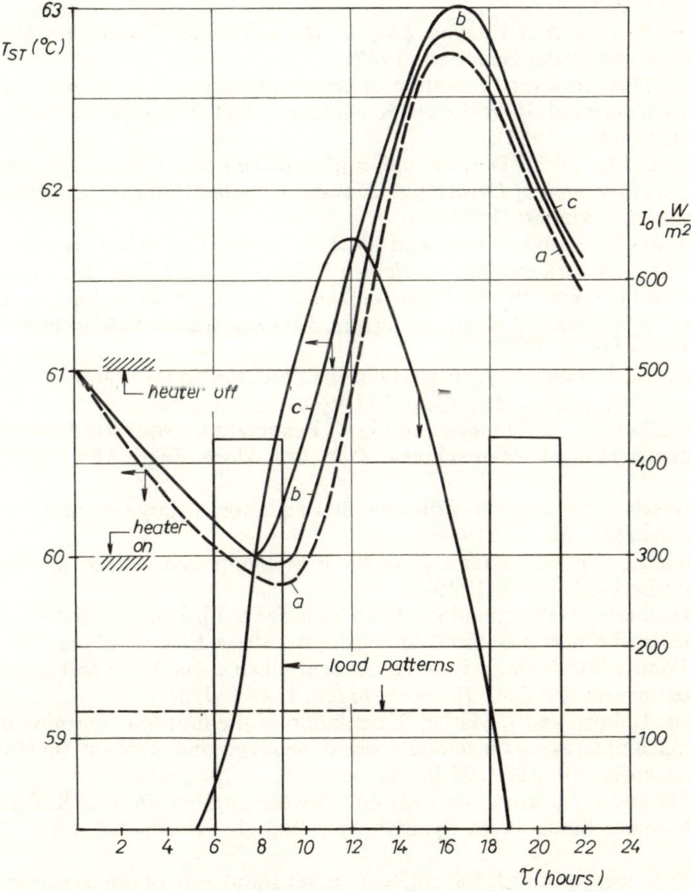

Figure 15.26 Daily temperature variations of storage tank on 16 September. Cases (a), (b), (c) are same as in Figure 15.25

higher at the end of the autumn day than in springtime. It can be seen that if during the day the total absorbed heat is the same, then the temperature course slightly differs during the day, but the global energy balance on the spring day and the autumn day presented is essentially the same. In the real case the collector has a surface of 7.5 m^2 and the storage is a water tank of 1000 kg; thus, if the desired temperature in the storage unit is a minimum of 60 °C, additional heating is needed on cloudless days in winter only. In the summer period there is overproduction.

REFERENCES

1. L. Imre, *Heat Transfer in Composed Devices* (in Hungarian), Akadémiai Kiadó, Budapest (1982).
2. L. Imre, *Heating and Cooling Electric Machines and Devices* (in Hungarian), Müszaki Könyvkiadó, Budapest (1982).
3. L. Imre, 'Heat transfer simulation of composite devices', in *Numerical Methods in Heat Transfer* (ed. R. W. Lewis, K. Morgan and O. Zienkiewicz) Vol. I, Chapter 3. Wiley, London (1981).
4. I. Szabó and L. Imre, 'Determination of expected operational characteristics of driers', in *Handbook of Drying* (ed. L. Imre) Chapter 18 (in Hungarian), Müszaki Könyvkiadó, Budapest (1974).
5. L. Imre and I. Szabó, 'Static and dynamic analysis of driers by the mass and energy flow network method', *Periodica Polytechnica, El. Eng.* **18**, 145 (1974).
6. L. Imre and I. Szabó, 'Role of relative transfer factors in controlling the operation of driers', *Proc. First Int. Symp. on Drying, Montreal* (ed. A. S. Mujumdar), Science Press, Princeton, p. 76 (1978).
7. B. Latto and L. Galloway, 'A study of the fluid flow in the stator of an electrical machine', *Proc. of the IME*, **3**E, p. 1 (1969).
8. S. B. Preston, M. A. Thomas and G. H. Pennington, 'Non-Steady state thermal performance of electrical machines', *Proc. Inst. Mech. Engrs*, **184** (3E), 9, (1969–79).
9. T. J. Roberts, 'The solution of the heat flow equations in large electrical machines', *Proc. of the IME*, **3**E, 70 (1969).
10. Gy. Istvánfy, 'On the asymmetry of the heat conduction matrix' (in Hungarian), *Elektrotechnika*, **63**, 393 (1970).
11. S. Krezeminski, 'Determination of the transient temperature variation of electric machines by the matrix method' (in Polish), *Archivum Elektrotechniki*, **20**, 4 (1971).
12. U. V. Popov, 'Modelling of directly cooled electric machines and calculation of their heating' (in Russian), *Elektrotechnika*, **1**, 44 (1973).
13. I. Farkas, L. Imre and I. Máthé, 'Calculation of the transient warming of electric rotary machines, taking the temperature-dependence into account' (in Hungarian), *Elektrotechnika*, **70**, 217 (1977).
14. A. Meyer and I. Meister, 'Verlust- und Erwärmungsberechnung Käfiganker von Asynchronmaschinen beim dynamischen Betrieb', *Bull. ASE/UCS*, **70**, 1247 (1979).
15. L. Imre, Z. Jagasits and J. Barcza, 'Computer simulation of the transient warming of rotary electric machines', *Proc. 2nd. Int. Conf. on Numerical Methods in Thermal Problems*, Venice, (ed. R. W. Lewis, K. Morgan and B. A. Schrefler) Vol. II, Pineridge Press, Swansea (1981).

16. W. E. Mason, jr., 'Finite element analysis of coupled heat conduction and enclosure radiation', *Proc. 1st Int. Conf. on Numerical Methods in Thermal Problems* (ed. R. W. Lewis and K. Morgan) Vol. I, Pineridge Press, Swansea (1979).
17. R. D. Karam, 'Optimum solution of linearized radiation equation', *Proc. 1st. Int. Conf. on Numerical Methods in Thermal Problems* (ed. R. W. Lewis and K. Morgan) Vol. I, Pineridge Press, Swansea (1979).
18. J. Crank and P. Nicolson, *Proc. Camb. Phil. Soc.*, **43**, 50 (1947).
19. O. C. Zienkiewicz and R. W. Lewis, 'An analysis of various timestepping schemes for initial value problems', *Earthquake Engineering and Structural Dynamics*, **1**, 407 (1973).
20. M. Lees, 'A linear three-level difference scheme for quasilinear parabolic equations', *Math. Comp.*, **20**, 516 (1966).
21. C. Bonacina and G. Comini, 'On the solution of the non-linear heat conduction equations by numerical methods', *Int. J. Heat Mass Transfer*, **16**, 581 (1973).
22. G. Comini, S. Del Guidice, R. W. Lewis and O. C. Zienkiewicz, 'Finite element solution of non-linear heat conduction problems with special reference to phase-change', *Int. J. Num. Meth. Engn.*, **8**, 613 (1974).
23. J. Donea, 'On the accuracy of finite element solutions to the transient heat conduction equation', *Int. J. Num. Meth. Engn.*, **8**, 103 (1974).
24. O. C. Zienkiewicz, *Finite Element Analysis in Engineering Science*, McGraw-Hill, New York (1978).
25. P. H. G. Allen and A. H. Finn, 'Transformer thermal design by computer', *Instn. El. Engrs. Conf. Publication*, No. 51, 589 (1969).
26. L. Imre, 'Determination of the steady-state temperature distribution of transformer windings by the heat flux network method', *Periodica Polytechnica, El. Eng.*, **20**, 461 (1976).
27. L. Imre, I. Szabó and A. Bitai, 'Determination of the steady-state temperature field in naturally oil-cooled disc-type transformers', *Proc. 6th. Int. Heat Transfer Conf. Toronto*, Vol. II, EC-20, 123 (1978).
28. V. Preiningerova and M. Pivrnec, 'Temperature distribution in a coil of a transformer winding', *Proc. Instn. El. Engrs*, **124**, 218 (1977).
29. L. Imre and A. Bitai, 'A simultaneous and hierarchic network modelling conception of simulating the steady-state warming of naturally oil-cooled transformers', *Periodica Polytechnica, Mech. Eng.*, **22**, 265 (1978).
30. L. Imre, A. Bitai and P. Csényi, 'Parameter sensitivity investigation on the warming of oil-transformers', *Proc. 1st. Int. Conf. on Numerical Methods in Thermal Problems* (ed. R. W. Lewis and K. Morgan) Vol. I, Pineridge Press, Swansea, p. 850 (1979).
31. M. Yamaguchi, T. Kumashaka, Y. Inui and S. Ono, 'The flowrate in a self-cooled transformer', *IEEE Trans. on Power Apparatus and Systems*, **PAS-100**, 1 (1981).
32. P. H. G. Allen, O. Szpiro and E. Campero, 'Thermal analysis of power transformer windings', *El. Machines and Electromechanics*, **6**, 1 (1981).
33. A. J. Oliver, 'Estimation of transformer winding temperatures and coolant flows using a general network method', *IEE. Proc.*, **127**, 395 (1980).
34. R. Baehr and J. Gerth, 'Verfahren zur Überprüfung der Wicklungserwärmung OD-gekühlter Transformatoren', *ETZ-A*, **96**, 572 (1975).
35. L. Kiss, 'Quasi-stationary warming of three winding transformers' (in Hungarian), *Elektrotechnika*, **69**, 241 (1976).
36. L. Imre, A. Bitai and P. Csényi, 'Thermal transient analysis of OFAF transformers by numerical methods', *Proc. 2nd Int. Conf. on Numerical Methods in Thermal Problems*, Venice (ed. R. W. Lewis, K. Morgan and B. A. Schrefler) Vol. II, Pineridge Press, Swansea, p. 899 (1981).

37. L. Imre, A. Bitai and P. Csényi, 'Computer simulation of the transient thermal performance of ONAN oil transformers', *Proc. 7th Int. Heat Transfer Conference, Munich* (in preparation) (1982).
38. J. A. Duffie and W. A. Beckman, *Solar Energy Thermal Processes*, Wiley, New York (1974).
39. I. Benkö and L. I. Kiss, 'An improved model for calculation of heat transfer due to solar radiation through windows', in '*Energy Conservation in Heating, Cooling and Ventilating Buildings*' (ed. C. J. Hoogendorn and N. H. Afgan), Hemisphere, Washington, Vol. 1, p. 251 (1978).
40. L. W. Butz, W. A. Beckman and J. A. Duffie, 'Simulation of a Solar Heating and Cooling System', *Solar Energy*, **16**, 129 (1974).
41. S. A. Klein and W. A. Beckman, 'A general design method for closed-loop solar energy systems', *Solar Energy*, **22**, 269 (1979).
42. O. Jørgensen, *Investigation of the Performance of Solar Heating and Cooling Systems. Modelling and Simulation.* Report of International Energy Agency (1979).
43. J. A. Manrique, 'Digital simulation of a solar water heating system under natural circulation conditions', *Proc. ICHMT Seminar on Heat and Mass Transfer in Buildings*, Dubrovnik (1977).
44. A. J. De Ron, 'Dynamic modelling and verification of a flat-plate solar collector', *Solar Energy*, **24**, 117 (1980).
45. L. Imre, L. I. Kiss, T. Környev and K. Molnár, 'Digital simulation of a solar hay drying system', in *Drying '80'* (ed. A. S. Mujumdar), Hemisphere, New York (1980).
46. L. I. Kiss, 'Simulation of the thermal characteristics of flat-plate solar collectors and glass shading structures', *Rep. of Tokai Univ. -techn. Ed. Tokai Univ.*, Tokyo, p. 30 (1980).
47. I. Szabó and L. I. Kiss, 'Long-term analysis of a solar hot-water supply system', Paper presented at *International Conference on Numerical Methods in Thermal Problems*, Venice (1981).

Numerical Methods in Heat Transfer, Volume II
Edited by R. W. Lewis, K. Morgan, and B. A. Schrefler

Chapter 16

The Influence of Creep and Transformation Plasticity in the Analysis of Stresses due to Heat Treatment

F. G. Rammerstorfer, D. F. Fischer, E. T. Till, W. Mitter, O. Gründler

SUMMARY

During the quenching of heated steel bodies stresses arise due to thermal and transformation-induced volume changes. There is no proper experimental way to determine the stress field during the heat treatment. Since the knowledge of this stress history is a prerequisite to avoid cracking during the heat treatment and to predict the residual stress field in the finished body a mathematical approach is developed which solves this problem with the aid of the finite element method. It is shown that the results of these thermoelastic-plastic analyses depend to a high degree on the accuracy by which the temperature-dependent material behaviour is modelled.

In the case of thick units the results are rather meaningless if creep effects are disregarded. Furthermore, it may happen that the results are correct only if the so-called transformation plasticity—which may be interpreted as a pseudoplasticity effect due to microstresses—is taken into account. The latter effect becomes important if the phase transformation leads to considerable volume changes, e.g. in the case of steel grades with low-temperature transformation.

In this chapter the thermal and the stress analysis procedures are presented and results of several analyses are discussed. A comparison with experimental results describing the residual stress field explains the above-mentioned effects of different material modelling.

16.1 GENERAL REMARKS ON THE ANALYSIS OF HEAT-TREATMENT PROCESSES

The calculation of the transient stress field in bodies during heat treatment is strongly nonlinear. The material behaves in a visco-elastic–plastic manner and all its properties are temperature dependent. The behaviour of the material during the heat-treatment process depends not only on the current temperature field but also on the history of plastic and creep deformation and on the real time. A very careful modelling of this complicated material in terms of continuum mechanics is necessary to obtain useful results. The numerical solution algorithm has to be controlled precisely to avoid errors caused by numerical inaccuracies.[1]

These sensitivities of the analysis of heat-treatment processes may be the reason for some uncertainties recorded in the related literature. A brief review of this literature is given in reference 1. An attempt to compare calculated and (rarely published) measured results often fails. Two essential aspects will be emphasized in the following chapters: the transformation plasticity and the creep behaviour. Neglecting one or both of these two effects in the material model leads to completely worthless results, as will be shown later.

16.1.1 The transformation plasticity in heat-treatment processes

During phase transformations connected with volume changes considerable plastic deformations are observed even at very low stress levels. This behaviour has been termed transformation plasticity.[2] There are different explanations for this phenomenon:

(a) During transformation, point defects are generated, favouring climb of dislocations which gives rise to creep.[3]

(b) Phase transformation involves reconstruction of the lattice accompanied by a partial loss of coherent bonding between the atoms.[4]

(c) The kinetic energy released during the formation of martensite plates is sufficient to move dislocations out of their sources and to force pile-up arrays against the grain boundaries. Therefore only small additionally applied stresses are necessary to activate slip in adjacent grains and produce considerable plastic deformations.[5]

(d) Local microstresses are induced by the transformations (internal stresses of the second kind, according to reference 6). These stresses act in combination with the stresses due to thermal or mechanical loading and cause reasonable plastic deformation at a low load level.[7] Hence, the 'effective yield stress' is very low during phase transformation.[1]

The latter explanation (d) is applicable to both diffusion controlled and martensitic transformations and, according to reference 8, will give a useful and realistic model for analyses. Based on explanation (d) as well as separate investigations and calculations as in reference 1, it is shown in Section 16.3 that a reduction of the yield stress within the transformation temperature interval is a simple, economic and effective way of accounting for the transformation plasticity.

16.1.2 Creep effects during quenching

Creep and relaxation effects become important especially if the heat-treated body is fairly thick. The chilling down period during which thin bodies are exposed to a high temperature is too short for the development of significant creep strains (see, for example, Section 16.3). On the other hand, in thick bodies, as for example the turbine disc investigated in Section 16.4, a considerable part of the irreversible strains is caused by creep, and completely wrong residual stresses will be calculated if creep effects are neglected. If the heat-treatment process includes long time periods at which the overall temperature is held at a high level, e.g. to reduce self-equilibrated stresses in a body, creep plays a significant role regardless of the size of the body.

The creep behaviour of the metal, i.e. the deformation rate under a certain stress and temperature, depends on the metallurgical situation which, in turn, is influenced by the heat-treatment history.[9] Hence, experiments for finding the uniaxial creep law must take account of the alteration of the metallurgical state. The uniaxial creep test data which are creep strain versus time curves at several temperatures and stress levels, are transformed into the creep law by curve fitting using a regression algorithm.

16.2 THE NUMERICAL ALGORITHMS USED IN THE ANALYSIS

If the heat generation due to deformation of the body and the influence of stresses on the phase transformation are neglected a separation of the thermal analysis from the stress analysis is allowed.

16.2.1 The thermal analysis procedure

The lumped parameter method,[10] subdividing the body into finite elements, see Figure 16.1, leads to the heat balance relation for element i:

$$\sum_j {}^tH_{ij}({}^tT_j - {}^tT_i)\,\Delta t = {}^tC_i({}^{t+\Delta t}T_i - {}^tT_i), \tag{16.1}$$

where ${}^tT_i$ is the element temperature at time t and

$${}^tC_i = {}^tc_{pi}\,{}^t\rho_i V_i, \tag{16.2}$$

$${}^tH_{ij} = \frac{A_{ij}}{\dfrac{l_{ij}}{{}^t\lambda_i} + \dfrac{l_{ji}}{{}^t\lambda_j}}. \tag{16.3}$$

Equation (16.1) is summed up over all elements j which are neighbours of element i. ${}^tc_{pi}$ is the temperature-dependent specific heat of the material in element i at temperature ${}^tT_i$, ${}^t\rho_i$ is the mass density, ${}^t\lambda_i$ the heat conduction coefficient, and V_i the volume of element i. The surface A_{ij} and the distance l_{ij} are explained in Figure 16.1.

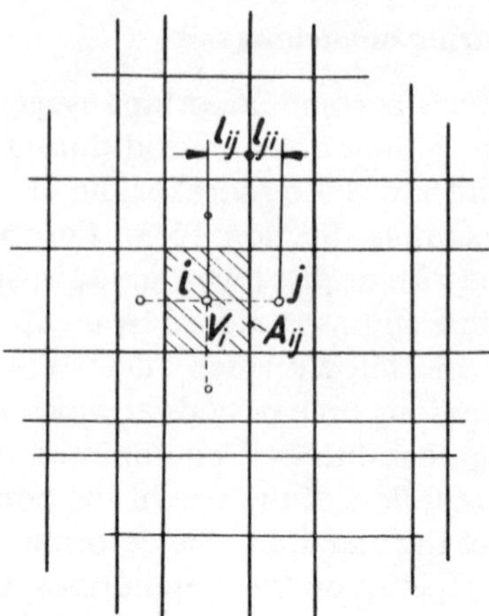

Figure 16.1 The lumped-parameter model

For each element ($i = 1, 2, \ldots, n$) in each time step ($t \rightarrow t + \Delta t$), Equation (16.1) yields an approximate linear relation. The resulting system of algebraic equations is solved iteratively because of the nonlinearities due to the temperature dependence of the thermal material properties (see, for example, reference 11). In Equation (16.1) no heat-source term is included. Hence, the transformation-induced heat generation (latent heat) is simulated by a modified temperature dependence of the specific heat, as shown in Section 16.3.

16.2.2 The stress analysis procedure

The stress-analysis problem is highly nonlinear due to the nonlinear temperature-, state-, and time-dependent material behaviour. The finite element method, in comparison to some finite difference approaches[12,13] and the method of successive elastic solution,[14] has the main advantage of being almost independent of the geometry of the considered body. Here the procedure summarized in references 1 and 11 using the finite element program ADINA[15] is applied. This procedure is based on the tangent stiffness matrix concept in combination with an incremental-iterative approach. Assuming only thermal loading, the governing finite element equations for an increment $t \rightarrow t + \Delta t$ are

$$^{\tau}[K]\{\Delta u\}^{i} = -^{t+\Delta t}\{F\}^{i-1}, \tag{16.4}$$

where $^{\tau}[K]$ is the tangent stiffness matrix at time $\tau \leq t$, which is updated at time steps predicted by the analysis control. $^{t+\Delta t}\{F\}^{i-1}$ is a vector of nodal-point forces corresponding to the internal element stresses in iteration $(i-1)$ at time $t+\Delta t$. $\{\Delta u\}^i$ is the increment in the nodal-point displacement vector in iteration i, such that

$$^{t+\Delta t}\{u\}^i = {}^{t+\Delta t}\{u\}^{i-1} + \{\Delta u\}^i. \tag{16.5}$$

Equation (16.4) corresponds to the modified Newton–Raphson iteration. The convergence of the iteration can be significantly improved using the BFGS updates as described in reference 16. The material model and numerical solution scheme for the thermo-elastic–plastic and creep analysis is described in detail in reference 17 and summarized in reference 1. The essence of the algorithm is the decomposition of the total strain tensor into the elastic, plastic, creep and thermal strain tensor:

$$^{\tau}e_{ij} = {}^{\tau}e^{\mathrm{E}}_{ij} + {}^{\tau}e^{\mathrm{P}}_{ij} + {}^{\tau}e^{\mathrm{C}}_{ij} + {}^{\tau}e^{\mathrm{TH}}_{ij}. \tag{16.6}$$

The constitutive law for the isotropic, thermo-elastic material is used:

$$^{\tau}\sigma_{ij} = {}^{\tau}C^{\mathrm{E}}_{ijkl}({}^{\tau}e_{kl} - {}^{\tau}e^{\mathrm{P}}_{kl} - {}^{\tau}e^{\mathrm{C}}_{kl} - {}^{\tau}e^{\mathrm{TH}}_{kl}), \tag{16.7}$$

where ${}^{\tau}C^{\mathrm{E}}_{ijkl}$ denotes the elastic constitutive tensor, defined by Young's modulus, $E({}^{\tau}T)$, and Poisson's ratio, $\nu({}^{\tau}T)$. The plastic strains are calculated using classical theory of time-independent plasticity with the von Mises yield criterion, the associated normality flow rule, and kinematic hardening is preferred to isotropic hardening.[1]

The von Mises yield function for nonisothermal kinematic hardening

$$^{\tau}F = \tfrac{1}{2}({}^{\tau}s_{ij} - {}^{\tau}\gamma_{ij})({}^{\tau}s_{ij} - {}^{\tau}\gamma_{ij}) - \tfrac{1}{3}{}^{\tau}\sigma^2_{\mathrm{y}} \tag{16.8}$$

contains the instantaneous yield stress, ${}^{\tau}\sigma_{\mathrm{y}} = \sigma_{\mathrm{y}}({}^{\tau}T)$, including the pseudoplasticity due to transformation effects, i.e. the transformation plasticity. ${}^{\tau}s_{ij}$ denotes the deviatoric stresses, ${}^{\tau}\gamma_{ij}$ is the yield surface translation tensor depending on the hardening modulus, ${}^{\tau}E_{\mathrm{T}}$, and on the history of plastic deformation and temperature.

The creep strain tensor is determined using the uniaxial creep law in combination with a flow rule. Strain hardening for variable loading and the Oak Ridge National Laboratory auxiliary hardening rules for cyclic behaviour are included in the procedure.

The thermal strains are determined by

$$^{\tau}e^{\mathrm{TH}}_{ij} = {}^{\tau}\alpha_{\mathrm{m}}({}^{t}T - T_{\mathrm{ref}})\delta_{ij}, \tag{16.9}$$

where δ_{ij} is the Kronecker delta and ${}^{\tau}\alpha_{\mathrm{m}}$ is the mean coefficient of thermal expansion

$$^{\tau}\alpha_{\mathrm{m}} = \int_{T_{\mathrm{ref}}}^{{}^{\tau}T} \alpha(\theta)\,\mathrm{d}\theta/({}^{\tau}T - T_{\mathrm{ref}}), \tag{16.10}$$

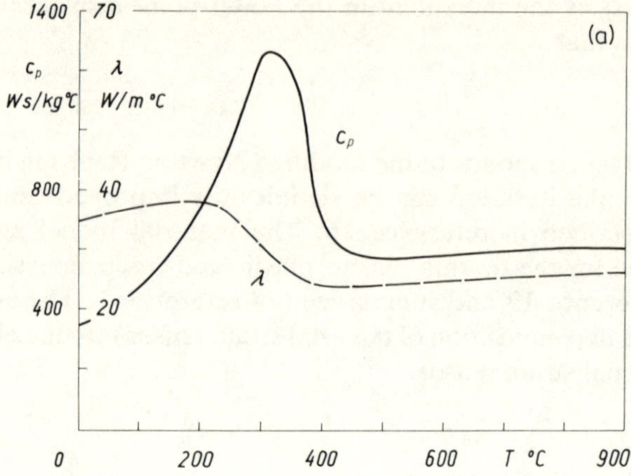

1400
70
(a)
c_p
Ws/kg°C
λ
W/m °C
c_p
800
40
λ
400
20
0
200
400
600
T °C
900

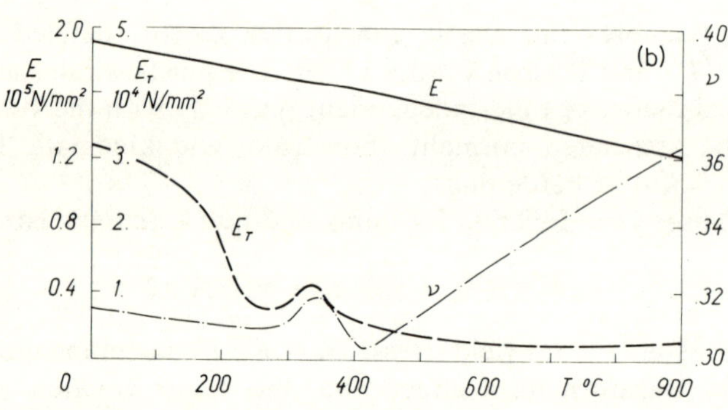

2.0
5.
.40
(b)
E
10^5N/mm²
E_T
10^4 N/mm²
E
ν
1.2
3.
.36
0.8
2.
E_T
.34
0.4
1.
ν
.32
.30
0
200
400
600
T °C
900

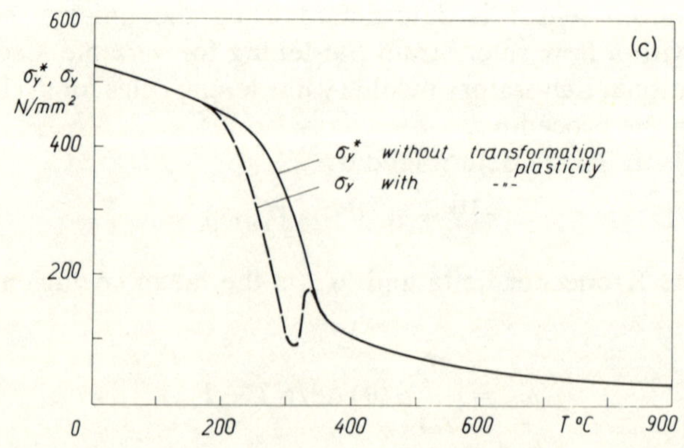

600
(c)
σ_Y^*, σ_Y
N/mm²
400
σ_Y^* without transformation plasticity
σ_Y with
200
0
200
400
600
T °C
900

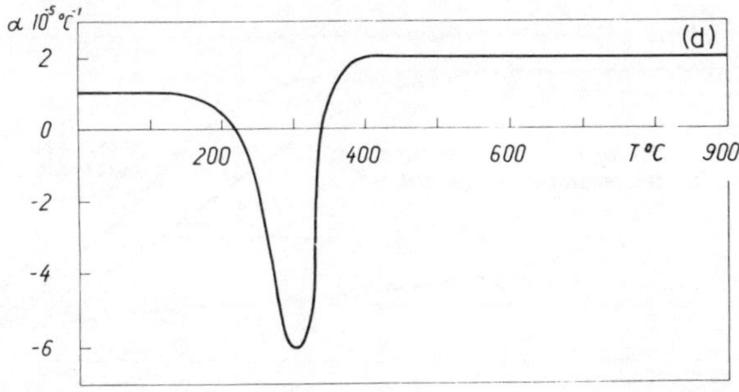

Figure 16.2 The thermal and mechanical properties as a function of temperature

where $\alpha(\theta)$ is the instantaneous linear coefficient of thermal expansion at temperature θ including both the temperature-induced and the transformation-induced volume changes. T_{ref} is the stress-free reference temperature.

16.3 QUENCHING ANALYSIS OF A STEEL CYLINDER

To demonstrate the influence of the transformation plasticity some results of an analysis described in detail in reference 1 are discussed now.

A long nickel–steel cylinder with a diameter of 50 mm is heated up to $T_{ref} = 900$ °C, held at this temperature and finally chilled in ice water. The transient temperature and stress fields are determined using the methods described above.

The temperature dependence of the thermal and mechanical properties is shown in Figure 16.2.

Since the cylinder is very long it is sufficient to analyse a typical cross-section using an axisymmetric model with no heat flux in the axial and circumferential direction and assuming generalized plane strain conditions.

Figure 16.3 shows the calculated temperature–time history for distances of $r = 0.0$, 0.125 and 25 mm from the axis. The phase transformation from austenite to nickel–martensite occurs in the temperature interval also indicated in Figure 16.3.

The influence of creep and, particularly, of transformation plasticity becomes clear by considering Figure 16.4 which shows the time history of the axial stress component at the cylinder axis.

Curve (a) represents the results disregarding both creep and transformation plasticity. The axial component of the residual stresses at $r = 0$ does not agree at all with the experimental value as shown in Figure 16.4. Including creep with the experimentally obtained uniaxial creep law for constant stress

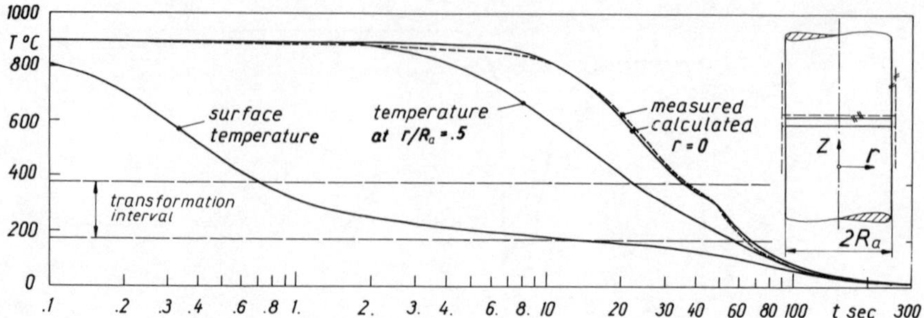

Figure 16.3 Temperature history for the surface, the axis and an intermediate position of the nickel–steel cylinder

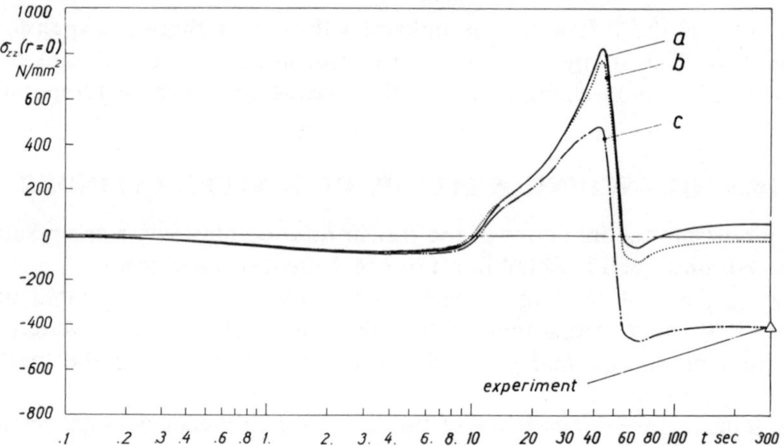

Figure 16.4 Histories of the axial component of the stress at the axis of the cylinder

level, σ,

$$^{\tau}e^{C} = 1.321 \times 10^{-4} \exp\left(-\frac{1.152 \times 10^{4}}{T^{*}}\right)\sigma^{3.142}\tau^{0.165} \qquad (16.11)$$

(T^{*} is the absolute temperature in K, τ the time in seconds, σ is measured in N/mm^2) renders curve (b). The effects of creep are nearly negligible because the cylinder is rather thin and is exposed to a high temperature for a time period which is too short for the development of significant creep strains.

Including transformation plasticity shows much more effect, as represented by curve (c): The transformation plasticity is taken into account by a reduction of the yield stress $^{\tau}\sigma_{y}^{*}$ as explained in Section 16.1.1 (see the dotted line in Figure 16.2(c)). The transformation from austenite to nickel–martensite is

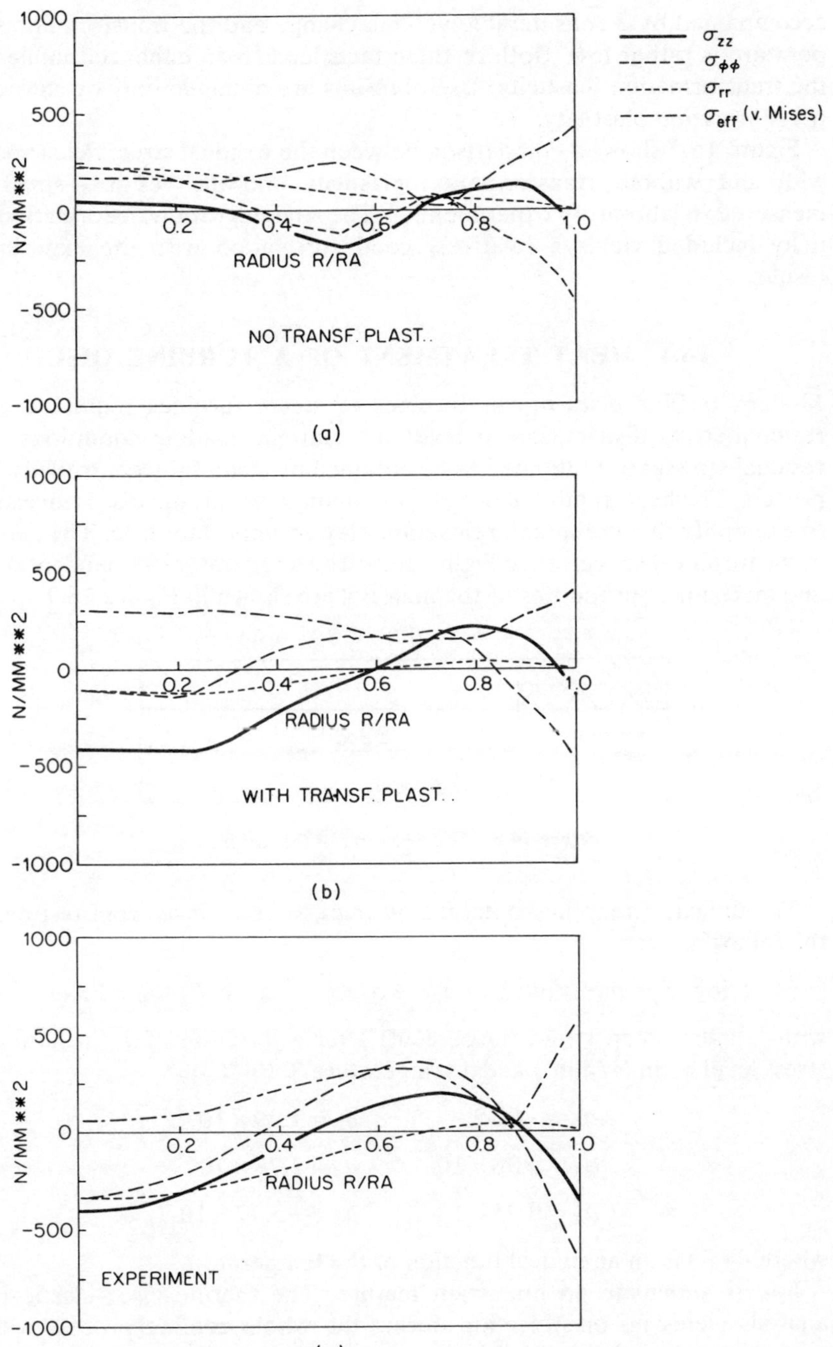

13096 **Figure 16.5** Residual stresses analysed and measured

accompanied by a considerable volume change and the transformation temperature is rather low. Both of these facts lead to an enhanced influence of the transformation plasticity. Useful results are obtained only by considering transformation plasticity.

Figure 16.5 shows a comparison between the residual stress fields analysed with and without transformation plasticity and the residual stress field measured in laboratory experiments.[18] The analysis with transformation plasticity included yields a relatively good correlation with the experimental results.

16.4 HEAT TREATMENT OF A TURBINE DISC

Modern turbine discs of gas turbines or steam turbines require a certain residual stress distribution to resist the extreme loading conditions. These residual stresses are intended to be obtained by a carefully controlled cooling process. The heat treatment of a rather voluminous turbine disc is investigated to exemplify that creep and relaxation play an important role. The geometry of the turbine disc is given in Figure 16.6. The temperature-dependent thermal and mechanical properties of the material are shown in Figure 16.7.

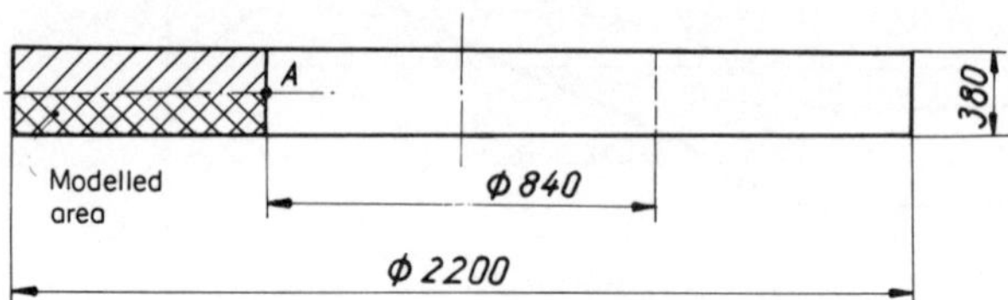

Figure 16.6 The geometry of the turbine disc

The uniaxial creep law obtained by regression analysis from test data has the following form:

$$\log {}^{\tau}e^{C} = a_0 + a_1\Phi(T) + a_2\sigma + a_3 \log \tau + a_4 \Phi(T) \log \tau + a_5\sigma^2 \quad (16.12)$$

with ${}^{\tau}e^{C}$, the creep strain (in percent) after a duration of τ (in hours) at a stress level σ (in N/mm^2) and a temperature T (in °C);

$$a_0 = -6.32, \qquad a_3 = 1.29 \times 10^{-1},$$
$$a_1 = 2.30 \times 10^{-2}, \qquad a_4 = 2.78 \times 10^{-3},$$
$$a_2 = 6.95 \times 10^{-3}, \qquad a_5 = -5.77 \times 10^{-7},$$

where $\Phi(T)$ is an analytical function of the temperature.

Let us anticipate an important feature: The thermo-visco-elastic–plastic analysis yields no plastification during the whole cooling process with the exception of one integration point at one single time. The residual stresses

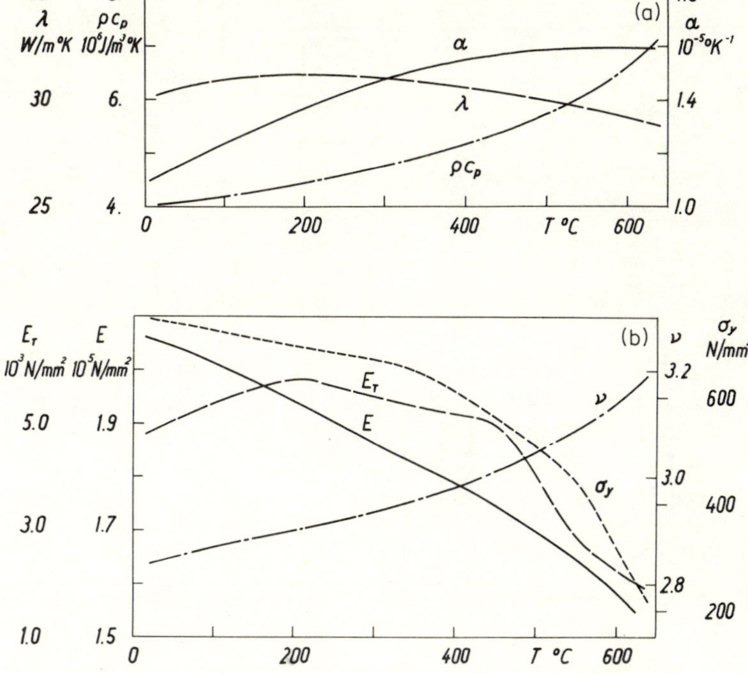

Figure 16.7 The temperature dependence of the material properties

are solely a consequence of irreversible creep deformations. Hence, a thermo-elastic–plastic analysis—with no creep included—cannot predict the stress development during this heat-treatment process. On the other hand, since no phase transformation occurs, the transformation plasticity effects that were important in the previous cylinder example are meaningless in this analysis.

In Figure 16.8 the residual stresses, once calculated with creep and once without creep, are shown. Significant differences can be observed. The analysis without creep results in a residual stress field which has significant stress values in a very local domain at point A only. This analysis yields irreversible (purely plastic) deformation in this local region exclusively.

In order to get a better understanding of the development of the stresses, the time history of the temperature and the equivalent stress at the point A (see Figure 16.6), which is the most stressed point, is shown in Figure 16.9.

In Figure 16.9 the yield stress corresponding to the current temperature is presented. Since no relaxation appears in the thermoelastic–plastic analysis, in which no creep effects are included, much more plastification occurs.

Finally, Figure 16.10 shows the time history of the axial component of the stress at point A. A comparison between the measured and the calculated

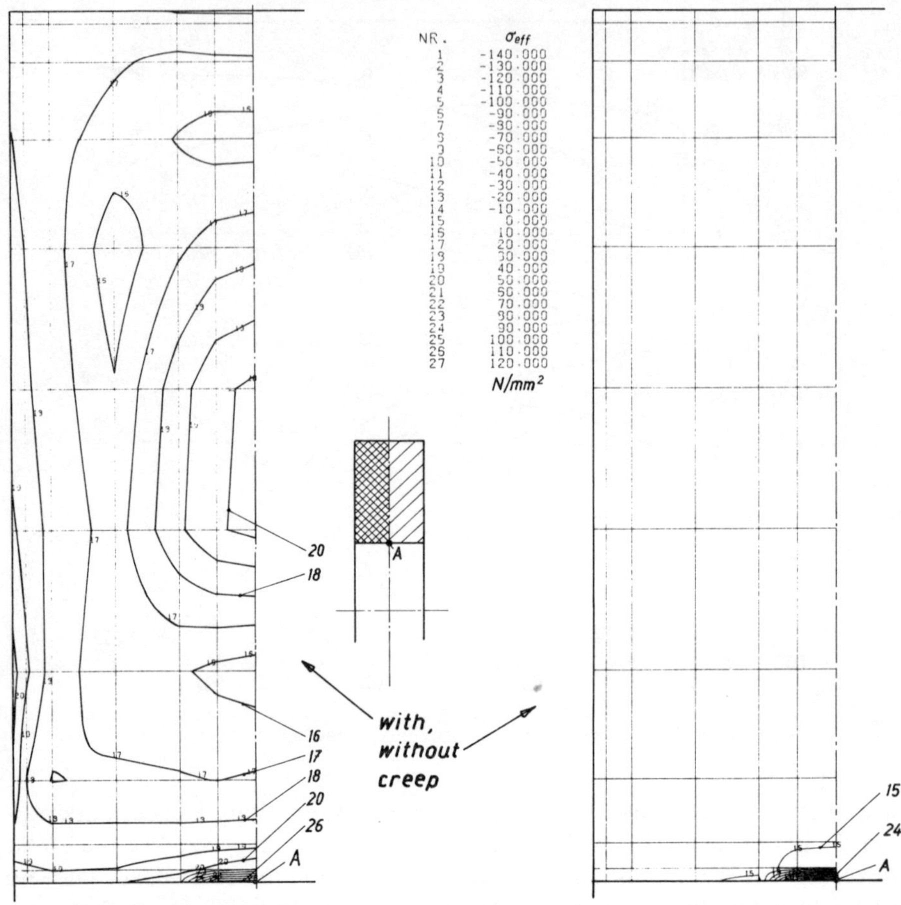

Figure 16.8 The residual stresses shown by equivalent stress distribution

residual stress values shows good correspondence if creep is included into the analysis.

CONCLUSIONS

Suitable material modelling is a prerequisite for useful results in heat-treatment analyses. Specifically in the case of phase transformation with considerable volume changes, a pseudoplasticity effect—the so-called transformation plasticity—must be taken into account. For thick or voluminous bodies or in the case of a slow temperature variation, creep and relaxation must be included into the analysis, otherwise rather meaningless results are obtained.

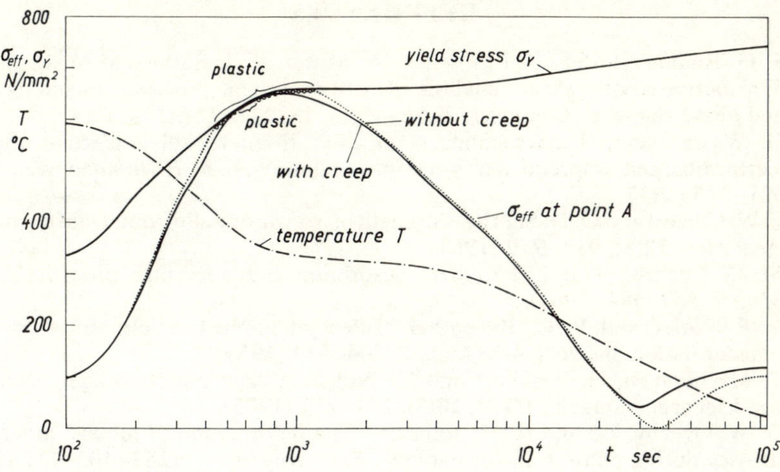

Figure 16.9 History of the temperature and the equivalent stress at point A

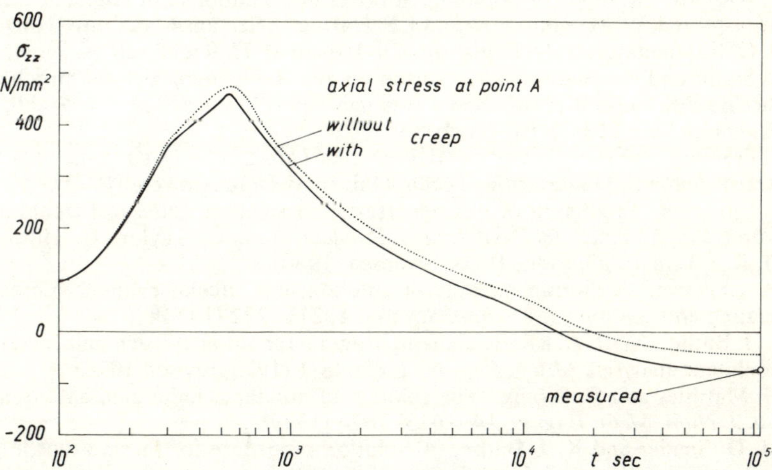

Figure 16.10 History of the axial stress component at point A—comparison between measured and calculated residual stress values

ACKNOWLEDGEMENTS

The financial support of the Austrian Forschungsförderungsfonds der gewerblichen Wirtschaft is gratefully acknowledged. We also thank Mr. Jaquemar and Mr. Bartosch of VOEST-ALPINE AG, Linz, Austria, for their efforts in calculating the transient temperature fields.

REFERENCES

1. F. G. Rammerstorfer, D. F. Fischer, W. Mitter, K. J. Bathe and M. D. Snyder, 'On thermo-elastic-plastic analysis of heat-treatment processes including creep and phase changes', *Computers & Structures*, **13**, 771–779 (1981).
2. G. Wassermann, 'Untersuchungen an einer Eisen–Nickel–Legierung über die Verformbarkeit während der γ-α-Umwandlung', *Arch. Eisenhüttenwes.*, **10**(7), 321–325 (1937).
3. F. W. Clinard and O. D. Sherby, 'Strength of iron during allotropic transformation', *Acta Met.*, **12**(8), 911–919 (1964).
4. M. G. Lozinsky, 'Grain boundary adsorbtion and 'superhigh plasticity' ', *Acta Met.*, **9**, 689–694 (1961).
5. L. F. Porter and P. C. Rosenthal, 'Effect of applied tensile stress on phase transformation in steel', *Acta Met.*, **7**, 504–514 (1959).
6. E. Macherauch, H. Wohlfart and U. Wolfstieg, 'Zur zweckmäßigen Definition von Eigenspannungen', *HTM*, **28**(3), 201–211 (1973).
7. G. W. Greenwood and R. H. Johnson, 'The deformation of metals under small stresses during phase transformations', *Proc. Roy. Soc.*, **A283**, 403–421 (1965).
8. K. A. Padmanabhan and G. J. Davies, *Superplasticity*, Springer-Verlag, Berlin, Heidelberg, and New York (1980).
9. G. Bernasconi and G. Piatti (eds.), *Creep of Engineering Materials and Structures*, Applied Science Publishers, London (1979).
10. P. Razelos, 'Methods of obtaining approximate solutions', in *Handbook of Heat Transfer* (ed. W. M. Rohsenow and J. P. Hartnett), Section 4, McGraw-Hill (1973).
11. F. G. Rammerstorfer, D. F. Fischer, Ch. Jaquemar, H. Kreulitsch, A. Scheinecker, J. Stulik and A. Niedermayr, 'Verfahren zur Berechnung der thermischen und thermo-visko-elasto-plastischen Vorgänge beim Stranggießen von Stahl', *Arch. Eisenhüttenwes.*, **51**(2), 61–66 (1980).
12. S. Franitza, *Zur Berechnung der Wärme- und Umwandlungsspannungen in langen Kreiszylindern*', Dissertation, Techn. University Braunschweig (1972).
13. S. Sjöström, 'Prediction of quench stresses in steel—a numerical treatment', in *Numerical Methods for Non-Linear Problems* (ed. C. Taylor, E. Hinton and D. R. J. Owen), Pineridge Press, Swansea (1980).
14. H. Ishikawa, 'A thermoelastoplastic solution of a circular cylinder subjected to heating and cooling', *J. Thermal Stresses*, **1**, 211–222 (1978).
15. K. J. Bathe, 'ADINA: a finite element program for automatic dynamic incremental nonlinear analysis', *M.I.T. Rep.* AVL 82448-1 (1975, revised 1978).
16. H. Matthies and G. Strang, 'The solution of nonlinear finite element equations', *Int. J. Num. Meth. Engng.*, **14**, 1613–1626 (1979).
17. M. D. Snyder and K. J. Bathe, 'A Solution procedure for thermo-elastic-plastic and creep problems', *J. Nucl. Engng. Des.* (1981).
18. O. Gründler, W. Mitter and G. Wiedner, 'Comparative measurements of residual stresses on quenched nickel-steel cylinders', *Techn. Messen*, **3**, 93–98 (1980).

Numerical Methods in Heat Transfer, Volume II
Edited by R. W. Lewis, K. Morgan, and B. A. Schrefler

Chapter 17

The $\dot{Q}$ Method—A Compact Technique for Describing the Heat Flux Present at the Mould-metal Interface in Solidification Problems

C. Wei, P. N. Hansen, and J. T. Berry

ABSTRACT

An initial effort at describing transient diffusion of heat at corners is presented. The technique is applied to finite difference simulations of the solidification and cooling of castings, where significant numbers of volume elements in the mould are replaced by the $\dot{Q}$ function. The $\dot{Q}$ method yields a more efficient use of computer capacity in calculating solidification profiles of shaped castings and therefore holds considerable promise in this type of application.

NOMENCLATURE

K	thermal conductivity of moulding materials
α	thermal diffusivity of moulding materials
$\dot{Q}$	heat removed from the mould–metal interface per unit time
c	specific heat capacity of moulding materials
$\frac{\partial \hat{C}}{\partial A}$	excess heat extraction per unit area at two-dimensional external corners of a casting
T	temperature
θ	modified temperature
H	heat content function
L	latent heat or length parameter
ρ	density of moulding materials
x, y, z	Cartesian coordinates
r	polar coordinate
A	surface area at mould–metal interface
t	time

17.1 INTRODUCTION

One of the principal problems in truly three-dimensional finite difference and finite element simulations of physical systems, such as the solidification and cooling of castings, is the great number of volume elements needed to ensure accuracy in the result of the computations. Since typically, say, 70–90% of these elements are in the mould, it is obvious that great savings may be obtained if the cooling effect of the mould is described in a compact analytical fashion. In this initial study only silica sand moulds are examined, and for this case it is reasonable to assume that the temperature in both the sand and the metal at the interface is identical.

17.2 ANALYTICAL MODELS

17.2.1 The planar, cylindrical and the spherical interfaces

The temperature in the mould may be schematically represented as shown in Figure 17.1, where T_{if} and T_i respectively represent the interface and initial mould temperatures. $\dot{Q}$ is the heat transfer rate at the interface from the metal into the mould, x is the distance measured from the interface into the mould, and r_0 is the distance from a fixed point to the interface.

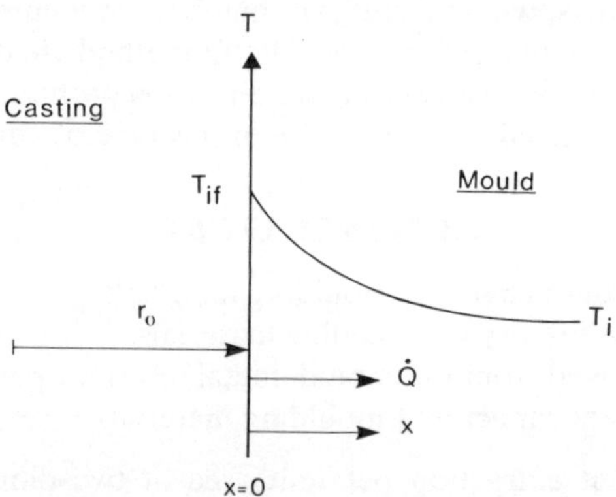

Figure 17.1 Temperature distribution in the mould

The temperature distribution at a specified time, t, may be described for the planar interface (Carslaw and Jaeger[1]) as

$$T = (T_{if} - T_i)\,\mathrm{erfc}\left(\frac{x}{2\sqrt{(\alpha t)}}\right) + T_i, \tag{17.1}$$

α being the thermal diffusivity and t the time.

The heat flux (per unit area and time) at the interface ($x = 0$) is given as

$$\frac{\dot{Q}}{A} = -K\frac{\partial T}{\partial x}\bigg)_{x=0} = (T_{\text{if}} - T_{\text{i}})\frac{K}{\sqrt{(\pi\alpha t)}}, \tag{17.2}$$

or

$$\frac{\dot{Q}}{A} = (T_{\text{if}} - T_{\text{i}})\sqrt{\left(\frac{K\rho c}{\pi}\right)}\frac{1}{\sqrt{t}}. \tag{17.3}$$

In Equations (17.2) and (17.3), K is the thermal conductivity and A is the actual surface area. The quantity $\sqrt{(K\rho c)}$ is the so-called coefficient of heat diffusivity for the mould material. Equation (17.3) represents the desired compact description of the heat flux.

For the case of a cylindrical interface the corresponding heat flux expression for small values of time may be expressed as

$$\frac{\dot{Q}}{A} = (T_{\text{if}} - T_{\text{i}})\sqrt{\left(\frac{K\rho c}{\pi}\right)}\left(\frac{1}{\sqrt{t}} + \frac{\sqrt{(\alpha\pi)}}{2r_0}\right), \tag{17.4(a)}$$

where r_0 is the distance from the centre of the cylinder to the interface. For relatively small castings this expression is applicable through the end of the solidification process.

For a spherical interface the corresponding heat flux expression is

$$\frac{\dot{Q}}{A} = (T_{\text{if}} - T_{\text{i}})\sqrt{\left(\frac{K\rho c}{\pi}\right)}\left(\frac{1}{\sqrt{t}} + \frac{\sqrt{(\alpha\pi)}}{r_0}\right), \tag{17.4(b)}$$

r_0 being the distance from the centre of the sphere to the interface.

Equations (17.4(a)) and (17.4(b)) may be expressed in the form

$$\frac{\dot{Q}}{A} = (T_{\text{if}} - T_{\text{i}})\sqrt{\left(\frac{K\rho c}{\pi}\right)}\left(\frac{1}{\sqrt{t}} + \frac{\sqrt{(\alpha\pi)}}{L}\right), \tag{17.5}$$

L being a length dimension of value $2r_0$ for a cylinder and r_0 for a sphere.

17.2.2 Internal corners of castings

Product solutions of one-variable problems may be used to formulate solutions of problems in several space variables for certain types of initial and boundary conditions.[1] Thus, the respective expressions for *local* heat flux along the plane $x = 0$ for the cases of two- and three-dimensional right-angled corners are

$$\left(\frac{\dot{Q}}{A}\right)_{\text{local}} = (T_{\text{if}} - T_{\text{i}})\sqrt{\left(\frac{k\rho c}{\pi}\right)}\left[\frac{1}{\sqrt{t}}\operatorname{erf}\left(\frac{y}{2\sqrt{(\alpha t)}}\right)\right], \tag{17.6}$$

y being the distance from the right-angled two-dimensional corner, and

$$\left(\frac{\dot{Q}}{A}\right)_{\text{local}} = (T_{\text{if}} - T_{\text{i}})\sqrt{\left(\frac{K\rho c}{\pi}\right)}\left[\frac{1}{\sqrt{t}}\operatorname{erf}\left(\frac{y}{2\sqrt{(\alpha t)}}\right)\operatorname{erf}\left(\frac{z}{2\sqrt{(\alpha t)}}\right)\right], \tag{17.7}$$

y and z being the coordinates measured from the three-dimensional corner on the plane normal to the heat flux direction.

By using cylindrical coordinates, the heat flux expression for the two-dimensional corner may be obtained by writing two expressions in the notation of Figure 17.2 as

$$\left(\frac{\dot{Q}_1}{A}\right)_{\text{local}} = (T_{\text{if}} - T_{\text{i}})\sqrt{\left(\frac{k\rho c}{\pi}\right)}\left[\frac{1}{\sqrt{t}}\operatorname{erf}\left(\frac{r - r_0}{2\sqrt{(\alpha t)}}\right)\right], \tag{17.8}$$

and

$$\left(\frac{\dot{Q}_2}{A}\right)_{\text{local}} = (T_{\text{if}} - T_{\text{i}})\sqrt{\left(\frac{K\rho c}{\pi}\right)}\left[\left(\frac{1}{\sqrt{t}} + \frac{\sqrt{(\alpha\pi)}}{2r_0}\right)\operatorname{erf}\left(\frac{z}{2\sqrt{(\alpha t)}}\right)\right]. \tag{17.9}$$

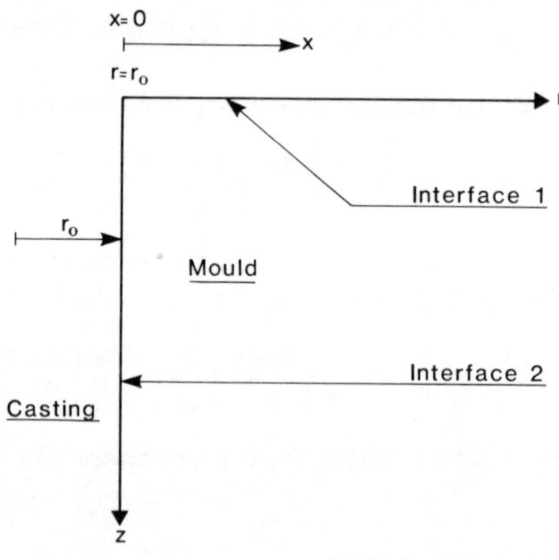

Figure 17.2 Representation of a two-dimensional internal corner—cylindrical coordinates

17.2.3 Proposed application of the $\dot{Q}$ method to external corners of a casting

The approach, inspired by the work of Ruddle and Skinner,[2] consists of comparing the 'extra cooling effect' due to the two-dimensional external corner with the expression for heat flux for the plane wall. This suggests that

Equation (17.3) may be modified as

$$\left(\frac{\dot{Q}}{A}\right)_{\text{local}} = (T_{\text{if}} - T_{\text{i}})\sqrt{\left(\frac{K\rho c}{\pi}\right)}\left[\frac{1}{\sqrt{t}} + \frac{\partial \hat{C}}{\partial A}\right], \tag{17.10}$$

which, when summed around the corner, gives an averaged heat flux expression for the corner as

$$\sum_{\text{corner}} \left(\frac{\dot{Q}}{A}\right)_{\text{local}} = \left(\frac{\dot{Q}}{A}\right)_{\substack{\text{average}\\\text{corner}}}$$

$$= (T_{\text{if}} - T_{\text{i}})\sqrt{\left(\frac{K\rho c}{\pi}\right)}\left[\frac{A}{\sqrt{t}} + \sum_{\substack{\text{corner}\\\text{area}}} \frac{\partial \hat{C}}{\partial A}\right], \tag{17.11}$$

where

$$\sum_{\substack{\text{corner}\\\text{area}}} \frac{\partial \hat{C}}{\partial A} = \text{constant}$$

$$= \text{'time-independent extra cooling effect'}$$

and A = total corner area.

In order to quantitatively establish the value of this constant, the actual distribution of $\dot{Q}$ around the external corner must be considered.

An approach such as that outlined in the preceding is currently being explored to include external/internal corner combinations. The compatibility of the results obtained by approximate analysis with those obtained by numerical computations using the nodes in the mould must be established to validate this approach.

As an example, a combined internal/external corner is proposed as shown in Figure 17.3. For the interface 1 and 3, the respective heat flux expressions proposed are

$$\left(\frac{\dot{Q}}{A}\right)_{\text{local}} = (T_{\text{if}} - T_{\text{i}})\sqrt{\left(\frac{K\rho c}{\pi t}\right)}\left[1 + 0.47\,\text{erfc}\left(\frac{x}{2\sqrt{(\alpha t)}}\right)\right], \tag{17.12}$$

and

$$\left(\frac{\dot{Q}}{A}\right)_{\text{local}} = (T_{\text{if}} - T_{\text{i}})\sqrt{\left(\frac{K\rho c}{\pi}\right)}\left[\frac{1}{\sqrt{t}}\,\text{erf}\left(\frac{y}{2\sqrt{(\alpha t)}}\right)\right]. \tag{17.6}$$

In Equation (17.12) the coefficient 0.47 results from a recently introduced theory on the transient edge effect upon heat conduction in wedges[3] and the experimental work by Ruddle and Skinner.[2] The complementary error function profile in Equation (17.12) is found to be a good representation of the 'extra cooling effect' of the external corner.[4]

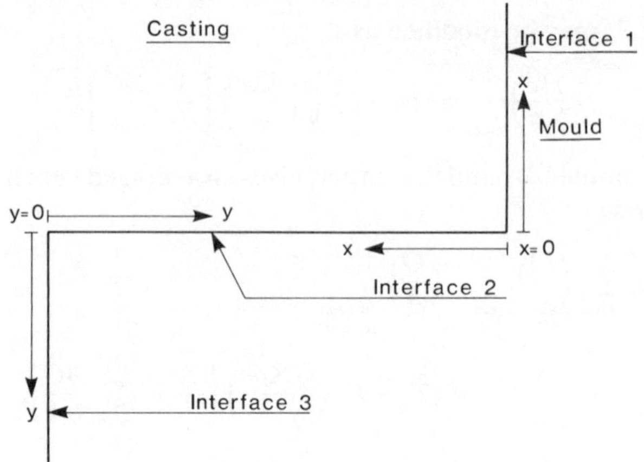

Figure 17.3 A two-dimensional casting geometry

To obtain the corresponding expression for the interface 2, Equations (17.6) and (17.12) may be combined to obtain

$$\left(\frac{\dot{Q}}{A}\right)_{\text{local}} = (T_{\text{if}} - T_{\text{i}})\sqrt{\left(\frac{K\rho c}{\pi t}\right)}\left[1 + 0.47\,\text{erfc}\left(\frac{x}{2\sqrt{(\alpha t)}}\right)\right]\left(\text{erf}\,\frac{y}{2\sqrt{(\alpha t)}}\right) \tag{17.13}$$

An analytical solution for the $\dot{Q}$ function for general corners (not necessarily rectangular) is currently being investigated.[4]

17.2.4 Limitations on the $\dot{Q}$ expressions

The derived expressions for $\dot{Q}$ are restricted by the fact of having assumed a constant interface temperature. The assumption is exactly valid only for planar wall solution when casting at the liquidus temperature. The effect of this assumption on other $\dot{Q}$ expressions is also being examined currently. This is done by setting (at every time step during computation of temperature field in the casting) T_{if} at an arbitrary point of the interface equal to the calculated temperature at that point at that time for use in the $\dot{Q}$ function. It is then necessary to compare the results of calculations using $\dot{Q}$ functions with those obtained by conventional techniques that use many net points in the mould.

17.3 NUMERICAL/COMPUTATIONAL TECHNIQUES

A standard finite difference scheme explained by Hansen[5,6] is being incorporated for the computations. Pure aluminium is used as the alloy. The heat content function[5] shown in Figure 17.4 and the thermal conductivity function

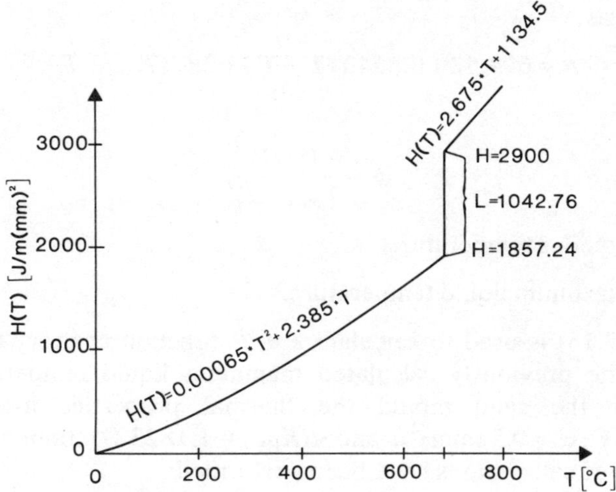

Figure 17.4 Heat content for pure aluminium

shown in Figure 17.5 are utilized. To overcome the effect of discontinuity in Figure 17.5 a modified temperature is defined as

$$\theta = \frac{1}{K_0}\int_0^T K \, dT, \tag{17.14}$$

K_0 being the conductivity at a temperature of 0 °C. An attempt is being made to use a modified thermal conductivity for liquid aluminium to accommodate

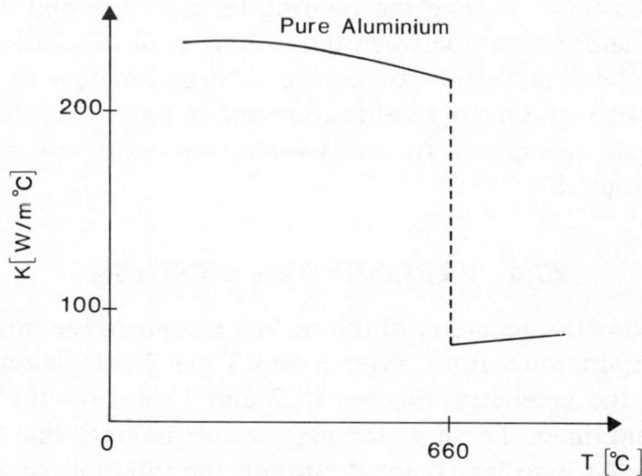

Figure 17.5 The coefficient of thermal conductivity for pure aluminium

convection,[7] as:

$$\theta = 604.52 + 0.3543\,(T - T_l) + 3\phi^7 (T_{max} - T_l), \tag{17.15}$$

with

$$\phi = \frac{T - T_l}{T_{max} - T_l},$$

and T_l = liquidus temperature

and T_{max} = maximum liquid temperature.

Equation (17.15) is used to calculate a θ–T function at every time step by also using the previously calculated maximum liquid temperature in the casting. For the sand mould the thermal properties used are $K = 0.836$ W/m °C, $\alpha = 0.5\ (\text{mm})^2/\text{s}$ and $\sqrt{(K\rho c)} = 1.1823\ \text{J/m(mm) °C(s)}^{1/2}$.

Two types of calculations have been performed:

Type 1: Calculations with net points in both the casting and the mould.
Type 2: Calculations with net points in the casting and the use of $\dot{Q}$ functions.

In both cases the net points in the casting have been spaced 1 mm apart. To ensure that Type 1 calculations are very accurate, the spacing of the net points in the mould has been chosen according to the following scheme:

Close to the interface: spacing $= \frac{1}{10}$ mm.

Away from the interface the spacing is gradually increased. (The moulds have been chosen so thick as to be considered as thermally infinite during the solidification process of the castings.)

Several initial computations have been performed for the planar, cylindrical and spherical castings, varying the pouring temperature and the volume of the castings. The difference between the two types of calculation procedures is expected to be small when comparing such parameters as the interface temperature transient, time of solidification and so forth. Calculations of Type 2 will most likely be superior for cases where the net points in the sand are not finely distributed.

17.4 PRELIMINARY RESULTS

Figure 17.6 shows the geometry of the casting examined for two-dimensional numerical computations. Both Type 1 and Type 2 calculations have been performed for the geometry. Figures 17.7 and 17.8 show the solidification front at different times. These preliminary results indicate that the effort is a promising one. As is to be expected, around the internal corner the results are almost the same due to the mathematical 'correctness' of the $\dot{Q}$ function. Around the external corners the agreement is still considered adequate.

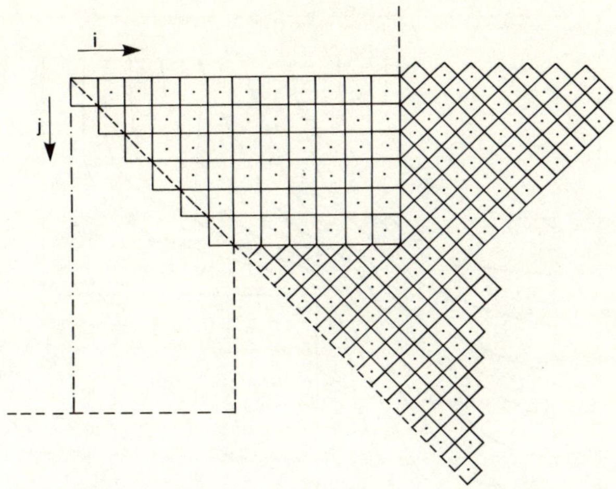

Figure 17.6 Quarter cross section of a two-dimensional casting. (The broken line indicates the casting. Due to geometric symmetry only $\frac{1}{8}$ of the casting needs to be used in the calculations. This part is enmeshed. (The dots are the net points.)

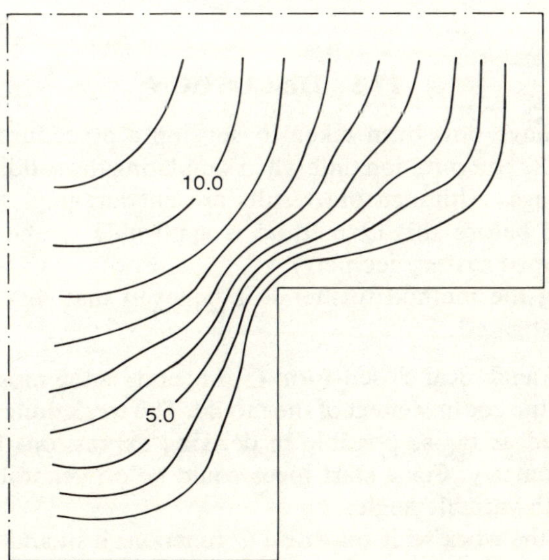

Figure 17.7 The solidification front at different times after pouring—Type 1 calculations. (5 and 10 on the curves indicate $\sqrt{(\text{time})}$ in $s^{1/2}$)

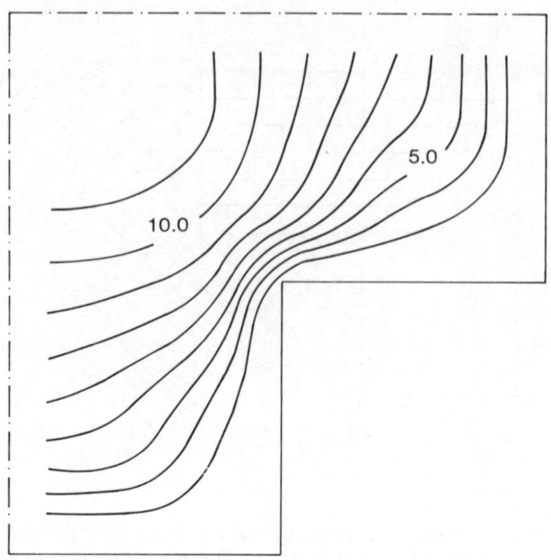

Figure 17.8 The solidification front at different times after pouring—Type 2 calculations. (5 and 10 on the curves indicate $\sqrt{(\text{time})}$ in $s^{1/2}$)

17.5 DISCUSSION

The first steps have now been taken to develop a procedure, which can be used to reduce the computation time when simulating the solidification profiles of shaped castings. Although the results are encouraging, still much more work is needed before this $\dot{Q}$ method is applicable to the general three-dimensional shaped casting geometry.

In developing the method further it is believed that several possibilities need to be investigated:

(a) The use of analytical closed-form $\dot{Q}$ functions is the most handy way of describing the cooling effect of the mould. The work done so far[3,4] should be extended as far as possible in deriving expressions for the general surface geometry. For a start focus could be on general 3-dimensional corners with variable angles.

(b) Parallel to the work with analytical $\dot{Q}$ functions it should be investigated of what help polynomial $\dot{Q}$ descriptions could be. These descriptions would have to be developed by the use of regression analysis of numerical results obtained by the use of type 1 calculations (netpoints in both the casting and the mould) for certain standard interface geometries.

(c) Still another possibility to take into consideration is an extended use of the product solution principle. In section 17.2.2 this principle was used in combining one variable analytical temperature curves to obtain the solution for a geometry of several space parameters. For cases where analytical descriptions are non-existent, numerically determined temperature curves could be used.

This could be done in the following way:

The general casting geometry is treated using type 2 calculations in the main part of the computer program. In every time step during the computations subprograms are operated to provide these one parameter type 1 calculated temperatures to be used in the $\dot{Q}$ expressions, which are linked to the main programme.

From this discussion it is seen, that the $\dot{Q}$ method might be generalized in several ways. Any combination of the three principles just mentioned should be feasible in constructing $\dot{Q}$ descriptions.

17.6 CONCLUSIONS

The $\dot{Q}$ method has been shown to hold considerable promise in computing the solidification profiles of shaped castings. It is believed that $\dot{Q}$ functions of the type proposed herein can provide valuable boundary information for many diffusion related problems. Further work on formulating a more general analytical $\dot{Q}$ function for nonrectangular corners is in progress. The method embodied should prove to be of great utility in applications of the type described.

17.7 ACKNOWLEDGEMENTS

The current work is sponsored by the National Science Foundation under the Grant No. DAR78-24301. The authors would also like to acknowledge facilities provided by the Georgia Institute of Technology through the School of Mechanical Engineering under its director, Dr. S. P. Kezios, and the valuable suggestions provided by Mr. R. W. Ruddle of Foseco, Inc., Dr. M. P. Stallybrass of the School of Mathematics at Georgia Tech., and their colleagues Drs. P. Desai and J. Hartley in the School of Mechanical Engineering.

REFERENCES

1. H. S. Carslaw and J. C. Jaeger, *Conduction of Heat in Solids*, Oxford University Press p. 11 (1973).
2. R. W. Ruddle and R. A. Skinner, 'Heat extraction at corners and curved surfaces in sand moulds' *J. Inst. Metals*, **79**(1), p. 35–56 (1951).

3. C. Wei and J. T. Berry, 'An analysis of the transient edge effect on heat conduction in wedges', to appear in *Int. J. Heat and Mass Transfer*.
4. C. Wei, 'An analysis of the transient corner effect of heat conduction and its application to casting solidification', *Ph.D. Thesis* (in preparation), School of Mechanical Engineering, Georgia Institute of Technology.
5. P. N. Hansen, 'Numerical simulation of the solidification process', in *Solidification and Casting of Metals*, The Metals Society, London, p. 350 (1979).
6. P. N. Hansen, 'Solidification and related structure as a function of the metal/mould boundary temperature', paper presented at the *International Conference on Solidification Technology in the Foundry and Casthouse*, held at the University of Warwick, Coventry, 15–17 September 1980, sponsored by the Metals Society, London.
7. P. H. Hansen and J. T. Berry, 'The simulation of heat transfer in castings and weldments .... some thoughts on needed research', in *Modelling of Casting and Welding Processes*, ed. H. D. Brody and D. Apelian, The Metallurgical Society of AIME, p. 497 (1981).

Numerical Methods in Heat Transfer, Volume II
Edited by R. W. Lewis, K. Morgan, and B. A. Schrefler

Chapter 18

Galerkin Finite Element Analysis of Temperature Distributions on Dissolving Coal

Duane R. Skidmore and Fred K. Fong

ABSTRACT

A newly developed finite element model was used to simulate heat transfer in *in situ* coal liquefaction. A modified Galerkin method was refined for use. New coordinates were derived from conventional Cartesian coordinates to simplify computer programming. A quadrilateral element was adopted to accommodate the cylindrical shape of the reaction zone and the resulting system was analysed numerically for several flow systems simulating underground coal liquefaction processes.

INTRODUCTION

Solvent temperature and heat flux estimations are important problems in *in situ* coal conversion. This study seeks to solve a system of coupled equations which have been found descriptive of the process.

The mathematical procedure is based on the Galerkin finite element method. The result is a 'strict' numerical solution to a formidable two-dimensional moving-boundary value problem with variable geometry.

Earlier investigators have often attempted to simplify a two-dimensional heat conduction problem into a one-dimensional problem. The reaction zone in underground coal gasification processing was simulated as bell-shaped[1] (see Figure 18.1) where axial heat conduction was negligible. This assumption rendered the problem amenable to numerical solution.

Finite difference schemes by Fong and Skidmore[2] combined central difference and ADI iteration to solve a moving boundary problem. Then boundary moving rates were uniformly distributed along the reacting channel length, i.e. did not vary with *Z*. Greatly simplified mathematical expressions

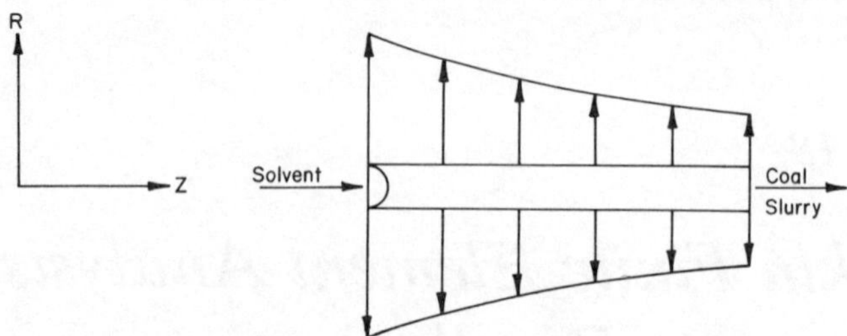

Figure 18.1 Cartesian coordinate system and idealized reactor zone

resulted because the finite difference method is awkward if variable boundary moving rates are encountered.

This study develops mathematically and analytically a strict numerical solution to a heat conduction problem with variable geometry and irregular boundary moving rates.

18.1 FORMULATION OF THE PROBLEM

This heat transfer problem is derived from *in situ* coal liquefaction and was described earlier[3] as follows:

Referring to Figure 18.1, at time $t = 0$, a solvent flows into a circular channel embedded in a coal medium. The temperature of the solvent decreases along the channel as the heat is dispersed to the surroundings. The governing equations can be written in dimensionless form as:

Fluid:

$$-F\frac{\partial V}{\partial Z}+H(U-V)=\frac{1}{\bar{D}^2}\frac{\partial}{\partial \tau}(\bar{D}^2 V) \tag{18.1}$$

Boundary:

$$K\frac{\partial U}{\partial N}=U-V \tag{18.2}$$

Surrounding:

$$\frac{1}{e}\frac{\partial^2 U}{\partial Z^2}+\frac{\partial^2 U}{\partial R^2}+\frac{1}{R}\frac{\partial U}{\partial R}=\frac{\partial U}{\partial \tau} \tag{18.3}$$

Boundary moving rate:

$$\frac{\mathrm{d}D}{\mathrm{d}\tau}=2\dot{R}, \qquad \bar{D}=\bar{D}_0+2\int \dot{R}\,\mathrm{d}\tau \tag{18.4}$$

Boundary conditions:

$$U = 0 \quad \text{at } R = \infty$$
$$V = 1 \quad \text{at } Z = 0 \tag{18.5}$$

Initial conditions:

$$U = 0,\ V = 0,\ \bar{D} = \bar{D}_0 \quad \text{at } \tau = 0$$

Where

$$U = (T_p - T_\infty)/(T_0 - T_\infty), \qquad V = (T_f - T_\infty)/(T_0 - T_\infty), \qquad e = \frac{z_0^2}{r_0}$$

$$Z = z/z_0 \qquad R = r/r_0 \qquad \tau = \alpha_p t/r_0^2$$

$$N = n/r_0 \qquad F = \bar{F}/\bar{D}^2 \qquad H = \bar{h}/\bar{D}$$

$$\dot{R} = \dot{r} r_0/\alpha_p \qquad K = k_p/r_0 h \qquad \bar{D} = D/r_0$$

$$\bar{v} = r_0^2 v/Z_0 \alpha_p \qquad F = \bar{v}\bar{D}^2$$

$$\bar{h} = 4 h r_0/\rho_f D_f \alpha_p$$

T_f, T_b, and T_p are the temperatures of fluid, boundary, and surroundings respectively (°C),

v	is the fluid linear velocity (m/s),
D	is the local diameter of the pipe (m)
D_0	is the initial channel diameter (m)
h	is the fluid heat transfer coefficient (cal/m^2.sec. °C)
n	is the normal vector to the boundary (m)
α_p	is the thermal diffusivity of the surrounding material (m^2/s)
k_p	is the thermal conductivity of coal (cal/m.s. °C)
ρ_f, ρ_p	are the densities of fluid and coal, respectively (kg/m^3)
C_f, C_p	are the heat capacities of fluid and coal, respectively (cal/kg. °C)
t	is time (s)
r, z	are the radial distance and longitudinal distance (m)
and $\dot{R}$	is the velocity of the moving boundary (m/s).

In Equation (18.4), the dimensionless boundary moving rate, $\dot{R}$, can be a function of longitudinal distance, Z, and time, t.

In the heat transfer problem, the boundary moving rate, $\dot{R}$, relates to the chemical reactivity between coal and solvents. Since the heat disperses from solvents to the surrounding coal, it may be expected that the solvent temperature, as the boundary moving rate, drops along the channel. As a consequence, the circular channel will gradually develop into a bell-shaped reaction zone.

Strict numerical solutions are obtainable by the introduction of new coordinates which can be defined as:

$$
\begin{aligned}
X &= Z \\
Y &= R - 1 - \int \dot{R}(Z)\,\mathrm{d}\tau \\
\tau' &= \tau
\end{aligned}
\tag{18.6}
$$

The coordinate system (X, Y, τ') was selected such that it simplifies the derivation. This will become clear after Equation (18.14).

The Galerkin interpretation of Equation (18.3) can be written as

$$\int_A \left(\frac{1}{e} U_{,ZZ} + U_{,rr} \frac{+1}{r} U_{,r} - U_{,\tau}\right)\phi_i \,\mathrm{d}A = 0, \tag{18.7}$$

where

$$U_{,ZZ} = \frac{\partial^2 U}{\partial Z^2}$$

$$U_{,rr} = \frac{\partial^2 U}{\partial r^2}$$

$$U_{,r} = \frac{\partial U}{\partial r}$$

and

$$U_{,} = \frac{\partial U}{\partial \tau}.$$

Symbols thereafter are defined in a similar fashion.

Using the relation $\mathrm{d}A = R\,\mathrm{d}R\,\mathrm{d}Z$, the Green–Gauss theorem, and introducing $U = \phi_i U_i$ yields

$$
\begin{aligned}
&U_{i,\tau} \int A\phi_i\phi_j R\,\mathrm{d}R\,\mathrm{d}Z + U_i \int_A \left(\frac{1}{e}\phi_{i,Z}\phi_{j,Z} + \phi_{i,R}\phi_{j,R} - \dot{R}\phi_i\phi_{j,R}\right) R\,\mathrm{d}R\,\mathrm{d}Z \\
&\quad = \int_S \frac{1}{K}(U - V)\phi_i R\,\mathrm{d}s.
\end{aligned}
\tag{18.8}
$$

Equation (18.8) can be expressed as

$$M_{ij}U_{i,\tau} + D_{ij}U_i = F_i \tag{18.9}$$

(a) $M_{ij} = \int_A \phi_i\phi_j R\,\mathrm{d}R\,\mathrm{d}Z$

Referring to Figure 18.2, the reference Cartesian coordinates (Z, R) are related to the isoparametric coordinates (f, g) as:

$$Z = \sum \Phi_M(f, g) Z_M \tag{18.10}$$

$$R = \sum \Phi_M(f, g) R_M \tag{18.11}$$

where M denotes the node number.

Term $\int_A \mathrm{d}R\, \mathrm{d}Z$ in M_{ij} can be written as

$$\int_A \mathrm{d}R\, \mathrm{d}Z = \int_A |J|\, \mathrm{d}f\, \mathrm{d}g, \tag{18.12}$$

where

$$J = \begin{vmatrix} \dfrac{\partial Z}{\partial f} & \dfrac{\partial R}{\partial f} \\ \dfrac{\partial Z}{\partial g} & \dfrac{\partial R}{\partial g} \end{vmatrix} = Z_{,f} R_{,g} - Z_{,g} R_{,f}$$

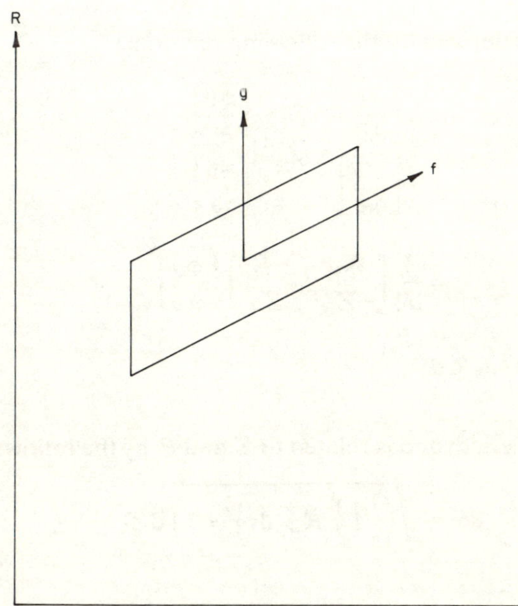

Figure 18.2 Transformed coordinate system

The new coordinates (Equation (18.6)) are so chosen that J can be simplified to the following expression after mapping:

$$J=\begin{vmatrix}\dfrac{\partial X}{\partial f} & \dfrac{\partial Y}{\partial f}\\ \dfrac{\partial X}{\partial g} & \dfrac{\partial Y}{\partial g}\end{vmatrix}, \tag{18.14}$$

where

$$X=\Phi_M X_M \tag{18.15}$$

$$Y=\Phi_M Y_M \tag{18.16}$$

$$X_M=R_M \tag{18.17}$$

$$Y_M=R_M-1 \qquad \text{at } t=0 \tag{18.18}$$

Equations (18.16) and (18.17) indicate that as time (τ) proceeds, nodes (X_M, Y_M) do not change and are equal to (Z_M, R_M-1) at $\tau=0$. Thus, J is not a function of time. A constant J implies that the areas of the elements are constant and do not vary with time. This greatly simplifies the derivation.

(b) $D_{ij}=\int\left(\frac{1}{e}\phi_{i,Z}\phi_{j,Z}+\phi_{i,R}\phi_{j,R}-\dot{R}\phi_i\phi_{j,R}\right)R\,\mathrm{d}R\,\mathrm{d}Z$

$\phi_{i,Z}$ and $\phi_{i,R}$ can be derived by the following relations:

$$\begin{bmatrix}\dfrac{\partial\Phi}{\partial Z}\\ \dfrac{\partial\Phi}{\partial R}\end{bmatrix}=[J^{-1}]\begin{bmatrix}\dfrac{\partial\Phi}{\partial f}\\ \dfrac{\partial\Phi}{\partial g}\end{bmatrix}$$

$$=\frac{1}{|J|}\begin{bmatrix}R_{,g} & -R_{,f}\\ -Z_{,g} & Z_{,f}\end{bmatrix}\begin{bmatrix}\phi_{,f}\\ \phi_{,g}\end{bmatrix}$$

(c) $F_i=\int_S\frac{1}{K}(U-V)\phi_i R\,\mathrm{d}S$

S denotes the curve length and is related to Z and R by the following relation:

$$\mathrm{d}S=\sqrt{\left(\frac{1}{e}\left(\int\dot{R}_{,Z}\,\mathrm{d}\tau\right)^2+1\right)}\,\mathrm{d}Z \tag{18.19}$$

where

$$\int\dot{R}_{,Z}\,\mathrm{d}\tau=\int\dot{R}_M\phi_{M,Z}\,\mathrm{d}\tau$$

SUMMARY OF THE FINITE ELEMENT SOLUTION

Upon the use of a quadrilateral element, the interpolation function is as follows:

$$\phi_i = 0.25(1 + f_i f)(1 + g_i g).$$

After introducing the interpolation function, M_{ij}, D_{ij} and F_i of Equation (18.1) can be expanded as

$$M_{ij} = \tfrac{1}{512} \int_{N=i,j} \pi(1 + f_N f + g_N g + f_N g_N f g) R_M \times (1 + f_M f + g_M g + f_M g_M f g)(a + bf + cg)\, df\, dg$$

$$D_{ij} = \tfrac{1}{16} \int (R_M \phi_M \phi_i G_{2j} + G_{1i} G_{1j})(J + G_{2i} G_{2j}) J \cdot \phi_M R_M\, df\, dg$$

$$F_i = \int \frac{1}{K}(U - V)\phi_i \phi_M R_M \sqrt{\left[\frac{1}{e}\left(\frac{\Delta R_M\, \Delta t}{\Delta X_M}\right)^2 + 1\right]} \Delta X_M\, df \tag{18.20}$$

and

$$G_{1i} = (g f_i g_i)(g_M R_M + f f_M g_M R_M)(g_i + f f_i g_i)(f_M R_M + g f_M g_M R_M)$$

$$G_{2i} = -(f_i + g f_i g_i)(g_M X_M + f f_M g_M X_M) + (g_i + f f_i g_i)(f_M X_M + g f_M g_M X_M)$$

$$a = (X_4 - X_2)(Y_1 - Y_3) - (X_1 - X_3)(Y_4 - Y_2)$$

$$b = (X_3 - X_4)(Y_1 - Y_2) - (X_1 - X_2)(Y_3 - Y_4)$$

$$c = (X_4 - X_1)(Y_2 - Y_3) - (X_2 - X_3)(Y_4 - Y_1).$$

As to the time domain of Equation (18.9), the central difference method can be applied.

18.3 NUMERICAL SOLUTIONS

The numerical solutions of Equations (18.1) and (18.9) are rewritten in accordance with the alternating variables method.[2]

Fluid ($\tau = n\, \Delta\tau$, $n = 1, 2, 3, \ldots$)

$$\left(\frac{-F\bar{D}^2}{2\,\Delta Z}\right)^{\tau-1/2} (V_i^\tau - V_{i-1}^\tau + V_i^{\tau-1} - V_{i-1}^{\tau-1})$$

$$-\left(\frac{H\bar{D}^2 + 4\dot{R}\bar{D}}{4}\right)^{\tau-1/2} (V_i^\tau + V_{i-1}^\tau + V_i^{\tau-1} + V_{i-1}^{\tau-1})$$

$$+\left(\frac{H\bar{D}^2}{2}\right)^{\tau-1/2} (U_i + U_{i-1})^{\tau-1/2}$$

$$= \left(\frac{\bar{D}^2}{2\Delta\tau}\right)^{\tau-1/2} (V_i^\tau - V_i^{\tau-1} + V_{i-1}^\tau - V_{i-1}^{\tau-1}). \tag{18.23}$$

Medium ($\tau = [n+1/2]\Delta\tau$, $n = 0, 1, 2, \ldots$)

$$\left(\frac{M_{ij}}{\Delta\tau}+D_{ij}\right)U_i^{\tau+1} + U_{1,s}^{\tau+1}F_{i,1}^{\tau+1} + U_{2,s}^{\tau+1}F_{i,2}^{\tau+1}$$

$$= V_{1,s}^{\tau+1}F_{i,1}^{\tau+1} + V_{2,s}^{\tau+1}F_{i,2}^{\tau+1} + \frac{M_{ij}}{\Delta\tau}U_i^{\tau}. \tag{18.24}$$

Equation (18.23) is written according to central difference while backward difference is applied for Equation (18.24). Figure 18.3 depicts the mesh points of Equations (18.23) and (18.24).

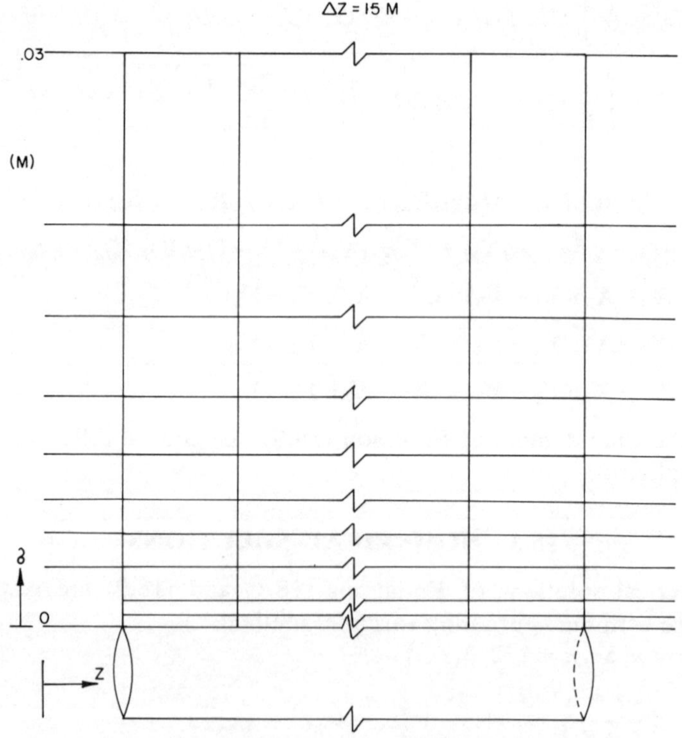

12204 **Figure 18.3** Mesh points

From previous experience,[2] it has been found that the coefficients of the medium equation are rather insensitive to change in temperature, and no noticeable error was introduced when the coefficients were evaluated at the previous time step. This study adopts the same assumption.

Table 18.1 Numerical values of the parameters in the model system. Model coal/anthracene oil systems

$\dot{R} = 5.55 \times 10^{-6}$ m/s (0.02 m/h)
$Z_0 = 15\ m$
$v = 3.05$ m/s
$\mu = 0.001$ kg/m.s
$\alpha_p = 1.89 \times 10^{-7}\ m^2/s$
$T_f = 427$ °C
$T_0 = 25$ °C
$r_0 = 0.15$ m
$\rho_f = 1099\ kg/m^3$
$\rho_p = 1289\ kg/m^3$
$k_p = 0.0560$ cal/m.s.C
$D_0 = 0.3$ m
$C_f = 800$ cal/kg.°C
$C_p = 230$ cal/kg.°C
$h = 277\ cal/m^2.s.°C$

18.4 NUMERICAL RESULTS

The model system of the coal/anthracene oil was chosen as a basis for the numerical studies. The numerical values of the parameters in the model are listed in Table 18.1.

The effects of the change in boundary moving rates on the heat dispersion had been investigated. Figure 18.4 demonstrates the various types of moving

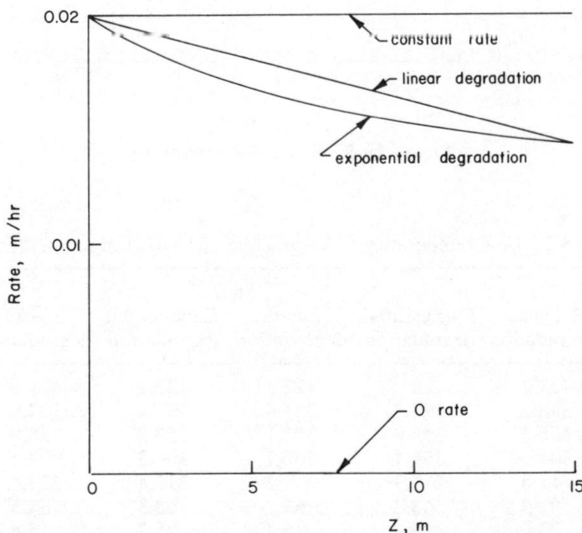

Figure 18.4 Various boundary moving rates along the channel

boundaries which had been studied. Figures 18.5 and 18.6; Tables 18.2 and 18.3 show the temperature distributions of solvent, boundary and coal medium.

Equations (18.1)–(18.5) had been solved previously by a finite difference method.[2] Figure 18.7 compares the solutions of the present study with the previous results.[2] It shows that the finite element method, while more flexible, provides relatively smaller temperature values at longer computation times. A copy of the computer program is available from the authors.

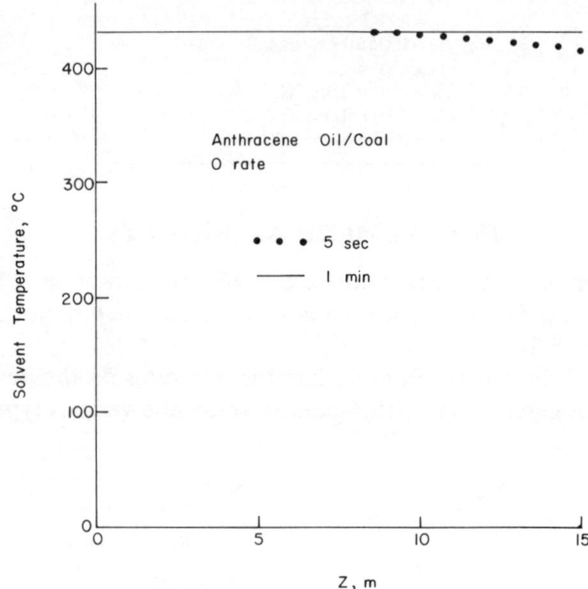

Figure 18.5 Solvent temperatures

Table 18.2 Coal temperatures. Anthracene oil/coal, coal temperature (°C)

z	1.5 m		6 m		12 m	
R	$\dot{R}$ Linear Degradation	Exponential degradation	Linear degradation	Exponential degradation	Linear degradation	Exponential degradation
0	423.7	423.7	422.1	422.1	421.9	421.9
0.002	398.6	398.7	397.4	397.4	397.6	397.6
0.006	358.8	358.8	358.1	358.2	358.8	358.8
0.01	304.3	306.4	306.1	306.2	307.1	307.2
0.02	211.8	211.9	212.2	212.4	213.6	213.7
0.05	63.3	63.3	63.7	63.8	64.5	64.6
0.1	25.6	25.6	25.4	25.7	25.6	25.7

* Initial coal temperature 25 °C; solvent inlet temperature 427 °C

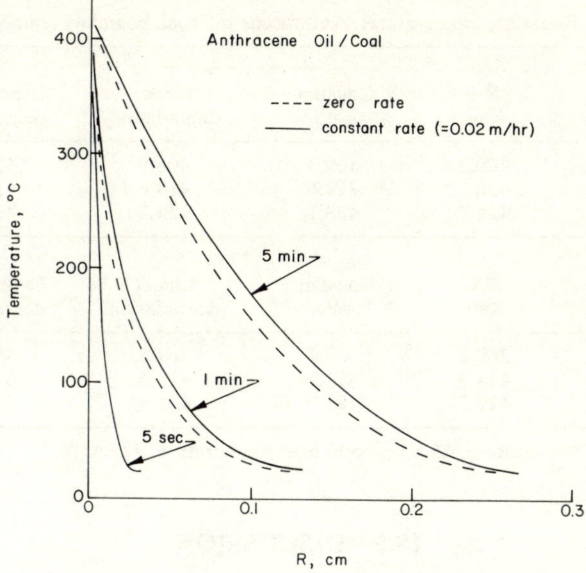

Figure 18.6 Coal temperatures

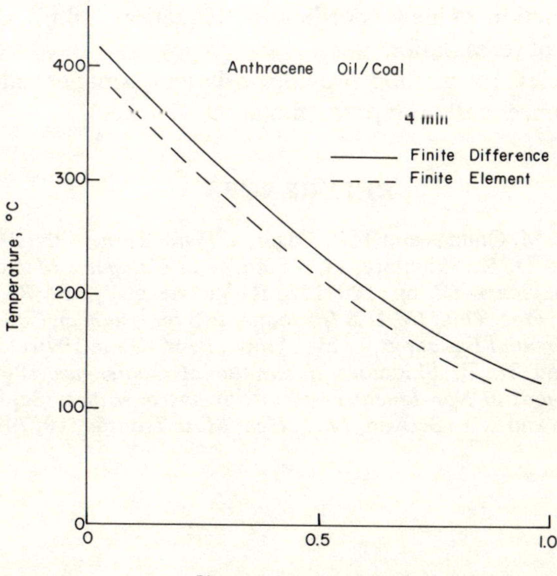

Figure 18.7 Comparison between finite difference and finite element methods

Table 18.3 Boundary temperatures. Anthracene oil/coal, boundary temperature (°C)

Z			1.5 m	
t	$\dot{R}$ 0 Rate	Constant rate	Linear degradation	Exponential degradation
5 sec	409.3	409.3	408.9	408.9
1 min	420	419.6	419.6	419.6
5 min	424.2	423.7	423.7	423.7
Z			**12 m**	
t	$\dot{R}$ 0 Rate	Constant rate	Linear degradation	Exponential degradation
5 sec	369.8	369.1	369.2	369.2
1 min	414.3	411.1	413.5	413.6
5 min	422.7	421.7	421.9	421.9

* Initial real temperature, 25 °C; solvent inlet temperature, 427 °C

18.5 DISCUSSION

(a) The method provides simple expressions which can be easily manipulated for variable coal comminution rates.
(b) The use of the weak form of Equation (18.3) greatly reduced the amount of mathematical manipulation required.
(c) The introduction of new coordinates (Equation (18.6)) simplified the finite element formulation.
(d) The finite element method provides relatively smaller values than the finite difference method (approximately 9 °C lower).

REFERENCES

1. B. Dinsmoor, J. M. Gallans and T. F. Edgar, *J. Petro. Tech.*, 695 (1975).
2. F. K. Fong and D. R. Skidmore, in *Advances in Computer Methods for Partial Differential Equations—III*, pp. 120–126, R. Vichnevetsky and R. S. Stepleman, IMACS (1979); *Proc. Third IMACS International Symposium on Computer Methods for Partial Differential Equations*, Lehigh Univ., Penn. (June 1979).
3. F. K. Fong and D. R. Skidmore, *International Conference* (Proceedings) *on Numerical Methods in Non-Linear Problems, Swansea, Wales*, (September 1979).
4. B. A. Finlayson and L. E. Scriven, *Int. J. Heat Mass Transfer*, **10**, 799–821 (1969).

Numerical Methods in Heat Transfer, Volume II
Edited by R. W. Lewis, K. Morgan, and B. A. Schrefler

Chapter 19

Mathematical Modelling of Heat and Mass Transfer in Iron Ore Packed Bed Industrial Processes

F. Hoislbauer and C. Jaquemar

ABSTRACT

Two mathematical models are described:

(a) the induration process of iron ore pellets;
(b) the iron ore sintering process.

Both models are based on the transient mass and energy balance, including heat transfer, combustion, decomposition of lime, etc. The temperature and humidity profiles of the bed are calculated versus height and length in the solid and the gas. Simultaneously for the sintering process concentrations of coke, limestone and melted ore of the solid, and of oxygen and carbon dioxide of the gas are incorporated in the mathematical model. For a given pressure drop across the bed, the gas-flow distribution along the moving grate is calculated. The zoning of the moving grate due to process parameters can be chosen freely.

The work has demonstrated that one has to distinguish between calculated data and results from laboratory tests.

NOMENCLATURE

a_{1T}, a_{2T}	thermal diffusivity	m^2/s
c_{p_c}	specific heat of coke	J/kg K
c_{PCaCO_3}	specific heat of lime	J/kg K
c_{PCO_2}	specific heat of CO_2	J/kg K
c_{p_g}	specific heat of gas	J/kg K

c_{p_s}	specific heat of solid (pellet or sinter)	J/kg K
c_{p_v}	specific heat of water vapour	J/kg K
c_{p_w}	specific heat of water	J/kg K
d_c	equivalent coke-grain diameter	m
d_{CaCO_3}	equivalent lime-grain diameter	m
d_s	equivalent solid (pellet or sinter) grain diameter	m
h_{H_2O}	enthalpy of water vapour	J/kg
h_{CO_2}	enthalpy of CO_2	J/kg
$\dot{m}_g$	mass flow density of gas	$kg/m^2 s$
$\dot{m}_s$	mass flow density of solid	$kg/m^2 s$
n_c	number of coke particles per unit volume	$1/m^3$
n_L	number of lime particles per unit volume	$1/m^3$
H_{uc}	combustion heat	J/kg
H_L	decomposition heat	J/kg
r	heat for vaporization	J/kg
r_c	radius of coke grain	m
r_L	radius of lime grain	m
r_m	heat for melting	J/kg
R	gas constant	J/mol K
R_{CO_2}	gas constant of CO_2	J/mol K
R_e	Reynolds number	—
T	gas temperature	°C
T_s	boiling point of water	°C
VF	distribution factor of combustion heat	—
$\bar{w}$	mean velocity	m/s
X_{br}	break point of moisture content	kg/kg
X_c	coke content	kg/kg
X_{CaCO_3}	lime content	kg/kg
X_{CO_2}	CO_2 content	kg/kg
X_m	content of melt	kg/kg
X_{O_2}	O_2 content	kg/kg
X_{sH_2O}	moisture content of solid	kg/kg
X''_{gH_2O}	saturation moisture of gas	kg/kg
X_{gH_2O}	moisture of gas	kg/kg
α	heat-transfer coefficient	$J/m^2 s K$
β	mass-transfer number of water	$kg/m^2 s$
δ	solid temperature	°C
η	dynamic viscosity	kg/ms
λ_g	thermal conductivity of gas	W/m K
λ_s	thermal conductivity of solid	W/m K
ν	kinematic viscosity	m^2/s
ρ_g	density of gas	kg/m^3
ρ_s	density of solid	kg/m^3

ψ	porosity	m^3/m^3
Ω	specific surface	m^2/m^3

19.1 INTRODUCTION

The mathematical modelling of processes in the iron and steel industry by use of modern computer technology has become a necessity in order to improve product quality and reduce product costs, e.g. energy costs. For the sintering and pelletizing process physical and chemical fundamentals are given in references 1–5. In this paper a comprehensive description of the modelling and solution procedure is presented.

19.1.1 Pelletizing process

The pelletizing process under consideration is one of the agglomeration processes which uses iron ore fines. The iron ore particles with a diameter of about 1–8 μm are wetted and placed on a rolling plate. There balls with a diameter of 10–12 mm (called green pellets) are made.

On the moving grate (Figure 19.1) is a hearth layer of fired pellets with a height of 10 cm and on top of it a layer of green pellets approximately 30 cm high.

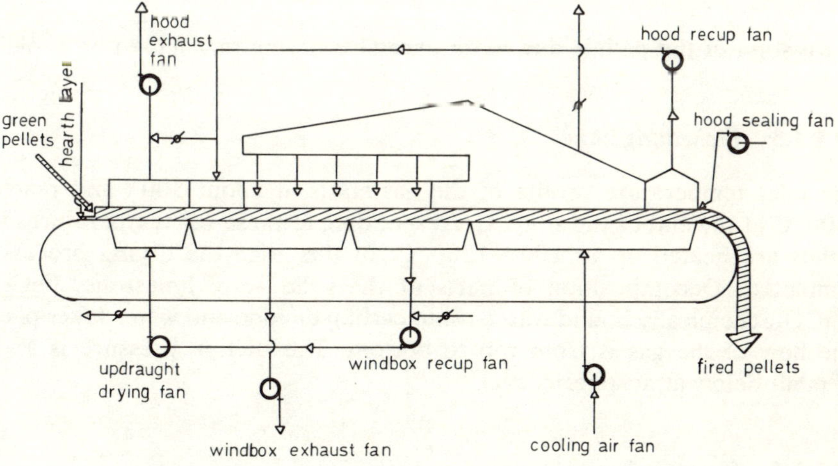

Figure 19.1 Arrangement of the pelletizing strand process

19.1.1.1 *Updraught drying zone*

Hot process gas at a temperature of 200–350 °C is blown through the grate and the packed bed with an approach velocity of about 2.8 m/s.

As the pellets are heated up, their moisture content decreases. At a certain equilibrium temperature an energy balance between convective heat transfer and heat consumption by vaporization is achieved (~90 °C). The temperature of the pellets starts to rise again when the humidity of the pellets has dropped below a certain critical value. Above 100 °C the pellets are dry and their temperature increases sharply and assumes the shape of an exponential function. The temperature difference between solid and process gas becomes very small.

If the temperature rise during the drying process is too high, the vaporization pressure may cause bursting of the pellets. This can be prevented by the use of the hearth layer. In the upper layers of the bed the gas temperature may fall below its dew point and rehumidification of the pellets occurs. About one third of the green pellets are dried in the updraught drying zone.

The pressure in the windbox is about 40 mbar above and in the hood 0.2–0.8 mbar below atmospheric pressure.

19.1.1.2 *Downdraught drying zone*

Hot process gas with a temperature of 200–300 °C is sucked down through the packed bed. In the hood there is a pressure of about 0.2–0.8 mbar below atmospheric level and in the windbox a negative pressure of 30 mbar.

Nearly all of the packed bed should be dried through when leaving this zone.

Bursting of the pellets due to an excessive drying rate has to be avoided.

19.1.1.3 *Preheating zone*

The inlet temperature profile of the gas starts at about 300 ° and reaches 1300 °C at the end of the zone. Corresponding to these gas temperatures the pellets are heated up to 1000–1200 °C. In this zone the drying process is completed. Decomposition of parts of the solid—esp. limestone, $FeCO_3$, $MnCO_3$, chemically bound water—into carbon dioxide and water takes place. The flow of the gas is from top to bottom. The suction pressure is about 40 mbar below atmospheric level.

19.1.1.4 *Firing zone*

The gas temperature has a constant temperature profile of 1300–1350 °C. The temperature of the solid has to reach at least 1250 °C in order that recristallization and sintering of the iron ore powder occurs. For haematite pellets only sensible heat needs to be considered, but for magnetite pellets additional heat due to oxidation has to be taken into account.

19.1.1.5 *Afterfiring zone*

Gas inlet temperature is reduced to about 900 °C. Heat is transported by convection from the hot upper layers to the cooler lower layers. Burnthrough (1250 °C) must be completed over the whole height of the bed at the end of this zone.

19.1.1.6 *Cooling zone*

The direction of the gas flow is reversed to an upward direction. Ambient air is used to cool the pellets down to about 120 °C.

19.1.2 Sintering process

The feed mixture consists of ore, return fine (sinter of very small pieces), coke, lime and water for moistening (in order that the small grains adhere to the big ones). As in the pelletizing process there is cross-stream for heat and mass transfer between the process gas and the solid (feed mixture). During the sinter process the packed bed passes through the following sections (Figure 19.2).

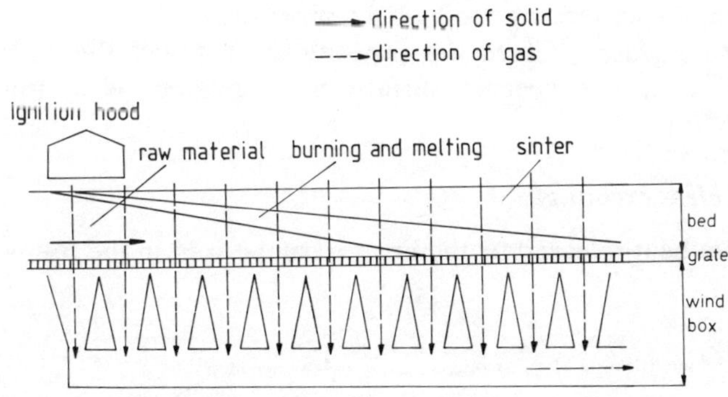

Figure 19.2 Arrangement of the sinter strand process

19.1.2.1 *Ignition zone*

Hot process gas at a temperature of 1100–1200 °C is sucked through the feed mixture from above at a free stream velocity of about 0.5 m/s. The solid is dried, heated up, the carbonates start to decompose between 700 °C and 1100 °C and ignition of the coke occurs above 900 °C. The sintering of the feed takes place at about 1450 °C (melting point).

19.1.2.2 *Sintering zone*

After a firing layer has been established further burning proceeds automatically. Over the height of the bed three zones can be distinguished. In the upper zone the burned sinter heats up the cold gas. In the middle zone burning of the sinter occurs and heat is released to the gas and lower layers of the feed. The shape of this zone is divergent in the direction of the moving bed. In the lower zone the gas dries and preheats the solid.

For optimum operation of a sintering machine the burning zone (middle zone) must reach the bottom of the bed just at the end of the moving grate.

19.2 FUNDAMENTALS FOR CALCULATION

19.2.1 Pelletizing process

19.2.1.1 *Geometric data of packed bed*

- *Green pellets.* Pellet spheres have a mean diameter of 12 mm, consist of iron ore dust with a grain size of 1–8 μm and have a corresponding surface blain from 1500–3500 cm^2/g. The height of the green pellet layer is 0.3 m.
- *Hearth layer.* A hearth layer of height 0.1 m consists of fired pellets and is necessary to prevent the green pellets from drying too quickly and the grate from reading too high a temperature.
- *Specific surface.* Given the porosity $\psi = 0.4$ and the pellet diameter $d_s = 12$ mm, the specific surface is calculated as $\Omega = 6(1-\psi)/d_s = 300\ m^2/m^3$.

19.2.1.2 *Heat exchange*

The sensible heat released by the hot gas can be split in the following way:

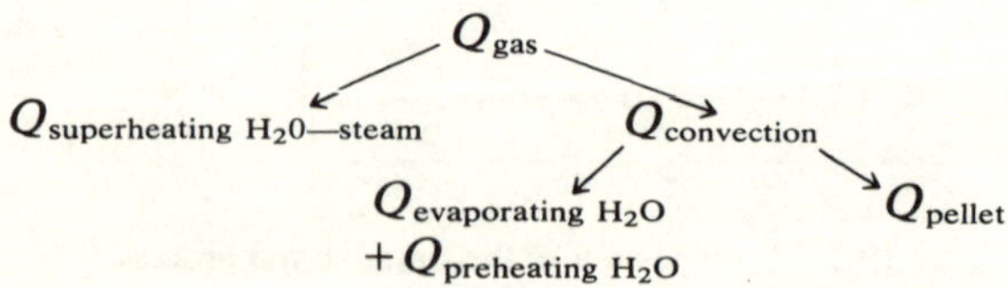

Heat transfer coefficient (α): According to reference 1 the heat transfer coefficient in a packed bed is calculated from

$$\alpha = \frac{\lambda_g(1-\psi)}{d_s\,\psi}\left(\frac{2\,\psi}{1-\psi} + \sqrt{R_e} + \frac{R_e}{200}\right), \tag{19.1}$$

where

$$R_e = \dot{m}_g \frac{d_s}{\rho_g \nu (1-\psi)}.$$

19.2.1.3 *Mass transfer*

For the mass transfer calculation of vapour it is assumed that mass transfer is proportional to the difference in concentration between two exchanging media. The coefficient of proportionality (mass-transfer number β) depends on the moisture concentration (X_s) of the pellet.

At high values of X_s the mass transfer number is a constant. When the humidity of the pellets drops below a certain limit (break away point) mass transfer is influenced by capillarity effects and the mass transfer number decreases linearly with the moisture content of the pellet:

$$\begin{aligned} \beta &= \beta_0 \frac{X_s}{X_{br}} \qquad \text{for } \frac{X_s}{X_{br}} < 1, \\ \beta &= \beta_0 \qquad \text{for } \frac{X_s}{X_{br}} \geqslant 1, \end{aligned} \tag{19.2}$$

with

$$\beta_0 = 7.2 \times 10^{-3}\ \mathrm{kg/m^2\,s}$$

$$X_{br} = 0.04\ \mathrm{kg/kg}.$$

19.2.1.4 *Pressure loss (gas stream)*

The pressure loss is given by

$$\frac{dp}{dy} = \varepsilon_s \frac{\rho_g \bar{w}^2}{d_s} \frac{1-\psi}{\psi^3} \tag{19.3}$$

after measurements of Ergun

$$\varepsilon_s = \frac{150}{R_e} + 1.75 \tag{19.4}$$

with

$$R_e = \frac{\bar{w} d_s}{\nu} \frac{1}{1-\psi}$$

$$\dot{m}_g = \bar{w} \rho_g$$

$$\eta = \nu \rho_g.$$

Equation (19.3) becomes

$$\frac{dp}{dy} = \underbrace{\frac{150\,(1-\psi)^2}{d_s^2 \psi^3} \nu}_{a_1} \dot{m}_g + \underbrace{\frac{1.75\,(1-\psi)}{d_s \psi^3 \rho_g}}_{a_2} \dot{m}_g^2,$$

$$\frac{dp}{dy} = a_1 \dot{m}_g + a_2 \dot{m}_g^2.$$

Temperature dependancy of the density and viscosity has to be considered.

19.2.2 Sintering process

19.2.2.1 *Geometric data of packed bed*

— *Raw material.* The grain of the feed mixture has an average diameter of 2 mm. The range of the grain size varies from 0 to 10 mm (ore), 0 to 5 mm (coke), 0 to 3 mm (lime) corresponding to the screen analysis. The height of the raw material layer is 0.35 m. Due to suction pressure and the sintering process this height is reduced by 5% at the end of the sinter machine.
— *Hearth layer.* To save the grate bar from heat a hearth layer of about 3 cm of sinter grains with a diameter of 15–25 mm is placed on the grate.
— *Grain diameter.* The grain diameter can be calculated by one of the following methods.

(a) mean value of screen analysis

(b) $$d_s = \frac{6(1-\psi)}{\Omega} \tag{19.5}$$

(c) from reference 1,

$$d_s = 1 \Big/ \frac{V_1}{V}\frac{1}{d_1} + \frac{V_2}{V}\frac{1}{d_2} + \cdots + \frac{V_n}{V}\frac{1}{d_n},$$

where V_n, d_n are fraction volume, fraction diameter respectively
— *Specific surface*
Assuming an average grain diameter $d_s = 2.0$ mm
and a porosity of $\psi = 0.33$
the specific surface becomes $\Omega = 1980\ \text{m}^2/\text{m}^3$

19.2.2.2 *Heat exchange*

Heat exchange for drying is analogous to the pelletizing process. Heat radiation from gas (CO_2 and H_2O) and dust can be neglected. Heat radiation between the ignition hood and the upper surface of the sinter bed is considered.

Heat generation by combustion of coke

$$q_c = R_c^* H_{uc} \tag{19.6(a)}$$

and by decomposition of carbonates

$$q_L = -R_L^* H_L \tag{19.6(b)}$$

19.2.2.3 *Mass transfer*

Mass transfer for water evaporation is analogous to the pelletizing process.

Mass transfer by burning coke. Sintering process is a forced burning process with a small content of combustible in the packed bed. CO content in flue gas is small (1–2%), therefore Equation (19.7) can be assumed to represent the total combustion process.

$$C + O_2 = CO_2 + H_{uc} \tag{19.7}$$

The amount of heat loss by CO production (measured) value is considered. The combustion rate of a coke particle depends on the chemical reaction rate (reactability), the mass tranfer coefficient (diffusion of O_2 to coke), the surface of the coke particle and the oxygen concentration in the gas.[2]

Combustion rate

$$R_c^* = 4\pi r_c^2 n_c K_c^* X_{O_2}. \tag{19.8}$$

Mass transfer number

$$K_c^* = \frac{k_p k_m}{k_p + k_m}. \tag{19.9}$$

Chemical reaction rate

$$k_p = 6.52 \times 10^5 \exp\left(-\frac{185{,}000}{R(\delta + 273)}\right) \surd(\delta + 273). \tag{19.10}$$

Mass-transfer coefficient

$$k_m = \frac{D_{ON}}{d_s} |A + BS_c^{1/3} R_e^n|, \tag{19.11}$$

$$A = \frac{2}{1 - (1 - \psi)^{1/3}},$$

$$B = \frac{0.61}{\psi},$$

$$R_e = \frac{d_s \dot{m}_g}{\mu}.$$

Gas viscosity

$$\mu = \mu_0 \left(\frac{T + 273}{273}\right)^{3/2} \frac{C + 273}{C + T + 273},$$

$$C = 113 \text{ for air},$$

$$\mu_0 = \text{viscosity at } 0\,^\circ\text{C}.$$

Diffusion coefficient of O_2 through N_2

$$D_{ON} = 1.8\ 10^{-5} \left(\frac{T + 273}{273}\right)^{1.75}.$$

Schmidt number

$$S_c = \frac{\mu}{\rho_g D_{ON}},$$

$$C_K = 4.65\, R_e^{-0.28},$$

$$n = \frac{2 + C_K}{3(1 + C_K)}.$$

Mass transfer by carbonate decompositon. Carbonate decomposition of $CaCO_3$ (lime), $FeCO_3$, and $MgCO_3$ occurs during the sintering process. Only the decomposition of lime is calculated in detail; the other carbonates are considered by adjusting the amount of lime by a factor.

The decomposition number of lime depends on the reaction rate, the mass-transfer coefficient, the surface of the lime particle, the CO_2 concentration on the surface, and the combustion time.

Decomposition number

$$R_L^* = 4\pi r_L^2 n_L K_L^* (X_{CO_2}^* - X_{CO_2}). \tag{19.12}$$

Mass-transfer number

$$K_L^* = \frac{k_{pL} k_{ML}}{k_{pL} + k_{ML}}. \tag{19.13}$$

Chemical reaction rate

$$k_{pL} = 91.2 \exp\left(-\frac{40{,}000}{R(\delta + 273)}\right). \tag{19.14}$$

Mass-transfer coefficient

$$k_{ML} = \frac{D_{CO_2}}{d_s} |A + B S_c^{1/3} R_e^{1/2}| = \frac{D_{CO_2}}{d_s} |2 + 0.75\, S_c^{1/3} R_e^{1/2}|. \tag{19.15}$$

Diffusion coefficient

$$D_{CO_2} = 1.38 \times 10^{-5} \left(\frac{T + 273}{273}\right)^2.$$

Surface concentration

$$X_{CO_2}^* = 0 \qquad \text{if } T < 650\,°\text{C},$$

$$X_{CO_2}^* = \frac{6.8 \times 10^{-6} (T - 650)^3}{R_{CO_2}(T + 273)\rho_{air}}.$$

Pressure loss (gas stream)

Pressure loss in the sinter bed is analogous to that in the pelletizing bed.

3 MODEL FOR CALCULATION

For the coordinate system and the direction of mass flow see Figure 19.3

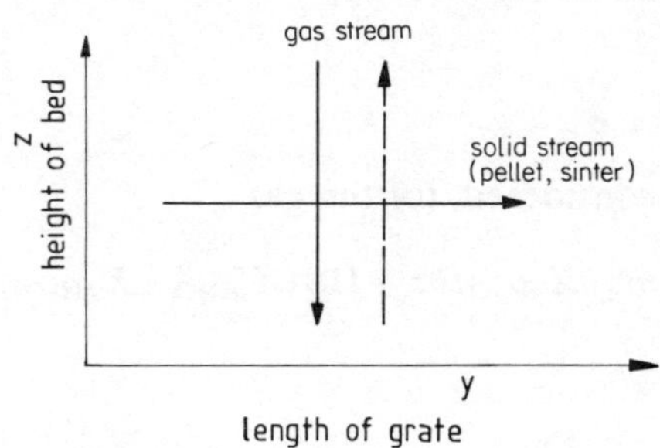

Figure 19.3 Coordinate system of model

19.3.1 Pelletizing process

19.3.1.1 *Energy balance*

For the gas

$$\dot{m}_g(c_{p_g}+c_{p_v}X_{gH_2O})\frac{\partial T}{\partial z}$$

$$= -\alpha\,\Omega(T-\delta)-\dot{m}_g\frac{\partial X_{gH_2O}}{\partial_z}(h_{H_2O}(T)-h_{H_2O}(\delta)). \qquad (19.16)$$

for the solid (pellets)

$$\dot{m}_s(c_{p_s}+c_{p_w}X_s)\frac{\partial \delta}{\partial y}=\alpha\,\Omega(T-\delta)+\dot{m}_s\frac{\partial X_{sH_2O}}{\partial y}r_{H_2O} \qquad (19.17)$$

In Equation (19.16) and (19.17) the term on the l.h.s. represents the energy change due to the sensible heat and the first term on the r.h.s. the heat transfer by surface convection. The last term in Equation (19.16) takes care of superheating of vapour and the last term in Equation (19.17) represents the heat of evaporation.

In the energy balance the expressions for heat conduction

$$\Delta T = a_{1T}\frac{\partial T}{\partial t}$$

$$\Delta \delta = a_{2T}\frac{\partial \delta}{\partial t}$$

are neglected because of

(a) small temperature gradient

(b) small thermal diffusivity (a_{1T}, a_{2T})

19.3.1.2 *Mass balance*

Only evaporation is considered, for the gas

$$\dot{m}_g\, \partial X_{gH_2O}/\partial z = \Omega\beta(X''_{gH_2O} - X_{gH_2O}), \tag{19.18}$$

and for the pellet

$$\dot{m}_s \frac{\partial X_{sH_2O}}{\partial y} = -\Omega\beta(X''_{gH_2O} - X_{gH_2O}). \tag{19.19}$$

19.3.2 Sintering process

Generally the geometrical data (d_s, ψ, Ω) are readjusted after the process of drying, decomposition of lime and burning. This has a direct influence on the heat and mass transfer and the pressure drop calculation.

19.3.2.1 *Energy balance*

For the gas

$$\dot{m}_g(c_{p_g} + c_{p_v}X_{gH_2O} + C_{pCO_2}X_{CO_2})\frac{\partial T}{\partial z}$$
$$= -\alpha\Omega(T-\delta) - \dot{m}_g\frac{\partial X_{gH_2O}}{\partial z}(h_{H_2O}(T) - h_{H_2O}(\delta))$$
$$-\dot{m}_g\frac{\partial X_{CO_2}}{\partial z}(h_{CO_2}(T) - h_{CO_2}(\delta)) - \dot{m}_s\frac{\partial X_c}{\partial y}H_{uc}(1-VF) \tag{19.20}$$

and for the solid (sinter)

$$\dot{m}_s(c_{p_s} + c_{pH_2O}X_{sH_2O} + c_{pCaCO_3}X_{CaCO_3} + c_{p_c}X_c)\frac{\partial\delta}{\partial y} + \frac{\partial X_M}{\partial_y}r_m\dot{m}_s$$
$$= \alpha\Omega(T-\delta) - \dot{m}_s\frac{\partial X_c}{\partial y}H_{uc}VF + \dot{m}_s\frac{\partial X_{CaCO_3}}{\partial y}H_{CaCO_3}$$
$$+\dot{m}_s\frac{\partial X_{sH_2O}}{\partial y}r_{H_2O}. \qquad 0 \leqslant VF \leqslant 1. \tag{19.21}$$

In the energy balance the expressions for heat conduction

$$\Delta T = a_{1T}\frac{\partial T}{\partial t}, \qquad \Delta\delta = a_{2T}\frac{\partial \delta}{\partial t}$$

are neglected due to the similar reasons as for the pelletizing process.

19.3.2.2 *Mass balance*

In addition to evaporation (Equations (19.18) and (19.19)) the following mass-transfer equations are considered.

Gas

For oxygen

$$\dot{m}_g\frac{\partial X_{O_2}}{\partial z} + R_C^* = 0 \qquad \text{for } X_{O_2} \geqslant 0 \tag{19.22}$$

and for carbon dioxide

$$\underbrace{\left(-\dot{m}_s\frac{\partial X_{CaCO_3}}{\partial y}\right)}_{R_L^*} + \underbrace{\left(-\dot{m}_s\frac{\partial X_c}{\partial y}\right)}_{R_C^*} = \dot{m}_s\frac{\partial X_{CO_2}}{\partial z}. \tag{19.23}$$

Solid

For carbon

$$\dot{m}_s\frac{\partial X_c}{\partial y} + R_C^* = 0 \tag{19.24}$$

and for lime

$$\dot{m}_s\frac{\partial X_{CaCO_3}}{\partial y} + R_L^* = 0. \tag{19.25}$$

19.3.3 Calculation method

To solve the system of differential equations the whole packed bed was subdivided into finite elements (Figure 19.4). The mass and energy equations were solved for each element in a finite difference form. As the problem is an initial value problem the calculation can be started either along the entrance of the solid stream or along the entrance of the gas stream.

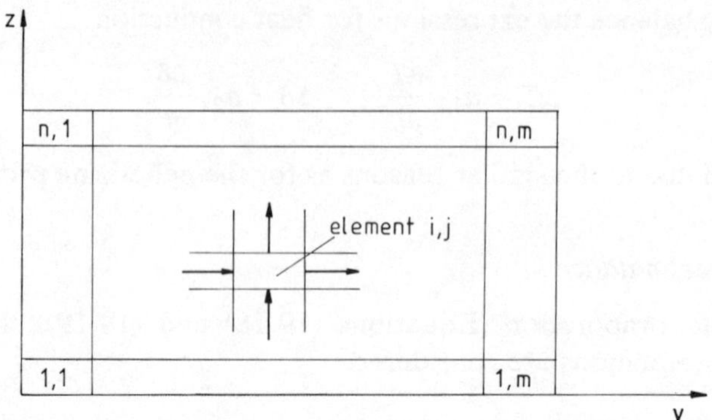

Figure 19.4 Subdivision of pellet and sinter bed into finite elements

For the pelletizing model an explicit (forward) form of solution was chosen (Equation (19.26)). For the sintering process, however, using the explicit method, instabilities showed up due to small grain sizes and high gas inlet temperatures at the start of the process. This problem can be overcome by drastically reducing the size of the elements or by using a semi-implicit method. The latter method was adopted, by applying the implicit form to the drying process and the change of sensible heat (Equation (19.27))

The energy balance for pelletizing in explicit form is as follows.

Gas

$$\dot{m}_g(c_{p_g}(T_1)+c_{p_v}(T_1)X_{gH_2O})\frac{T_2-T_1}{\Delta z}$$
$$= -\alpha\Omega(T_1-\delta_1)-\Omega\beta(X''_{gH_2O}(\delta_1)-X_{gH_2O})(h(T_1)-h(\delta_1)). \quad (19.26(a))$$

Solid

$$\dot{m}_s(c_{p_s}(\delta_1)+c_{p_w}(\delta_1)X_s)\frac{\delta_2-\delta_1}{\Delta y}$$
$$= \alpha\Omega(T_1-\delta_1)-\Omega\beta(X''_{gH_2O}(\delta_1)-X_{gH_2O})T_{H_2O}(\delta_1). \quad (19.26(b))$$

The energy balance for sintering in semiimplicit form is as follows.

Gas

$$-\dot{m}_g(c_{p_g}(T_1)+c_{p_v}(T_1)X_{gH_2O}+c_{pCO_2}(T_1)X_{CO_2})\frac{T_2-T_1}{\Delta z}$$
$$= -\alpha\Omega(T_2-\delta_2)-\Omega\beta(X''_{gH_2O}(\delta_2)-X_{gH_2O})(h_{H_2O}(T_1)-h_{H_2O}(\delta_1))$$
$$-(R^*_L+R^*_c)(h_{CO_2}(T_1)-h_{CO_2}(\delta_1))+R_c(T_1,\delta_1)H_{uc}(1-VF). \quad (19.27(a))$$

Solid

$$\begin{aligned}
&\dot{m}_s(c_{p_s}(\delta_1)+c_{pH_2O}(\delta_1)X_{sH_2O}+c_{pCaCO_2}(\delta_1)X_{CaCO_3}\\
&\quad +c_{p_c}(\delta_1)X_c)\frac{\delta_2-\delta_1}{\Delta y}+\frac{\partial X_m}{\partial y}r_m\dot{m}_s\\
&=\alpha\Omega(T_2-\delta_2)+R_c^*(T_1,\delta_1)H_{uc}VF-R_L^*H_{CaCO_3}\\
&\quad -\Omega\beta(X''_{gH_2O}(\delta_2)-X_{gH_2O})r_{H_2O}(\delta_1) \qquad (19.27(b))
\end{aligned}$$

The nonlinearities of the material properties like the saturation line of vapour (X''), combustion- and decompostiion rate etc. have been considered by calculating these values at the beginning of each step i based on the results of step $i-1$.

Simultaneously with the calculation of the energy and mass balance the pressure-drop calculation is performed in each element. An initial gas flow is assumed, which is re-evaluated iteratively until the given pressure-drop distribution along the strand is achieved.

19.4 MODEL VALIDATION

19.4.1 Pelletizing process

19.4.1.1 *Results of calculation*

In Figures 19.5–19.7 the calculated temperature and moisture distribution in the solid are presented graphically. The parameter for each curve corresponds to a distinct height of the solid above the grate. For the calculation the bed was subdivided into 100 horizontal layers. The results of every second of these layers are shown.

Updraught drying zone. Across the height of the hearth layer (dried pellets) there is a fast increase of the temperature in the pellets. Heating of the green pellets stops for some time at a temperature of about 90 °C until the moisture of the pellets reaches the breakaway point. Temperature now increases slowly up to 100 °C, at which point the pellets are dried completely. Then the pellet temperature rises very sharply. From the plot of moisture in the pellets rehumidification of the upper pellet layers by gas, which is saturated with water, can be seen. At the end of updraught drying zone 64% of the packed bed is dried.

Downdraught drying zone. By the inversion of the gas-flow direction the topmost pellet layers are dried quickly, while in the lower layers drying is slower. This leads to a cross over of the temperature–moisture curves as

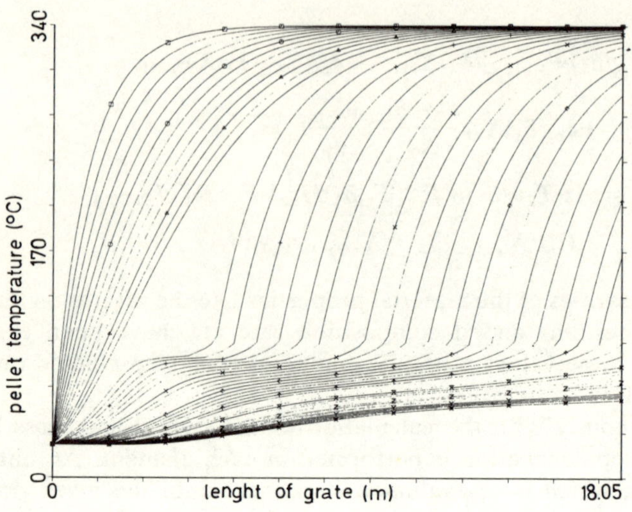

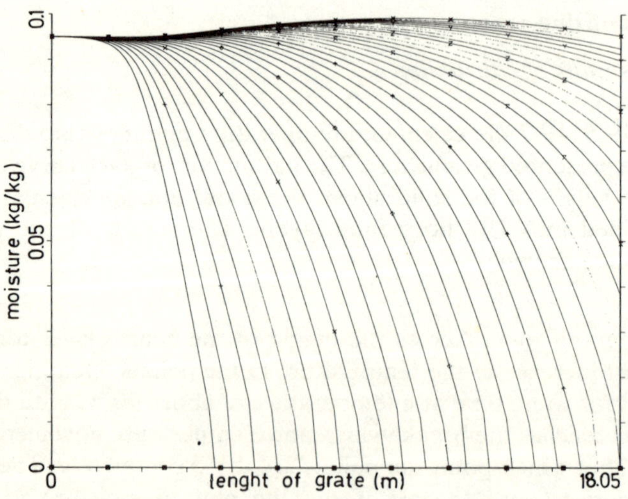

Figure 19.5 Pellet temperature and moisture, updraught drying zone

shown on Figures 9.6 and 9.7. The hearth layer is strongly heated in the updraught drying zone and cooled in the downdraught drying zone. At the end of downdraught drying there is still residual moisture of 64%–72% in the bed.

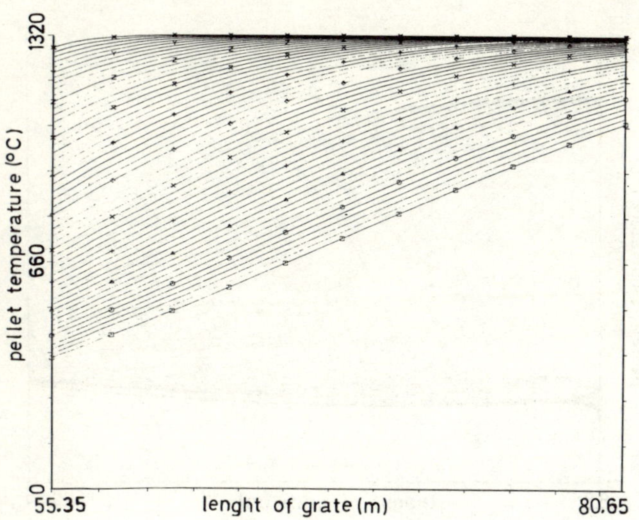

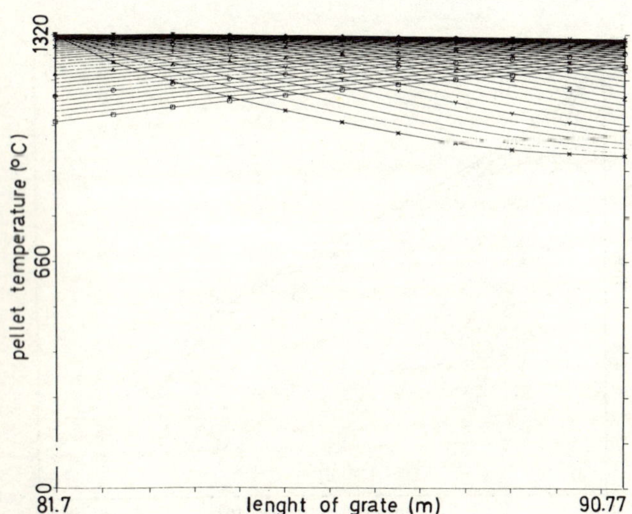

Figure 19.7 Pellet temperature, firing and afterfiring zone

19.4.1.2 *Comparison of calculation with measurements on a pot grate*

Experiments to compare the measurements with calculation were done on a pot grate which is sketched in Figure 19.8. In Figure 19.9, temperature in the middle of the bed height, fairly good agreement is achieved between calculation and measurement in the downstream zones of the strand. The

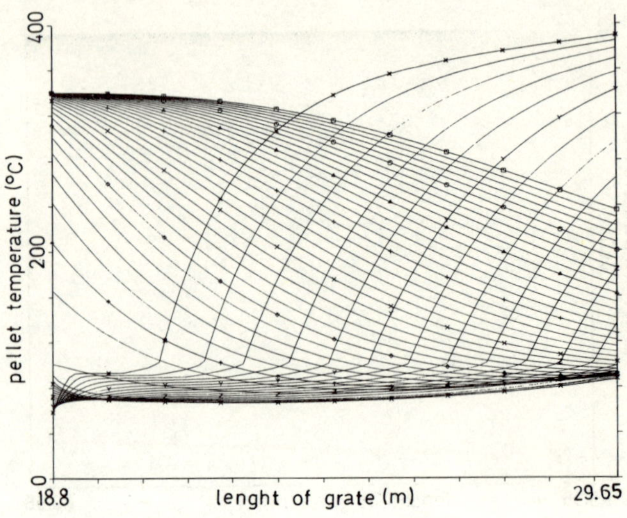

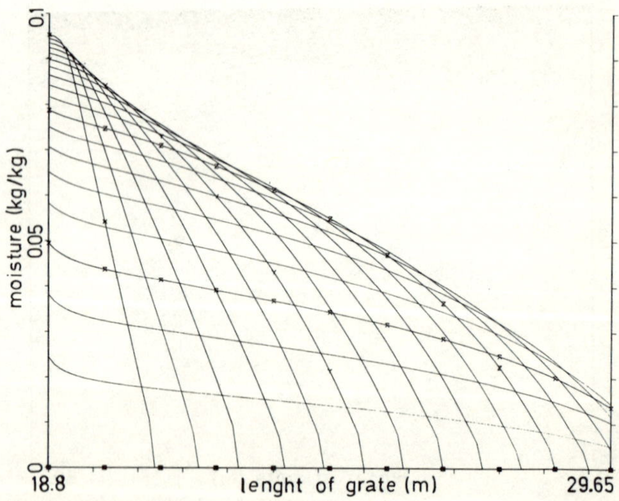

Figure 19.6 Pellet temperature and moisture, downdraught drying zone

Amount of gas. The distribution of gas flow along the grate can be seen Table 19.1. There is a maximum change of gas-flow density of about This is mainly due to the varying gas-temperature field along grate.

Table 19.1

Zone	Mass-flow density (kg/m^2 s) Start	End
Updraught drying	2.52	1.99
Downdraught drying	1.46	1.37
Preheating	1.855	1.23
Firing I	1.624	1.20
Firing II	1.423	1.169
Afterfiring	1.19	1.27
Cooling	1.279	2.74

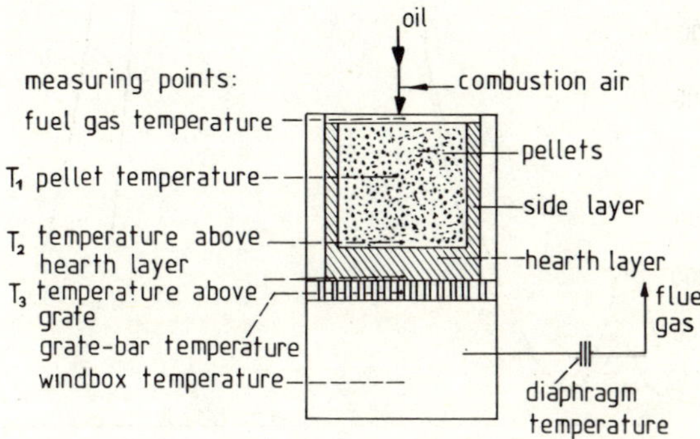

Figure 19.8 Schematic of pellet pot-grate experiment

difference in the upstream zones can be attributed either to a time lag of the thermocouple or to a not exactly defined position of the thermocouple, i.e. there is no control whether the gas or the solid temperature is actually measured.

From Figures 19.10 and 19.11 a very good agreement of T_2 and T_3 between measured and calculated results can be seen along the whole strand.

19.4.2 The sintering process

19.4.2.1 *Results of calculation (Figures 19.12 and 19.13)*

The height of the packed bed is again the parameter for temperature and moisture curves.

Ignition hood. The heating of the moist sinter mixture stops temporarily at a temperaure of 99 °C until the moisture content of the sinter mixture has reached the breakaway point. After complete drying of the sinter (100 °C)

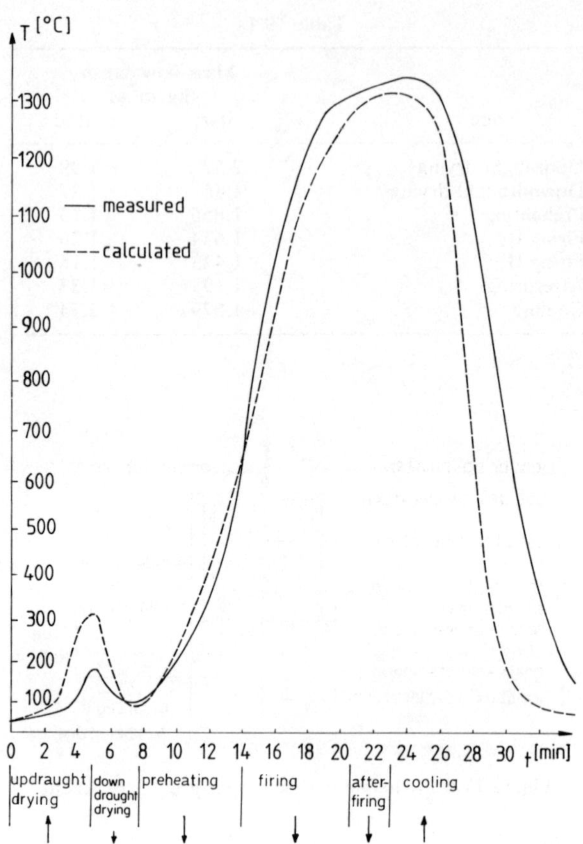

Figure 19.9 Pellet temperature in the middle of bulk height of the pot (T_1)

there follows a very steep increase of the temperature by heating of the mixture. Near 900 °C lime decomposition starts and the temperature curve gets almost horizontal. With increasing temperature the coke combustion accelerates and the temperature curve becomes steeper again until melting temperature is reached. Up to total combustion of the coke the content of molten material increases at constant temperature. Then the molten material solidifies again and cools down.

Humidification of the sinter from the saturated gas occurs downward from 50% of the height of the bed. At the end of the ignition hood the sinter mixture is dried to 13% of the height.

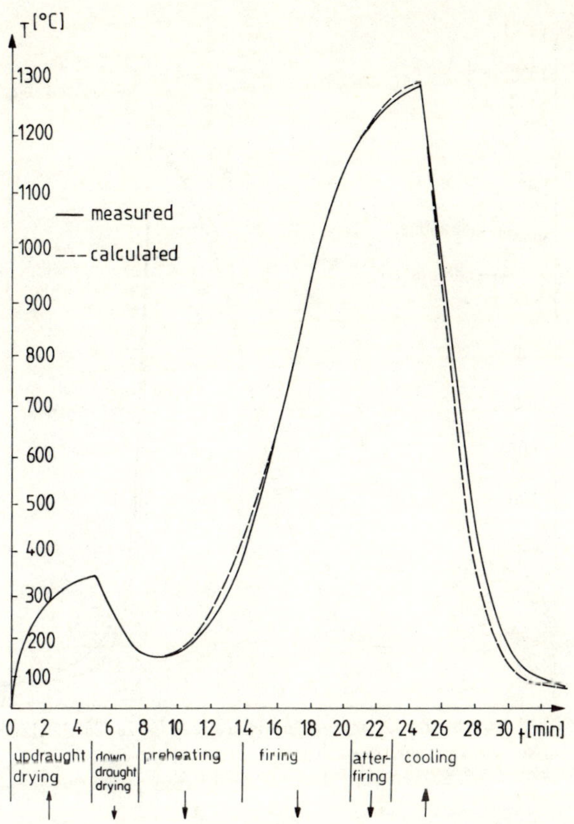

Figure 19.10 Temperature above hearth layer of the pot (T_2)

Sintering zone. The cold gas gets heat from the upper layers of the sinter so that it can bring the coke of the deeper layers to reaction temperature. The time the sinter stays at melting temperature gets longer in the direction of the moving strand.

The plot also shows the temperature curves of the hearth layer, grate bar, and moving grate.

Amount of gas. The distribution of the gas mass-flow density can be seen in Table 19.2. The strong influence of the different permeabilities of the feed mixture and processed sinter to the gas mass-flow distribution can be seen easily.

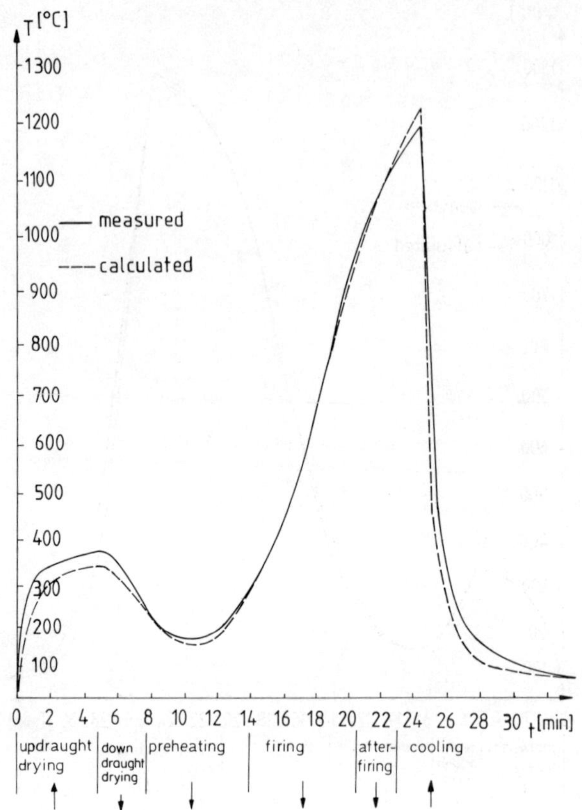

Figure 19.11 Temperature above the grate of the pot (T_3)

Table 19.2

Zone	Mass-flow density (kg/m² s) Start	End	Mean Value
Ignition hood	0.85	0.78	0.81
Sinter zone	0.77	1.84	1.14

19.4.2.2 *Comparison of calculation with measurements on a potgrate*

The experiments for comparison of the measurements with calculation was done on a sinter pot which is sketched in Figure 19.14. Measurements and

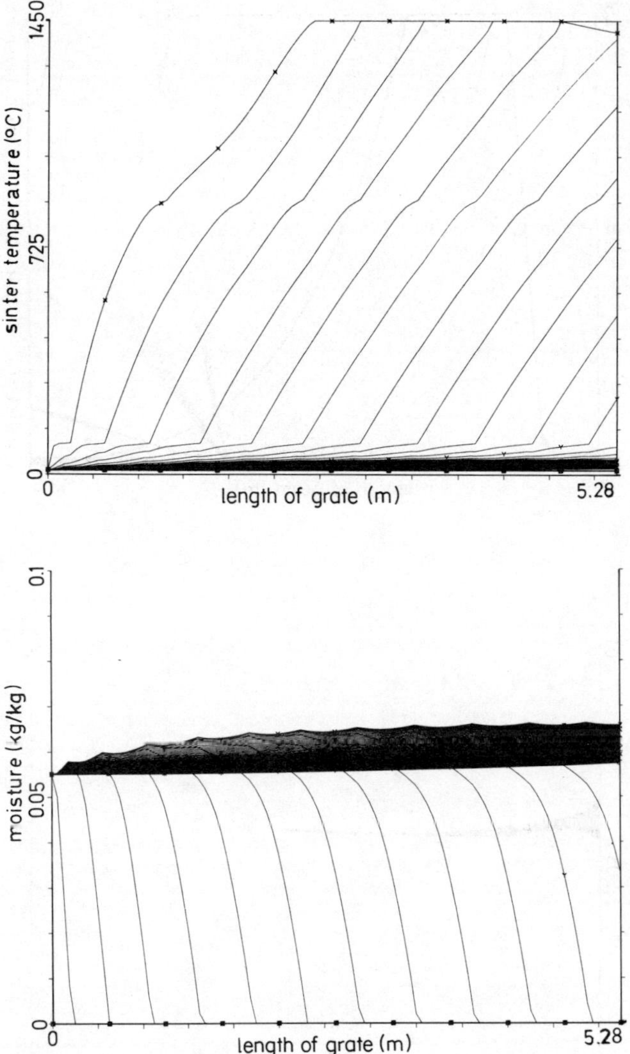

Figure 19.12 Sinter temperature and moisture, ignition zone

calculations were made in the following heights in centimetres above the grate bar.

$$T_1 = 30, \quad T_2 = 20, \quad T_3 = 16, \quad T_4 = 3.$$

The calculations and the measurements (Figure 19.15) agree well considering a scattering of the measurements of 25%.[6] The peak in the calculated gas temperatures is due to the numerical discretization.

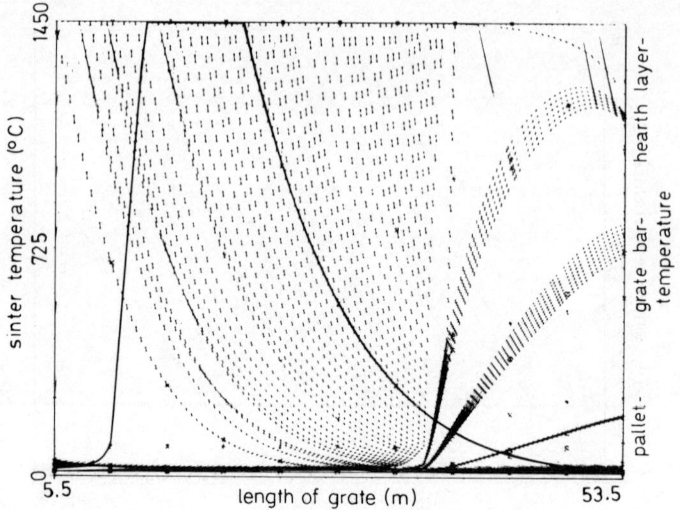

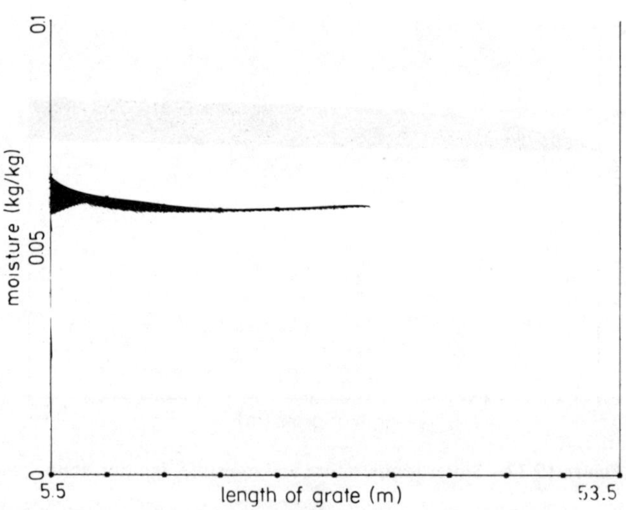

Figure 19.13 Sinter temperature and moisture, sintering zone

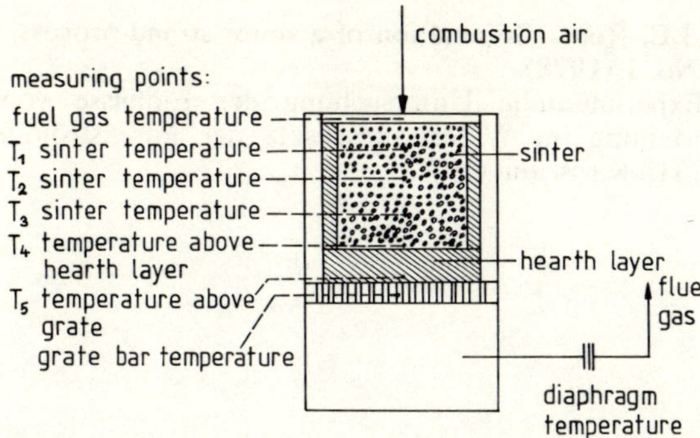

Figure 19.14 Schematic of a sinter pot grate experiment

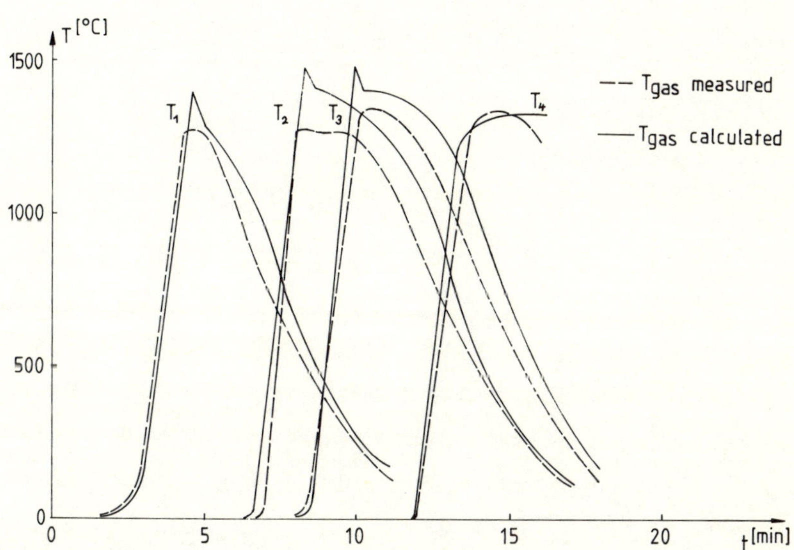

Figure 19.15 Calculated and measured temperature in a sinter pot

REFERENCES

1. R. Jeschar 'Wärmeübergang in Mehrkornschüttungen aus Kugeln,' *Archiv. für das Eisenhüttenwesen*, **6** (1964).
2. I. Dash, C. E. Carter and E. Rose, 'Heat wave propagation through a sinter bed; a critical appraisal of mathematical representations', Pro. *SIMAC* **74**, 8/1 (1974).
3. H. Bauer and D. Mewes, 'Strömungswiderstand sowie Stoff und Wärmeübergang in ruhenden Füllkörperschichten', *Chemie Ingenieur Technik*, **44**(1) (1972).
4. R. W. Young. 'Dynamic mathematical model of sintering process', *Ironmaking and Steelmaking*, No. 6 (1977).

5. I. R. Dash and E. Rose. 'Stimulation of a sinter strand process, *Ironmaking and Steelmaking*', No. 1 (1978).
6. A. Gupte, 'Experimentelle Untersuchung der Einflüsse von Porosität und Korngrößenverteilung im Widerstandsgesetz der Porenströmung', *Dissertation Universität* (T. H.) Karlsruhe (1970).

Numerical Methods in Heat Transfer, Volume II
Edited by R. W. Lewis, K. Morgan, and B. A. Schrefler

Chapter 20

Heat and Moisture Movement in Wood Composite Materials During the Pressing Operation—a Simplified Model

Ferhan Kayihan and Jay A. Johnson

NOMENCLATURE

c	specific heat (kJ/kg. K)
D_s	effective bound water (liquid) diffusivity (m^2/s)
D_v	effective vapour diffusivity (in air) (m^2/s)
α	thermal diffusion coefficient (m^2/s)
$\mathcal{M}$	molecular weight (kg/kmol)
P	pressure (N/m^2)(Pa)
R	gas constant (8314 N m/kmol K)
T	temperature (K)
u	moisture content (kg/kg) (dry basis)
ε	porosity (m^3/m^3)
λ	heat of evaporation + desorption (kJ/kg)
δ	thermal gradient coefficient (1/K)

Subscripts

a	air
v	vapour
l	liquid
s	solid
B	bound water
F	free water
m	maximum

20.1 BACKGROUND

Wood is used in a vast array of products. An extensive system of breakdown and reconstitution processes exist for converting trees, the raw material source of wood, into these products (Figure 20.1). Production of lumber, veneer and pulp involve age-old technologies while the emergence of plywood, laminated beams and other laminated elements evolved with the development of the synthetic glue technology. Production of composition boards for building components was stimulated by the desire to use waste fibre from other processes. The manufacturing technology has now progressed to the point where wood particle sizes are engineered and carefully controlled to produce three basic types of wood composite, panel products: fibreboard, particleboard and flakeboard. There is a trend toward producing structural components with demanding end-use requirements. Higher-quality products require tighter control over the manufacturing processes, consequently the processes must be better understood.

All composition boards are subjected to a pressing operation during manufacturing. The furnish (fibres, particles or flakes) is blended with resin (glue, approximately 5–10% by weight), formed into a low-bulk-density mats and compacted to form thin, relatively dense sheet materials. The technology of pressing is very demanding, the hardware impressive. Heated platens as large as 2.5×10 m are used to compact the mats. The platens are squeezed together by huge hydraulic systems. Multi-opening presses (Figure 20.2), some with up to 24 openings, can be as big as a two- or three-storey building. Yet to maintain uniform product thickness, the position of the platens must be controlled to within 0.25 mm. More details concerning manufacturing may be found in the works of Maloney,[1] Moslemi,[2] and Johnson and Montrey.[3]

Thermal energy is an important ingredient in the recipe for manufacturing composition boards. During pressing heat is transferred into the interior of the mat by surface contact with the heated platens. This heat activates polymerization and 'sets' the resin. The heat also affects the compaction process by accelerating stress relaxation mechanisms in the wood substance. Water, bound in the wood substance, is driven into the void spaces and ultimately forced out of the mat by vapour-pressure gradients induced by the rise in temperature. It is clear that three simultaneous and interrelated processes are occurring in the mat during the pressing operation:

(a) simultaneous heat and mass transfer;
(b) stress relaxation;
(c) resin polymerization.

If the press is opened too soon the board will 'blow' (delaminate and fly apart) due to incomplete resin cure and unrelieved steam pressure. Since it is extremely difficult to obtain experimental data of the type necessary to

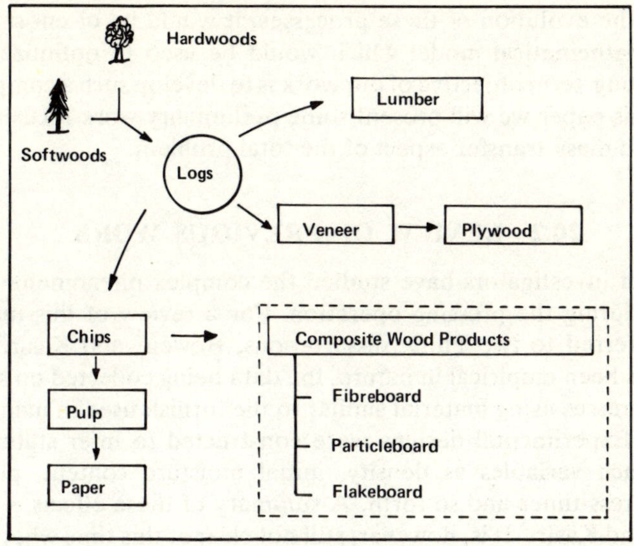

Figure 20.1 Breakdown and reconstitution processes for wood products

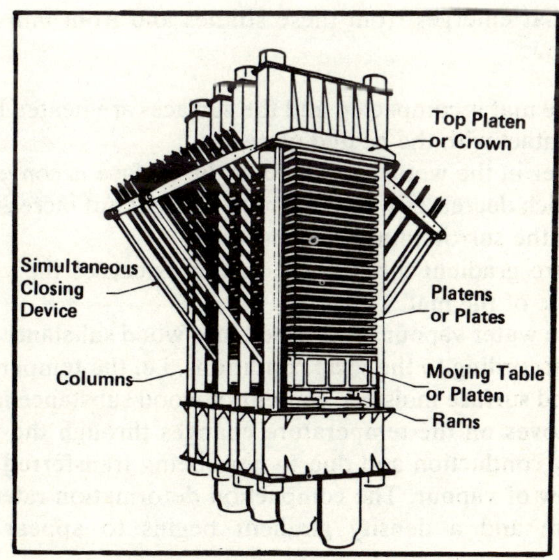

Figure 20.2 Multi-opening press for manufacturing wood composition boards
(Courtesy of T. M. Maloney)

understand the evolution of these processes, it would be of enormous value to have a mathematical model which would be used to optimize the press cycles. The long-term objective of our work is to develop such a comprehensive model. In this paper we will present some preliminary work focusing only on the heat- and mass-transfer aspect of the total problem.

20.2 REVIEW OF PREVIOUS WORK

A number of investigators have studied the complex phenomena that occur in the mat during the pressing operation. For a review of this material the reader is referred to two Ph.D. dissertations, Bowen[4] and Kasir.[5] Much of the work has been empirical in nature, the data being collected on small-scale laboratory presses using material similar to the furnish used in manufacturing operations. Experimental designs were constructed to infer statistically the effects of such variables as density, initial moisture content, platen temperatures, press times and so forth. A summary of these effects is also given in Bowen[4] and Kasir.[5] It is, however, still not clear at this time what difference exists between laboratory-scale, experimental results and behaviour exhibited during full-scale production runs. This problem, parenthetically, is one of the reasons why mathematical modelling is so important: there is a need to calibrate easy-to-perform laboratory test results with actual operating procedures in the plants.

The picture that emerges from these studies and from mill experience is summarized below.

(a) Initially, the mat is compacted and the surfaces are heated by conduction through contact with the heated platens.
(b) Bound water in the wood substance at the surface is converted to water vapour, which decreases wood moisture content but increases the vapour pressure in the surrounding void spaces.
(c) The pressure gradient through the thickness causes the vapour to flow into the core of the mat.
(d) Some of the water vapour is adsorbed into wood substance at intermediate layers according to the local conditions, i.e. the temperature, vapour pressure and surface moisture content of wood substance in these zones.
(e) As time moves on the temperature changes through the thickness due to ordinary conduction and due to heat being transferred into the core by bulk flow of vapour. The compaction deformation rates change with temperature and a density gradient begins to appear through the thickness.
(f) The polymerization reaction rates also change with temperature and the resin starts to 'kick over' at different times depending on thickness position.

(g) A lateral vapour-pressure gradient starts to develop and water vapour begins to flow in this direction eventually escaping to the atmosphere through the edges.

(h) At an intermediate time an interesting phenomenon occurs: the centre line temperature levels off, typically near 100 °C, and maintains a quasi-equilibrium condition for some time before rising slowly to the platen temperature level.

(i) During this period of quasi temperature equilibrium water is driven from the mat apparently in such a way that a steadily decreasing parabolic moisture distribution exists through the thickness which corresponds with a 'steady-state' flow of water vapour escaping through the edges.

(j) At some point the compaction stress in the mat decays away. If sufficient water is removed the shrinkage of the wood causes a gap between the mat and platens. Excess steam in the mat escapes through the faces and along this gap. A sudden cooling through the mat thickness results as a quantity of thermal energy is swept away.

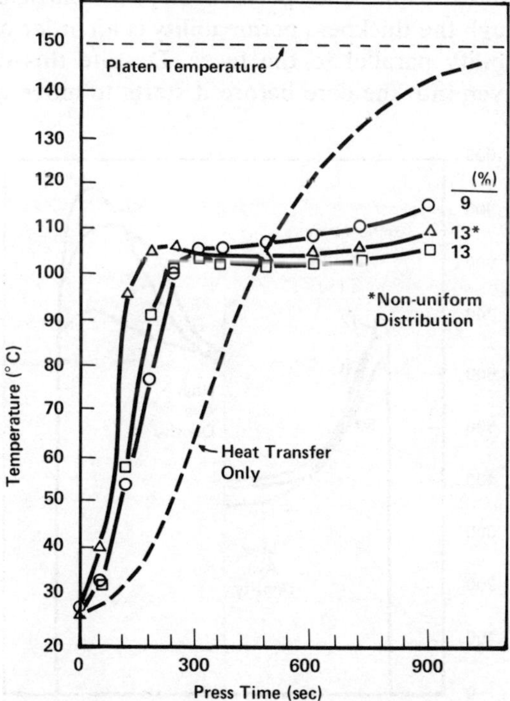

Figure 20.3 Temperature at the centre of a particleboard mat during pressing at various moisture contents compared to a theoretical calculation based on heat transfer alone (adapted from W. A. Kasir, Ph.D. dissertation[5])

A comprehensive model must be able to predict this behaviour in the proper sequence. Moreover, it must account for variations resulting from parameter changes. Before discussing the model, we will illustrate a few empirical relationships in more detail.

That heat and moisture mass movement in the mat are inseparably intertwined can be seen in Figure 20.3. The particleboard core temperature should follow the theoretical curve if the heat transfer process were independent. Obviously it does not. A typical value of thermal diffusivity for particle board was used in this calculation. It appears then that bulk flow of water vapour is responsible for the initial increased heating rate of the core. This vapour, however, is adsorbed and/or condensed, which leads to an intermediate temperature plateau in later stages. A similar type of calculation for independent, diffusive moisture flow out of an axisymmetric particleboard disc with impermeable faces (the platens) leads to drying times much too long compared with observations. This is true even for large diffusion coefficients corresponding to elevated temperatures.

Bowen[4] has shown that the permeability of particleboard is quite anisotropic. Through the thickness permeability is an order of magnitude less than the permeability parallel to the faces. Despite this difference, water vapour is first driven into the core before it starts to move out to the edges.

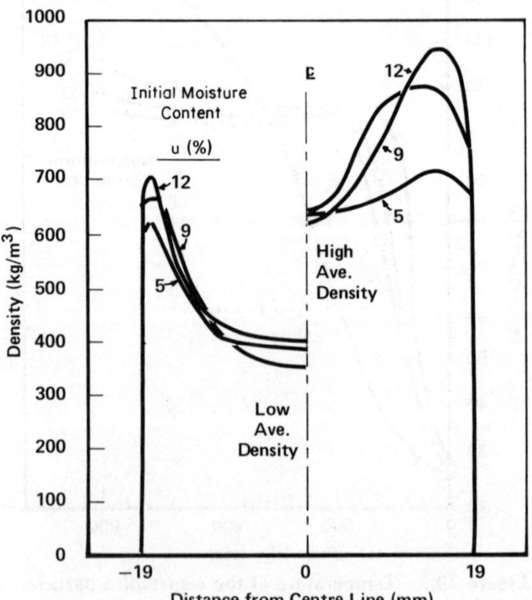

Figure 20.4 Density profiles through the thickness of particleboards with various initial moisture contents (adapted from M. E. Bowen, Ph.D. dissertation[4])

The permeability in both directions is a strong function of density. Given various initial conditions, it is possible to compact the mat at differential rates through the thickness. Density profiles reveal this phenomenon (Figure 20.4). A balance is struck in the intermediate stages between the flow of thermal energy to the core, the subsequent re-evaportation of the core water and the outward flow of vapour. Although the vapour flow pattern has never been measured, Bowen[4] measured and classified six types of temperature patterns, Figure 20.5 existing in the mat during the press cycle. These are undoubtedly related to the extent of vapour flow to the edges.

In a series of experiments conducted at the Weyerhaeuser Technology Center, we have found similar experimental trends with fibreboard. We controlled platen temperature, initial moisture content and pressed the mat for very long times to obtain fundamental information about the heat- and mass-transfer process.

The effect of platen temperature can be seen in Figure 20.6. In all but the low-temperature cases, an intermediate plateau was observed which, however, was followed by an unusual 'dip' in core temperature. This dip actually occurred simultaneously throughout the thickness as shown in Figure 20.7.

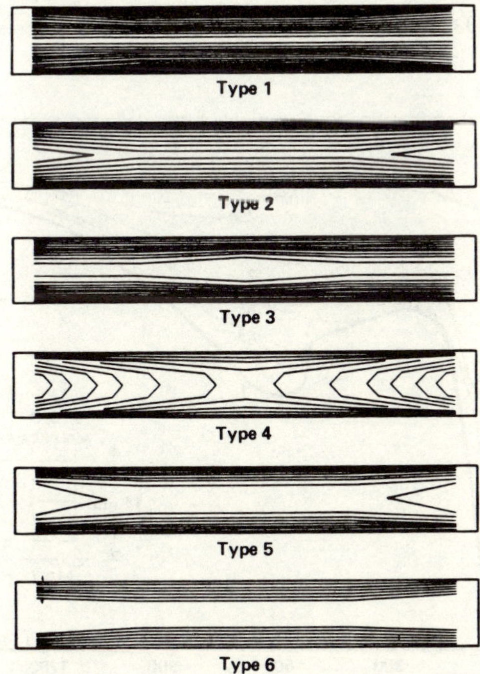

Figure 20.5 Six types of temperature contours in particleboard measured during the pressing cycle (adapted from M. E. Bowen, Ph.D. dissertation[4])

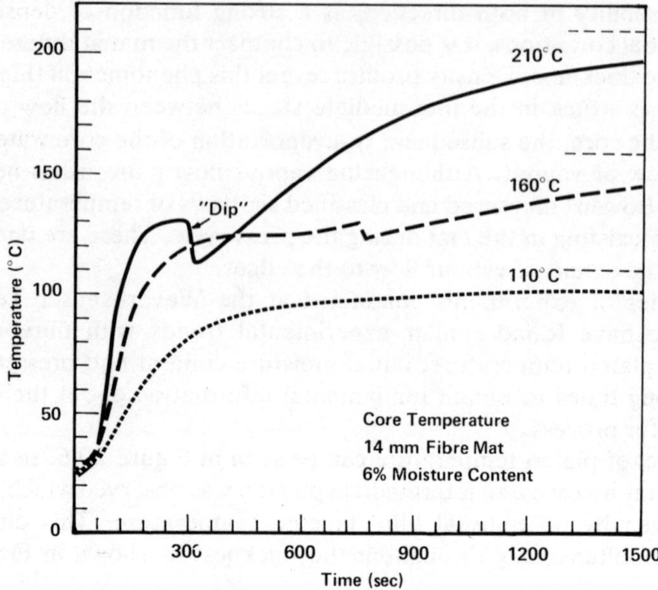

Figure 20.6 Temperature profiles of fibreboard during pressing with different platen temperatures

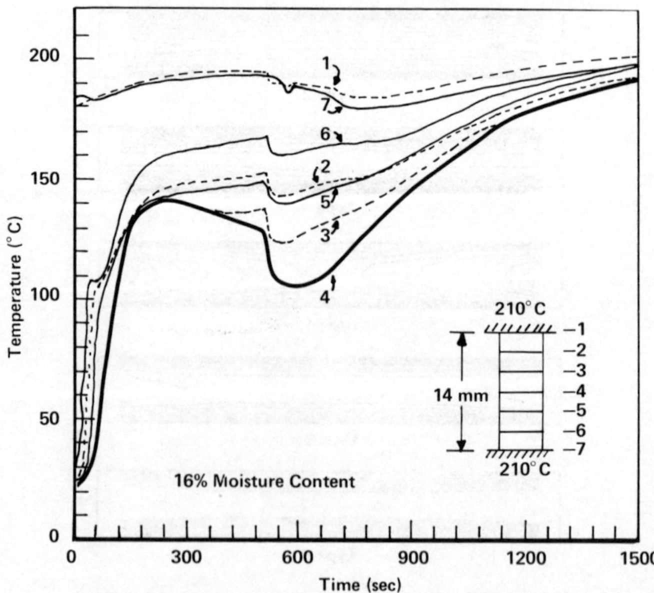

Figure 20.7 Temperature profiles at various locations through the thickness of fibreboard

The sudden temperature drop must be related to a quick release of vapour, but we did not observe anything out of the ordinary, like a surge of steam from the press, during the experiment.

The average moisture content at various thickness levels are shown in Figure 20.8 at different times along with corresponding temperature profiles. Both temperature and moisture fields appear to be parabolic in shape, indicating that a quasi, steady-state condition prevails during the intermediate phase. The heat flux from the platens balances the heat being driven off in the form of vapour as the moist fibre in the central region of the mat dries out.

20.3 THE ONE-DIMENSIONAL LEAKING COMPARTMENT MODEL

There have been some attempts at calculating the dynamic variation of the transport process variables during the hot pressing of a wood particulate or fibrous mat. Bowen[4] used a finite difference scheme to determine temperatures. Only a heat-transfer process was considered and the difference between the calculated and measured results was used to infer the amount of energy consumed in the other processes. Denisov and Juskov[6] purport to predict the time of press opening from a few theoretical and empirical results. (Shortening of the press cycle tends to bc a very popular subject for obvious

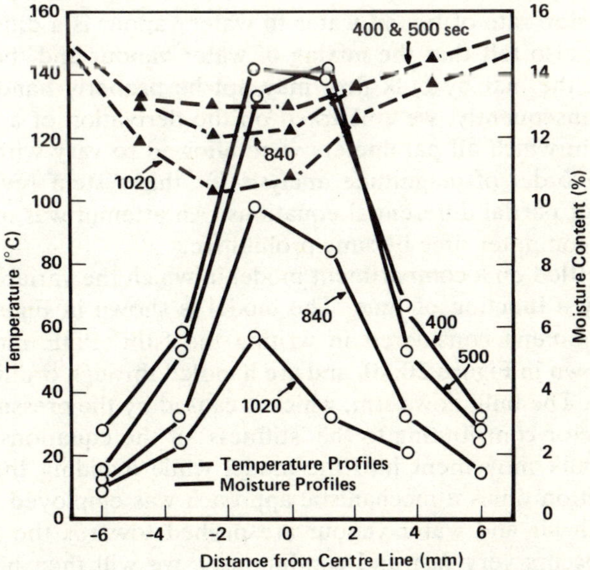

Figure 20.8 Average moisture content and temperature distributions through the thickness of fibreboard

economic reasons.) Gefahrt[7] has developed a one-dimensional model based on a moving boundary where moisture is assumed to be 'boiling·off'. As far as we can determine only the thermal field has been considered in modelling work thus far.

We are attempting to develop a comprehensive model which will be able to predict the transient behaviour of the following variables: (a) temperature, T, (b) moisture content, u, (c) pressure, P, and (d) density, ρ. In this chapter we will report on a one-dimensional model that dealt with the first three variables. Initially the set of three simultaneous systems of equations derived by Luikov[8] were thought to be applicable to the pressing problem. A generalized form of these equations can be expressed as

$$\begin{Bmatrix} \partial T/\partial t \\ \partial u/\partial t \\ \partial P/\partial t \end{Bmatrix} = \begin{bmatrix} a_{11} & a_{12} & a_{13} \\ a_{21} & a_{22} & a_{23} \\ a_{31} & a_{32} & a_{33} \end{bmatrix} \begin{Bmatrix} \nabla^2 T \\ \nabla^2 u \\ \nabla^2 P \end{Bmatrix} + \begin{Bmatrix} s_1 \\ s_2 \\ s_3 \end{Bmatrix},$$

where t is time, ∇ the gradient operator, $[a_{ij}]$ and $\{s_i\}$ (some elements are zero) are the coefficient matrix (thermal diffusivity, moisture diffusivity, permeability, thermal gradient coefficient, etc.) and a vector of 'sources', respectively.

We had difficulty determining whether this system of equations would be applicable to the problem of hot pressing a mat. At the microscopic level, we envisioned two phases: void space and wood substance. An *a priori* estimate of the conversion rate of bound water to water vapour is a difficult value to come by. We also felt that the mixing of water vapour and the subsequent drying out of the mat by bulk flow may not be properly handled by these equations. Consequently, we embarked on the derivation of a complete set of equations in which all parameters were allowed to vary with the process variables. An order-of-magnitude analysis for this system revealed a very 'stiff' system of partial differential equations. An attempt was made to solve them but the computer time became prohibitive.

We then settled on a compartment model in which the variables at a given level are only a function of time. The model is shown in Figure 20.9. The physical phenomena considered in writing the 'stiff' PDE model are still present as shown in Figure 20.10, and are handled through ordinary differential equations. The bulk flow term, which is caused by the pressure gradients, is the only factor contributing to the 'stiffness' of the equations. In order to include this bulk movement into the model while avoiding the excessively long computation times a mechanistic approach was employed. As pressure builds up both air and water vapour are pushed towards the low-pressure side. If this occurs very fast and continuously, we will then have a mixing effect among the vapour chambers. It is this idea of mixing which is used as a measure to simulate the bulk movement from high- to low-pressure regions.

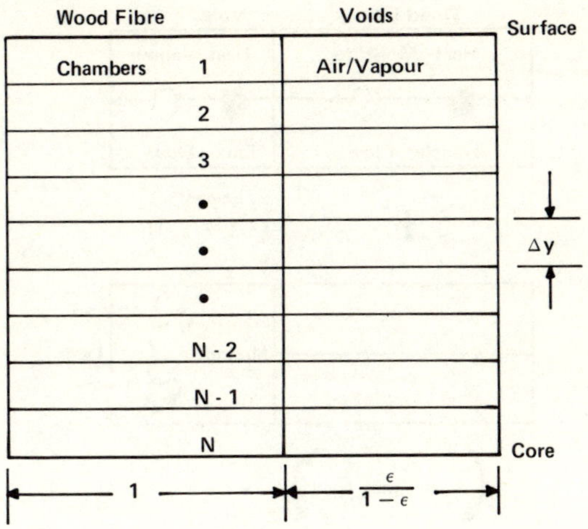

Figure 20.9 The compartment model

Another phenomenon taking place as a part of the changes due to concentration, temperature, and pressure gradients is the equilibrium between the water vapour, adsorbed water, and the possible condensed (free) water. Because this is an algebraic problem rather than a time-dependent one, it is also treated separately from the diffusion model. The computational steps are shown in Figure 20.11. For one time intergration step, there are three main mechanisms taking place in the model: (a) diffusion, (b) mixing, and (c) evaporation/condensation. An assigned number of mixing and equilibrium iterations within a time step are prescribed arbitrarily. The mixing is done by chamber pairs; i.e., (1, 2), (2, 3), . . . , $(N-1, N)$.

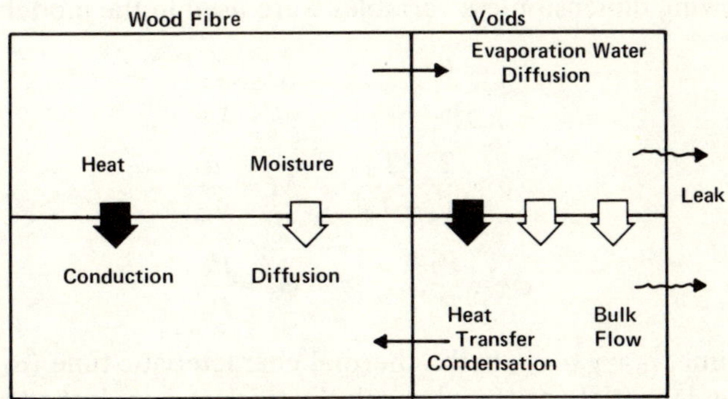

Figure 20.10 Physical phenomena for the one-dimensional leaking compartment model

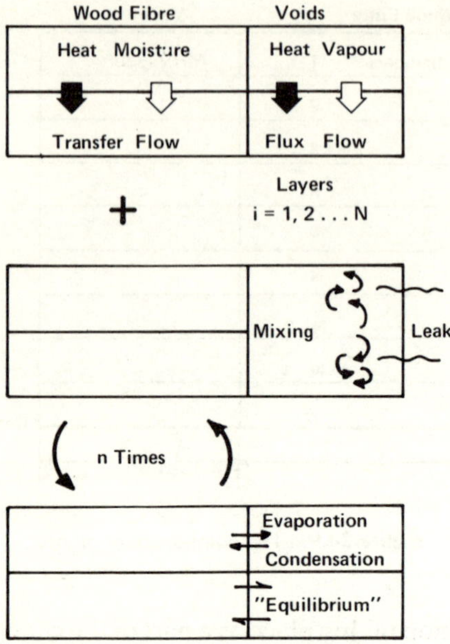

Figure 20.11 Computation steps of the model

As apparent from Figure 20.11, there is also a leak term which accounts for the transfer of vapour moisture between adjacent lateral sections. Of course, we are deviating from the one-dimensional arrangement of the original model. However, this leaking effect was believed to be the major cause for the temperature drops observed in the experimental boards. Thus the model accepts a user-defined leak term to account for sudden heat losses (and simultaneous pressure drops) which were observed experimentally.

The following dimensionless variables were used in the model:

$$\tau = \frac{t}{t_0}, \qquad Y = \frac{y}{y_0},$$

$$\theta = \frac{T - T_0}{T_1 - T_0}, \qquad M = \frac{u}{u_m},$$

$$\phi_T = \frac{P_T}{P_a}, \qquad \phi_v = \frac{P_v}{P_a},$$

where t is time, $t_0 = y_0^2/\alpha_{LS}$ is the thermal characteristic time (α_{LS} is thermal diffusivity), y is spatial position through the thickness, y_0 is the half thickness, T_0 and T_1 are initial and final temperatures respectively, u_m is the saturated moisture content and P_v, P_a, P_T are the vapour, atmospheric and total pressures respectively. Dimensionless coefficients required for the model were computed as follows:

$$a_0 = T_0/(T_1 - T_0), \quad a_1 = D_s/\alpha_s(\Delta Y)^2, \quad a_2 = \delta(T_1 - T_0);$$

$$a_3 = \mathcal{M}_a\mathcal{M}_v D_v/\mathcal{M}_{av}\alpha_v(\Delta Y)^2, \quad a_4 = \mathcal{M}_v/\mathcal{M}_{av},$$

$$a_5 = [(1-\varepsilon)(\Delta Y)^2]^{-1}, \quad a_6 = \varepsilon\lambda u_m/c_s(T_1 - T_0);$$

$$a_7 = \varepsilon \mathcal{M} P_a c_v/(1-\varepsilon)\rho_0 c_s R(T_1 - T_0),$$

$$a_8 = \varepsilon\mathcal{M}_v P_a/u_m\rho_0(1-\varepsilon)R(T_1 - T_0),$$

where ΔY is the dimensionless thickness of a compartment and the definitions for the other parameters are found in the 'Nomenclature' section.

The following equations were developed for an intermediate compartment. The computer code is such that the first and the last compartment equations are adjusted to reflect the boundary conditions.

$$\frac{\partial M_{BF,n}}{\partial\tau} = a_1\{M_{BF,n-1} - 2M_{BF,n} + M_{BF,n+1}\}$$
$$+ a_1 a_2\{\theta_{n-1} - 2\theta_n + \theta_{n+1}\},$$

$$\frac{\partial M_{v,n}}{\partial\tau} = a_3\{M_{v,n-1} - 2M_{v,n} + M_{v,n+1}\},$$

$$\frac{\partial\theta_n}{\partial\tau} = a_4\{\theta_{n-1} - 2\theta_n + \theta_{n+1}\},$$

$$M_{BF} = M_B + M_F, \quad M_{BF_v} = M_v + M_{BF},$$

where M_F = dimensionless free (condensed) water moisture.

These 'diffusion equations' (finite difference form) cause bound water, water vapour, and thermal energy to flow from one component to another. Weighted averaging is invoked to calculate water vapour pressure, total pressure and temperature between the chambers n and $n+1$ (for $n = 1, 2, \ldots, N$).

$$\phi_{v,n}^{(m)} = \tfrac{1}{2}(\theta_n^{(m)} + a_0)\left\{\frac{\phi_{v,n}}{\theta_n + a_0} + \frac{\phi_{v,n+1}}{\theta_{n+1} + a_0}\right\},$$

$$\phi_{T,n}^{(m)} = \tfrac{1}{2}(\phi_{T,n} + \phi_{T,n+1}),$$

$$\theta_n^{(m)} = \frac{2\phi_{T,n}^{(m)}}{\left\{\dfrac{\phi_{T,n}}{\theta_n + a_0} + \dfrac{\phi_{T,n+1}}{\theta_{n+1} + a_0}\right\}} - a_0.$$

The adjacent chamber temperature, total pressure and water-vapour pressure are then equated to simulate 'mixing'.

$$\theta_n = \theta_{n+1} = \theta_n^{(m)},$$

$$\phi_{T,n} = \phi_{T,n+1} = \phi_{T,n}^{(m)},$$

$$\phi_{v,n} = \phi_{v,n+1} = \phi_{v,n}^{(m)}.$$

The temperature between the two phases in each chamber are equilibrated.

$$\theta_n^{(m)} = \frac{a_7\phi_{T,n} + (1 + u_m M_{BF,n})(\theta_{s,n} + a_0)}{\left(\dfrac{a_7\phi_{T,n}}{\theta_{v,n} + a_0}\right) + (1 + u_m M_{BF,n})} - a_0,$$

$$\theta_{s,n} = \theta_{v,n} = \theta_n^{(m)},$$

where the superscript 'm' denotes conditions after mixing.

Pressures are adjusted to correspond to the new temperatures.

$$\phi_{T,n}^{(m)} = \phi_{T,n}\left(\frac{\theta_n^{(m)} + a_0}{\theta_n + a_0}\right),$$

$$\phi_{v,n}^{(m)} = \phi_{v,n}\left(\frac{\theta_n^{(m)} + a_0}{\theta_n + a_0}\right).$$

At any time in any chamber there may be up to three phases: bound water, free water (condensed), and vapour water. If there is any condensed water, then the vapour must be saturated. Likewise, if the calculated vapour pressure is above the saturation value, then some condensation must take place to reduce the (vapour) partial pressure to the saturation level. Simultaneously with these conditions, the adsorbed water is always in equilibrium with the humid air and vapour mixture.

Therefore, an algorithm was developed and used in the code, going through the following steps:

(a) satisfy adsorption equilibrium with no condensation/evaporation;
(b) if humidity is less than 100%, go to (d); otherwise,
(c) satisfy adsorption and vapour/liquid equilibrium with humidity = 100%;
(d) calculate net condensation/evaporation.

The adsorption isotherm used in this algorithm is the Hailwood and Horrobin sorption isotherm with the constants given by Skaar[9] on p. 187.

After equilibrium conditions are satisfied, Step (d) of the described procedure indicates a net change of vapour water content. Let us call this ΔM_v. Because of this net evaporation ($\Delta M_v > 0$) or condensation ($\Delta M_v < 0$) there is an associated temperature change due to the heat of evaporation. An enthalpy balance indicates that the change in temperature is:

$$\Delta\theta = -a_6\left(\frac{\Delta M_v}{1+u_m M_{BF_v}}\right).$$

This, and a relation between M_v, ϕ_v, and θ,

$$M_v = a_8\left(\frac{\phi_v}{\theta + a_0}\right),$$

are used in an iterative scheme with the equilibrium relations to find the correct temperature change associated with the sensitive temperature vapour/bound water/free water balance.

20.4 RESULTS

As discussed earlier, the model treats the mathematical solution in two sections. One, where only the diffusion terms are solved; and two, where the vapour–liquid equilibrium and the mixing effects are computed. To help in the testing of the program and to understand the relative contributions of each of the two aspects on temperature–time history, simulations were made where temperatures at different layers were predicted using only diffusion terms and also complementing the values of the dimensionless parameters used in the calculations:

$$a_0 = 2.4, \qquad a_1 = 0.5N^2$$

$$a_2 = 2.5, \qquad a_3 = 20N^2$$

$$a_4 = 0.72, \qquad a_5 = 1.7N^2$$

$$a_6 = 0.40, \qquad a_8 = 5.2\times 10^{-3}$$

The results can be observed in Figure 20.12 where temperature profiles for four layers of a ten-layer (surface-to-core) board are plotted. The profiles predicted by diffusion–conduction terms all underestimate the temperature increase in each layer and this becomes more pronounced towards the core. What we see here is that if we are only interested in temperature profiles we will probably be not very far away from the actual behaviour if we consider only the diffusion terms. However, the predictions of moisture, condensation and evaporation are completely dependent on the equilibrium and mixing calculations and for that reason only, if not any other reason, the model needs these additional effects considered. A final observation from this figure is that all temperature profiles reach the surface value. This is expected because, by nature, the one-dimensional model does not have the heat-transfer terms to outside (ambient) or the continuous moisture diffusion in the lateral direction.

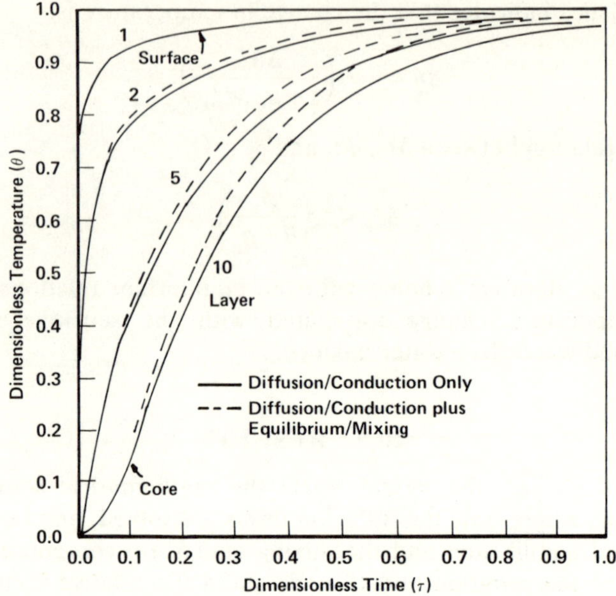

Figure 20.12 Temperature variation with time in different layers

Figures 20.13 and 20.14 show the behaviour of moisture distributions through the thickness of the board at different times and the predictions for the condensation of water. It is important to observe that towards the end of a press cycle moisture accumulated at the core and there is a significant amount of condensation. However, this is the extreme case because there is continuous migration of moisture towards the edges in the real press.

The observation of inflection points and dips in the experimental temperature profiles imply that there must be some heat-sink effects influencing the temperature rise in the board. One explanation can be the following: as the board heats, the moisture adsorbed in the fibres must be released into the vapour phase, thus increasing the pressure. With increased pressure air and water vapour are driven to outside through a lateral movement because the sides of the board are open to atmosphere. The speed of this movement is controlled by the board permeability. While the temperatures are moderate, it is easier for vapour to escape because the physical structure of the fibres is not deformed. However, with time and higher temperatures, the fibres plasticize and reduce the permeability, and thus practically trap the formed vapour in the available void space. With increasing temperature the pressure of the trapped vapour increases and eventually overcomes the resistance to flow out. When this happens, a significant amount of vapour escapes and in turn reduces the pressure. Because vapour–liquid equilibrium must always be

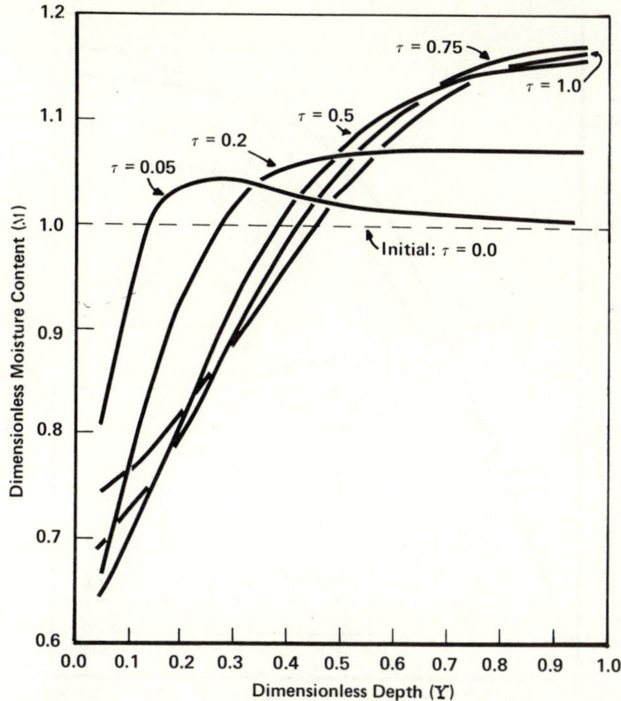

Figure 20.13 Moisture content as a function of position for different dimensionless times

satisfied, as a reaction to the pressure drop more liquid (adsorbed or free) evaporates and thus reduces the temperature due to the significant heat of evaporation.

In the mathematical model the only way to see the effects of lateral movement is to have the second (or preferably the third) dimension in the space characterization. Because of the initial development on the one-dimensional description we decided to incorporate a 'leak' term into the model which essentially simulates the lateral migration of vapour. It is important to realize that the objective in including this 'leak' term is to check the hypothesis that vapour migration is responsible for the temperature dips, but not to reproduce experimental results quantitatively. For that we definitely need to extend this model to three dimensions and deal with migration through porous media with changing permeability rather than with the artificial 'leak' term.

Figures 20.15–20.17 show that with arbitrarily selected 'leaks' we can create either simple inflection points or major dips in the temperature profiles. These effects indicate that our understanding of the physical process is in the right direction and that future efforts should include the incorporation of lateral moisture movement into the model.

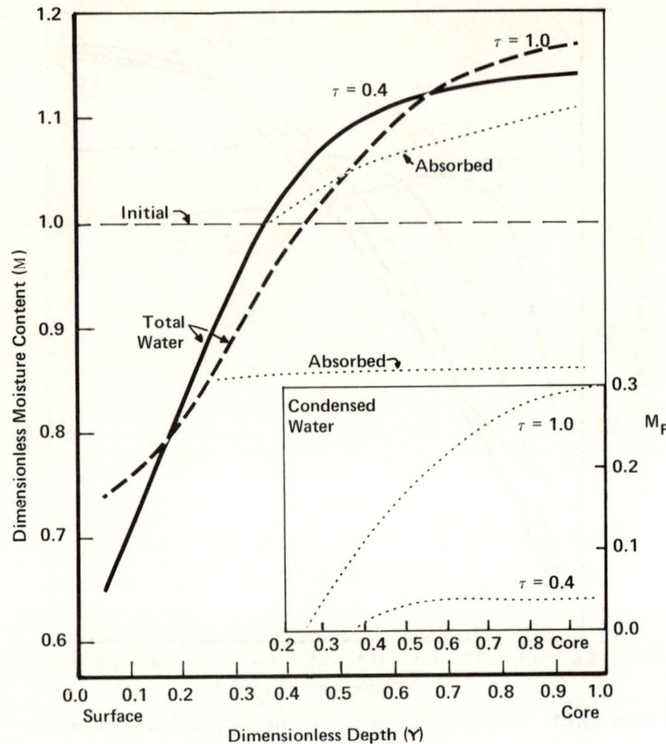

Figure 20.14 Condensation as a function of position

20.5 CONCLUSIONS

A mathematical model for the simultaneous heat and mass transfer during the hot pressing of a particleboard (or fibreboard) board has been developed in terms of a simplified one-dimensional picture which considers property changes with depth and time. The model predicts temperature, bound moisture, condensed (free) moisture, vapour moisture, density, vapour partial pressure, and total pressure profiles at different depths.

Due to the mathematical nature of the problem certain mechanistic assumptions are made to overcome excessive computation requirements with the computer. Pressure differences play an important role in forcing vapour to move towards the core of the board (in the one-dimensional picture) which causes numerical difficulties (a 'stiff' system of differential equations) and can be handled only at the expense of very long computation times. This difficulty is overcome in the model by introducing a mixing concept to simulate migration due to pressure differences.

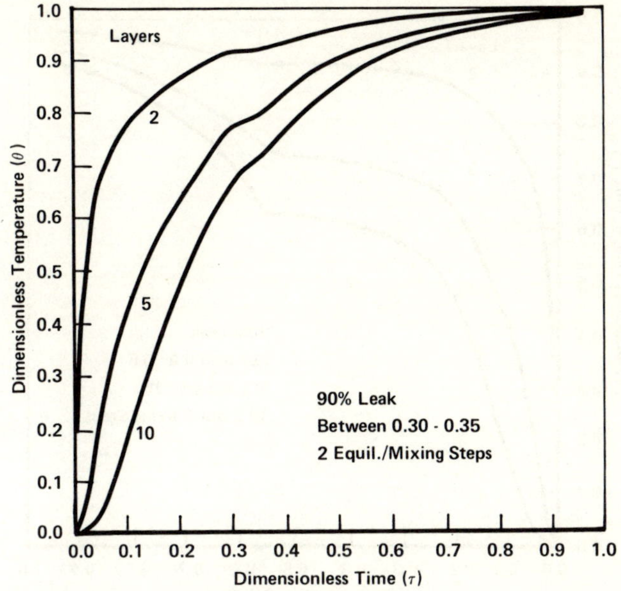

Figure 20.15 Temperature variation with time showing effect of vapour/air leak

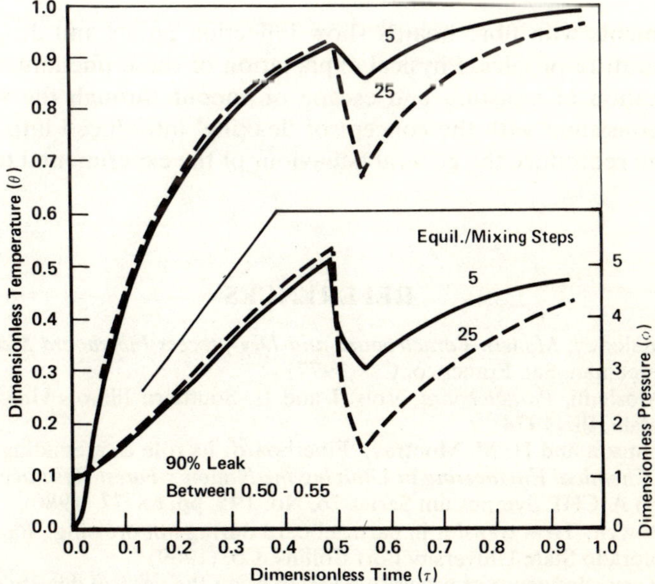

Figure 20.16 Effect of leaks on temperature and pressure profiles for layer 5

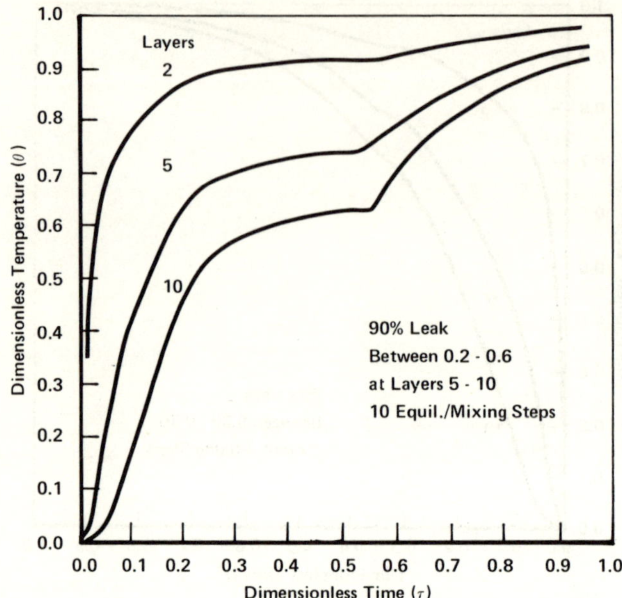

Figure 20.17 Effect of continuous leaking on temperature profiles

Experiments with fibre boards show inflection points and drops (dips) in the temperature profiles. Physical explanation of these phenomena in terms of evaporation of moisture and escape of vapour through the sides of the board is consistent with the concept of 'leaking' introduced into the model. Theory can reproduce the general behaviour of the experimental temperature profiles.

REFERENCES

1. T. M. Maloney, *Modern Particleboard and Dry-process Fiberboard Manufacturing*, Miller Freeman, San Francisco, Ca. (1977)
2. A. A. Moslemi, *Particleboard*, Vols. I and II, Southern Illinois University Press, Carbondale, Ill. (1974)
3. J. A. Johnson and H. M. Montrey, 'Fiberboard: its role as a building material' in *Role of Chemical Engineering in Utilizing the Nation's Forest Resources*, (ed. G. R. Lightsey) AICHE Symposium Series 76, No. 195, pp. 68–77 (1980).
4. M. E. Bowen, 'Heat transfer in particleboard during hot pressing', *Ph.D. Dissertation*, Colorado State University Fort Collins, Co. (1969)
5. W. A. Kasir, 'Influence of processing variables on the vertical density gradient and properties of particleboard', *Ph.D. Dissertation*, North Carolina State University Raleigh, N.C. (1979).

6. O. B. Denisov, and V. V. Juskov, Calculating the pressing time in particleboard production (translation) *Holztechnologie*, **15**(3) (1974)
7. J. Gefahrt, A contribution to the subject of preheating particles with high-frequency energy: model for computing the temperature curve in the center of the mat during hot-pressing (translation).
8. A. V. Luikov, *Heat and Mass Transfer in Capillary-porous Bodies*, Pergamon Press, London (1966)
9. C. Skaar, *Water in Wood*, Syracuse University Press, Syracuse, N.Y. (1972)

Author Index

The page numbers in ordinary type indicate where the author's work is mentioned in the text. The numbers in **bold** type refer to the pages where the journal or book reference is given.

Subject Index